Skeletal Muscle

Form and Function

Second Edition

Brian R. MacIntosh, PhD
University of Calgary

Phillip F. Gardiner, PhD
University of Manitoba

Alan J. McComas, MB
McMaster University (Emeritus)

Human Kinetics

Library of Congress Cataloging-in-Publication Data

MacIntosh, Brian R., 1952-
    Skeletal muscle : form and function / Brian R. MacIntosh, Phillip F. Gardiner,
Alan J. McComas.
        p.   cm.
    Includes bibliographical references and index.
    ISBN 0-7360-4517-1 (hardcover)
    1. Striated muscle--Physiology.   2. Striated muscle--Anatomy.   I. Gardiner,
Phillip F., 1949-   II. McComas, Alan J.   III. Title.
    QP321.M127   2006
    612.7'4--dc22   2005003557

ISBN-10: 0-7360-4517-1
ISBN-13: 978-0-7360-4517-9

The Web addresses cited in this text were current as of May 25, 2005, unless otherwise noted.

Acquisitions Editor: Loarn D. Robertson, PhD; **Developmental Editor:** Elaine H. Mustain; **Assistant Editor:** Lee Alexander; **Copyeditor:** Joyce Sexton; **Proofreader:** Erin Cler; **Indexer:** Nancy Ball; **Permission Manager:** Dalene Reeder; **Graphic Designer:** Fred Starbird; **Graphic Artist:** Yvonne Griffith; **Photo Manager:** Kelly J. Huff; **Cover Designer:** Keith Blomberg; **Photographer (cover):** Dr. Russ Hepple; **Art Manager:** Kelly Hendren; **Illustrators:** Alan J. McComas and Brian R. MacIntosh, unless otherwise noted; **Printer:** Sheridan Books

Printed in the United States of America   10 9 8 7 6 5 4 3 2

Human Kinetics
Web site: www.HumanKinetics.com

United States: Human Kinetics, P.O. Box 5076, Champaign, IL 61825-5076
800-747-4457
e-mail: humank@hkusa.com

Canada: Human Kinetics, 475 Devonshire Road Unit 100, Windsor, ON N8Y 2L5
800-465-7301 (in Canada only)
e-mail: orders@hkcanada.com

Europe: Human Kinetics
107 Bradford Road, Stanningley, Leeds LS28 6AT, United Kingdom
+44 (0) 113 255 5665
e-mail: hk@hkeurope.com

Australia: Human Kinetics
57A Price Avenue, Lower Mitcham, South Australia 5062
08 8277 1555
e-mail: liaw@hkaustralia.com

New Zealand: Human Kinetics
Division of Sports Distributors NZ Ltd., P.O. Box 300 226 Albany, North Shore City, Auckland
0064 9 448 1207
e-mail: info@humankinetics.co.nz

Contents

Part II Putting Muscles to Work 85

Part III The Adaptable
Neuromuscular System 224

Preface

There was a time, perhaps only 35 years ago, when a neurophysiologist could claim to "know" the entire field of neuroscience, including muscle. Not any longer. So great has been the expansion of knowledge that previously coherent areas of scientific inquiry have become completely fragmented. In skeletal muscle, for example, the expert on ion channels may neither know nor care about the mechanism of contraction. Even a subject such as muscle fatigue has become so large that a specialist in Ca^{2+} kinetics may be totally unfamiliar with the changes in motoneuron firing rate or with the difference in susceptibility to fatigue among the various types of muscle fiber. And then, as in other tissues, there is the heavy imprint of molecular biology to be absorbed.

Yet, despite the fragmentation of knowledge, skeletal muscle continues to fascinate. Perhaps no other tissue in the body is so amenable to study, and it was muscle and its nerve supply that gave the first clues to how synapses work and also revealed the twin phenomena of programmed neuronal death and transient hyperinnervation in the embryo. Again, it was muscle that provided the basis of our understanding of trophic dependencies among cells, and other examples of the primacy of discovery are not difficult to find.

The present book is an attempt to bring this huge diversity of information together so that almost all aspects of muscle can be considered and, where appropriate, related to each other. While it is natural to reveal what is new, it is important to remember the classic experiments upon which much of our knowledge rests. So elegant and fundamental has been this work that it is hardly surprising that several lifelong or erstwhile myologists have won Nobel Prizes: Charles Sherrington, Edgar Adrian, A. V. Hill, Andrew Huxley, Alan Hodgkin, Jack Eccles, Bernard Katz, Bert Sakmann, and Erwin Neher.

This textbook provides a detailed look at motoneuron and muscle structure and function and is intended for all those who need to know about skeletal muscle, from the undergraduate student gaining advanced knowledge in kinesiology to the physiotherapist, the physiatrist, the graduate student, and the electromyographer. The book is appropriate for a course designed to advance the understanding and knowledge of students in these fields but could also serve as a valuable resource for scientists and clinicians doing research in these fields. A unique feature of this book is that it combines the basic sciences—anatomy, physiology, biophysics, and chemistry—with clinical applications and interesting notes on applied aspects of this field of study.

To assist learning, the text has been interspersed with declarative key points, presented in the margins, and a large number of figures, many of them original. Points of special interest have been incorporated in special graphic elements, and an extensive glossary has been provided to explain various terms, symbols, and abbreviations. An important and novel feature of the book is the applied physiology section at the end of each chapter. The intention here was to show how specific defects of muscle or nerve cells can result in certain clinical disorders that are known, at least by name, to most laypeople. For example, chapter 1 describes the consequence of a genetic aberration leading to the absence of a specific protein, dystrophin. This condition results in Duchenne muscular dystrophy. Additional diseases presented in the applied physiology sections include Charcot-Marie-Tooth disease, malignant hyperthermia, Bell's palsy, McArdle's syndrome, familial hyperkalemic periodic paralysis, Guillain-Barré syndrome, and amyotrophic lateral sclerosis. In addition to diseases, the "Applied Physiology" sections include discussion of the mechanisms of action of some biological toxins that affect neuromuscular function, adaptations to specific training programs, the consequences of unfamiliar exercise, and the ergogenic benefits of glycogen loading. In many of these presentations, the special role of the electromyography lab for diagnosis or quantification is described.

The book is divided into three parts. Part I presents the structures of the neuromuscular system: muscle, motoneurons, neuromuscular junctions, and sensory receptors, as well as the development of these structures. This survey provides the framework for understanding part II, "Putting Muscles to Work," which covers the function of these parts. This section begins with a consideration of the ion channels and pumps, then goes on to axoplasmic transport and finally to the membrane properties underlying excitability. The next three chapters deal with the mechanism of contraction and the organization, properties, and recruitment of the motor units. The final chapter of part II presents the metabolic aspects of muscle; this includes the energetic cost of muscle contraction as well as the means by which this energy is provided. Part III considers the adaptability of our muscles. This section includes the acute changes associated with fatigue, the chronic adaptations associated with denervation and other forms of disuse or inactivity, and the improved function associated with training. The special trophic influence of nerve on muscle, and vice versa, is covered in chapter 18. Changes in muscle due to injury, and the remarkable property of regeneration, are then covered. The final chapter presents the diminution of muscle function associated with aging.

This is the second edition of this book. The basic format of the first edition has been retained, but new information has been added. Some of the chapters (11, 14, and 15) have received major rewrites, while others have retained much of the original presentation. The glossary of this edition has grown substantially from that of the original version. Terms defined in the glossary are italicized the first time they appear in any chapter. For terms that appear in only a few chapters, those chapter numbers are listed in the glossary. This will enable readers to identify quickly the sections where they may learn more about the less common terms used in the text. Abbreviations are also presented in the glossary.

It is our hope that the reader will be intrigued by the wonder that is skeletal muscle, as we have been.

I

Structure
and Development

The structure of a tissue reflects its function. In the case of skeletal muscle, that function is to generate force or to produce movement. To achieve these mechanical goals effectively, the muscle fiber is packed with contractile elements in the form of thick and thin filaments. In part I of this book, we will examine the structure of the filaments and that of the special molecules that hold the filaments in place, thereby giving skeletal muscle its characteristic banded appearance.

How is a signal brought to the filaments to make them interact with each other and produce movement or force? To understand the excitation of muscle, we must begin by examining the structure of the motoneurons (motor nerve cells) in the spinal cord. We must then look at the structure of the nerve fibers (axons) that enable impulses to be transmitted rapidly and repetitively from the motoneuron cell body to the nerve terminals. However, a nerve impulse cannot simply leap across the narrow gap that separates an axon and a muscle fiber. Instead, the structure of the neuromuscular junction allows the impulse in the axon to release a chemical (acetylcholine) that then initiates an impulse in each muscle fiber associated with that motoneuron. This impulse in each muscle fiber must somehow be conducted not only along the fiber, but also deep into the muscle cell to provide a signal in the region of the contractile filaments. Once again, there is a structure appropriate for this function in the form of the transverse or T-tubules. We will discover that the T-tubules make special contacts with sacs inside the fiber so that another chemical messenger, this time Ca^{2+}, can act as a link in the excitation pathway. Thus, at every step in the chain of events, the structure of the muscle or nerve fiber appears appropriately suited to the task that must be performed. In the survey of muscle structure, we must not forget the organelles and other cell parts that are common to all tissues—the nuclei, mitochondria, and cell membranes. From the familiar light microscope of the last century, the investigative tools have progressed to the interference, electron, and confocal microscopes, to the technique of freeze-fracture, and to X-ray diffraction. These techniques have provided us with knowledge of the structures of skeletal muscle at the macromolecular and molecular levels.

After the anatomy of the muscle fiber, motoneuron, and neuromuscular junction has been reviewed in the first three chapters, there comes a chapter devoted to muscle sensory receptors. For convenience, the physiology of the muscle sensory receptors is dealt with at the same time. Although the muscle receptors intrude from time to time in our survey of skeletal muscle and its nerve supply, the real place for the receptors lies in the study of motor control and is therefore outside the scope of the present book. The last two chapters of part I (chapters 5 and 6) consider how muscle fibers and motoneurons develop in the embryo and how the growing nerve fibers can establish synaptic connections with the muscle fibers. This is an exciting and topical line of scientific inquiry, requiring the skills of the microscopist and those of the molecular biologist too. It is for the molecular biologist to identify the signals responsible for activating the genes of the muscle fiber in the correct sequence. Doubtless many signals remain to be discovered, but enough are now known to give some idea of the way in which the genetic instructions are formed and processed during development.

In this part of the book, as throughout, a section on applied physiology concludes each chapter. The intention here is to show how disease or dysfunction can result from a precise alteration in structure or from aberrant embryological development.

1

Muscle Architecture and Muscle Fiber Anatomy

It is amazing to think that muscle tissue can generate force, execute movements of the skeletal components, adapt to a wide variety of needs, contribute to temperature regulation of the body, and repair itself when the need arises. The capability to generate force can be finely regulated, with voluntary control of the rate and extent of motor unit activation. Each muscle of the body has a general structure that includes various connective tissue components that give the muscle shape and organization; bundles of individual muscle cells (each muscle cell is referred to as a fiber) are called fascicles. Each cell or fiber has a complex structure that relates to specialized functions like contractile properties, metabolic provision of energy, and regulation of protein synthesis. Within each muscle of the body there is a vascular network that permits the muscle fibers to obtain oxygen and chemical substrates for energy transduction, and to dispose of heat and chemical products resulting from this metabolism. This vascular network permits maintenance of a constant environment in the immediate vicinity of each muscle fiber during rest and helps minimize the disturbance to that *homeostasis* during exercise. The arrangement of individual muscle fibers within a muscle, the neural connections, and the contractile properties of the individual fibers all contribute to the vast movement capabilities we have.

In this chapter, the sizes and shapes of muscles are considered first. Next comes an account of the connective tissue that, among other functions, provides a scaffolding for the muscle fibers. These general aspects of the muscle can be referred to as its architecture. The structure of one of the many thousands of fibers in a muscle belly is then described, making use of observations from both the light and electron microscopes. The description begins with the thin membrane that envelops the fiber and then examines in turn the various features in the fiber interior. All of the parts of a fiber are concerned in some way with the production of movement or force that results from contraction, but it is the *myofibrils* that ultimately produce these effects, and it is the myofibrils that make up the bulk of the fiber (see figure 1.1).

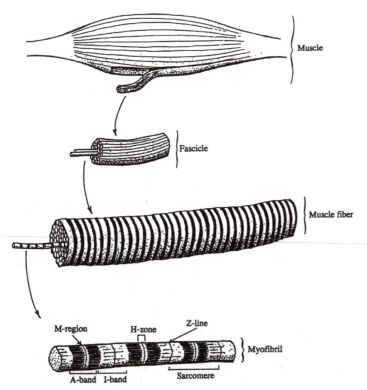

Figure 1.1 The structure of a skeletal muscle. The muscle is made up of bundles (fascicles) of fibers, each bundle being lined by perimysial *connective tissue* (see figure 1.4). The muscle fibers and the myofibrils appear striated due to alternate dark (A) and light (I) bands; the bands are "in register" between all the myofibrils in the same fiber and across the fibers within a muscle.

Occupying less volume within the muscle, but of no less importance, are the structures that allow a signal from the motoneuron to trigger contraction and the components that provide the energy required within the myofibrils to generate force. These parts of a muscle fiber involved in signal transduction include two networks of tubular structures: the *transverse (T-) tubules* and the *sarcoplasmic reticulum* (SR). The T-tubules permit transmission of the signal into the depths of the muscle fiber to initiate contraction, and the SR is the intracellular storage site for Ca^{2+} that is released to trigger the contractile response and taken back up to permit relaxation. The *mitochondria* are appropriately distributed within the fiber to provide adenosine triphosphate (ATP), the source of energy for all energy-requiring processes within the fiber. Last, but by no means least, are the nuclei of the muscle fiber that, through the expression of their genetic information, provide the instructions for the synthesis of a remarkable variety of proteins of the fiber: those needed in the membrane, myofibrils, tubular systems, cytoplasm, and the organelles. The structures of all these components of the muscle fiber will be described in this chapter, but only after an account has been given of the general features of a muscle.

Muscle Architecture

For humans, as for most members of the animal kingdom, to move is to survive. Apart from thinking, every human activity requires a movement, or at least a muscle contraction, whether for walking or running, catching or letting go, shouting or whispering, looking or listening, or even standing still. And, of course, no muscle contractions are more important than those responsible for the simple task of breathing. This enormous diversity of muscle function is reflected in the variety of sizes and shapes of the muscles of our body.

Muscles are as diverse as the tasks they perform.

Muscle Diversity

Let us consider the sizes of the muscles first. The tensor tympani, a small muscle responsible for adjusting the tension on the eardrum, contains only a few hundred muscle fibers, while the medial gastrocnemius muscle, one of the principal calf muscles employed for walking and running, has over a million fibers. The number of fibers in these two examples and in a selection of other human muscles is given in table 1.1.

Table 1.1   Numbers of Muscle Fibers in Various Human Muscles

Muscle	Number of muscle fibers
First lumbrical	10.250[a]
External rectus	27,000
Platysma	27,000
First dorsal interosseous	40,500
Sartorius	128,150[a]
Brachioradialis	129,200[a]
Tibialis anterior	271,350
Medial gastrocnemius	1,033,000

Note. Results given to nearest 50. [a]Average values. Value for sartorius from MacCallum 1898; all others from Feinstein *et al.* 1955.

The shapes of the muscles also vary. Some muscles are relatively thick—for example, the vastus muscles, which form most of the quadriceps, or the gluteal muscles, which abduct or extend the hip. Other muscles, however, form long, slender straps—such as the gracilis and sartorius. Yet other muscles, such as the flexors and extensors of the fingers and toes, are characterized by their very long tendons. This variation is suited to the different types of tasks that muscles must perform. The longer a muscle is, the more it can shorten and the higher its velocity of shortening; in contrast, the thicker a muscle, the more force (tension) it can develop. The long tendons of the finger flexors run beneath fibrous bands (*retinacula*), which serve as pulleys, and are inserted into the bases of the phalanges. This arrangement enables modest excursions of the tendons to be translated into large grasping movements of the fingers.

Muscle Fiber Pinnation

Muscles of the human body can be categorized as parallel fibered, *fusiform*, or *pinnate*. In parallel-fibered and fusiform muscles, the fibers lie in the longitudinal axis of the muscle belly (figure 1.2). For these muscles, the shortening of individual muscle fibers can be translated to shortening of the muscle. In most human muscles the fibers have a fusiform shape (figure 1.2). In some of the larger muscles, however, the fibers are obliquely inserted into the tendon (or *aponeurosis*) and, because of the resemblance to a feather, this arrangement is termed pinnated or pennated.

The muscle fibers in a pinnate muscle are obviously shorter than those in a fusiform belly and, because they pull on the tendon at an angle, the effective force transferred to the tendon (in the direction of tendon orientation) is less than the full force of contraction; the proportion of force transmitted is the cosine of the angle formed between the muscle fiber and the line of action of the muscle (see "Function and Muscle Architecture"). The advantage of pinnation is that the effective cross-sectional area of the muscle is increased, since there are many more muscle fibers in a pinnate muscle than in a fusiform muscle belly of the same size. The force developed by the muscle, which depends on the combined cross-sectional area of all the fibers, is increased proportionately. The compromise that is made to achieve this greater cross-sectional area is that the individual cells are shorter, and therefore the effective length range over which they can operate and the maximum velocity of fiber shortening are decreased. A factor that counteracts this consequence of pinnation is that the distance moved by a central tendon during a contraction is actually greater than that apparently contributed by the muscle fibers. This is achieved by a change in the angle of pinnation during the contraction (see "Function and Muscle Architecture"). This effect can provide an additional advantage. It enables the muscle fibers to function over the optimum part of their length–tension curves (Gans & Gaunt, 1991; also figure 11.7 in chapter 11), achieving a greater length change than is evident from the fiber length change.

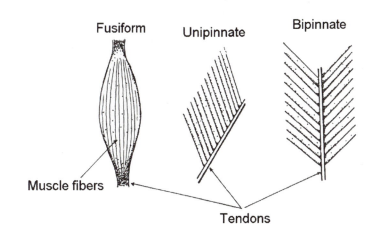

Figure 1.2 Three different arrangements of muscle fibers. Fusiform muscles are composed of tapered fibers. Pinnate fibers are arranged in parallel, but the direction of fiber orientation is not in line with the action of the muscle.

Muscle Striation

When viewed in longitudinal sections under the microscope, each muscle fiber has a series of alternating light and dark striations. These striations are more apparent when viewed with polarized light. The light bands have a consistent index of refraction, at any incident angle. This property is referred to as isotropic, so this light band has been named the *I-band*. The dark band is anisotropic and is called the *A-band*.

At the myofibril level, further subdivisions within these bands can be observed (see figure 1.1 and 1.9). At the middle of the I-band is the *Z-line* or Z-disk. In the middle of the A-band is an area that is less dark, the *H-zone*. The H-zone is bisected by the *M-region*. The part of the fiber between successive Z-disks is termed a *sarcomere,* and in a relaxed fiber each sarcomere has a length of approximately 2.2 μm. Adjacent myofibrils are attached at the Z-disks by *costameres*. This arrangement aligns the banded pattern of each myofibril across the full fiber, and indeed between fibers, and gives skeletal muscle its characteristic striated appearance. Specific proteins are responsible for the bands and zones within each sarcomere. Details of these protein structures will be given later in this chapter.

Individual muscle fibers are composed of a number of sarcomeres in series. The length of the fiber depends on the number of these sarcomeres. Table 1.2 gives estimates of the numbers of sarcomeres arranged end to end in single fibers of various human muscles. Note that the numbers of sarcomeres in some of the longer muscles can account for their length. It has been reported that the longest human muscle fibers are about 12 cm. Muscles that appear to be longer than this

Function and Muscle Architecture

Muscles in the human body can have any of several different shapes: parallel fibered, fusiform, unipinnate, and multipinnate (see figure 1.2). Each shape of muscle has specific advantages and compromises.

To illustrate the functional characteristics of muscle design, consider a unipinnate muscle, about the size of the vastus lateralis. This hypothetical muscle has a volume of 300 cm^3. Muscle fiber length is 12 cm, and the fibers are oriented at 15° relative to the aponeurosis. Figure 1.3 illustrates this muscle in its rested (precontractile), contracted, and shortened states.

This pinnate muscle would have a physiological cross-sectional area of 25 cm^2. Assuming a maximum isometric force of 20 N · cm^{-2}, and considering effective force transmission to be COS θ · 20 N · cm^{-2} · 25 cm^2, maximum isometric force would be 492 N. The angle θ is that between the line of action of the muscle and the fiber orientation.

The muscle illustrated in figure 1.3 undergoes a fiber shortening of 33% (4 cm) and a fiber angle change from 15° to 22°. Because of the alteration in fiber angle, the change in muscle length is from 36.8 cm to 28.8 cm. Assuming a maximal velocity of 10 fiber lengths · s^{-1}, this shortening (4 cm) could be achieved in 0.033 s. The maximal velocity of muscle length change would then be 240 cm · s^{-1}.

A parallel-fibered muscle of similar size such as the sartorius, or one of the hamstrings, would have substantially different properties. Suppose such a muscle had the same volume as in the previous example (300 cm^3) and a length of 36.8 cm. The cross-sectional area would be 300 · 36.8^{-1} = 8.15 cm^2, permitting a maximal isometric force of 8.15 cm^2 · 20 N · cm^{-2} = 163 N. The maximal velocity of shortening (at 10 fiber lengths · s^{-1}) would be 368 cm · s^{-1}.

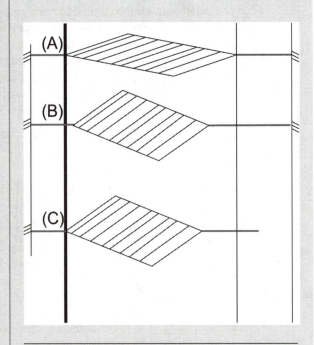

Figure 1.3 Schematic representation of a pinnate muscle: (A) relaxed, (B) isometrically contracting, and (C) shortened. During contraction and shortening, fiber length decreases and the angle of pinnation increases. Note the stretch of the connective tissue in series with the muscle (the tendon) in B.

Table 1.2 Numbers of Sarcomeres in Human Muscles

Muscle	Number of sarcomeres per fiber (× 10^4)		
	I[a]	II[a]	III[a]
Tibialis posterior	1.1	1.5	0.8
Soleus	1.4	—	—
Medial gastrocnemius	1.6	1.5	1.5
Semitendinosus	5.8	6.6	—
Gracilis	8.1	9.3	8.4
Sartorius	15.3	17.4	13.5

[a]Refers to individual limbs analyzed. The values for the three thigh muscles at the bottom of the table do not take into account the fibrous inscriptions in the belly, and the true values will be rather lower than those stated.

Data from "Muscle Architecture of the Human Lower Limb" by T. L. Wickiewicz, R. R. Roy, P. L. Powell, and V. R. Edgerton, 1983, *Clinical Orthopaedics and Related Research, 179*, p. 277.

(i.e., gracilis, sartorius, etc.) have fibers arranged in series to account for this long length. This feature of the strap muscles is described next.

Muscle Compartments

The fibers in the long strap muscles (e.g., sartorius) are often said to run from one end of the muscle to the other. Under the microscope, however, the muscle bellies of these strap muscles are seen to be divided into compartments by one or more transverse fibrous bands *(inscriptions)*; the sartorius muscle has three such inscriptions, giving four compartments; the semitendinosus has three compartments, and the biceps femoris and gracilis each have two. These compartments restrict the length of the longest human muscle fibers to approximately 12 cm, which corresponds to 5.5×10^4 sarcomeres in series. Each compartment must necessarily have its own nerve supply, and individual nerve fibers do not appear to supply muscle fibers in adjacent compartments (see review by Monti *et al.,* 2001). In the case above, where individual motor units are confined to a single compartment of the muscle, selective activation of one motor unit, and therefore only one compartment, will result in stretching of the muscle fibers in the adjacent compartment because they will not be activated. Effective muscle contraction would rely on coordinated activation of motor units in each compartment of the muscle. Only by ensuring that contraction occurs fairly synchronously along the muscle belly can the compartmental innervation permit the muscle belly to shorten very rapidly. A possible advantage of compartmentation is that this allows a more effective distribution of *neurotrophic factors* from the motoneurons to the muscle fibers (see chapter 18).

Sheard (2000) has observed another fiber arrangement. Muscle fibers can apparently be arranged in an interdigitating manner. This arrangement permits a muscle that may be 20 to 40 cm in length to be composed of fibers that are each only a few centimeters in length. In this case, there is evidence that a single motoneuron can innervate fibers that are collectively situated along the entire length of the muscle, enabling effective shortening to occur. Lateral transfer of force is another important issue in muscles that are arranged in this manner. Lateral transfer of force means that the fibers of a given motor unit do not have to be arranged directly in series, yet they can transmit force along the length of the muscle. The review by Monti *et al.* (2001) provides a summary of this property of the architecture of muscle.

Motor Unit Distributions

The functional unit of a muscle is considered to be the motor unit. A *motor unit* is a single motoneuron and all the muscle fibers innervated by that motoneuron. It is a functional unit in that it is the smallest part of a muscle that can be activated at any given instant. Details of motor unit structure and contractile properties are presented in chapter 12. However, there are a few key features of motor unit architecture that will be presented now. It is important to realize that each muscle fiber is innervated by a single motoneuron and that a given motoneuron can innervate a number of muscle fibers. This number can range from just a few fibers (as in the external ocular muscles) to approximately 2,000 fibers (as in the large limb muscles). It should also be realized that mammalian motor units can be categorized by their contractile and metabolic properties into fiber types. The muscle fibers of a single motor unit are rarely located adjacent to each other. They are distributed widely over a large portion of the volume of the muscle. This issue is further discussed in chapter 12 and is illustrated in figure 12.5.

Muscle Connective Tissue

The *connective tissue* of a muscle is almost as important as the muscle fibers, for without it there would be no structure to the muscle belly and no way for the movements and forces produced in the fibers to be transmitted to the tendon. It will be seen that the connective tissue in the muscle belly has several anatomical parts, associated with different sizes and orientations of *collagen* fibers.

Muscle connective tissue is organized at three levels: *epimysium, perimysium,* and *endomysium.*

Muscle Connective Tissue Organization

Like connective tissue elsewhere in the body, that of muscle consists of fibers embedded in an *amorphous ground substance*. Most of the fibers are collagen, of which there are at least five (I-V) immunologically distinct types; the remaining fibers are *elastin*. The connective tissue in muscle has three anatomical parts, the epimysium, perimysium, and endomysium (figure 1.4; see also Borg & Caulfield, 1980; Rowe, 1981).

The epimysium is a particularly tough coat that covers the entire surface of the muscle belly and separates it from other muscles. It contains tightly woven bundles of collagen fibers, from 600 to 1,800 nm in diameter, that have a wavy appearance and are connected to the perimysium.

The perimysium is also tough and relatively thick; it divides the muscle into bundles, or *fascicles* of fibers, and it also provides the pathway for the major blood vessels and nerves to run through the muscle belly. Some of the large collagen bundles lie alongside the outer muscle fibers in the fascicles; others run around the fascicles so as to form a crisscross pattern. Underneath the coarse perimysial sheets of connective tissue is a looser and more delicate network in which collagen fibrils run in all directions, some being connected to the endomysium (discussed next). The arterioles and venules are found in these regions, often with intramuscular nerve branches (figure 1.4A). A cross section through a fascicle shows that the muscle fibers have polygonal rather than circular outlines (figure 1.4B); the polygonal configuration enables the greatest number of fibers to be contained within a fascicle. The interstitial spaces separating the fibers are usually no wider than 1 μm, and individual fibers in an adult muscle have a diameter of about 50 μm. Fibers are much smaller in infancy, and specific training can cause cellular *hypertrophy* such that fibers are more than double this diameter.

The endomysium envelops each muscle fiber and is composed of a dense feltwork of collagen fibrils, from 60 to 120 nm in diameter, some of which are continuous with the fine fibrillar mesh of the perimysium. It is probable that the endomysium also makes connections to the *basement membrane,* a glycoprotein layer that lies on the outside of the muscle fiber membrane.

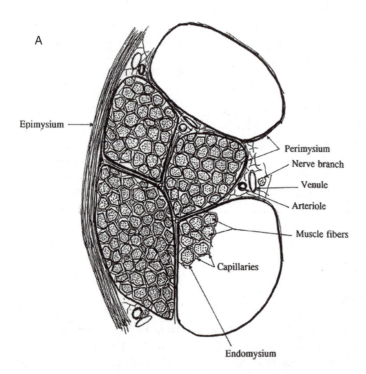

A

Epimysium

Perimysium
Nerve branch
Venule
Arteriole
Muscle fibers
Capillaries
Endomysium

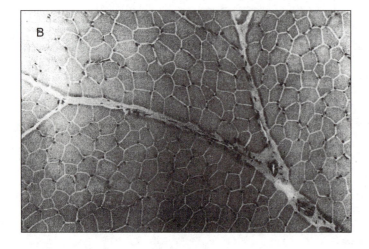

B

Figure 1.4 (A) Cross section of a muscle drawn schematically to show the three types of connective tissue sheath. (B) Cross section of human quadriceps muscle, stained with hematoxilin and eosin. Note the tight packing of muscle fibers within each fascicle and the broader gap where the perimysium is located. The dark spots identify capillaries.

B: Courtesy of Dr. John Maguire.

Muscle Connective Tissue Functions

Muscle connective tissue serves three major functions:

- Development: The connective tissue serves as a scaffolding upon which the muscle fibers can form. When muscle development is complete, the connective tissue holds the muscle fibers together and largely determines the gross structure of the muscle belly.
- The loose connective tissue of the perimysium provides a conduit for the blood vessels and nerves supplying the muscle fibers (figure 1.4).
- The connective tissue resists excessive passive stretching of the muscle and distributes forces to minimize damage to the muscle fibers. Further, the elasticity of the elastin fibrils and the wavy collagen bundles enable the muscle belly to regain its shape when external forces are removed.

The endomysium, through lateral connections between the muscle fibers, conveys part of the contractile force to the tendon and to adjacent muscle fibers. For example, Street and Ramsey (1965) crushed single muscle fibers, causing the myofibrils to retract; when the intact portion of the fiber was stimulated, the full force was still transmitted through the empty "tube" to the far end of the fiber. Although some of the force could have been transferred through the cytoskeletal elements *(spectrin, dystrophin, actin)* under the *plasmalemma,* the endomysium and basement membrane would also have been involved through structures (costameres) at the level of the Z-disks. The costameres connect myofibrils to each other and to the plasmalemma and basement membrane, and probably also to the endomysium.

Tendons Transmit Force

The ends of the muscle fibers are specialized for transmitting force to the tendon.

As the muscle fibers approach their tendon of attachment, they narrow considerably, decreasing their diameters by as much as 90% (Loeb *et al.,* 1987), giving the muscle belly its typical fusiform shape. Young *et al.* (2000) have reported that the tapering in the guinea pig sternomastoid muscle is achieved in step-like fashion, presumably because the myofibrils end abruptly. These step-like changes in diameter are accompanied by "a fine spray of finger-like projections" (p. 136) that probably provide connections through the basement membrane and endomysium to adjacent muscle fibers. This structure would allow lateral transmission of force.

At the very ends of each fiber there is extensive folding of the plasmalemma, the folds interdigitating with connective tissue processes. This folding ensures that the contractile force will be distributed over a larger area, reducing the stress on the surface of the fiber. Also, because the force is transmitted at an angle, it causes a shearing stress; and structures to which the muscle is attached, such as the different layers of tissue at the fiber ends, are more resistant to shearing stresses than they are to orthogonally applied forces. With the electron microscope, it can be seen that the myofibrils do not extend all the way to the plasmalemma; instead, the thin filaments (actin) are inserted immediately beneath the plasmalemma into a dense layer of material that contains the attachment proteins: *vinculin, talin, paxillin,* and *tensin.* Figure 1.5 shows a hypothetical model for

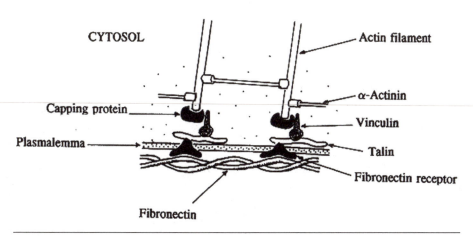

CYTOSOL — Actin filament — α-Actinin — Vinculin — Talin — Fibronectin receptor
Capping protein — Plasmalemma — Fibronectin

Figure 1.5 A possible way in which the intracellular attachment proteins are connected to the plasmalemma and to the extracellular supporting matrix. Not all the known attachment proteins are shown. Note how the α-actinin molecules brace the actin filaments.

Adapted from Alberts et al. 1989.

the connection of the actin filaments to the plasmalemma, incorporating talin and vinculin. In this model, talin is joined to the *fibronectin receptor;* this receptor is situated in the plasmalemma and is connected to bundles of *fibronectin* outside the cell. The fibronectin receptor belongs to a class of proteins, the *integrins,* that span the surface membrane.

On the outer side of the surface membrane is a well-developed basement membrane, which consists of several layers and is considered in more detail in the following section. Just as there are specialized connecting proteins on the internal face of the plasmalemma, so there are on the outer surface. These include fibronectin and *tenascin,* and they link the integrin molecules in the plasmalemma to the type I collagen fibers in the tendon.

Basement Membrane

The basement membrane is a glycoprotein complex that surrounds the muscle fiber, lying between the endomysium and the plasmalemma. Although thin, the basement membrane consists of different layers, each with a distinct structure. Further, there are a number of specialized proteins associated with the basement membrane. Thus, in addition to providing, with the endomysium, a scaffolding for the muscle fiber, the basement membrane has enzymatic actions and also "trophic" functions during development and innervation.

The basement membrane has several layers.

Basement Membrane Structure and Composition

With the electron microscope, it can be seen that the basement membrane consists of two parts, a basement lamina and a *reticular lamina* (figure 1.6). The reticular lamina is composed of collagen and other fibrils within an amorphous ground substance; toward the ends of the muscle belly, it is connected to the collagen fibers in the muscle tendon. The basement lamina is itself made up of two layers, the thinner of which is 2.5 nm thick and, because of its electron translucence, is termed the *lamina lucida;* the other layer, the *lamina densa,* is from 10 to 15 nm thick and is more easily seen with the electron microscope. The plasmalemma and the various layers of the basement membrane are collectively referred to as the sarcolemma (figure 1.6). Some authors, however, prefer to use the term sarcolemma to describe the skeletal muscle plasmalemma alone.

The basement membrane contains several types of protein and carbohydrate.

The composition of the basement membrane has been investigated with immunohistochemical techniques. Among the proteins identified are the following:

- *Acetylcholinesterase (AChE).* This is the only enzyme found so far in the basement membrane; it splits *acetylcholine* released from the motor nerve terminals (see chapter 10). Except in special circumstances (i.e., *denervation;* see chapter 10), AChE is localized to the end-plate region of the sarcolemma.

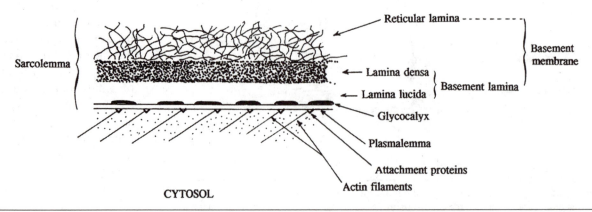

Figure 1.6   The different components of the basement membrane. The basement membrane and the plasmalemma constitute the sarcolemma, while the basement membrane is itself made up of the basement and reticular laminas.

- Collagen. Several types of collagen are present, including the relatively thick type I collagen fibers. The collagen fibers and the thinner fibrils give strength to the basement membrane and also help to connect the muscle fiber to the endomysium.
- Fibronectin and tenascin. Both these proteins link the integrin molecules of the plasmalemma to the collagen fibers of the endomysium.
- Agrin. This protein is involved in the formation of the neuromuscular junction by instructing the motor nerve and muscle fiber membranes to undergo the necessary structural specializations (see chapter 18).

Some of the carbohydrates in the lamina lucida consist of the sugar residues of various glycoproteins that are embedded in the plasmalemma. However, most of the carbohydrate is in the form of *glycosaminoglycans.* These molecules are long, unbranched saccharide polymers and include hyaluronic acid, chondroitin sulfate, and *heparin sulfate.* These compounds coat the proteins in the basement membrane and also form the amorphous ground substance.

Basement Membrane Functions

The basement membrane has several important functions.

Far from being an inert covering for the muscle fiber, as was once thought, the basement membrane has a number of important functions.

- Termination of *synaptic transmission,* through hydrolysis of the transmitter, acetylcholine, by the enzyme acetylcholinesterase (see chapter 10).
- Attachment of the muscle fiber to the endomysium, the motor nerve terminal, and, at the ends of the fiber, to the muscle tendons.
- Scaffolding for muscle fiber *regeneration,* by ensuring that the *satellite cells* multiply within the confines of the damaged fiber (see chapter 21).
- Development of neuromuscular junction. Not only does the basement membrane somehow guide a regenerating axon to the site of the former neuromuscular junction, but it also provides a signal for the growth cone to develop the specialized structures characteristic of a motor nerve terminal. Similarly, the basement membrane, even in the absence of a motor nerve terminal, is able to stimulate the muscle fiber to develop synaptic folds and to incorporate *acetylcholine receptors* into the plasmalemma (see chapter 17).

Plasmalemma

The contents of all living cells are bounded by a plasma membrane, the plasmalemma, some 7.5 nm thick; the cell contents themselves comprise an aqueous solution of inorganic ions, sugars, amino acids, peptides, and proteins that is termed the *cytosol.* The cytosol and the filaments and organelles within it are the *cytoplasm;* in the case of the skeletal muscle fiber, the cytoplasm is sometimes referred to as the *sarcoplasm.*

Apart from giving each cell its own anatomical identity, the main function of the plasmalemma is to enable the cytosol to have a chemical composition markedly different from that of the fluid surrounding the cell. The plasmalemma achieves this last function through populations of protein channels and pumps, which are embedded in the two lipid layers of the membrane. In nerve and muscle fibers, as distinct from most other cells, the plasmalemma has the additional property of excitability, enabling impulses to be transmitted over the length of the cell (see chapter 9).

The Plasmalemma Has Fluid Properties

The cell membrane is a lipid bilayer and has fluid properties.

The plasmalemma is not a smooth structure over the entire muscle fiber. Rather, in the region of the innervation zone (motor end-plate), the plasmalemma is highly convoluted due to the presence of *junctional folds* (see chapter 3). Elsewhere along the surface of the fiber the membrane displays much shallower folds; these result from slackness of the membrane when the fiber is in its short resting or contracted state, and they disappear if the fiber is stretched. There are also numerous small in-pocketings of membrane, the *caveolae,* which are connected to the surface membrane by narrow necks; their function is uncertain, though they can act as reserve sources of membrane during stretching of the fiber (Dulhunty & Franzini-Armstrong, 1975). Ignoring the tapered ends, it can be estimated that a muscle fiber of 10-cm length with a diameter of 50 μm has a surface area of 1.57×10^{-5} m^2. When the fiber shortens to 6 cm, the volume remaining constant, the surface area decreases to 1.22×10^{-5} m^2, or to 77.5% of its previous value.

Biochemical and ultrastructural investigations indicate that the plasmalemma is largely composed of *phospholipid* molecules arranged perpendicularly to the surface of the fiber and forming two layers (figure 1.7). The *hydrophilic* "heads" of the phospholipid molecules form the internal and external surfaces of the membrane, while the *hydrophobic* "tails" make up the interior of the membrane. The heads are composed of *choline, phosphate,* and *glycerol;* the tails consist of fatty acid chains, and there are two chains associated with each head. The plasmalemma also contains relatively large amounts of *cholesterol* (approximately 50% of the lipid), the molecules of which are interposed between the phospholipids and serve to stiffen the membrane. Since much of the membrane lipid has a melting point below body temperature, the plasmalemma would be expected to have fluid properties. By labeling a spot of membrane with a fluorescent dye and then observing its enlargement under the microscope, Fambrough *et al.* (1974) were able to show that this was indeed the case. The fluidity is due to the ability of the lipid molecules to exchange places with each other within their layer of the membrane. Increased cholesterol content reduces this fluidity.

Proteins Are Embedded in the Plasmalemma

The membrane also contains proteins, of which two types are generally recognized; these are referred to as extrinsic and intrinsic. *Extrinsic ("peripheral") proteins* are exposed only at the internal or external surface of the membrane and can be dislodged relatively easily by chemical means. In contrast, *intrinsic ("transmembrane") proteins* penetrate the full thickness of the membrane and are difficult to remove. Many of the proteins are glycosylated; that is, they have sugar residues that may extend as far as the lamina lucida of the basement membrane (figure 1.6). Freely ending sugar residues may trap molecules in the extracellular fluid and make these molecules more accessible

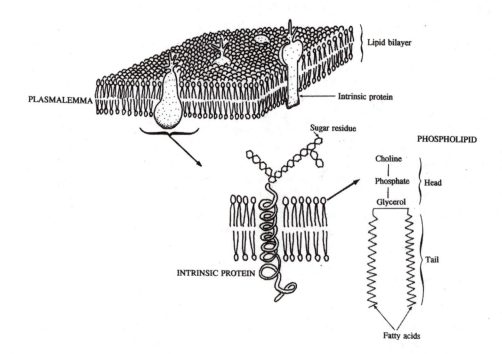

Figure 1.7 The plasmalemma of the muscle fiber, consisting of two closely applied lipid leaflets in which proteins are embedded. A glycosylated transmembrane protein, formed by a single α-helix of amino acids and a "sugar residue," is shown in the lower part of the figure. At bottom right is an enlargement of a single phospholipid molecule.

to the protein in the membrane. Among the special proteins known to be localized in the plasmalemma are the following:

• Transport systems for sugars, lipids, amino acids, and ions. The transport proteins fall into several classes. *Uniports* carry only one solute across the membrane. *Symports* function by moving two solutes in the same direction; for example, sugar molecules are often transported with Na$^+$ in the same direction. *Antiports* act by transferring two solutes in opposite directions; the best known example is the Na$^+$-K$^+$ pump, which extrudes Na$^+$ from the cell in exchange for K$^+$. Other transport proteins are the *ion channels,* which allow a specific ion to cross the membrane down its *electrochemical gradient.* Ion channels can be opened by a change in voltage across the membrane, or by the binding of a triggering substance *(ligand).* The acetylcholine receptor is an example of this latter type of ion channel.

• *Adenylate cyclase.* This enzyme is responsible for synthesizing one of the second messengers, cyclic adenosine monophosphate. A *guanosine triphosphate*-binding protein *(G protein)* on the cytoplasmic surface of the plasmalemma is involved in the activation of adenylate cyclase.

• *Kinases.* Kinases phosphorylate various other proteins. The kinases are critically important in a number of signaling cascades (see chapter 14).

• *Hormone receptors.* The ability of the muscle fiber to respond to such varied hormones as thyroid hormone, insulin, and epinephrine depends on the presence in the plasmalemma of specific receptors for each hormone. Hormone receptors are often linked to G protein, adenylate cyclase, or specific kinases.

• Integrins. These proteins link the basement membrane and endomysium to the plasmalemma and to cytoskeletal structures. Two of the integrins are the fibronectin receptor and the dystrophin-associated glycoproteins (see figures 1.5 and 1.8). Integrins can have important cell signaling functions (Carson & Wei, 2000), in conjunction with focal adhesion kinase, an enzyme that associates with the sarcolemma (Flück *et al.,* 2002).

In comparison with the lipid molecules, the ability of proteins to move within the membrane is restricted, because they are anchored to intracellular or extracellular filaments through binding proteins such as *ankyrin* and *desmin.*

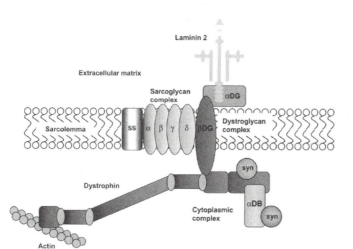

Figure 1.8 Schematic representation of connecting proteins allowing lateral transmission of force across the sarcolemma. Cytoskeletal actin is connected to laminin via the dystroglycan complex.

Reprinted, by permission, from D. J. Blake et al., 2002, "Function and genetics of dystrophin and dystrophin-related proteins in muscle," *Physiological Reviews* 82: 291-329.

The Cytoskeleton

The cytoskeleton strengthens the plasmalemma and anchors intracellular structures.

Like other cells in the body, the muscle fiber has a cytoskeleton that gives it shape as well as integrity and holds the different intracellular structures in place. At the periphery of the fiber the cytoskeleton reinforces the plasmalemma, preventing it from tearing during contraction and relaxation, much the way an internal frame supports a tent. Actin and spectrin are especially important in this role, as is dystrophin. Other intermediate filaments, composed of desmin, synemin, and vimentin, are wrapped around the myofibrils at the Z-disks and bind the myofibrils together. The same types of filament serve to hold some of the intracellular organelles, such as the nuclei and mitochondria, in position. These intermediate filaments are considered again later.

Myofibrils

The most obvious structures within the muscle fiber are the myofibrils, which are the units responsible for contraction and relaxation of the fiber. Each myofibril is from 1 to 2 μm in diameter and is separated from its neighbors by mitochondria and the sarcoplasmic and transverse tubular systems; in a fiber of 50 μm diameter, there are up to 2,000 myofibrils. It will be seen that the myofibrils contain two primary types of protein filament, actin and *myosin,* and that it is the regular disposition of these filaments along the myofibril that gives the myofibril, and the muscle fiber, its striated appearance under the microscope. Ultimately the filaments, through the supporting structures, are connected to the endomysial connective tissue and thence to the muscle tendon.

Actin and myosin filaments correspond to light and dark bands.

Myofilaments

It is the light and dark bands of the myofibrils that give the muscle fiber its striated appearance (figure 1.9A). The dark, anisotropic, A-bands correspond to the presence of myosin (thick) filaments, while the light, isotropic, I-bands contain the actin (thin) filaments (figure 1.9B). The latter filaments are, of course, separate from the actin filaments, already discussed, that help to form the cytoskeleton of the muscle fibers. The molecular structures of the actin and myosin filaments are considered again in chapter 11, in relation to the contractile mechanism of the myofibrils.

In a resting muscle fiber, the actin filaments overlap the myosin filaments to some extent, in an interdigitating manner; the central region without filament overlap constitutes the relatively pale H-zone in the A-band. In the middle of the H-zone is a dark region, the M-region, formed by the presence of fine, filamentous structures that cross-connect the myosin filaments and give them a regular spacing from each other; the structure of the M-region is considered in more detail in the following section. There is another array of fine filaments, composed of *titin,* that stabilizes the myosin filaments in the longitudinal axis (see p. 15). Each of the myosin filaments is surrounded by a hexagonal lattice of actin filaments. During a contraction, the myosin filaments contact the actin filaments by molecular cross-bridges which attempt to slide opposing actin filaments toward each other within each sarcomere, thereby diminishing the widths of the H-zone and the I-band. The thin filaments contain, in addition to actin, two regulatory proteins, *troponin* and *tropomyosin,* and also a strengthening protein, *nebulin.* Nebulin is thought to be the template for thin filament formation. Troponin is composed

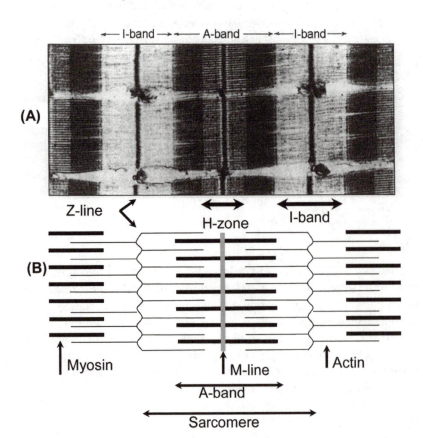

Figure 1.9 (A) Longitudinal electron micrograph of several myofibrils, showing the characteristic striations. (B) The overlapping arrangement of the thick (myosin) and thin (actin) filaments responsible for the striated appearance.

Adapted, by permission, from H. E. Huxley, 1972, Molecular basis of contraction in cross-striated muscles. In *The structure and function of muscle* (New York: Academic Press), 309.

of three subunits: troponin I (inhibitory), troponin C (calcium binding), and troponin T (tropomyosin binding). Tropomyosin spans seven actin monomers and is thought to occupy a position at rest that prevents actin–myosin interaction. The interaction between the actin and myosin filaments through the cross-bridges is considered in chapter 11.

Myosin Filament Array

Even in the early studies with the light microscope, a dark line could be distinguished in the center of the A-band. With the electron microscope, however, it became apparent that there are actually several *M-lines,* and hence it is more appropriate to refer to the *M-region* rather than to the M-line. A further complication of the M-region was the finding of fine filaments *(M-filaments),* approximately 5 nm in diameter, that run parallel to the myosin filaments and appear to be connected to the latter as well as to each other. It is the cross-links between the two types of filament that are responsible for the fixed structural array of the thick filaments and the appearance of multiple M-lines in each A-band. Antibody labeling and extraction studies have identified two of the proteins in the M-region, in addition to myosin. One of these proteins is the M-protein, or myomesin; it has a molecular weight (MW) of 165 kD and is thought to correspond to

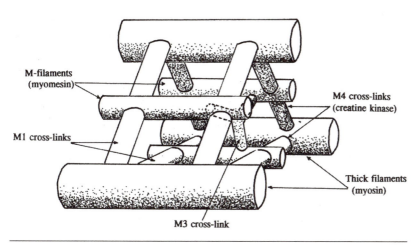

Figure 1.10 Model of the M-region, showing the thick (myosin) filaments and the M-filaments, together with the cross-links holding them in position.

Adapted from Luther and Squire 1978 and Strehler *et al.* 1983.

the M-filaments, as shown in figure 1.10. The other protein is *creatine kinase,* and it is likely that in addition to having a role in metabolism, these molecules form the struts holding the myosin filaments in a tight lattice and in register with each other. Other proteins, yet to be identified, evidently brace the M-filaments and produce additional M-lines.

Titin keeps the myosin filaments in the middle of the sarcomere.

If a solvent such as gelsolin is used to dissolve the actin filaments in a skinned muscle fiber, it can be seen that the thick filaments of myosin are connected to the Z-disks by fine strands. These strands, which are approximately 5 nm in diameter and span a full half sarcomere, consist of an extremely large protein, titin (MW 3,000 kD); it is probable that 6, or possibly 12, titin (also known as connectin) molecules make up each of the strands (Trinick, 1991). Part of the titin strand overlaps the myosin filament and is firmly attached to it, while the remainder extends through the I-band and is likely to have elastic properties (figure 1.11). From its structure and position, each pair of titin strands is thought to function as a longitudinal stabilizer for the myosin filament, keeping it in the center of the sarcomere during contraction and relaxation. It is likely that the titin strands contribute much of the elasticity of the muscle fiber when the latter is stretched, either passively by forces applied to the ends of the muscle, or actively by unequal sarcomere shortening in the long axis of the fiber. Figure 1.11 also shows the strands of nebulin, which support the actin filaments.

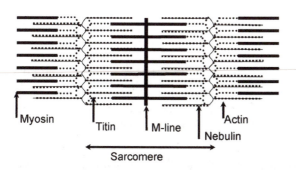

Figure 1.11 The myosin-stabilizing protein, titin, connects the thick filament to the Z-line, with connection to the thin filaments; titin in the I-band is elastic, being stretched when the muscle fiber is lengthened. Also shown is nebulin, which follows and strengthens the actin filaments.

In addition to its structural role, it has been suggested (Gregorio *et al.,* 1999) that titin serves as a template or organizer for myofibril assembly. Such a role would require capacity to bind to the two primary filaments of the sarcomere, actin and myosin. This has been shown to be the case. Titin also binds to *α-actinin* at the Z-disk and myomesin in the M-region.

The I-Band

The filaments of actin, like those of myosin, require positional support; they achieve this partly from nebulin and partly from their insertions into the Z-disk in the center of each I-band. The Z-disk is composed mainly of the proteins α-actinin, desmin, *vimentin,* and *synemin.* α-Actinin, a protein dimer of 180 kD, is thought to attach the ends of the actin filaments on one side of the Z-disk to those on the opposite side. The actin filaments, which do not themselves extend across the Z-disk, undergo a transition from a hexagonal lattice around the myosin filaments to a square lattice at their insertions into the Z-disk. A further feature of the Z-disk is that the point at which an actin filament is attached on one side is midway between two attachment points on the opposite side of the disk. Desmin, vimentin, and synemin, the other main protein components of the Z-disk, form intermediate filaments that are wrapped around the disk and also link adjacent disks together in the transverse axes (figure 1.12). Each transverse array of intermediate filaments constitutes a costamere. It is these intermediate filaments that, through their transverse connections from one Z-disk to another, keep all the myofibrils "in register" within a single muscle fiber. The Z-disks are also attached, through the intermediate filaments, to the cytoskeleton beneath the plasmalemma and to the plasmalemma itself. Through the attachment proteins of the plasmalemma, the Z-disks are ultimately connected to the basement membrane and the endomysium ensheathing the muscle fiber (see figure 1.8).

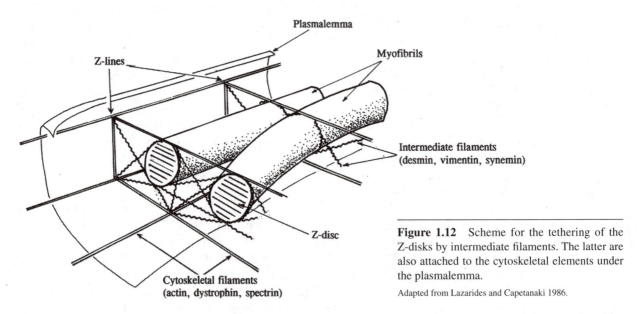

Figure 1.12 Scheme for the tethering of the Z-disks by intermediate filaments. The latter are also attached to the cytoskeletal elements under the plasmalemma.

Adapted from Lazarides and Capetanaki 1986.

Tubular Systems

Early in the last century, Veratti (1902) was able to stain a fine, interlacing network within the muscle fiber. It was only many years later, with the aid of the electron microscope, that the details of this structure could be resolved into tubular systems comprising two parts (see, for example, Franzini-Armstrong & Porter, 1964): the SR and the transverse tubules (T-tubules).

•   **The sarcoplasmic reticulum.** An elaborate meshwork of channels surrounds individual myofibrils (figure 1.13). These channels appear anchored at the T-tubules in an enlarged form of the network, the terminal cisternae. The remaining parts of this meshwork are referred to as longitudinal reticulum, because they surround the myofibril along the length of the sarcomere. Considering that the volume of this reticulum does not change when the muscle shortens, it can be imagined that the longitudinal reticulum becomes shorter in proportion to the change in sarcomere length and that the reticulum becomes wider to accommodate the same volume in a shorter length. The function of the

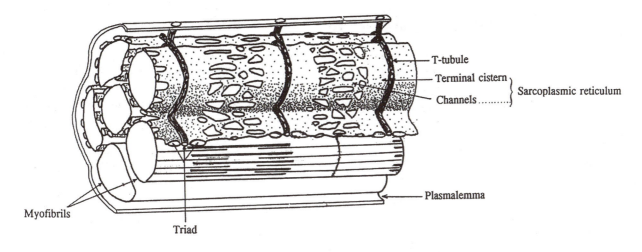

Figure 1.13   Three-dimensional representation of myofibrils, and enveloping sarcoplasmic reticulum and T-tubules.

SR is to release Ca^{2+} into the cytosol around the myofibrils, where it combines with troponin C and allows contraction to take place. This release function is accomplished by the terminal cisternae. When Ca^{2+} is pumped back into the longitudinal SR, the contraction is terminated. In keeping with its role in "activating" the fiber, the SR has a membrane that contains both Ca^{2+} release channels (ryanodine receptors primarily at the terminal cisternae) and Ca^{2+} ATPase (adenosine triphosphatase) pumps, primarily associated with longitudinal reticulum.

• **The transverse tubular system.** The T-tubules lie perpendicular to the long axis of the muscle fiber in the form of narrow channels that encircle the myofibrils at regular intervals (figure 1.13). In mammalian muscle fibers, including those of humans, there are two zones of transverse tubules in each sarcomere; they lie at the junctions of the A- and I-bands, at each end of the thick filaments. In cardiac muscle fibers and in the skeletal muscle fibers of frogs, in contrast, there is only one zone of T-tubules in each sarcomere, and this is situated at the Z-line. Here are the basic facts about T-tubules:

-As the T-tubules encircle the myofibrils, they interrupt the longitudinal channels of the SR. At these points of contact, the SR is dilated to form terminal cisternae (or *lateral sacs*); neighboring sacs are connected. Figure 1.13 shows that each of the relatively narrow T-tubules is embraced on either side by a cistern; these three elements, which surround the myofibril, are referred to as a *triad*.

-Electron microscopy reveals that the membranes of the T-tubules and SR, although closely apposed at the triads, remain separate. According to Peachey (1965), approximately 80% of the transverse tubular system in a frog muscle fiber is surrounded by SR.

-At the surface of the muscle fiber, the T-tubules form small openings, and their membranes are continuous with the plasmalemma. Through examination of muscle fibers bathed in solution containing electron-dense material or fluorescent dyes, it has been shown that the T-tubules contain extracellular fluid.

-The principal function of the T-tubules is to conduct impulses from the surface to the interior of the muscle fiber, thereby initiating the release of Ca^{2+} from the lateral sacs of the SR. For this reason, the continuity with the surface membrane is important. The regular position of the T-tubules permits conduction of the impulses into the interior at regular intervals along the length of the muscle fiber and to all depths of the muscle fiber.

-Protein structures called *junctional feet* straddle the gap between T-tubule and SR membranes. Junctional feet are most likely part of the dihydropyridine receptors, embedded in the T-tubule membrane. They are *voltage sensors* that function to detect depolarization of the T-tubule membrane (see chapter 11) and trigger opening of the Ca^{2+} channels in the terminal cisternae.

Nuclei and Mitochondria

The muscle fiber contains numerous nuclei (myonuclei) that in healthy muscle fibers are dispersed along the inner surface of the plasmalemma, particularly in the region of the motor end-plate; each *nucleus* is bounded by two membranes: the nuclear membrane and the sarcolemma (figure 1.14). The function of the nucleus is to prepare and send instructions for protein synthesis out into the cytoplasm. The instructions come from the genes contained in the chromosomes.

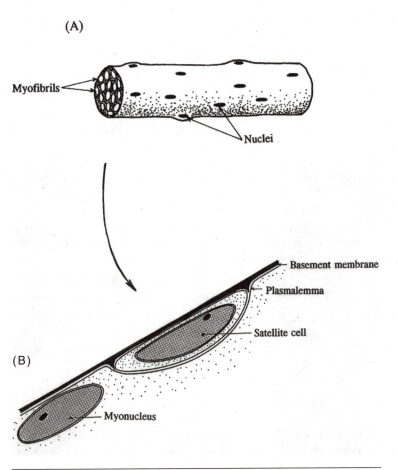

Figure 1.14 (A) Single muscle fiber with myonuclei at periphery. (B) *Myonucleus* and satellite cell. The satellite cell is separated from the fiber by its own plasmalemma and that of the fiber, but lies within the basement membrane of the skeletal muscle fiber.

Other than the nuclei, myofilaments, and tubular systems, the main structures in the cytoplasm are the mitochondria. During evolution, these organelles have become specialized to form ATP, the major energy-yielding compound of living cells. Although most of the cell enzymes of metabolism are housed in the mitochondria, others are in the cytosol. In addition, the cytoplasm contains two of the chemical fuels that the mitochondria can degrade to produce ATP; these are the glycogen granules and the lipid droplets. The nuclei and the mitochondria will now be considered.

Nuclear DNA Is Transcribed to RNA

In humans there are 23 pairs of chromosomes, which may hold 30,000 or so genes. Each gene contains the blueprint (code) for a specific protein. Although many proteins are coded here, only the ones specifically required in muscle cells will be expressed in muscle cells. When the deoxyribonucleic acid (DNA) of a gene is to be expressed, the message is first transcribed into primary *ribonucleic acid* (RNA). While still in the nucleus, the RNA is then processed, with noncoding sequences of nucleic acids being cut out. Within a special part of the nucleus termed the nucleolus, the messenger RNA is then packaged with proteins to form relatively large ribosomes; the latter are made smaller and then exported through pores in the nuclear membrane into the cytosol of the muscle fiber. The ribosomes, aided by transfer RNA, then start to assemble the sequence of amino acids that is characteristic of a particular protein; for each of the amino acids there is a genetic code in the form of a triplet of nucleotides (codon). Particularly in developing or regenerating muscle fibers, the ribosomes are attached in the cytoplasm to sheets of membrane enclosing a flattened cavity; this is the endoplasmic reticulum.

With the light microscope, the myonuclei are indistinguishable from the nuclei of the satellite cells. The latter cells probably account for less than 1% of the muscle fiber nuclei in the adult. They can be differentiated from the myonuclei only by the electron microscope, which reveals the presence of twin membranes separating the cytoplasm of the satellite cell from that of the muscle fiber (figure 1.14B). The satellite cells are of particular importance for the regeneration of muscle following disease or injury (see chapter 21).

Mitochondrial Distribution and Structure

The mitochondria (sarcosomes) appear in light and electron microscope sections as ovoid structures, measuring from 1 to 2 μm in their longest diameters. Mitochondria are found especially between the myofibrils in the region of the Z-line and also in relation to the nuclei and the motor end-plate. Although the mitochondria appear oval in longitudinal sections of muscle fibers, some form extended reticula that should be referred to as mitochondrial reticulum. Each mitochondrion has a double membrane, the inner one being repeatedly folded to form *cristae;* these folds bulge into the central compartment of the mitochondrion (figure 1.15). The matrix between the cristae houses the enzyme systems required for the Krebs tricarboxylic acid cycle, which breaks down pyruvate to carbon dioxide and water. This oxidative metabolic system is dealt with further in chapter 14. The matrix also contains several copies of the mitochondrion's own DNA.

Before finishing the description of the cytoplasmic components, it should be pointed out that glycogen granules are distributed throughout the cytoplasm. The granules, some 25 to 40 nm in diameter, are the major source of energy for the muscle fiber. Lipid droplets are often to be found close to mitochondria and contain fatty acids or triglycerides; they form important supplementary sources of energy.

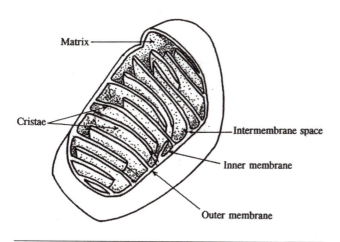

Figure 1.15　A muscle fiber mitochondrion, with part of the membrane casing removed so as to expose the cristae.

Applied Physiology

The sections on applied physiology in various chapters provide an opportunity to put the contents of the chapter into the context of real life. In the case of the present chapter, we first consider how the absence of a single type of protein at the surface of the muscle fiber results in a fatal disease. Next we will consider how simple modifications to these proteins can impair muscle function.

Duchenne Muscular Dystrophy

In the middle of the 19th century there were two reports of a disease, apparently confined to male children, in which progressive weakening of skeletal muscles leads to death (Duchenne, 1861; Meryon, 1852). This disease, subsequently given the eponym *Duchenne muscular dystrophy (DMD),* affects one in 3,000 males; on average, one of three cases is the result of a new mutation in the ovum of the mother or grandmother. In many instances the diagnosis of DMD is made in the fourth or fifth year of life because the boy is clumsy and late in walking, while attempts to run are frustrated by frequent falls. It is interesting that the calf muscles may be considerably enlarged, as Duchenne clearly recognized (figure 1.16). Walking, when acquired, is rather ponderous as weakness of the paraspinal muscles causes the pelvis to sag when the foot is raised, which is compensated for by tilting of the body to the opposite side.

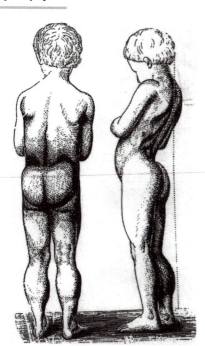

Figure 1.16　A drawing by Duchenne (1861) of a boy with muscular dystrophy. The drawing emphasizes the curious combination of muscle wasting in the arms and shoulders with muscle enlargement in the calves and buttocks. There is increased anteroposterior curvature of the lumbar spine (lordosis). Not only was Duchenne a fine artist, but he was probably the first medical photographer and electromyographer; he also invented the muscle biopsy needle and several ingenious orthotic devices (Reicke & Nelson, 1990).

Usually, by the age of 12 years, walking is no longer possible, and a wheelchair existence begins. Weakness, having started in the larger, more proximal muscles, spreads to the smaller, more distal ones. Eventually only respiration and movements of the fingers, face, and tongue may be left. Death occurs in the late teens or early 20s from bronchopneumonia or else from heart failure, since cardiac and smooth muscle are affected in the same way as the skeletal muscles.

For many years, the pathogenesis of DMD was a mystery. Some authorities postulated primary involvement of the intramuscular blood vessels or, since a third of the cases are mentally retarded, of the nervous system. The first real clue to the correct mechanism was the finding, with the electron microscope, of small discontinuities of the plasmalemma at a time when the remainder of the muscle fiber appeared normal (Mokri & Engel, 1975). It is now thought that these tears in the membrane allow the entry of Ca^{2+} from the interstitial fluid, with subsequent disruption of enzymatic process in the fiber and local necrosis (see figure 1.17). The necrotic area, usually wedge shaped, is invaded by macrophages that ingest and remove the decomposing debris. The damaged sector may then be sealed off from the remainder of the fiber by newly formed plasmalemma, as in figure 1.17. Satellite cells may also become active and aid in the repair process. Alternatively, the necrosis may progress

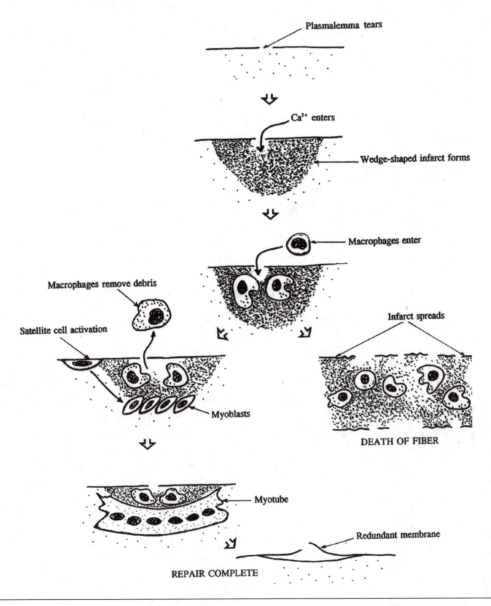

Figure 1.17   Possible sequence of events following damage to the plasmalemma of a DMD muscle fiber.

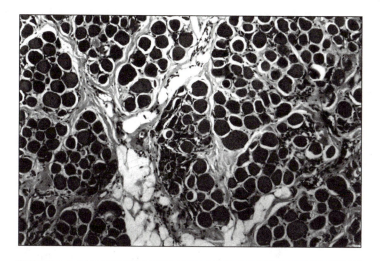

Figure 1.18 Microscopic changes in DMD muscle. In this cross section, the muscle fibers are rounded and of very different sizes; some fibers have central nuclei. There are several infiltrations of inflammatory cells where fibers are undergoing necrosis or, in some cases, starting to regenerate. The perimysial and endomysial connective tissues are more prominent and there is considerably more fat (white areas) than is typical. Hematoxilin and eosin stain.

along the length of the fiber, causing death of most or all of the muscle fiber. Some of the microscopic changes in DMD are shown in figure 1.18. In an advanced case of DMD, all the fibers within a large region of the muscle belly may be destroyed.

Although the membrane tears provided a clue to the nature of this disease, this observation raised an important question. What caused the membrane tears? The answer to this question came independently from two molecular biology laboratories. Since only males are affected by DMD, the gene for the condition must be recessive, residing on the X-chromosome. In the puzzling condition of a Belgian woman with what appeared to be DMD, it was shown by Worton and colleagues that part of the X-chromosome had switched places (translocated) with part of chromosome 21 (Ray *et al.*, 1985). In the other laboratory, Kunkel's group (Hoffman & Kunkel, 1989) found that, in approximately 5% of DMD cases, there were major deletions in the X-chromosome; in one patient, in whom DMD was associated with two other inherited diseases, the deletion was large enough to be seen with the light microscope. It was argued that the deletions, like the translocation in the female case of DMD, must have involved the site of the DMD gene; and by the technique of chromosome walking, the gene was quite rapidly identified and cloned. The DMD gene proved to be the largest yet discovered, with 2×10^6 base pairs, and it coded for a previously unrecognized protein, dystrophin, having a MW of 400 kD (see figure 1.8).

At first it was not clear where dystrophin was located in the muscle fiber, but it is now agreed that the molecule not only lies close to the plasmalemma but is attached to the latter through glycoproteins; outside the fiber, the glycoproteins are connected to *laminin* in the basement membrane (Ibraghimov-Beskrovnaya *et al.*, 1992; Matsumura & Campbell, 1994; see figure 1.8). Most of the subplasmalemmal portion of the molecule consists of a rod, 120 nm long, formed by a triple helix of 2,700 amino acids; it is possible that the rod molecules form an interlocking lattice (Hoffman & Kunkel, 1989). Due to its location under the membrane and its resemblance to the cytoskeletal protein spectrin, it is reasonable to assume that dystrophin binds to cytoskeletal actin, in association with the costameres. It is, of course, quite possible that such a large molecule serves more than one function, particularly since it is found in many neurons of the central nervous system (see Matsumura & Campbell, 1994) and is found in increased amounts at the neuromuscular junction.

In Becker dystrophy, the abnormal dystrophin is partly effective.

Becker Muscular Dystrophy

Although *Becker muscular dystrophy* (BMD) resembles DMD in many respects, it has a later onset and runs a much slower course. The gene for dystrophin is also the culprit in this condition; however, hardly any dystrophin-like proteins can be found in DMD whereas significant amounts of abnormal, smaller molecules are found in BMD. At the level of the gene, the DMD deletions probably cause "frame shifts" that result in premature termination of messenger RNA translation and the synthesis of very small, unstable fragments. In BMD, however, the reading frame is preserved, allowing the production of apparently useful portions of dystrophin (Monaco *et al.*, 1988).

Now that the structure of a muscle cell and its conglomeration into a muscle have been described, we can move on to consider a very different type of cell. This cell, which interacts so closely with the muscle fiber, is the motoneuron.

2

The Motoneuron

When a muscle is required to contract, it is sent the necessary instructions in the form of nerve *impulses (action potentials)* by large cells lying in the ventral gray matter of the spinal cord (or in a corresponding region of the brainstem). These cells are the *motoneurons,* of which there are usually a hundred or more for each muscle. A motoneuron consists of a cell body, or *soma,* and special processes termed the *dendrites* and axon (figure 2.1).

As we shall see later (chapter 8), the motoneurons also send chemical signals to the muscle fibers, via a relatively slow, energy-dependent mechanism called *axoplasmic transport.* For the

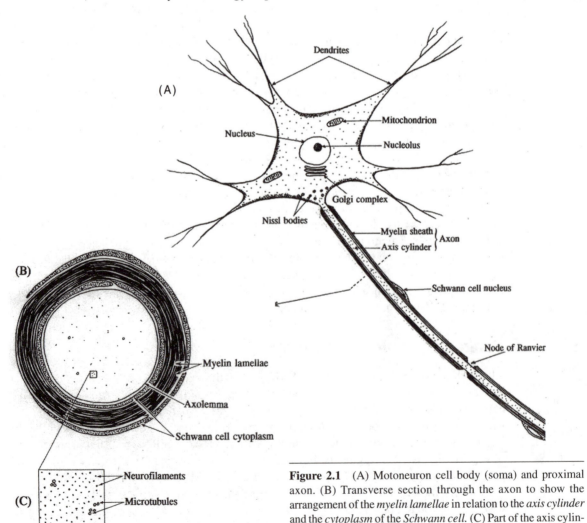

Figure 2.1 (A) Motoneuron cell body (soma) and proximal axon. (B) Transverse section through the axon to show the arrangement of the *myelin lamellae* in relation to the *axis cylinder* and the *cytoplasm* of the *Schwann cell.* (C) Part of the axis cylinder, enlarged to show the *microtubules* and *neurofilaments.*

moment, however, it is necessary to consider the general features of the motoneuron, including its processes. Following this is a more detailed examination of the structure of the axon, including the fatty coat that surrounds it.

General Features of Motoneurons

The soma of the motoneuron contains the *nucleus* of the cell together with various structures that will be considered later. The function of the dendrites is to receive signals from other neurons, while the major role of the axon is to transmit a resulting message to the muscle fibers in the form of action potentials. An action potential is a triggered and patterned exchange of ions across the membrane of an excitable cell. This exchange propagates along the length of the fiber (see chapter 9). The axons also transport chemical messages and organelles between the motoneurons and the muscle fibers in both directions: *anterograde* (toward the muscle fibers) and *retrograde* (toward the soma).

Some human axons are extremely long.

The lengths of the axons vary considerably. In the case of a man 180 cm tall, an axon running from the lumbosacral region of the spinal cord to one of the plantar muscles would be about 125 cm long. In contrast, the fibers passing in the cranial nerves to the external ocular muscles, or to the muscles of the face or tongue, would measure only about 1/10 of this length. As the axon nears the muscle, it splits, with each branch dividing further, so that eventually a single parent axon may contact several hundred muscle fibers. The contact region between an axonal twig and a muscle fiber is a specialized *synapse* called the neuromuscular junction. The synapse possesses special structural and functional features that enable the impulse in the nerve fiber to be translated into an impulse in the muscle fiber through the action of a chemical link, the transmitter *acetylcholine (ACh)*. These are described in chapter 3. In the normal adult mammal, most muscle fibers have only one neuromuscular junction and are innervated by a single motoneuron. Sherrington (1929) pointed out that a single motoneuron and the colony of muscle fibers that it supplies could be considered as a functional entity, because each time the nerve fiber discharges an impulse, the muscle fibers of the colony are excited together; Sherrington described this entity as a *motor unit* (see chapter 12).

Several methods have been used to study the structure of the motoneuron.

Haggar and Barr (1950) examined serial sections of cat spinal cord and were able to make three-dimensional models of the soma and dendrites. Another technique, devised by Chu (1954), has been to dissect pieces of ventral horn from recent autopsy specimens of human spinal cord and to make them into a crude suspension with physiological saline. Within this suspension, some motoneurons are completely dissociated from other tissue and yet retain considerable lengths of axons and dendrites suitable for examination. Somewhat later, a technique for injecting the dye procion yellow through a micropipette into the soma of a neuron was applied to motoneurons (Kellerth, 1973). The dye diffuses into the dendrites and proximal axon, enabling the full extent of the cell to be visualized. *Horseradish peroxidase (HRP)* can be used for the same purpose; even the fine dendritic branches are permeated by this material, which can then be stained (figure 2.2).

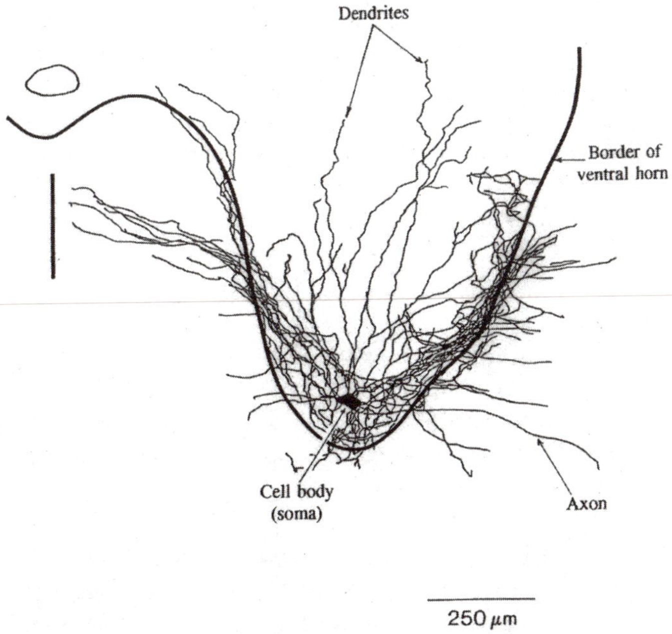

Dendrites

Border of ventral horn

Cell body (soma)

Axon

250 μm

Figure 2.2 Cell body and processes of a motoneuron. The motoneuron is from the cervical region of the cat spinal cord and has been injected with horseradish peroxidase. Note the extensive territory covered by the dendritic branches, several of which extend into the white matter surrounding the ventral horn.

Courtesy of Dr. Ken Rose.

More recently, a number of substances have been developed that can be applied to cut nerves or injected into muscles in small volumes in order to identify the location of the cell bodies in the cord, the three-dimensional structures of the individual neurons, and distributions and densities of motoneuron pools. Some are fluorescent (Fluorogold, True Blue, Fast Blue, fluorescent latex microspheres, dextran/rhodamine and dextran/fluorescein conjugates), allowing more than one tracer to be used concurrently, enabling examination of the distributions of motoneurons innervating different targets in the same preparations.

For information concerning the fine structure of the motoneuron, it has been necessary to turn to the electron microscope, and several comprehensive reports are available (see, for example, Bodian, 1964). The various parts of the motoneuron will now be considered in more detail.

Within the ventral gray matter of the spinal cord, the cell bodies of the motoneurons are arranged in columns parallel to the long axis of the cord. The cells in each column are usually connected to the same muscle, and, as a consequence, the muscle receives its nerve supply from two or more adjacent segments of the spinal cord (figure 2.3). The cells supplying the *extrafusal muscle fibers* are termed α-*motoneurons* in order to distinguish them from the γ-*motoneurons* (fusimotoneurons) innervating the small fibers within the *muscle spindles*.

Muscle is innervated by a column of motoneurons in the spinal cord.

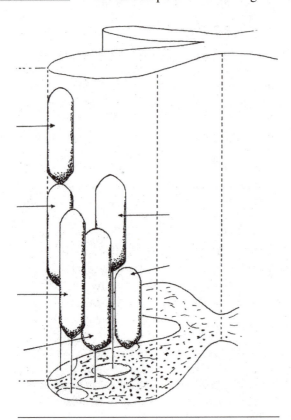

Figure 2.3 Longitudinal columns of α- and γ-moto-neurons in the lumbosacral region of the spinal cord. The diagram is schematic and does not show the overlaps in the columns or the longitudinal curvature of the ventral horn.

Based on the data of Romanes 1951.

Motoneuron Soma

The cell bodies of the α-motoneurons are much larger than those of γ-motoneurons and may have diameters of 100 μm; an average value for an adult human is about 70 μm. Large though these cell bodies are, their volumes are very much less than those of their axons. Thus, the axon of a lumbosacral motoneuron contains approximately 100 times as much *cytoplasm* as the soma. Indeed, if the motoneuron soma were enlarged to the size of a tennis ball, the axon would stretch the length of seven football fields!

The most prominent feature of the soma is the relatively large nucleus, which is bounded by a wavy double membrane and contains nucleoplasm and a *nucleolus*. Within the nucleoplasm are the 23 pairs of chromosomes, which together contain the 30,000 or so genes characteristic of the human genome. Only a fraction of the genes are expressed in the motoneuron, however, and many of these will be different from those expressed in the muscle fiber. It is the genes that are ultimately responsible for directing the protein synthesis in the neuron. They do so by sending instructions, in the form of messenger *RNA,* out into the cytoplasm of the cell, possibly through the pores in the nuclear membrane, which can be seen with the electron microscope. Since human motoneurons do not multiply after birth, the chromosomes cannot be distinguished in the nucleoplasm of the motoneuron; all that can be seen in the nucleoplasm are irregularly distributed particles, from 1 to 2 μm in diameter. Within the nucleus, the nucleolus appears as a large structure, measuring about 4 μm across and consisting mostly of ribosomal RNA. This RNA is made on the *DNA* templates of the genes *(transcription)* and is eventually passed out from the nucleolus into the cytoplasm to take part in the synthesis of proteins.

Genetic instructions are sent from the nucleus via the nucleolus.

The cytoplasm of the motoneuron soma contains several different types of organelles, the most prominent of which are the *Nissl bodies.* These bodies, which are especially numerous near the base of the axon, range from 0.5 to 3 μm in diameter, and electron micrographs show them to be composed of tightly packed arrays of *endoplasmic reticulum.* The membranes of the reticulum are studded with ribosomes, giving them a rough appearance. Acting on instructions received through messenger RNA, the ribosomes engage in the synthesis of proteins *(translation)*. Their behavior during the reactions of the cell following injury to the axon is of great interest (see chapter 18).

In addition to the Nissl bodies, the cytoplasm of the motoneuron contains many mitochondria. There are also *microtubules, microfilaments,* and *neurofilaments,* most of which protrude into the dendrites and the axon. The *Golgi apparatus* is the name given to a system of flattened tubular channels found near the nucleus; one of its functions is thought to be the packaging of proteins prior to their delivery into the axon. Finally, with advancing age, the motoneuron cytoplasm contains increasingly large masses of pigment, the *lipofuscin granules;* the significance of this material is unknown.

Cytoskeletal Proteins in the Motoneuron

The motoneuron, like other cells in the body, contains cytoskeletal proteins. In the motoneuron, three of these are especially plentiful and serve important functions; they are *actin, tubulin,* and the neurofilament proteins.

Functions of Actin, Tubulin, and the Neurofilament Proteins

The actions of these three cytoskeletal proteins are the following:

- By forming the cytoskeleton of the motoneuron, they not only strengthen the cell and give it a characteristic shape, but they also anchor the nucleus and some of the major structures within the interior of the cell.
- By a process of elongation, they help to create new segments of *axis cylinder,* and so play a critical role in nerve *regeneration.*
- By mechanisms described in chapter 8, actin and tubulin provide a pathway for the conveyance of cellular material from the soma to the axon, dendrites, *Schwann cells,* and muscle fibers; this process is axoplasmic transport.

Actin is a pear-shaped globular protein with a MW of 42 kD and a diameter of 4 nm; each molecule is stabilized by being tightly bound to one Ca^{2+} ion. In the presence of *ATP,* the globular (G) actin polymerizes to form filamentous (F) actin. Each actin filament appears as if two strands of globular molecules are twisted round each other to form a helix, though, in fact, actin strands cannot exist on their own. The spacing between successive turns of the helix is approximately 37 nm (figure 2.4A). In motoneurons, microfilaments are formed from actin monomers. Actin is also considered in chapter 11, in which a description is given of its role in the sliding filament mechanism of muscle contraction.

Tubulin is another globular protein and has a MW of 50 kD; slightly different forms of the protein, α-tubulin and β-tubulin, link together to form a dimer. The tubulin dimers can remain in solution in the cytoplasm or can polymerize under the influence of ATP to form protofilaments, with the α-tubulin of one dimer attached to the β-tubulin of another. Characteristically, 13 of the protofilaments then link up to form each helical loop of a hollow structure, the microtubule (figure 2.4B).

The three neurofilament (NF) proteins are NF-L, NF-M, and NF-H, with MWs of 70, 160, and 200 kD, respectively. Each filament protein molecule has an amino head and a carboxy-terminal tail; in between is the rod domain, in the form of an α-helix. The three types of protein line up with each other to produce neurofilaments (figure 2.4C), 7 nm in diameter. The carboxy-terminals of the NF-M and NF-H molecules have large numbers of sites that can be phosphorylated; *phosphorylation* causes the tails to stick out, where they may connect to other neurofilaments and form a lattice (figure 2.4C). At the same time, the projecting tails will delay the movement of the neurofilaments down the axon, hence the slow component of axoplasmic transport (see chapter 8). The effect of phosphorylating the amino head terminal is quite different, for it promotes the disassembly of the filament network by depolymerization (see Nixon & Sihag, 1991).

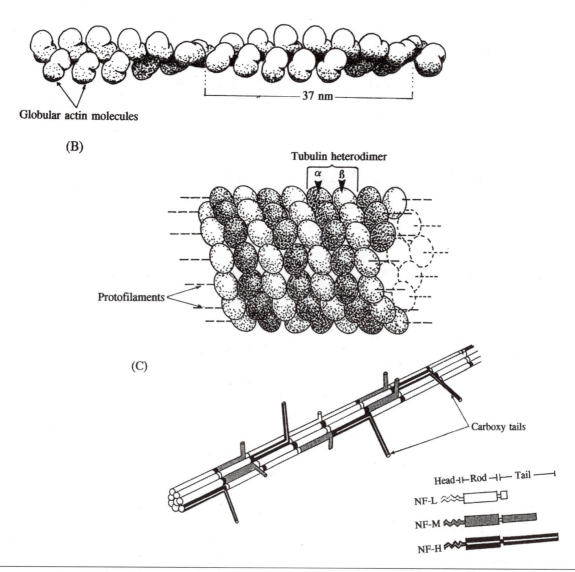

Figure 2.4 (A) Part of an actin filament, formed by globular actin molecules packed into a tight helix, giving the appearance of two strands twisted around each other. (B) A short length of microtubule; the microtubule is formed by 13 parallel protofilaments, each of which consists of a series of αβ-tubulin dimers. (C) A single neurofilament, consisting of three types of interlocking subunits. The carboxy (COOH) tails of the subunits are protruding from the neurofilaments.

A and B: Adapted from Amos *et al.* 1976; C: From R. A. Nixon and R. K. Sihag, 1991, "Neurofilament phosphorylation: A new look at regulation and function," *Trends in Neurosciences* 14: 502. Copyright 1991 by Elsevier Trends Journals. Adapted with permission.

Micro-filaments and microtubules can lengthen or shorten.

How Microfilaments and Microtubules Work

Both microfilaments and microtubules are polar structures, since their respective subunits have specific orientation within the polymers. The ends of the actin filaments and microtubules can grow by *polymerization* or shorten by depolymerization. The + (plus) end can lengthen much more rapidly than the – (minus) end. Capping proteins can bind to one end or the other of the polymer, stabilizing that end and perhaps also attaching it to another structure at the surface or in the interior of the cell. This anchoring of the microfilaments takes place via plasma membrane proteins such as *spectrin, ankyrin, vinculin, talin,* and *integrin.*

Several drugs are known that bind to actin and tubulin and thereby interfere with their structure and function. *Colchicine* causes depolymerization of the microtubules by inhibiting the addition of further tubulin molecules, while the *cytochalasins* have a similar effect on actin. *Vinblastine* and *vincristine,* two drugs derived from the periwinkle and used in cancer treatment, induce the formation of tubulin crystals. In contrast, *taxol* stabilizes the microtubules, and *phalloidin* does the same to the actin filaments. These drugs are useful in evaluating the functions of these cytoskeletal proteins.

Axon, Dendrites, and Glia

We shall now consider in more depth the special processes that radiate from the cell body of the motoneuron and also the *glial cells* that surround the cell body in the spinal cord. Of the special processes, it is the axon that will receive the most attention, in part because it has been studied more extensively than the dendrites. Of particular interest are the structure and physiological properties of the fatty layers (myelin sheath) that form the insulating coat of the axon.

Dendrites of Motoneurons

The dendrites of motoneurons branch extensively.

The motoneurons possess several dendrites that radiate from the cell body in dorsal, superior, and inferior directions; the dendrites become progressively narrower and also divide into several branches. The remarkable complexity of the dendritic arborization is evident in figure 2.2, which shows a single motoneuron injected with dye. It can be seen that the dendrites extend over considerable distances in the gray matter of the spinal cord, enabling them to receive information from the axons of a variety of other neurons. The connections between the axon and dendrites are made at synapses, which, in their fine structures, have certain features in common with the neuromuscular junction (chapter 3); the terminations of the axon twigs are expanded to form boutons, inside which many synaptic vesicles can be distinguished. The packing density of the boutons on the dendritic membrane increases as the dendrites narrow; more proximally, there are gaps into which processes from the glial cells project. Some synapses are also found on the membrane of the soma.

The boutons are the loci for contact from excitatory and inhibitory neurons, and their summed influence governs the state of excitation or inhibition of the motoneuron. While ACh is the only neurotransmitter at the neuromuscular junction, the principal neurotransmitters at synapses onto motoneurons include *gamma-aminobutyric acid* (GABA) and *glycine,* which are inhibitory, and *glutamate,* which is excitatory. Receptors to these neurotransmitters are thus found on the postsynaptic (i.e., motoneuron) side of these synapses.

Role of Glial Cells

Glial cells provide structural and metabolic support of the motoneurons.

The glial cells of the central nervous system do not conduct impulses and therefore have no direct role in signaling. Their importance lies in various supportive functions; they provide a structural matrix for the neurons and also control the passage of substances from the capillaries into the neuronal milieu. In addition, it appears that the metabolism of the glial cells is at least partly linked to that of the neurons. In the case of the motoneuron, several oligodendroglial cells cluster around its periphery and occupy any spaces available between the synaptic knobs (boutons) of the incoming fibers. One of the metabolic activities of glial cells becomes particularly important when the motoneuron is discharging impulses: the removal of K^+ from the narrow interstitial spaces by means of the Na^+-K^+ pump (see chapter 7). The glial cells can also assist in the removal of excitatory and inhibitory transmitter molecules diffusing from the narrow *synaptic clefts* that separate the membranes of the boutons and dendrites. It has been suggested that, within the central nervous system, the oligodendroglia provide the myelinated axons of neurons with Na^+ channels at the *nodes of Ranvier* (Bevan *et al.,* 1985).

Motoneuron axons have myelin sheaths interrupted by nodes.

The Motoneuronal Axon

The axon arises from a conical protrusion of the motoneuron soma known as the *axon hillock* and extends to the muscle as a long axis cylinder. The cylinder is bounded by a membrane, the *axolemma,* that has structural and functional features similar to those of the muscle fiber *plasmalemma.* Within the cytoplasm of the axis cylinder are arrays of small microtubules and neurofilaments that run in the long axis of the fiber between the soma and the many neuromuscular junctions. The microtubules are about 20 nm wide, while the neurofilaments are smaller, being approximately 7 nm thick (see p. 26). An interlacing system of wider tubules forms the *endoplasmic reticulum.*

The axon hillock, also known as the *initial segment,* is the most excitable region of the motoneuron and is thus the locus from which action potentials are generated. It is in this region where one finds the highest concentration of voltage-regulated sodium conductance channels, and relatively few synaptic contacts from other neurons. This specialized region of the motoneuron is considered an integrator and transducer, in that it sums the excitatory and inhibitory influences from the soma and dendrites and transduces this information into producing action potentials (if the excitatory balance is high enough) of a certain frequency (depending on the continuing strength of the excitatory "wave").

All the motor axons have lipid coverings surrounding the axis cylinders; these are the *myelin sheaths,* and they are derived from a type of *satellite cell* known as the Schwann cell. The motor and sensory axons in humans are not quite as thick as those in the cat and monkey; if the myelin sheaths are included in the measurements, their diameters range from approximately 2 to 14 μm (figure 2.5). In the *ventral roots,* which contain motor fibers only, the axon diameters fall into a bimodal distribution with a separation at about 7 μm. The population of fibers with large (7-14 μm) diameters corresponds to the α-motor axons supplying the extrafusal muscle fibers; the smaller axons are the γ-axons innervating the *intrafusal muscle fibers* inside the muscle spindles.

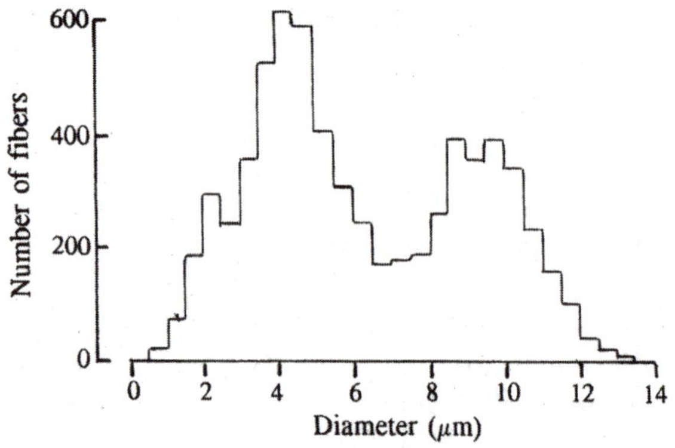

Figure 2.5 Diameters of myelinated nerve fibers in a specimen of sural (cutaneous) nerve from a 14-year-old boy. The range of diameters and the bimodal distribution of values are typical of both motor and sensory nerves in humans.

Adapted, by permission, from P.J. Dyck, P.K. Thomas, and E.H. Lambert, 1975, "Sensory potentials of normal and diseased nerves," *Peripheral Neuropathy* 1: 444.

Myelin sheaths are formed by Schwann cells.

Myelin and Nodes of Ranvier

The myelin sheath has been studied with the electron microscope and has been shown to consist of spiral wrappings of tightly packed membranes laid down by the Schwann cell (Geren, 1954). The myelin sheath is not a continuous structure but is interrupted at regular intervals by the *nodes of Ranvier.* It will be seen in chapter 9 that the nodes of Ranvier are the sites at which the action potential is generated as it travels down the axon. The distance between two successive nodes is an *internode* and corresponds to the territory of a single Schwann cell. At birth, the internodes measure about 230 μm in human motor nerves; as the limb grows, the lengths of the internodes increase, because the number of Schwann cells remains the same. Vizoso (1950) found that the longest internodes in the ulnar and anterior tibial nerves of an 18-year-old subject were 1,100 and 980 μm, respectively, indicating that there had been a fourfold increase in their lengths. If, during the course of a disease process, the myelin sheath is destroyed and then remade, or if the axon is damaged and then regenerates, many of the newly formed internodes are much shorter than in the normal adult (figure 2.6). The explanation for this discrepancy is that during the recovery process the Schwann cells divide and thereby increase their number; each new cell occupies a smaller length of axon.

Schwann cell fingers and gap substance surround the axis cylinder at the nodes.

The electron microscope has been invaluable in revealing the ultrastructure of the nodes of Ranvier. Figure 2.7 summarizes, in diagram form, the findings of Williams and Langdon (1967). On the left side of the figure, the axon has been bisected by a vertical incision and a segment has been removed; to the right of the node, some of the *basement membrane* has been peeled away. In the bisected segment, it is apparent that each *myelin lamella* (several of which form the tightly packed

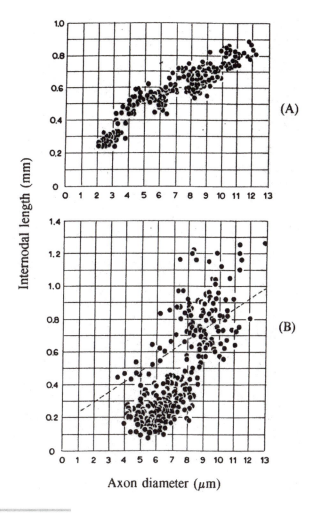

Figure 2.6 Internodal distances in teased anterior tibial nerve fibers of (A) an 18-year-old woman and (B) an 80-year-old man. The normal linear relationship between internodal length and axon diameter is disrupted in the older subject due to chronic nerve damage.

Adapted, by permission, from A.D. Vizoso, 1950, "The relationship between internodal length and growth in human nerves," *Journal of Anatomy* 84: 343-353.

membranes referred to in the previous paragraph) is attached to the axolemma, or plasma membrane covering the axon, by a small loop, evident in the figure by a slight enlargement at each lamella–axis cylinder interface. On the other side of the node *(paranodal region)* the myelin sheath is seen to be indented by columns of Schwann cell cytoplasm. As the columns approach the node, they first become confluent and then send finger-like processes toward the nodal portion of the axolemma.

The processes are embedded in an amorphous extracellular material known simply as *gap substance.* Studies by Landon and Langley (1971), among others, have demonstrated that the gap substance is composed of mucopolysaccharides, and that the anionic groups of these molecules exert a powerful electrostatic attraction for cations. Landon and Langley suggest that the gap substance may serve to maintain a high concentration of Na^+ ions available for flow across the axolemma during the action potential. Similarly, the gap substance might limit the diffusion of K^+ from the vicinity of the node following an impulse and hold it in readiness for active transport back into the fiber.

Cytoplasmic material is exchanged between the axis cylinder and Schwann cells.

It is tempting to assign a functional role to the Schwann cell processes that project to the nodal axolemma. It is possible that these convey ATP to the nerve membrane from the mitochondria in the columns of Schwann cell cytoplasm; the ATP might then be used to supply the energy required for Na^+ and K^+ pumping (see Landon & Langley, 1971). Another possible role for the Schwann fingers, analogous to that suggested by Ritchie (see Bevan *et al.,* 1985) for glial cells in the central nervous system, is to provide the axolemma with Na^+ channels.

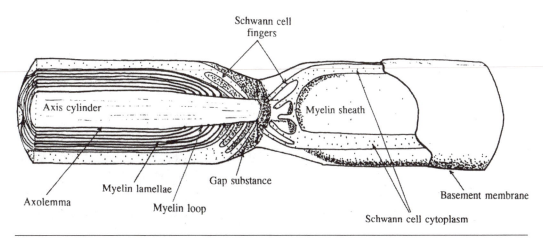

Figure 2.7 Structure of a single node of Ranvier, shown in longitudinal section.

An interesting feature of the myelin sheath is the *Schmidt-Lanterman incisures* (clefts). At each of these regions a narrow cytoplasmic process is sent from the Schwann cell to the axis cylinder. The projection does not actually penetrate the myelin wrappings but instead winds its way around the axon, following the plane of separation between the myelin lamellae. Using time-lapse cinematography, Gitlin and Singer (1974) have observed that the Schmidt-Lanterman clefts are not static structures but can be open or closed at different times. It is possible that the open phase enables materials, including *trophic* molecules (see chapter 18), to be passed between the axis cylinder, on the one hand, and the Schwann cell and the endoneurial space of the nerve fiber, on the other. Even the remainder of the myelin sheath is not stationary but can be seen to develop small indentations that then regress; possibly this type of movement is related to some component of axoplasmic flow (see chapter 8 and Weiss, 1969).

Structure of Peripheral Nerves

Axons are held together by connective tissue.

An account of the axon would not be complete without consideration of the way the axons (nerve fibers) are supported by the *connective tissue* within a peripheral nerve. Under the microscope, each nerve fiber is seen to be enclosed in a sheath of connective tissue, or *endoneurium*. Within the nerve trunk, the fibers are collected into a number of bundles or fasciculi, each being bounded by a condensation of connective tissue termed the *perineurium*. Last, there is a relatively thick coat of connective tissue that invests the whole nerve trunk; this is the *epineurium* (figure 2.8). Surrounding the nerve trunk are the vessels conveying blood to and from the fibers; these are the *vasa nervorum*. Small arterioles and venules pierce the epineurium at intervals, giving rise to an intraneural system of capillaries.

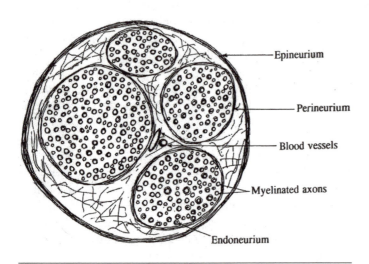

Epineurium

Perineurium

Blood vessels

Myelinated axons

Endoneurium

Figure 2.8 Transverse section through a small nerve trunk to show the different types of connective tissue wrapping. The general arrangement is reminiscent of that for muscle (see figure 1.4). The nonmyelinated nerve fibers have not been depicted.

Applied Physiology

There are literally hundreds of causes of peripheral nerve degeneration, but the disorder we shall consider now is hereditary. To make matters more interesting, the abnormal gene has been located, even though the function of the protein product is not yet known.

In 1886 two neurologists in France and one in England described patients with what appeared to be the same condition—severe *weakness* and wasting of muscles below the knees and in the hands, together with loss of sensation in the hands and feet. This condition, known as Charcot-Marie-Tooth (CMT) disease or peroneal muscular *atrophy,* is hereditary and produces its effects through degeneration of the peripheral nerves.

Charcot-Marie-Tooth Disease

Charcot-Marie-Tooth nerves show demyelination and Schwann cell hypertrophy.

Usually clinical evidence of CMT disease appears in the late teens, and the first signs are in the feet, which lose muscle tissue and toe movement and are highly curved (pes cavus). In most cases, the legs have been likened to inverted champagne bottles or to stork legs because of the extreme thinning below the knee, together with some wasting of the lower thigh (figure 2.9A). Several forms of the condition are now recognized, the distinctions depending on the mode of inheritance, the presence or absence of sensory nerve fiber involvement, and, especially, whether or not there is loss of myelin (demyelination) in the axons. In the latter cases, the nerves are often thickened and can be readily felt by an examiner; nerve conduction studies in the *electromyography (EMG)* laboratory make the diagnosis easy, for the impulse conduction velocities are greatly reduced (see chapter 9).

Under the microscope, many nerve fibers can be seen to have abnormal myelin sheaths, and some are completely *demyelinated*. A striking observation is *hypertrophy* of the Schwann cells, so that individual axons are surrounded by concentric layers of Schwann cell cytoplasm, giving the impression of an onion bulb (figure 2.9B). The bulb is thickened further by connective tissue in the form of *collagen* fibrils that run in the same direction as the nerve fiber. As the disease advances, more and more axons disappear, until some muscles become totally denervated.

Search for the Primary Defect in Charcot-Marie-Tooth Disease

Is the primary defect in the axis cylinder or Schwann cell?

In the demyelinating form of CMT disease, the question arises: Is the genetic abnormality one that affects the motor and sensory neurons primarily, with secondary involvement of the Schwann cells and their myelin sheaths, or is it the other way around? One experimental approach has been to study a strain of mouse (Trembler) that also has a demyelinating peripheral neuropathy. It was possible to transplant a segment of nerve from a Trembler mouse into the sciatic nerve of a normal littermate. When the nerve fibers of the host animal regenerated through the graft, they became ensheathed by Schwann cells from the donor animal and developed abnormal myelin sheaths (Aguayo *et al.,* 1977). Clearly, in the Trembler animals it is the Schwann cell rather than the axis cylinder that is at fault. Subsequently, grafts of nerve from CMT patients into immunodeficient but otherwise normal mice gave similar results (Aguayo *et al.,* 1978; see, however, Dyck *et al.,* 1978).

More recently, molecular genetics studies have identified a gene on chromosome 17 as responsible for demyelinating CMT in most of the families. Further, it is now known that this gene codes for one of the proteins associated with the myelin sheath, peripheral myelin protein (PMP-22; see figure 2.9C). Since the involved region of chromosome 17 is duplicated in CMT, it would appear likely that there is too much PMP-22, but the function of the protein in relation to the myelin sheath is not understood at present (Suter *et al.,* 1993).

It is time now to consider the structural features of the neuromuscular junction. Here, there are specialized structures and unique proteins that allow the propagating nerve action potential to trigger a chemical signal (release of neurotransmitter) to activate the muscle fiber.

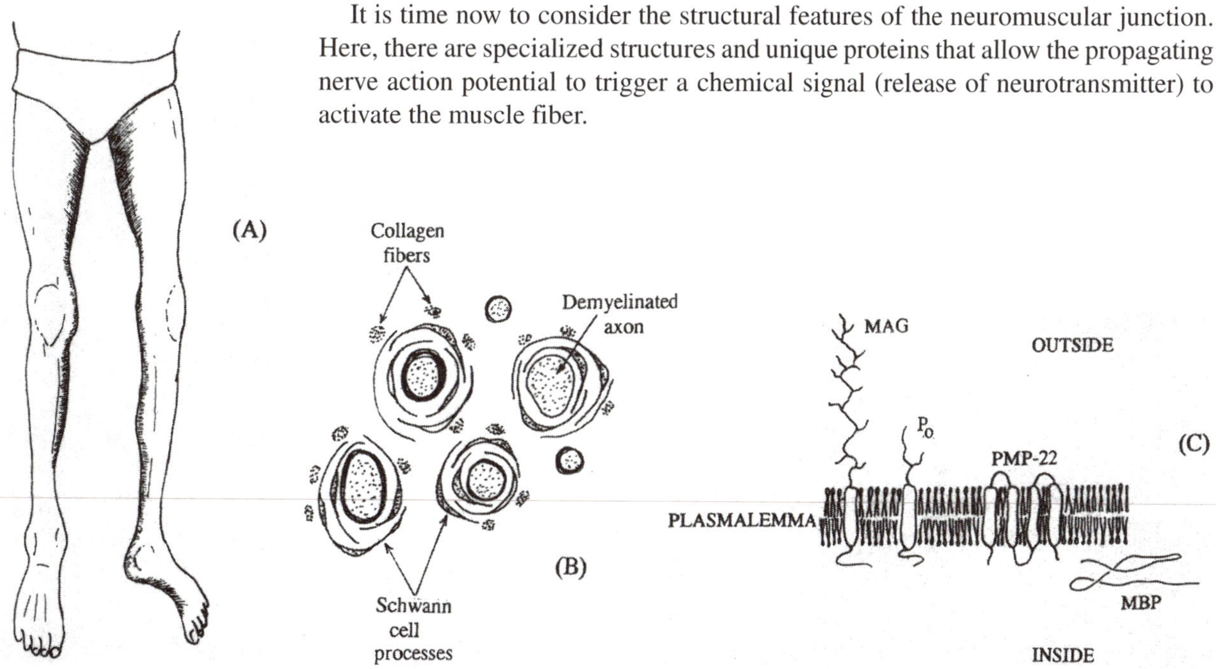

Figure 2.9 (A) Typical thinning of muscles in lower part of leg, and prominent curvature of feet, in demyelinating CMT disease. (B) Cross section through peripheral nerve in demyelinating CMT disease. Some axons are lost, and others have thin or absent myelin sheaths; some axons have an "onion bulb" appearance (see text). (C) Protein components of normal myelin sheath. MAG = myelin-associated glycoprotein; MBP = myelin basic proteins; PMP-22 = peripheral myelin protein-22; P_o = protein zero. See also Suter *et al.* (1993).

3

The Neuromuscular Junction

Now that the *motoneuron* has been described, we can explore the area of contact that its axon makes with the muscle fiber. This small but highly developed region is called the neuromuscular junction, or *synapse,* and it is here that excitation spreads from the axon to the muscle fiber. Since insufficient current flows from the axon to stimulate the muscle fiber directly, the nerve ending releases a chemical, *acetylcholine (ACh)*, to bring about the necessary transmission of the axonal signal.

General Features of the Neuromuscular Junction

The structure of the nerve ending reflects its ability to release ACh, and the underlying muscle fiber is constructed so as to capture as many of the ACh molecules as possible. In keeping with the organization of the first part of this book, this chapter deals with the structure of the neuromuscular junction; the opportunity to visualize the activity of the junction will come in chapter 10.

As the motor axon approaches the muscle fiber, it loses its *myelin* sheath and divides into several small twigs, the outermost ones in mammals often forming an incomplete ring (figure 3.1A). The twigs lie in grooves, or gutters, on the surface of the muscle fiber and run short distances before terminating; the twigs have irregular expansions (boutons) corresponding to the sites of transmitter release. The region of the muscle fiber under these twigs is termed the motor *end-plate;* it includes the *plasmalemma* and also a mound of *sarcoplasm,* the *sole-plate.* Within the sole-plate are collected a number of muscle fiber nuclei as well as many mitochondria, ribosomes, and pinocytic vesicles. Lying over the whole end-plate are membrane-covered cytoplasmic processes derived from the *Schwann cells.* With the electron microscope, it can be seen that the membrane of the axon terminal is separated from the muscle fiber plasmalemma by a distinct gap measuring about 70 nm; this is the *primary synaptic cleft* (figures 3.1 and 3.2). The primary cleft is interrupted by repeated invaginations of the plasmalemma into the sole-plate; each *junctional fold* forms one *secondary synaptic cleft.* The secondary clefts are approximately 0.5 to 1 μm deep and are rather wider at their terminations than at their necks. In amphibians, the junctional folds are consistently perpendicular to the long axis of a motor nerve twig, but in mammals, the arrangement is less regular because of the frequent bending of the twigs (figure 3.3).

The muscle membrane exposed to the primary cleft and the upper parts of the secondary clefts is thickened due to the presence of closely packed *acetylcholine receptors (AChRs).* The AChRs are held in position by 43-kD protein molecules that link them, via *spectrin,* to a network of *actin* filaments in the superficial *cytoplasm* of the muscle fiber. The crests of the junctional folds contain other proteins that also serve to tether the plasmalemma to the fiber cytoskeleton and to the overlying

A narrow space separates the membranes of the muscle fiber and nerve ending.

Synaptic clefts accommodate more receptors for ACh.

basement membrane, thereby maintaining the structure of the junctional folds. *Rapsyn, utrophin,* and α-dystrobrevin-1 are found near the tops of folds, while *ankyrin,* α-dystrobrevin-2, and *dystrophin* serve this function at the bottoms of the folds (Sanes & Lichtman, 1999). The functional benefit in having a folded plasmalemma is that the arrangement increases the surface area of the muscle fiber that can combine with ACh released from the nerve terminals. Although the surface area is expanded 8 to 10 times, the paucity of AChRs in the deeper parts of the secondary clefts makes for only a threefold increase in the AChR-bearing portion. The folding of the plasmalemma also increases the amount of *acetylcholinesterase (AChE)* that can be accommodated at the synapse, since this enzyme is located throughout the primary and secondary clefts, including the farthest reaches of the latter.

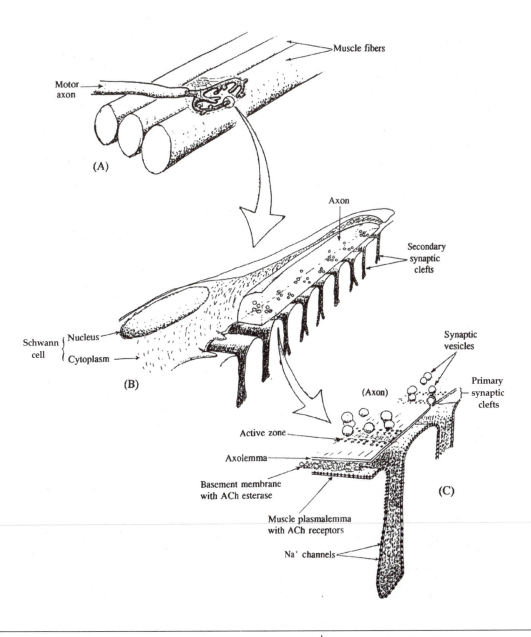

Figure 3.1   Mammalian neuromuscular junction, shown at progressively higher magnification. The different structures in the junction have not been drawn to scale. Note the synaptic vesicle emptying in C. See also figure 9.11.

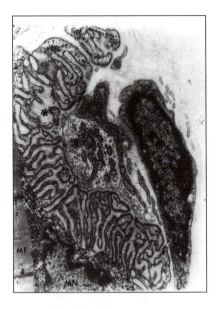

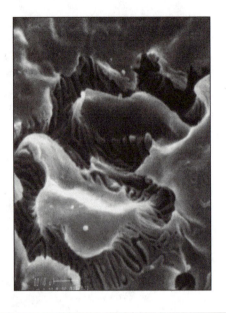

Figure 3.2 Electron micrograph of a human neuromuscular junction showing an axon terminal (Ax) lying beneath a Schwann cell (S, Schwann cell *nucleus*) and covering an array of secondary synaptic clefts (*); the latter are quite deep and often branch. The muscle fiber also displays a *myonucleus* (MN) and a superficial *myofibril* (MF). The arrow indicates the primary synaptic cleft. The various structures in the junction can be identified by referring to figure 3.1. Note the synaptic vesicles in the axon terminal. 30,600×.

Courtesy of Dr. A. G. Engel.

Figure 3.3 Junctional folds and secondary synaptic clefts in a mouse neuromuscular junction. The axon terminal has been removed and the exposed muscle fiber surface examined with a scanning microscope. 10,000×.

Journal of Neurocytology, Vol. 12, 1983, pg. 13-15, Scanning the light microscopic study of age changes at a neuromuscular junction in the mouse, M.A. Fahim, J.A. Holley, and N. Robbins, panel 6, with kind permission of Springer Science and Business Media.

Muscle Fiber Acetylcholine Receptors

The AChR was not only the first transmembrane ion channel to be visualized and purified, but also an ion channel that could be studied in detail by a variety of electrophysiological techniques. One of the factors facilitating the early investigations was the availability of naturally occurring toxins that bind to the receptor and can be labeled for microscopic identification.

The most frequently used AChR-binding toxin is *α-bungarotoxin,* derived from the highly poisonous venom of an Asian snake, the banded krait. This toxin, labeled with either [³H] or *horseradish peroxidase,* has been used to map the distribution of the AChRs at the neuromuscular junction and to determine their prevalence in the muscle plasmalemma. In the receptor-rich areas of the muscle plasmalemma, electron microscopy has shown that the latter contains particles some 6 to 12 nm in diameter occurring at intervals of 10 to 15 nm (Rosenbleuth, 1974). These particles are the AChRs themselves, and there are approximately $10,000 \cdot \mu m^{-2}$ of membrane; since each receptor has two sites for binding ACh, the density of the binding sites is about $20,000 \cdot \mu m^{-2}$.

Acetylcholine receptors are seen with the electron microscope when labeled with α-bungarotoxin.

High-power electron micrographs show the receptor to have a central pore with several rounded structures around it; at extreme magnification, assisted by computer processing, the receptor can be seen to span the muscle plasmalemma and to project into the *synaptic cleft* (figure 3.4A).

The purification of the AChR protein was first accomplished using the electric ray, *Torpedo californica.* This fish has an organ on each side of its head that is capable of delivering shocks of a thousand volts or so, sufficiently strong to stun prey or aggressors. The electric organ is itself composed of modified muscle fibers with an abundance of nerve–muscle contacts and AChRs. The purified receptor protein was found to have a MW of 268 kD and to consist of five glycopeptide subunits, α, α, β, γ, δ (Raftery *et al.,* 1980); the two α subunits are identical, and each contains a

The AChR has five subunits, two of which are identical.

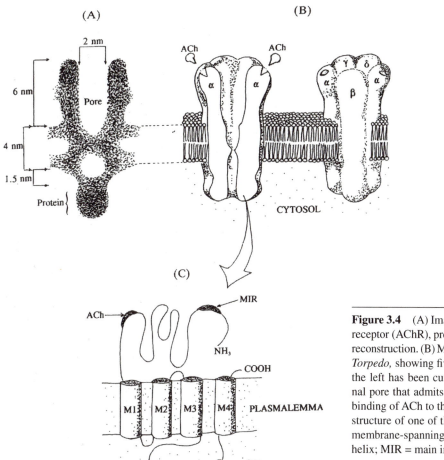

(A)

2 nm

6 nm

Pore

4 nm

1.5 nm

Protein {

(B)

ACh   ACh

γ   δ

α   α

α   α

β

CYTOSOL

(C)

ACh →

MIR

NH₃

COOH

M1 M2 M3 M4 PLASMALEMMA

MA

Figure 3.4 (A) Image of a single acetylcholine receptor (AChR), prepared by three-dimensional reconstruction. (B) Model AChR from electric ray, *Torpedo,* showing five subunits. The receptor on the left has been cut through to show the internal pore that admits Na⁺ and K⁺ ions, following binding of ACh to the α subunits. (C) Molecular structure of one of the α subunits, showing four membrane-spanning regions. MA = amphipathic helix; MIR = main immunogenic region.

A: Adapted from Toyoshima and Unwin 1988; C: Adapted from Beeson and Barnard 1990.

single ACh-binding site (figure 3.4B). The messenger RNAs (mRNAs) corresponding to the four different types of subunit are, like the AChRs, abundant in the electrical organ of *Torpedo,* and provided one approach to cloning the four receptor genes. Another technique was to construct and label a nucleotide coding for part of the AChR and to allow it to hybridize with fragments in the complementary DNA library (Noda *et al.,* 1984). Once the genes had been cloned, the complete amino acid sequences of the subunits could be easily determined, and inspection of these suggested that each subunit contained four regions that spanned the muscle fiber membrane (figure 3.4C). One of these regions, M2, is thought to form the lining of the central pore. Three rings of negatively charged amino acid residues, aspartate or *glutamate,* flank the M2 region and make the pore selective for cations.

In mammalian muscle cells, the α, α, β, γ, δ subunit composition is characteristic of fetal and noninnervated muscles, whereas in normally innervated adult muscle fibers the ε subunit replaces the γ subunit. Acetylcholine receptors with the ε subunit have a larger single-channel conductance and a shorter burst duration than their fetal counterparts.

Acetylcholine receptors are replenished in the plasmalemma.

It was something of a surprise to discover that AChRs are not fixed in the muscle fiber membrane but are being continually renewed and replaced. The receptors are first prepared on internal membranes and are then transported to the fiber plasmalemma where, in cultured chick *myotubes,* they remain for an average period of 22 hr (Devreotes & Fambrough, 1975); in adult neuromuscular junctions, however, they stay in place for considerably longer, the average half-life being 8 to 11 days (Fambrough, 1979). They are then taken from the membrane and carried back into the interior of the fiber within secondary *lysosomes* (degradation *vacuoles*).

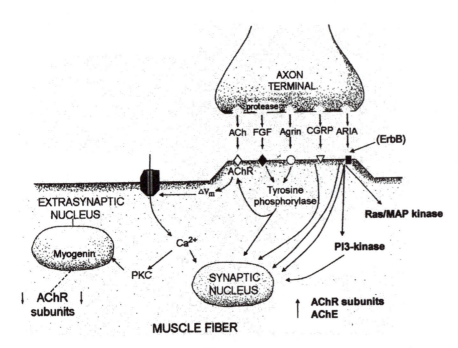

Figure 3.5 Possible means by which AChR genes are controlled in the myonuclei. The combination of ARIA and CGRP with their receptors and the influx of Ca^{2+} cause the genes to be expressed in the synaptic nuclei. Elsewhere in the muscle fiber, the AChR gene is repressed by protein kinase C (PKC) and the muscle *transcription* factor *myogenin*. *Agrin* and *fibroblast growth factor (FGF)* make the newly formed AChRs cluster in the synaptic clefts.

Adapted, by permission, from Z. W. Hall and J. R. Sanes, 1993, "Synaptic structure and development: The neuromuscular junction," *Cell* 72: 112.

The synthesis of the subunits of AChRs is regulated by those myonuclei lying in the sole-plate of the muscle fiber, and the nuclei, in turn, appear to respond to *neurotrophic factors* released from the motor nerve terminal; two such factors are *calcitonin gene-related peptide* (*CGRP;* New & Mudge, 1986) and *ACh receptor-inducing activity* (*ARIA;* Usdin & Fischbach, 1986). Similarly, mRNAs encoding for rapsyn, *neural cell adhesion molecule (N-CAM), muscle-specific kinase (MuSK),* sodium channels, and the catalytic subunits of AChE are found preferentially in the perijunctional region, giving more evidence of the specialized nature of subsynaptic nuclei (Burden, 2002). In contrast to the synaptic myonuclei, those nuclei outside the neuromuscular junction have their AChR genes repressed by muscle activity, probably through a chain of molecular signals that include Ca^{2+} and *protein kinase* C (PKC; figure 3.5).

The Basement Membrane at the Neuromuscular Junction

The basement membrane of the muscle fiber was considered in chapter 1. In the neuromuscular junction, the basement membrane overlies both the primary and the secondary synaptic clefts; at the edges of the junction it becomes continuous with the basement membrane over the rest of the muscle fiber and with that covering the Schwann cell. In electron micrographs, the basement membrane appears as a fuzzy structure from 10 to 15 nm thick; fine processes connect this to the plasmalemmae of the axon terminal and muscle fiber.

In chapter 1, attention was drawn to the different types of protein that the basement membrane contains in addition to *heparin sulfate proteoglycan*. In the neuromuscular junction, these proteins include *agrin*, AChE, *collagens* III and IV, *laminin, FGF,* and a *protease* (see "Location of Protein Molecules Concentrated at the Neuromuscular Junction"). Of these, agrin has been shown to promote the clustering of AChRs in the muscle fiber plasmalemmae of newly formed neuromuscular junctions during embryogenesis and presumably during *reinnervation.* Indeed, agrin fulfills the description of a *trophic* substance (see chapter 18) in that it is synthesized in the motoneuron *soma* and transported to the nerve endings before being inserted into the basement membrane where it is known to exert its influence on AChRs (Magill-Solc & McMahan, 1990b).

In normal adult neuromuscular junctions, agrin interacts with two proteins expressed by the muscle fiber, the agrin signaling receptor and α-dystroglycan, which in turn is connected to the transmembrane protein β-dystroglycan. These last two components are members of the large *dystrophin-glycoprotein complex (DGC;* see figure 1.8). Since AChRs are connected to the DGC via

The basement membrane contains a variety of proteins.

rapsyn, agrin helps keep all these components together. Therefore, once agrin has played its role during differentiation, it probably serves primarily as a structural component of the neuromuscular junction (Denzer *et al.,* 1997).

Fibroblast growth factor, also found in the basement membrane, has been shown to have a similar action to agrin in controlling the positions of AChRs at the neuromuscular junction (Baker *et al.,* 1992). The actions of agrin and FGF help to explain how the basement membrane exerts such a powerful influence on the morphological differentiation of the nerve terminal and the muscle fiber at the site of a prospective neuromuscular junction (Sanes *et al.,* 1978; see chapter 17). Of the other proteins associated with the basement membrane, the collagens provide strength, while the laminins appear to have a trophic role that includes axon guidance during embryogenesis. Dystrophin is one of the proteins that anchors the basement membrane to the muscle fiber plasmalemma, and it does so through attachments to laminin (Ibraghimov-Beskrovnaya *et al.,* 1992).

Location of Protein Molecules Concentrated at the Neuromuscular Junction

Postsynaptic membrane

AChR $\alpha_2\beta\gamma\delta$ and $\alpha_2\beta\epsilon\delta$

N-CAM

N-acetylgalactosaminyl transferase

Integrin β1 subunit

Voltage-sensitive Na^+ channel

GM2-like lipid

Cytoskeleton

43 K protein

Vinculin

β_2-Syntrophin

Talin

Paxillin

Filamin

Utrophin

α-Actinin

Rapsyn

Tropomyosin 2

58 K protein

87 K protein

Dystrophin

Dystroglycan $\alpha\beta$

MuSK

Acetylated tubulin

Ankyrin

Laminin B

Desmin

Actin

β-Spectrin

3G2 antigen

Nerve terminal membrane

Latrotoxin-R

Ca^{2+} channel

Syntaxin

Large or small synaptic vesicles

Choline acetyltransferase[a]

CGRP

Synaptophysin

Synapsin

P65B

SV2

Basement membrane

Agrin

Collagens α3(IV), α4(IV)

Laminins

AChE

Heparan sulfate proteoglycan

N-acetylgalactosaminyl-terminated carbohydrate

Protease nexin 1

Fibroblast growth factor (FGF)

[a]May be present in nerve teminal cytoplasm and/or associated with small vesicles.

Adapted, by permission, from Z. W. Hall, and J. R. Sanes, 1993, "Synaptic Structure and Development: The Neuromuscular junction," *Cell* 72: 103.

Situated within the basement membrane that runs through the primary and secondary synaptic clefts is an enzyme that hydrolyzes ACh and thereby terminates *synaptic transmission*. This enzyme, AChE, is specific for ACh, in contrast to *cholinesterase* (ChE), which splits additional *choline* esters and is also found in the basement membrane. The functional forms of AChE are composed of globular units of the enzyme. For example, the primary form of AChE traditionally associated with the hydrolysis of ACh at the neuromuscular junction is termed A12, since it is asymmetric, composed of 12 globular AChE units, and is attached to the basement membrane via a collagen tail. G4, another form of the enzyme composed of four globular units, is loosely associated with the basement membrane and exists throughout the peri-junctional region, where it is thought to function as a "scavenger" for ACh that might accumulate in the junction and cause receptor desensitization. There is good evidence that the AChE in the axon terminal has been transported there from the motoneuron soma by axoplasmic flow. Although the muscle fiber is also able to synthesize this enzyme, it is apparent from *denervation* experiments that much of this synthesis is under the trophic control of the motoneuron. Using labeled substrate, Barnard and colleagues (1975) have found that there are about 3,000 AChE molecules $\cdot \mu m^{-2}$ in the *postsynaptic membranes* at mouse end-plates. Unlike the AChRs, the AChE molecules are distributed evenly throughout the primary and secondary synaptic clefts, in both attached and solubilized form. The AChE molecules also differ in that they can be easily removed from the synaptic clefts following digestion with proteolytic enzymes. It is therefore probable that the AChE molecules lie within the cleft substance and that their protein tails have only weak attachments to the muscle plasmalemma. Like the AChRs, the AChE molecules are replaced with newly synthesized ones, their half-life being approximately 3 weeks (Kasprzak & Salpeter, 1985).

Recently, it has been suggested that AChE molecules, in addition to hydrolyzing transmitter, may be the precursors of protein-splitting enzymes (proteases) and have a trophic role (Small, 1990).

Axon Terminal

Now that descriptions have been given of the muscle fiber membrane and basement membrane at the neuromuscular junction, it remains to examine the structure of the axon (motor nerve) terminal. As stated in the introduction to this chapter, the terminal is designed for the release of ACh and, of course, for preparing and storing the transmitter as well.

Synaptic Vesicles and Acetylcholine Release

The axon resembles the muscle fiber in being highly specialized at the neuromuscular junction. The axon terminations contain mitochondria, which provide the energy necessary for synthesis and release of neurotransmitter, and also contain large numbers of small spheres with membranous linings; the majority measure some 50 to 60 nm across and are typically clustered opposite the secondary synaptic clefts. These spheres are the synaptic vesicles and contain the transmitter substance, ACh, as well as *ATP* and a *proteoglycan* (figures 3.1 and 3.2). It has been suggested that the vesicles are derived from *microtubules* or endoplasmic reticulum within the axon terminal (Blumcke & Niedorf, 1965). However, it is now clear that most synaptic vesicles are formed by a recycling process in which the plasmalemma *(axolemma)* of the nerve terminal becomes invaginated and then pinched off (Heuser & Reese, 1973).

At first the new vesicles are encased by small polygonal particles of the protein *clathrin* (Ungewickell, 1984) and are referred to as the coated vesicles that can be seen in electron micrographs. The clathrin heavy and light chains are then dissociated by an ATPase in the axoplasm, giving the smooth appearance characteristic of most of the vesicles in the nerve ending (Schmid *et al.,* 1984). Occasionally, much larger smooth-walled vesicles can be seen (giant vesicles), and these are thought to contain proportionately greater amounts of ACh. A fourth type of vesicle is in a minority but is of great theoretical interest; this is the dense-cored vesicle, which is from 70 to 110 nm in diameter. Similar vesicles are a prominent feature of adrenergic nerve endings in the central and autonomic nervous systems, and it is possible that, at the neuromuscular junction also, they contain monoamine transmitter molecules. Other possible contents are the neurotrophic factors secreted by the motoneuron; one such factor could be CGRP, which is known to be released from motor nerve terminals. The different types of vesicle contain special docking proteins in their walls, which enable the vesicles to attach themselves to the axolemma and then discharge their contents into the synaptic cleft (see chapter 10).

Spontaneous evacuation of synaptic vesicles causes MEPPs.

The possibility that the vesicles contain ACh was first considered when electrophysiological experiments indicated that the transmitter was released from the axon terminal in small packets, or quanta. Even at rest, there is a spontaneous random release of ACh, each quantum producing a small *depolarization* of the muscle fiber membrane (figure 3.6); the depolarizations were referred to as *miniature end-plate potentials (MEPPs)* by Fatt and Katz (1952), who described them first. By using raised concentrations of Mg^{2+} in the bathing fluid to reduce the release of ACh from the axon terminal, it can be shown that the *end-plate potential* following a nerve *impulse* is normally the summation of many MEPPs. It has also been shown that the spontaneous release of ACh probably reflects the resting level of Ca^{2+} ions within the nerve terminal and that this is regulated by Ca^{2+} uptake and release from the mitochondria (Alnaes & Rahamimoff, 1975).

The Vesicle Hypothesis of Neurotransmitter Release

The vesicle hypothesis of ACh release is also supported by evidence.

The vesicle hypothesis of synaptic communication suggests that nerve terminal vesicles each contain a quantum of neurotransmitter, and that transmitter release occurs via *exocytosis,* during which transmitter is released as the vesicle membrane is fused with the presynaptic plasma membrane. This is further dealt with in chapter 10.

The validity of the vesicle hypothesis is strengthened by the results of combined biochemical and electron microscopical studies on tissue homogenates; thus, both in the brain and in the electric organ of the electric eel, it has been possible to prepare cell fractions that are rich in vesicles and also contain large amounts of ACh (Whittaker *et al.,* 1964). Unfortunately, a similar experiment cannot be performed on motor nerve terminals because of the difficulty in obtaining satisfactory preparations of vesicles without contamination from subcellular components of muscle.

A second important test of the vesicle hypothesis has been to demonstrate a reduction in their number following massive release of ACh by the nerve terminal. In the experiments of Jones and

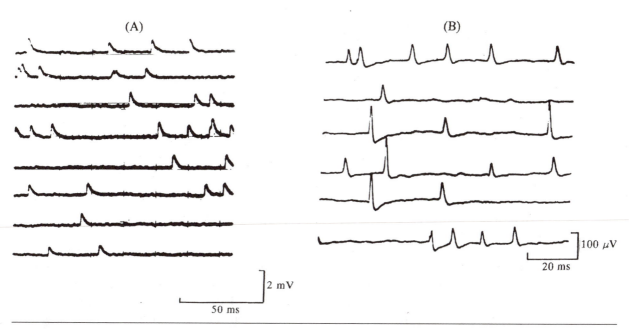

(A) (B)

2 mV

50 ms

100 µV

20 ms

Figure 3.6 Miniature end-plate potentials (MEPPs) recorded *in situ* (A) with an intracellular microelectrode from a mouse gracilis muscle fiber and (B) with a coaxial needle electrode from the vastus medialis muscle of a normal subject. Because in the latter instance the recording electrode is effectively extracellular, the potentials appear much smaller (note amplification).

Reprinted from A. J. McComas, 1977, *Neuromuscular Function and Disorders* (London: Butterworth Publishing Co.), 32.

Kwanbunbumpen (1970), the rat phrenic nerve was stimulated repetitively, and resynthesis of ACh was prevented by *hemicholinium*. The authors found that a significant depletion of vesicles had occurred in the zone of axoplasm bordering the synaptic cleft and that the residual vesicles were reduced in size.

Another, and rather interesting, technique for depleting the nerve terminals of ACh is to expose them to the venom of the black widow spider; with this method also, a reduction in the number of vesicles within the nerve terminal can be demonstrated (Clark *et al.*, 1970). A still more demanding technique is to freeze the muscle very rapidly, at a time when unusually large amounts of ACh are being released from the nerve endings, as, for example, following electrical stimulation of the nerve twigs in the presence of 4-aminopyridine. It is then possible, in electron micrographs, to see vesicles that have been captured while opening up after fusing with the axolemma (Heuser & Reese, 1981).

Quantal Size

Each synaptic vesicle contains approximately 10,000 molecules of ACh.

The number of ACh molecules inside a single vesicle has been determined in various ways. For example, Katz and Miledi (1972) analyzed *membrane potential* "noise" during the delivery of ACh from a micropipette; they estimated that a single molecule of ACh produced a depolarization of about 0.3 μV and that at least 1,000 molecules would be necessary to account for a MEPP of 0.3 mV in a frog muscle fiber. However, the number of molecules in a vesicle must be rather larger than this, for additional molecules will be lost by diffusion out of the synaptic cleft and by hydrolysis with AChE. Potter (1970), using radioactively labeled ACh, calculated that a single impulse in a phrenic motor nerve fiber released about 4×10^6 molecules of transmitter from the axon terminal; he considered that a typical vesicle might contain from 4,000 to 9,000 molecules of ACh.

An especially accurate (and elegant) estimate of *quantal size* was made by Kuffler and Yoshikami (1975), who devised a highly sensitive bioassay system to determine the amount of ACh delivered from a micropipette onto the exposed subsynaptic membrane in the frog and snake. The authors concluded that a quantum of transmitter contained fewer than 10,000 molecules of ACh. This value agrees well with that of Miledi *et al.* (1983), who collected the ACh released spontaneously in frog sartorius muscles in the presence of an anticholinesterase drug and compared the amount with the numbers of MEPP currents in the same period of time; they estimated that each vesicle contained, on average, approximately 12,000 molecules of ACh.

The functional advantage of containing the transmitter in a vesicle, rather than in the axoplasm, is probably twofold. First, the ACh is protected from the small amount of ChE that is normally to be found within the nerve terminal. Second, the influx of Ca^{2+} during the impulse is able to activate only a certain number of sites on the inner surface of the axolemma, and it is therefore advantageous to deliver an optimal quantity of ACh to each site (see the following section).

"Active Zones" at Neuromuscular Junctions

The axolemma contains active zones for the release of ACh.

The axolemma facing the synaptic cleft contains regular structures that can be seen by electron microscopy in regular ultrathin sections, but are best visualized after the membrane has been sheared into two leaflets by the freeze-fracture technique. These structures, termed active zones, consist of parallel rows of small particles arranged in two pairs. In the amphibian neuromuscular junction, in which the motor nerve terminals are oriented in the long axis of the muscle fiber, the active zones run transversely across the terminals and lie opposite the secondary synaptic clefts (figure 3.1); each pair of rows is separated from the other pair by a ridge. In mammalian neuromuscular junctions, the active zones have a more variable orientation to each other and to the axis of the nerve terminal; in addition, the rows contain fewer particles and there are no intervening ridges between the pairs of rows. The particles in the active zones are thought to be voltage-dependent potassium and *calcium channels* (Sanes & Lichtman, 1999). Their presence, together with that of the docking proteins, enables the synaptic vesicles to fuse with the axolemma (see chapter 10).

The nerve
ending
contains
cellular
organelles.

Other Constituents of the Nerve Terminal

In addition to the synaptic vesicles, the nerve ending contains large numbers of mitochondria, the main function of which is to provide ATP, through oxidative metabolism, for various synthetic and pumping activities. Thus, ATP is required for the conjugation of acetate and choline to form ACh and for fueling the Na^+-K^+ pump in the axolemma. In addition, ATP is consumed when clathrin is removed from the coated vesicles, as part of the membrane recycling process (see p. 38). Massive stimulation of the motor nerve endings can produce degenerative changes in the mitochondria (Jones & Kwanbunbumpen, 1970).

Electron microscopy of the nerve ending shows that it also contains *smooth endoplasmic reticulum,* microtubules, and *neurofilaments.* The microtubules and neurofilaments act as channels for the *axoplasmic transport* of material between the motoneuron soma and nerve terminals (see chapter 8). It is possible that some of the neurofilaments direct the synaptic vesicles to the active zones in the axolemma.

Applied Physiology

There are many fascinating clinical disorders that result from alterations in function at the neuromuscular junction. However, these disorders are better understood after the physiology of the junction has been studied and have therefore been deferred until chapter 10.

The account of muscle structure is almost complete. In the next chapter, we examine the final element—the sensory nerve endings in the muscle and tendon. It is these endings that signal the state of the muscle to the brain and spinal cord.

4

Muscle Receptors

For muscles to work effectively, the *motoneuron* must be able not only to commence discharging at the right instant, but also to adjust its firing rate from moment to moment, depending on the nature of the task and the changing loads imposed on the moving part. These adjustments in firing rate, in turn, depend on the continuous flow of information to the central nervous system from the muscles, joints, and skin. It is the nature of the receptors (sense organs) in the muscles that must now be considered.

It is convenient to classify muscle receptors into four types:

- Muscle spindles
- Golgi tendon organs
- Paciniform corpuscles
- Free nerve endings

Of these receptors, the *muscle spindles* and *Golgi tendon organs* are by far the most important in signaling information about mechanical events. The muscle spindles are exquisitely designed to sense the lengths of the muscle fibers during contraction, while the tendon organs are concerned more with the forces developed by the fiber. The *paciniform corpuscle* is another morphologically distinct receptor; it has a nerve ending that is ensheathed in thin wrappings of *connective tissue,* rather like the layers of an onion. The corpuscle is exquisitely sensitive to rapid small deformations and is widely distributed in such tissues as skin, periosteum, and abdominal viscera; in muscle it probably plays a minor role in providing information about contraction and relaxation. The free nerve endings are important in that one of their functions is to signal the metabolic status of the interstitial fluid and hence of the muscle fibers; these endings respond to changes in oxygen tension, $[H^+]$, $[K^+]$, and probably other signals as well.

The Muscle Spindle

The first of the muscle receptors that we will consider in detail is the muscle spindle. The spindle is remarkable for several reasons, one being the complexity of its structure and another being the way its responsiveness can be adjusted by the central nervous system. Muscle spindles are sensory organs that monitor the length of a muscle, and they are also sensitive to the rate change of length. These properties make the spindles important for reflex responses and coordination of patterned and learned movements, as well as for proprioception.

The Structure of the Muscle Spindle

The muscle spindle contains two types of muscle fiber.

A muscle spindle is many times longer than it is wide. Thus, although a spindle is only 80 to 250 μm in diameter at its thickest part, the capsule, it is usually several millimeters long and may even extend for 10 mm. The spindles are surrounded by, and are situated in parallel with, the *(extrafusal) muscle fibers* that make up most of the muscle belly. Most spindles occur singly, but others are found

Table 4.1   Values of Spindle Densities in Certain Human Muscles

Muscle	Number of spindles	Weight (g)	Density (spindles/g)
Obliquus capitis superior	141	3.3	42.7
Rectus capitis posterior major	122	4.0	30.5
Abductor pollicis brevis	80	2.7	29.3
First lumbrical, foot	36	1.7	21.0
Second lumbrical, hand	36	1.8	19.7
Opponens pollicis	44	2.5	17.3
First lumbrical, hand	51	3.1	16.5
Masseter, deep portion	42	3.8	11.2
Biceps brachii	320	164	2.0
Pectoralis major	450	296	1.5
Triceps brachii	520	364	1.4
Latissimus dorsi	368	246	1.4
Teres major	44	123	0.4

Values rounded off. Data from Voss (1937, 1956, 1958), Schulze (1955), Freimann (1954), Körner (1960).

From P.B.C. Matthews, 1972, *Mammalian muscle receptors and their central actions* (London: Edward Arnold), 48. Copyright 1972. Reproduced by permission of Edward Arnold.

in an end-to-end "tandem" arrangement. Muscle spindles are clearly not as long as extrafusal muscle fibers; however, the capsule is continuous with the connective tissue matrix of the muscle, and therefore changes in muscle length are transmitted to the intrafusal fibers. The number of spindles in a muscle belly varies greatly from one muscle to another, even when allowances are made for the size of the muscle (table 4.1). The muscles at the back of the neck and the small muscles of the hand have the richest supply of spindles, and the large muscles of the arm and leg are least well endowed. This difference in density is probably related to the ability to carry out small movements of the head and fingers rapidly and accurately.

Muscle Spindle Fiber Types, Innervation, and Electrical and Contractile Properties

The characteristic fusiform shape of the spindle is due to the presence, in its middle part, of a connective tissue capsule containing fluid. The composition and function of the fluid are still unknown, though it has long been thought that the fluid is lymph. In the longitudinal axis of the spindle, the central and outer regions are often referred to as the equatorial and polar regions, respectively.

A remarkable feature of the muscle spindle is that its sensitivity as a mechanoreceptor can be adjusted; this is possible because the sensory nerve endings are attached to intrafusal (inside the spindle) muscle fibers that can be made to contract and relax. Two types of intrafusal fiber can be recognized on the basis of the numbers and distributions of their *nuclei* (figure 4.1). The *nuclear bag fibers* have an extraordinary number of nuclei in their middle part, within the capsule; indeed, in this region the muscle fiber appears to consist of little else but nuclei. The *nuclear chain fibers*, in contrast, have their nuclei distributed more evenly along their lengths, and the nuclei are found in the centers of the fibers. The two types of fiber differ in other respects, for the bag fibers are usually thicker and rather longer than the chain fibers. In addition, the bag fibers are less numerous, for there are usually only two (between one and four) in each spindle, whereas there may be as many as 12 chain fibers (table 4.2). Although the two nuclear bag fibers appear very similar, they can be distinguished by histochemical staining for *myosin ATPase* and presumably have other differences in contractile proteins; the two fibers are referred to as bag$_1$ and bag$_2$.

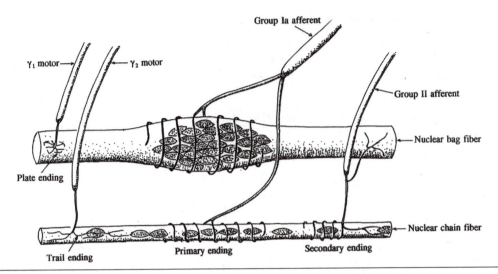

Figure 4.1   Diagram of central (equatorial) regions of a nuclear bag and a nuclear chain fiber, showing the sensory and motor *axons* that are characteristic of the two types of fiber.

Adapted from Matthews (1964, p. 232).

Table 4.2   Characteristics of Nuclear Bag and Nuclear Chain Intrafusal Fibers

	Nuclear bag	Nuclear chain
No. per spindle	2	2-12
Diameter (μm)	20-25	10-12
Motor nerve endings	Plate	Trail
Sensory response	Dynamic	Static
Sensory axon	Ia, II (bag$_2$)	Ia, II

The intrafusal muscle fibers have their own motor nerve supply.

The contractile and electrical properties of the intrafusal fibers differ.

Two types of motor nerve ending are found on the *intrafusal muscle fibers* in the polar regions (figure 4.1). One type, the *plate ending,* resembles that found at the neuromuscular junction of an extrafusal fiber; this type of ending is associated mainly with the nuclear bag fibers. The other type of motor ending is more extensive, running along the greater part of the fiber outside the central region; this type is the *trail ending* and is found mostly on the nuclear chain fibers.

Both types of ending are attached to γ-motor (fusimotor) axons, which can be identified in the *ventral roots* of the spinal cord by their small diameters. For most spindles there are two γ-motor axons, one for the nuclear bag fibers and one for the nuclear chain fibers (Porayko & Smith, 1968). A refinement of the situation is that, while the bag$_1$ fiber has its own γ-motor axon, the bag$_2$ fiber shares the other γ-axon with the chain fibers (Kucera, 1985). In a minority of spindles, however, there may be only a single γ-axon, and this is then responsible for innervating all the intrafusal muscle fibers. In addition to the γ-motor axons, a spindle may receive a branch from an α-motor axon. Such branches, termed β-axons, appear to supply only plate endings on the nuclear bag fibers.

On the basis of the differences in structure and innervation, it might be expected that the intrafusal muscle fibers would respond differently to excitation through their *axons,* and such is the case. By stimulating intrafusal fibers directly (Smith, 1966) or through their motor axons (Boyd, 1966; Bessou & Pagès, 1972), it is possible to show that the contractions of the bag fibers are localized and appear slower and weaker than those of the chain fibers; this distinction is especially true of the bag$_1$ intrafusal fibers. The electrical properties of the two types of fiber also differ. Bessou and Pagès (1972) collaborated with Barker (1974) in showing that the bag fibers usually developed graded, nonpropagated electrical potentials, while the nuclear chain fibers responded to nerve stimulation with propagated *action potentials,* like extrafusal muscle fibers.

Since the intrafusal fibers receive their motor innervation on either side of their central regions, and since, in the case of the bag fibers, the central region consists mainly of nuclei rather than *myofibrils,* it is the central region that is stretched by the contracting ends of the fibers. This stretching increases the sensitivity of the primary sensory nerve endings (see p. 45).

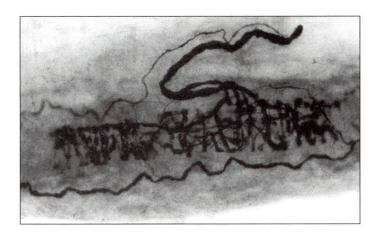

Figure 4.2 The central (equatorial) region of a human muscle spindle, showing the branching of the Ia axon (arrow) to supply annulospiral-like endings on the (unstained) nuclear bag and chain fibers.

Courtesy of Dr. Dennis Harriman.

Primary and secondary sensory endings are on the intrafusal muscle fibers.

Each spindle has one large *afferent* nerve fiber, the group Ia axon, which terminates in spirals around each of the intrafusal fibers within the equatorial region; this type of ending is known as the *primary ending* (figures 4.1 and 4.2). In human spindles the primary endings take the form, not of spirals, but of multiple C-shaped bands that partly surround the intrafusal fibers (Kennedy, 1970).

There is also a rather thinner sensory nerve fiber for each spindle, the group II axon, and this terminates in *flower spray endings* on the nuclear chain and bag$_2$ fibers; these constitute the *secondary endings*.

Activation of the Muscle Spindle

It is convenient to deal with the functional properties of the muscle receptors at this stage so that, apart from a small section in chapter 15, the examination of these sense organs can be completed.

Response of the Spindle to Passive Stretch

Passive spindle stretch produces both a rapid firing and a sustained discharge.

In the laboratory, the responses of the muscle spindles to passive stretch can be studied by recording the *impulse* discharges in the Ia and II sensory axons; in most experiments, stretch is applied in the form of a ramp. The impulse discharge in the two types of axon differs. The Ia axon fires a high-frequency burst of impulses while the stretching takes place *(dynamic response)* and a steady discharge of impulses when the muscle spindle is held at its new length *(static response,* figure 4.3); the II axon has only a static response. A further distinction is that the primary endings of the Ia axons have much lower thresholds

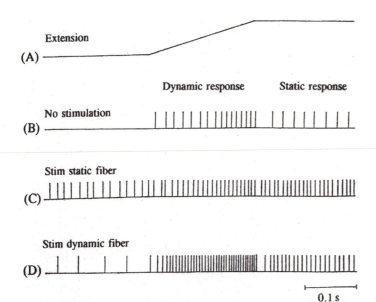

Figure 4.3 The effect of stimulating fusimotor axons on the impulse discharge in a Ia axon, during and following muscle stretch. (A) Muscle length is illustrated. (B) No stimulation is given, and the modest discharge has both dynamic and static components. (C) The static axon is stimulated, causing an increase in background activity and in the static component of the discharge. (D) Dynamic axon stimulation causes a great increase in the dynamic phase of the response and a smaller change in the static component.

Adapted, by permission, from A. Crowe and P. B. C. Matthews, 1964, "The effects of stimulation of static and dynamic fusimotor fibres on the response to stretching of the primary endings of muscle spindles," *Journal of Physiology* 174: 112.

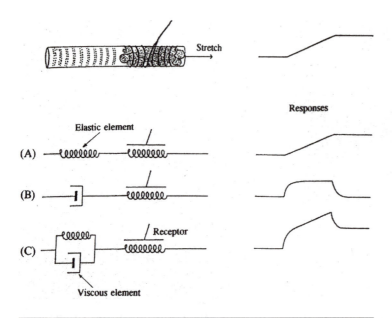

Figure 4.4 Models to illustrate how the Ia impulse discharge would be affected by the elastic and viscous properties of the intrafusal muscle fibers. The annulospiral (primary) ending (receptor) is shown as being in a purely elastic region of the spindle, and its response will depend on the stretch transmitted to this region. (A) Stretching the spindle evokes a discharge that is entirely proportional to the instantaneous length of the receptor region; that is, there is only a static component. (B) Replacement of the elastic element by a viscous one introduces a dynamic component in the discharge but abolishes the static one. (C) The introduction of elastic and viscous elements in parallel would be expected to give both dynamic and static phases of the response. Because these are the responses actually observed, this model is the best one.

Adapted, by permission, from P.B.C. Matthews, 1964, "Muscle spindles and their motor control," *Physiological Reviews* 44: 248.

to stretching than the secondary endings of the II axons. Matthews (1972) suggested that the complex response of the primary endings was due to the Ia axon's being connected to both a nuclear bag and a nuclear chain fiber (figure 4.1); the bag fiber could contribute the dynamic discharge and the chain fiber the static one. The two types of response are reflected in the *generator potential,* that is, the *depolarization* of the sensory nerve endings caused by the opening of the stretch-activated *ion channels* (Hunt *et al.,* 1978). There is still uncertainty as to how these channels are made to open, but presumably this has something to do with the extension of the nerve spirals around the intrafusal muscle fibers.

There is also uncertainty about the mechanism of the dynamic response and why the impulse frequency should be proportional to the rate of stretching. One factor seems to be the unusual viscous properties of the nuclear bag fiber, a factor that was first appreciated when single muscle spindles were dissected out and inspected under the light microscope while being stretched. Thus, if a bag fiber is pulled to a new length, the equatorial region (containing annulospiral endings) is first extended and then slides back to its original length over several hundred milliseconds (Smith, 1966). This behavior could be due to the fact that the equatorial region is predominantly elastic and the polar regions are viscous (Matthews, 1964; figure 4.4).

Types of γ-Motor Axons

Just as there are two types of sensory discharge for the spindle, dynamic and static, so there are two types of fusimotor axons. Stimulation of the dynamic fusimotor axon enhances the impulse frequency in the primary endings during the time that the spindle is being lengthened (Matthews, 1962), while stimulation of the static fusimotor fibers increases the impulse frequency in both primary and secondary endings once the spindle is at its new length. The simplest explanation for the two types of motor response is that the static fusimotor axons supply the nuclear chain fibers within the spindle and the dynamic fusimotor axons innervate the nuclear bag fiber (Matthews, 1962). In both cases, however, the fusimotor fibers produce their effects in the same way. By causing the intrafusal fibers to contract, especially in their polar regions, they simultaneously extend the equatorial regions in which the annulospiral endings are stretched and enhance the sensitivity of the latter to passive stretch. The spindle can therefore be regarded as an exquisitely engineered length sensor; not only can it signal both the rate and the extent of stretching, but its sensitivity can be enhanced by the central nervous system, as the need arises, by contraction of the intrafusal muscle fibers. The coactivation of *α-motoneurons* and fusimotor neurons ensures that information regarding muscle length is always available to the nervous system whether the muscle is at rest, moving, shortened or lengthened, or contracting isometrically. The associated information processing by the central nervous system requires monitoring not only of spindle output, but of fusimotor output as well.

Projection of Muscle Spindle Afferents

Muscle spindle axons project to the spinal cord and brain.

Part of the flow of impulses from the muscle receptors is directed to the brain, where it may or may not be appreciated at a conscious level. In addition, there are powerful synaptic connections to motoneurons, especially those innervating the same muscle *(homonymous motoneurons)*. Mendell and Henneman (1971) demonstrated that, in the cat hindlimb, each Ia afferent fiber makes contact with over 90% of homonymous motoneurons. Indeed, there is good evidence that the strongest connections of all are made with motoneurons supplying *motor units* in the vicinity of the spindles. This arrangement allows for functional compartmentalization of large muscles, and may subserve a motor scheme whereby certain compartments can be preferentially recruited over others.

Group Ia axons excite homonymous motoneurons.

The group Ia axons that project to motoneurons supplying the same muscle end in knobs, or boutons, on the *dendrites;* sometimes an additional neuron is involved in the pathway (figure 4.5). Each Ia afferent makes from two to six contacts with each individual motoneuron, in its dendritic tree (Brown & Fyffe, 1981). Regardless of the number of dendritic contacts or length of the neural pathway, the effect of the Ia axons is to excite the moto-neurons. These connections provide a means for rapidly executed reflex adjustments in muscle length following any stretching or change in muscle load. One can think of many situations when such adjustments might come into play—for example, restraining a fractious dog on the lead, holding a tray while objects are added or removed, standing in an accelerating or slowing bus, and so on. The mainte-nance of posture is also largely under the direction of the muscle spindles. Since the neural pathways are in the spinal cord and are relatively short, the change in motoneuron firing can be made more rapidly than if intervention by the brain were needed. Although the Ia axons are excitatory to motoneurons serving the same muscles or its synergists, they are inhibitory, through an interneuron, to motoneu-rons of the antagonist muscles (figure 4.5). Group II axons resemble the Ia axons in being excitatory to homonymous motoneurons, though the connections are usually through one or more interneurons.

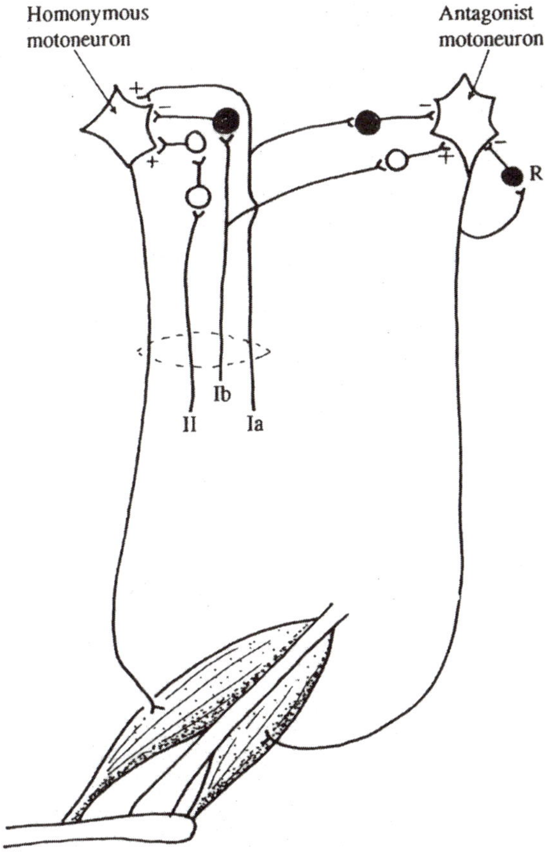

Figure 4.5 Synaptic connections of group Ia and Ib axons in the spinal cord. The inhibitory interneurons are shown by filled circles. Ia connections onto homonymous moto-neurons are mono- and bi-synaptic (the latter not shown for purposes of clarity).

Role of Muscle Spindles in Muscle Tone and Tendon Reflexes

It is the muscle stretch receptors that are responsible for the impression of "tone" that can be elicited by passive manipulation of a limb; in this case, extension of a joint will activate Ia afferent nerve endings in the flexor muscle and will evoke a reflex contraction of the muscle. If the nerve endings are compromised by disease or injury, so as to be relatively inexcitable, then the reflexes will be diminished, and the tone perceived by an examiner will be decreased. Conversely, an increase in spindle activity, as in patients with certain types of brain or spinal cord lesion, will give an impression of greater resistance when the limb is moved *(spasticity)*.

Muscle spindles are involved in muscle tone and in tendon jerks.

Muscle spindles are also responsible for the so-called tendon reflexes. A sharp tap on a tendon will stretch the muscle belly and the spindles contained within it, evoking an impulse volley in the Ia axons; these axons, in turn, evoke a brisk reflex contraction of the muscle (figure 4.6B), causing the limb to move. The *H-reflex,* named after Paul Hoffmann (1918), is the electrical equivalent of the tendon reflex; in this case, the impulse volley in the Ia axons is set up by an electrical stimulus to the motor nerve, rather than by a mechanical perturbation (figure 4.6C).

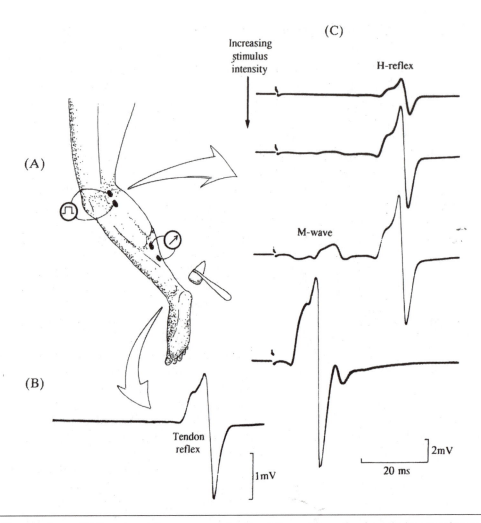

Figure 4.6 Muscle reflexes. (A) Experimental arrangements for stimulating and recording from the human soleus muscle. (B) The tendon reflex. An *EMG* response has been evoked in a human soleus muscle by a tap to the Achilles tendon. (C) H-reflexes elicited in the same muscle by progressively stronger electrical stimuli to the tibial nerve (from above downward). Beyond a certain value, an increase in stimulus strength is associated with a decline in the H-reflex, due to collision between antidromic and reflexly elicited impulses in the motor axons. Eventually the H-reflex disappears altogether, and only the indirectly evoked muscle compound action potential *(M-wave)* remains (lowest trace).

The Golgi Tendon Organ

Named after their discoverer, Camillo Golgi (1903), the Golgi tendon organs are fewer in number than the muscle spindles; a further difference is that the tendon organs are found at the musculotendinous junction rather than in the muscle belly. Also, a tendon organ has a much simpler structure than a muscle spindle, and its physiological properties are easier to comprehend. Finally, unlike muscle spindles, Golgi tendon organs are not innervated by motor axons. The characteristics of Golgi tendon organs are summarized as follows:

    • **Golgi tendon organs are connected to extrafusal muscle fibers.** The tendon organs are encapsulated structures, typically from 0.5 to 1.0 mm long and from 0.1 to 0.2 mm in diameter. A single nerve axon, of relatively large diameter, conducts the sensory impulses from one or more tendon organs to the spinal cord and is termed the Ib axon; it appears to arise from nonmyelinated sprays in contact with the tendon *fascicles* (figure 4.7A). High-powered microscopy reveals that the fine nerve twigs weave their way between the bundles of *collagen* fibers (figure 4.7B). When the muscle fibers contract, the collagen bundles straighten and compress the nerve twigs, setting

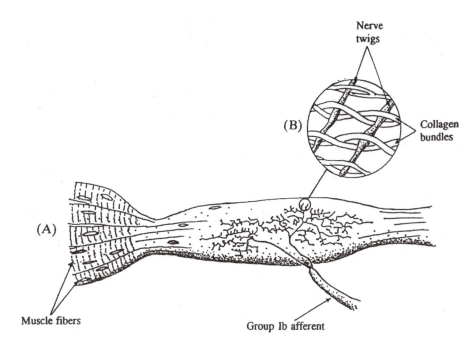

Figure 4.7 (A) Structure of a Golgi tendon organ; (B) the organ has been dissected from the tendon and the capsule has been removed. Interrelationship of nerve fiber twigs and collagen bundles in part of a tendon organ, seen in higher magnification than in A.

A: Adapted from Cajal 1909.

up an impulse discharge in the latter (Swett & Schoultz, 1975). Through the collagen bundles, a tendon organ is connected to 4 to 25 muscle fibers.

• **Tendon organs are excited by motor unit contractions.** In contrast to muscle spindles, tendon organs are not very sensitive to passive stretch of the muscle and its tendon, though there is some variation among muscles. When muscles contract, however, the tendon organs discharge, and their sensitivity is such that they can respond to a single motor unit through the one or two muscle fibers that attach the unit to the tendon organ. Tendon organs appear to be slightly more sensitive to changes in contractile force, as opposed to the absolute force level. Each tendon organ is influenced by 4 to 15 motor units (Houk & Henneman, 1967). Sensory information from the tendon organs, transmitted along Ib axons differs from that of spindles; this inhibits motoneurons of the same muscle and excites those of the antagonists (figure 4.5). It was once thought that Golgi tendon organs were relatively insensitive to low forces and provided for a protective inhibitory mechanism during high forces. However, it is now known that Ib afferents are activated during locomotion (Loeb, 1981), and thus Golgi tendon organs constitute an additional source of information on muscle forces during normal activities.

Free Nerve Endings

Free nerve endings may respond to specific stimuli or to tissue damage.

One last type of muscle receptor remains to be discussed, a type that is both the simplest in its structure and the most abundant within the muscle. This is the free nerve ending.

The free nerve endings relay their information to the spinal cord through the smallest myelinated nerve fibers (group III afferents) and the nonmyelinated fibers (group IV afferents). The free nerve endings innervate almost all structures within the muscle belly: the extrafusal fibers, the different types of connective tissue including the muscle sheath *(epimysium)*, the larger blood vessels, and even the muscle spindles and tendon organs (Stacey, 1969). Electrophysiological recordings from group III and IV afferents show that many of the free nerve endings are sensitive to mechanical stimuli, such as those associated with muscle contraction or with pressure or stretching (Mense & Meyer, 1985). Other free nerve endings respond specifically to changes in temperature or in their chemical environment; for example, elevations in K^+ or *lactic acid* concentrations may provoke discharges (Thimm & Baum, 1987). Yet other free nerve endings have high thresholds and discharge impulses only in response to stimuli that can cause tissue damage (Mense & Stahnke, 1983). These are the *nociceptive endings,* and it is possible that they are sensitized by the release of neuropeptides (e.g., bradykinin) or other stimulants (e.g., arachidonic acid) at the site of injury.

Some nonmyelinated endings are quiescent when a muscle is resting but become increasingly active during the course of ischemic contractions. These fibers, and those responsive to rises in K^+ or lactic acid concentrations, could be responsible not only for muscle discomfort during exercise, but also for cardiovascular and respiratory reflexes (Mitchell *et al.,* 1983). The endings could also provide the afferent arm of the reflex that has been postulated by Bigland-Ritchie *et al.* (1986b) to depress motoneuron excitability during sustained *effort* (see chapter 15).

Role of Muscle Receptors During Locomotor Activity

It should be kept in mind that, while the role of information from muscle (and other) receptors during voluntary activities is significant, it is difficult to determine this from studies of isolated reflex responses at rest. The organization of the spinal cord is such that most reflex pathways receive input from different types of primary afferent, as well as from a variety of supraspinal systems. The various inputs usually influence the motoneurons through spinal interneurons and, since there are several classes of interneuron, the same afferents can produce contrasting effects. One example is the "reflex reversals" involving Ib afferents; at rest, activity in these fibers inhibits extensor motoneurons, but during locomotion there is excitation of the same motoneurons instead (Hultborn, 2001). Not only are some interneurons excited during locomotion and others inhibited, but there is presynaptic suppression of afferent activity (McCrea, 2001). The role of muscle receptors, as well as that of other sources of afferent information, in controlling locomotor activities is a very active research area (for reviews see Burke, 1999; Hultborn, 2001; McCrea, 2001).

Applied Physiology

Instead of attending the neuromuscular clinic, we shall visit the physiology laboratory and try to answer a question that is as puzzling as it is fascinating, namely, *Do signals from muscle receptors reach consciousness?*

There is no doubt that impulse messages from the nociceptive free nerve endings can be perceived; an example is the momentary pain experienced by a patient when an electromyographer penetrates the epimysial sheath of a muscle with a needle electrode. Also, during and after severe exercise, there is a discomfort, poorly localized in the muscle, that is accentuated by pressure over the belly. But what about the messages from the muscle spindles and Golgi tendon organs—can these be perceived? Over the years, scientific opinion has tilted first in one direction and then in the other; and even today, despite many ingenious experiments, the answer to the question is still uncertain. The earliest view was that information from receptors in the muscles and joints was indeed required for awareness of the position of the limb (Sherrington, 1900). This view then succumbed to the results of experiments in which passive movements of a toe or finger could no longer be appreciated after impulses in joint nerve fibers had been blocked by local anesthesia of the digit (Browne *et al.,* 1954; Provins, 1958), despite the same changes in muscle length taking place as before. An even more direct experiment was that of Gelfan and Carter (1967), who exposed tendons in the ventral forearms of conscious subjects. These authors found that there was no sensation of movement when a tendon was pulled so as to stretch the muscle; when the tendon was pulled in the opposite direction so as to flex the finger, without disturbing the muscle belly, the new positions of the fingers were instantly detected.

Against these observations is the clear demonstration that impulse volleys in large-diameter muscle afferents evoke responses in the mammalian cerebral cortex, including that of the monkey (Albe-Fessard *et al.,* 1965). Further, when the tendons of the human biceps brachii or triceps brachii are vibrated, so as to excite the muscle spindles, the position of the elbow joint is misjudged; this discrepancy suggests that the Ia axon discharge has been misinterpreted by the brain as muscle stretch (Goodwin *et al.,* 1972).

More recently still, the pendulum appears to be swinging back toward the lack of conscious awareness. For example, when single muscle spindle afferents are stimulated electrically through tungsten microelectrodes inserted into the peripheral nerve, there is no detectable sensation; in contrast, electrical stimulation of most joint afferents does evoke sensation (Macefield *et al.,* 1990).

The situation cannot be regarded as closed, however (Gandevia *et al.,* 1992). For one thing, the excitation of a single spindle afferent by itself does not occur under natural circumstances. Also, as Gandevia and McCloskey (1976) showed, the sensations of movement and of joint position are normally required by the body during active contractions. Correspondingly, these authors have observed that after a finger has been anesthetized, there is a marked improvement in the detection of passive movements when a subject is allowed to tense the long finger flexor muscle.

The examination of the structure of skeletal muscle and its nerve supply is finished. In the next two chapters we shall see how the muscles and motoneurons develop in the embryo.

5

Muscle Formation

The muscles of the human body are formed long before birth. Indeed, fetal imaging has shown that the muscles of the trunk start to contract as early as the eighth week of gestation. By 20 weeks, a remarkably rich repertoire of apparently purposeful movements involving the hands, feet, and head has become established (table 5.1).

At least five major interrelated processes are involved in muscle development. The gross anatomical features of muscle development in both humans and animal species have been known for many years. Special insights have come from tissue transplants and other experiments in animal embryos, as well as from dissections, microscopic examination of embryonic tissue, and studies of muscle cells in culture. Recently, our understanding of muscle development has reached a new level, following the identification of some of the genes and peptide growth factors that are involved in muscle cell determination and differentiation as well as in the laying down of the body plan. It will be shown that the formation of muscles depends on five interrelated processes:

- Induction of *mesoderm* from presumptive *ectoderm*
- Commitment of a portion of the mesoderm to develop into skeletal muscle
- Proliferation and differentiation of myogenic (muscle-forming) cells
- Laying down of the body plan, including the outgrowth of the limb buds
- Assembly of muscles

Table 5.1 Appearance of Movements in Human Fetus

Gestational age (weeks)	Movement
8	Quick flexion and extension movements of the trunk.
10	Extension of arms and legs accompanies trunk movements.
12	Head rotation. Hands brought up to face. Sudden jerks elicited by mechanical stimuli ("startle" response).
13-14	Creeping and climbing movements. Mouth opening and tongue protrusion. Swallowing and breathing movements appear.
15	Thumb put in mouth and sucked.
16	Limb movements now well coordinated. Hands "explore" uterine and placental surfaces.
18	Exploration extended to parts of body (head, trunk, and legs).
20	More discrete movements, involving individual fingers, eyelids, feet.

Another particularly detailed study is that by de Vries and her collaborators (de Vries, 1987; de Vries *et al.,* 1982).

Based on Ianniruberto and Tajani (1981).

Step 1: Mesoderm Is Induced From Ectoderm

As the oocyte develops in the ovary, it not only becomes larger and accumulates yolk but also forms animal and vegetal "poles" (figure 5.1), which can be distinguished by the uneven distribution of their cytoplasmic inclusions (e.g., lipid droplets and yolk granules). Within the *cytoplasm* of the oocyte are large amounts of mRNA; although some of the mRNAs, such as those for histone proteins, remain freely distributed, other mRNAs start to move along the cytoskeleton and to accumulate at one or the other of the two poles. Some of the mRNAs collecting at the vegetal pole will subsequently produce inductive signals; as the oocyte has not yet been fertilized, these mRNAs are entirely maternal in origin.

Following invasion of the oocyte by a sperm, a dorsoventral polarity is imposed on the egg, in addition to the animal-vegetal one already present. The fertilized egg now divides repeatedly to form the multicellular blastula; the cells in the animal pole are destined to become ectoderm, and those in the vegetal pole *endoderm* (figure 5.1).

In the equatorial (marginal) zone of the blastula, the vegetal and animal cell masses are in contact, and inductive signals now pass from the vegetal mass, instructing the adjacent "animal" cells to develop into mesoderm (Nieuwkoop, 1973). The internal cavity in the blastula, the *blastocoel,* prevents the inductive signals from reaching the main part of the animal cap, which, as already stated, will become ectoderm (figure 5.1).

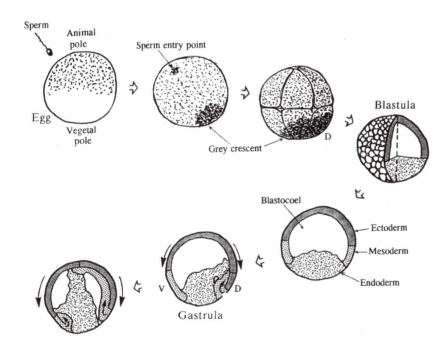

Figure 5.1 Early steps in embryonic development. After the egg has been penetrated by a sperm, a gray crescent forms on the opposite side and becomes the dorsal side of the embryo. In the blastula, mesoderm is formed by inductive signals passing from vegetal to animal cells, and then the mesoderm invaginates the embryo (gastrulation). The ectodermal cells spread from the animal pole to the vegetal pole, as indicated by the arrows in the last two drawings of the series. The formation of the nervous system is initiated by genetic instructions from the migrated mesoderm to the overlying ectoderm. For convenience, the drawing does not indicate the gradual increase in the size of the embryo during development.

Adapted from Ruiz i Altaba and Melton (1990, p. 57).

Peptide growth factors provide inductive signals.

The inductive signals are carried by peptide growth factors (PGFs); one of these, *activin* A (MW, 24 kD), was first identified in the medium bathing a *Xenopus* cell line (Smith *et al.,* 1990). Activin A appears to be the most important of the inducing PGFs in amphibians, but its effects in mammalian development are less striking (Matzuk *et al.,* 1995). Other inducers, possibly acting later and in a cascade, include *fibroblast growth factor (FGF)* and *transforming growth factor-β* (TGF-β).

The next major event in the development of the embryo is gastrulation, during which the blastula is invaginated by the expansion of the dorsal and ventral mesoderm so as to form a gastrula (figure 5.1; Ruiz i Altaba & Melton, 1990). As the dorsal mesoderm comes to lie beneath the dorsal ectoderm, it induces the ectoderm to form the central nervous system (figure 5.2, Gurdon *et al.,* 1989). The anteroposterior axis of the embryo is related to the extent to which the dorsal

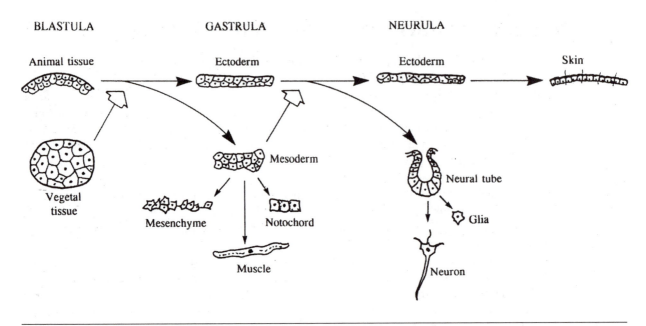

BLASTULA   GASTRULA   NEURULA

Figure 5.2   Sequential transformations and possible destinies of embryonic tissues following inductive signals (open arrows). The ectoderm is also responsible for forming the lens of the eye, under the influence of part of the *neural tube*.

From J.B. Gurdon et al., 1989, "Embryonic induction and muscle gene activation," *Trends in Genetics* 5: 51. Copyright 1989 by Elsevier Trends Journals. Adapted with permission.

mesoderm has migrated under the ectoderm. The cells that have moved farthest will induce the brain and eyes, while the posterior structures will be induced by the mesodermal cells that have traveled least. It now appears that the PGFs activins A and B, in addition to inducing mesoderm as previously described, are responsible for establishing the anteroposterior axis in the embryo. For example, if activin B mRNA is injected into a *Xenopus* embryo, it causes a second body axis to form (Thomsen *et al.,* 1990). The inducers appear to travel among the tissues of the embryo by passive diffusion and, within a few hours, can exert effects on many cells distant from their sites of origin (Gurdon *et al.,* 1994).

Step 2: A Portion of Mesoderm Forms Somites, Then Develops Into Skeletal Muscle

The cells in the equatorial zone of the blastula, which have been instructed to become mesoderm, are influenced in their further development by their positions in the dorsoventral axis. Following gastrulation, the dorsal cells ultimately give rise to notochord and skeletal muscle, while the ventral cells form blood and mesenchyme (embryonic *connective tissue*) (figure 5.3). The part of the dorsal mesoderm that is responsible for muscle formation divides, under the influence of activin B and the homeobox genes (see p. 60), into segmental blocks of tissue called somites (figure 5.4). It is now known that all the skeletal muscle in the body, whether in the trunk or in the limbs, is derived from the *somites* (for review, see Kenny-Mobbs, 1985).

The somitic mesoderm is responsible for certain other tissues besides muscle: the vertebral column and ribs, and the deeper part of the skin (dermis). Nonsomitic mesoderm gives rise to the notochord and to the heart and blood vessels (figure 5.4). The decision of somitic mesoderm to form muscle, rather than one of the other tissues, is made by the expression of one or more *myogenic master regulatory genes* in the mesodermal cells, to produce myogenic regulatory factors (MRFs; for review, see Olson, 1990; Buckingham, 2001). These genes code for the following MRFs: *myo D, myogenin, myf-5,* and *myf-6* (MRF 4). The similarities in the amino acid sequences among these

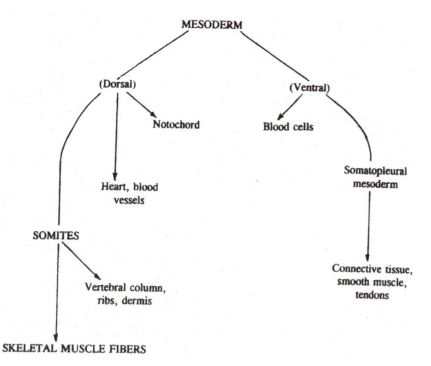

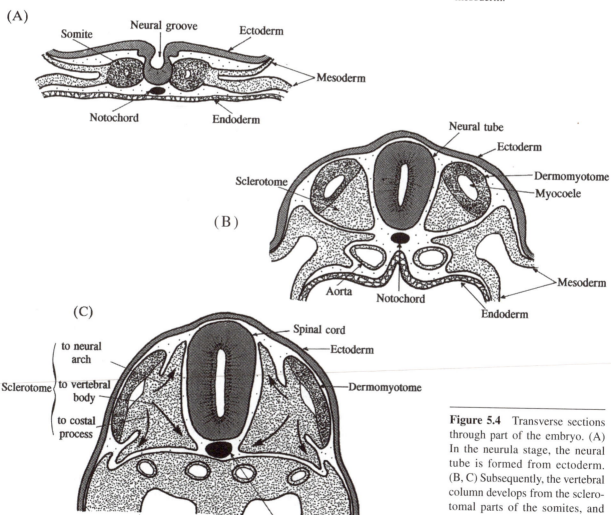

Figure 5.3 Eventual destinies of cells arising from embryonic mesoderm.

Figure 5.4 Transverse sections through part of the embryo. (A) In the neurula stage, the neural tube is formed from ectoderm. (B, C) Subsequently, the vertebral column develops from the sclerotomal parts of the somites, and skeletal muscle is seen to originate in the dermomyotomal parts.

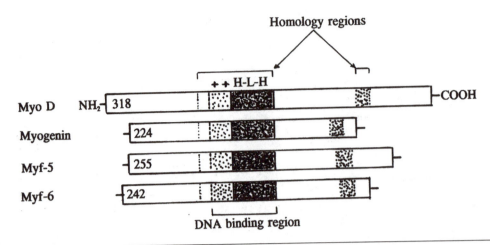

Figure 5.5 Structural similarities between myogenic regulatory factors. For each factor, the number of amino acids in the polypeptide is given at the left. The two regions of homology are shown at the top of the figure; the larger region contains a basic sequence (++) and a helix-loop-helix (HLH) arrangement. Only part of the larger homology region is necessary for the myogenic action of each polypeptide; this is the *DNA*-binding segment shown at the bottom.

Adapted, by permission, from E.N. Olson, 1990, "MyoD family: A paradigm for development," *Genes and Development* 4: 1455. Copyright 1990.

polypeptides are shown in figure 5.5. Each of these factors, once expressed, combines with myogenic genes in the nuclei of the mesodermal cells. Of the four genes, that for myogenin appears to be the most important. Very little skeletal muscle forms in experimental embryos in which this gene has been disrupted (Toyoshima & Unwin, 1988; Nabeshima *et al.*, 1993). However, the fact that the other myogenic genes appear to be less indispensable (Braun *et al.*, 1992; Rudnicki *et al.*, 1992) does not exclude the possibility that they have important roles to play in muscle formation.

Indeed, it is now thought that the genes act in a cascade, with myo D and myf-5 regulating the expression of myogenin (Emerson, Jr., 1993), and myf-6 (MRF 4) being involved at a later stage. The myf-5 gene appears to be the first MRF activated, as a result of signals generated by the adjacent neural tube and notochord, and it commits a fraction of the somitic cells into the myogenic lineage. When this gene is inactivated, the result is a delay in the formation of *myotubes* as well as aberrant migration of the muscle precursors into the adjacent tissues, where they assume nonmuscle identities. However, the expression of myo D, which occurs later, can compensate for this deficit, demonstrating the redundancy existent in the MRF system. Similarly, disruption of the myo D gene alone has very little effect on muscle outcome. When both genes are disrupted, a complete absence of differentiated skeletal muscle is the result. As for the later expression of myogenin and MRF 4, inactivation of the former gene has devastating effects, as already pointed out, while inactivation of the latter gene causes only a mild muscle deficiency (Arnold & Winter, 1998; Seale & Rudnicki, 2000).

So powerful is this family of regulatory factors that they have the ability to cause differentiated cells of other types (*fibroblasts,* adipocytes, liver and melanoma cells) to switch partially or completely to muscle development (Weintraub *et al.*, 1989).

Pax3 is a member of a different family of *transcription* factors. It is an upstream regulator of myf-5 and is thought to be involved in migration of myogenic cells to limbs and diaphragm. Inactivation of the Pax3 gene results in a lack of limb musculature (Arnold & Winter, 1998). Other factors are involved in the muscle differentiation process, such as members of the *myogenic enhancer factor 2 (MEF2)* family and *muscle LIM protein (MLP).* These factors are thought to act as essential cofactors for muscle-specific gene activation (Arnold & Winter, 1998).

While the factors referred to so far are all positive regulators of the myogenic program, there are also negative regulators. TGF-β, bone morphogenetic proteins (BMPs), FGFs, and *myostatin* have all been identified as factors that can inhibit myogenesis. Other nuclear factors inhibit myogenesis by sequestering myogenic factors into non-DNA-binding complexes, blocking target upstream sites, or forming complexes with myogenic factors and thus preventing their entry into the *nucleus*. These nuclear factors include Id (inhibitor of DNA binding) HLH proteins, Mist1, I-mfa protein, and ZEB (zinc finger/E box binding; Arnold & Winter, 1998).

Step 3: The Myogenic (Muscle-Forming) Cells Proliferate and Then Differentiate

The formation of a single human muscle requires that millions of cells must be created before joining together into muscle fibers. This enormous cellular proliferation starts in the somites, where the mesodermal cells multiply under the influence of activin A or B, or both, and the other PGFs (figure 5.6). The proliferation continues among the *myoblasts,* which are the earliest muscle-forming cells to appear, following the expression of the master regulatory muscle genes in a proportion of the mesodermal cells (see earlier). Whether an individual myoblast will continue replicating or will start to differentiate further depends, on the one hand, on the local concentrations of the PGFs expressed by the inducing genes (activins, FGF, TGF-β; see p. 56) and by the "immediate-early" gene family (fos, myc, and jun).

On the other hand, a myoblast will come out of the proliferative cycle and will continue its differentiation into a muscle cell if there is a relatively high concentration of the proteins expressed by myo D, other myogenic master regulatory genes, and proteins p21 and p57, two apparently redundant cell cycle inhibitors (Friday *et al.,* 2000). Inactivation of the latter two genes results in a phenotype nearly identical to that following the inactivation of the myogenin gene: a severe reduction in the appearance of muscle-specific proteins (Friday *et al.,* 2000). The ability of the growth factors to suppress differentiation may also be partly mediated by a protein called Id (inhibitor of DNA binding), which is found in many cell types and is known to be down-regulated during muscle differentiation (Benezra *et al.,* 1990).

Myoblasts are termed embryonic or primary myoblasts, fetal or secondary myoblasts, and adult myoblasts or *satellite cells,* depending on the developmental stage at which they migrate from the somites to the limb buds. These myoblasts differ in their morphological characteristics and in the expression of certain proteins, such as *myosin heavy chains.*

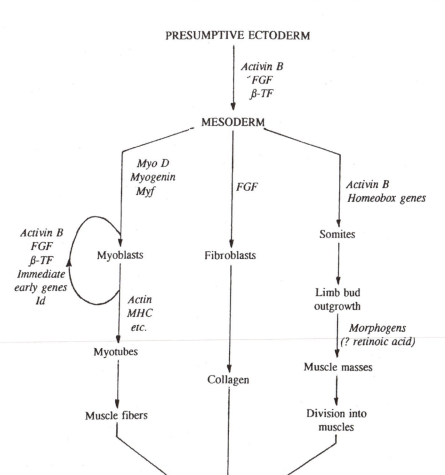

Figure 5.6 Transformation of presumptive ectoderm into mesoderm and subsequently into complete muscles. The genes and regulatory factors involved in the various steps are indicated by italics.

Formation of Myoblasts

Myoblasts resemble mesodermal cells and have simple structures.

When first formed, the myoblast is a spindle-shaped cell with a central nucleus and is indistinguishable in appearance from its mesodermal ancestors. Within the ovoid nucleus is a prominent nucleolus, indicating that the cell is actively synthesizing *RNA*. The cytoplasm of the cell is also packed with RNA in the form of ribosomes. Scattered mitochondria are present, and a small *Golgi apparatus* is evident; the endoplasmic reticulum is poorly developed. Electron microscopy also reveals thick (60 nm) filaments lying in the periphery of the cell under the plasma membrane. Some of these filaments are *actin* and are probably involved in movements of the myoblast and its appendages. Thick filaments are not seen in mammalian myoblasts except just prior to cell fusion, and there is therefore no evidence of *myofibrils* in most of the cells.

From microelectrode studies of muscle developing in tissue culture, it appears that the resting *membrane potential* of the myoblast is low (around –20 mV; Fambrough *et al.,* 1974). Since the internal K^+ concentration is normal, the low potential presumably results from leakiness of the plasma membrane for Na^+ (see chapter 9); the fact that a *basement membrane* has not yet developed may also be relevant in the control of *permeability.* Functional *acetylcholine receptors (AChRs)* do not appear to be present in the plasma membranes of most myoblasts (Fambrough *et al.,* 1974).

Fusion of Myoblasts to Form Myotubes

Structures resembling gap junctions form between fusing myoblasts.

The myoblasts divide repeatedly by mitosis and are seen to be mobile both in tissue culture and in regenerating muscle. After these proliferative mitoses, each cell undergoes a final division, the quantal mitosis, and then prepares for fusion into myotubes (figure 5.7). In culture the cells now extend cytoplasmic processes that explore the surfaces of other myoblasts and of myotubes lying nearby. Once the cells have recognized each other as myoblasts (or as myoblast and myotube), they line up together with their long axes parallel. It is probable that *fibronectin,* one of the extracellular glycoproteins involved in cell guidance, serves as a substrate upon which the myoblasts can move during these preparatory events. After alignment the myoblasts adhere to each other and eventually develop electron-dense structures under their respective *plasmalemmae,* which resemble the *gap junctions* found in other cell types (Rash & Fambrough, 1973). At this stage, connexin 43, a gene that codes for one of the gap junction proteins in various tissues of the body, is expressed in the developing skeletal muscle.

Factors Promoting Myoblast Fusion

Ca^{2+} entry, following depolarization, promotes cell fusion.

The formation of new proteins is also suggested by studies in which antibodies have been raised against membrane preparations made from cells prior to fusion. It is known that the entry of Ca^{2+} into the myoblasts is essential if fusion is to proceed, because the process is halted if these ions are removed from the culture medium (Shainberg *et al.,* 1969) or if Ca^{2+} entry is blocked pharmacologically. Conversely, precocious fusion occurs if a Ca^{2+} ionophore is included in the medium (David *et al.,* 1981). In turn, the entry of Ca^{2+} is dependent on transient *depolarization* of the myoblast, which increases the permeability of the plasmalemma to Ca^{2+}.

Several naturally occurring agents can induce depolarization of myoblasts in culture; and Entwistle and coworkers (1988a; 1988b) suggest that, *in vivo,* the developing muscle may have several independent mechanisms available in order to provide the fusion process with a large safety factor. One of these mechanisms is likely to be the binding of prostaglandin E with *prostanoid receptors.* In those myoblasts that have already formed AChRs, another mechanism may be the combination of the receptors with *ACh* released from these motor nerve terminals (figure 5.7).

Events During and Following Myoblast Fusion

Apposed cell membranes break down during myoblast fusion.

While the recognition and adhesion phases may last several hours, in culture at least, the fusion process itself is relatively brief, occupying 8 to 10 min (Bischoff & Holtzer, 1969) or even less (Rash & Fambrough, 1973). Experiments using cycloheximide show that although new proteins may be synthesized during the adhesion phase, this is not a requirement for the fusion stage (Bischoff &

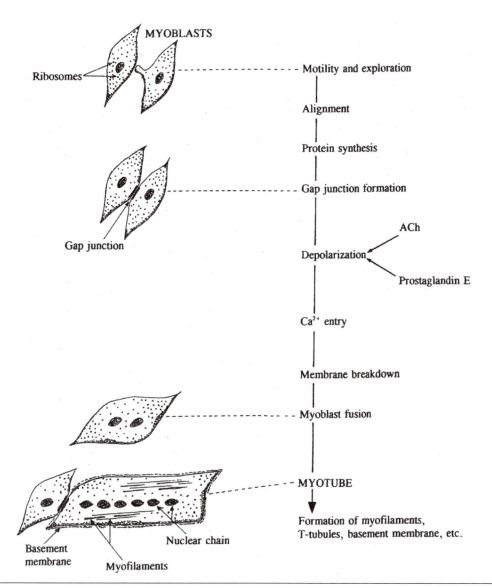

Figure 5.7 Stages in the formation of myotubes from myoblasts.

Lowe, 1974). It is interesting that in tissue culture, fusion may occur between myoblasts derived from different species and between cardiac and skeletal myoblasts, but not between myoblasts and other types of cell (e.g., from liver, kidney, or cartilage).

As part of the fusion process there is, at the surfaces of the cells, breakdown of the contiguous membranes and possibly the incorporation of membrane from closely applied cytoplasmic vesicles (Kalderon & Gilula, 1979). The nuclei come into apposition in the center of the newly formed myotube, and adjustments are made to the cytoskeleton so as to strengthen the structure of the syncytium. Additional details of myoblast fusion may be found in the reviews of Inestrosa (1982) and Wakelam (1985).

Once a myotube has been formed, further myoblasts will add themselves to it as described on page 58. The nuclei of the new cells continue to line up with preexisting nuclei in the center of the myotube so as to form a chain, while they become somewhat larger and rounded. These nuclei will not partake in any further mitoses. Meanwhile, the cytoplasm of the myotube increases in amount and contains increasing numbers of thick and thin filaments (myosin and actin) that have been synthesized by polysomes. The *myofilaments* are grouped in bundles; these become thicker and form myofibrils. Soon the *A-* and *I-bands* of the myofibrils can be distinguished, indicating that the component myosin and actin filaments are now in register; the *Z-lines* subsequently appear.

New proteins, including those for myofila- ments, are synthesized after fusion.

The sequence in which the special proteins of muscle appear during embryogenesis has been determined by *immunohistochemistry;* the first protein is *desmin,* and then follow *titin,* muscle-specific actin, myosin heavy chains, and, last, *nebulin* (Furst *et al.,* 1989). Some of these proteins can already be detected at the myoblast stage, though in small amounts. In the myotube, the myofibrils are formed in two parts; one of these is the Z-disk, with filaments attached at either side, and the other structural unit is the thick filaments of the A-band. Both structural units appear to form on bundles of *microfilaments,* which act as templates; once formed, the thick filaments and Z-disk units are linked together by the long titin filaments (see figure 1.11; Furst *et al.,* 1989).

Aside from the appearance of myofibrils, other changes become evident in the myotube. Mitochondria become more numerous and are elongated with densely packed *cristae.* A rudimentary transverse tubular system (T-system) forms from invaginations of the plasma membrane; similarly, *sarcoplasmic reticulum* is derived from the outer nuclear envelope. The development of a basement membrane is evident in electron microscopy as a fuzziness on the surface of the myotube.

Developments in Myotube Plasmalemma and Size

The myotube plasma-lemma becomes more permeable to K+ and develops ACh sensitivity.

Intracellular microelectrode studies of cultured cells show that the properties of the plasmalemma are changing. The resting membrane potential rises as the myotube becomes larger and reaches maximum values of about -60 mV (Fischbach *et al.,* 1971); this change is probably caused by a rise in the relative permeability of the membrane for K^+ as opposed to Na^+ (see chapter 9). The membrane is now capable of firing *action potentials* either spontaneously or following direct stimulation. In tissue culture, the myotubes can be seen to twitch spontaneously and synchronously; it is possible that the cells are electrically coupled to each other. The presence of tight junctions between adjacent myotubes, visible upon electron microscopy, would provide a possible mechanism.

By iontophoretic application of ACh or the use of ^{125}I-labeled *α-bungarotoxin,* it can be shown that the membrane now contains AChRs, some of which have been formed in the cytoplasm and then transported to the surface of the myotube. Although the whole surface of the cell is sensitive to ACh, there are certain "hot spots" corresponding to particularly high receptor densities. Fambrough *et al.* (1974) have estimated that in 1 hr, roughly 100 AChRs may be formed per square micrometer of membrane. The myotubes also contain *cholinesterase;* and, as with the ACh receptors, much of this appears to be manufactured in the vicinity of the nuclei and carried to the surface, where it is incorporated in the basement membrane.

As the myotube enlarges, it forms more myofilaments, and both the sarcoplasmic reticulum and T-tubular systems become better differentiated. At the same time, for most species, the nuclei move along the cytoskeleton from the center of the cell to take up new positions under the plasmalemma. The myotube has now completed its transition into a muscle fiber.

Step 4: The Body Plan Is Laid Down

While the somitic cells are pursuing their myogenic destinies, the shape of the embryo becomes increasingly well defined. The genes that control the shape of the embryo and the appearance of the major body features are called homeotic.

Role of Homeotic Genes

Homeotic genes regulate early embryonic development.

The homeotic genes are the master regulators of development and code for proteins that, in turn, bind to chromosomal DNA and thereby control the activities of other genes. The latter are those regulating the expression of yet other genes, including those involved in such cellular processes as signaling between cells, programmed cell movements, and the production of surface proteins necessary for cell-to-cell recognition. Some homeotic genes contain a common region of DNA; this region is the homeobox, and it codes for a 60-amino acid sequence that binds the remainder of the homeotic protein to the DNA strand.

It appears that the first expression of the homeotic genes is brought about by the same growth factors responsible for inducing mesoderm from the animal cap (i.e., activins A and B, FGF, TGF-β). Some of the homeotic genes are involved in the inducing action of mesoderm on ectoderm that results in the development of the head, body, and tail in the anteroposterior axis (see p. 54). Other homeotic genes control the segmentation of the embryo and the creation of the limb buds. If the embryo is to form properly, the homeotic genes must be switched on in the correct sequence; such an arrangement implies that each gene affects the transcription of the next one in the developmental program (Ruiz i Altaba & Melton, 1990).

Role of Morphogens

Different concentrations of morphogens provide positional cues.

Other cues for gene expression depend on the positional values of the cells. Thus, throughout the embryo, each cell can be defined in terms of its relationship to three orthogonal axes: dorsoventral, anteroposterior, and proximal-distal. Several workers, including Wolpert (1969) and Crick (1970), have proposed that for each axis there is a chemical *morphogen;* the morphogen would diffuse from the site of secretion establishing a concentration gradient in that axis. A cell would recognize its position in the embryo by sensing the strength of the different morphogens and would then express its genome appropriately. In the limb bud, the morphogen that determines the anteroposterior axis (thumb to little finger in the human hand) is derived from a small region on the posterior margin, the zone of polarizing activity. Tissue grafting experiments indicate that the proximal-distal axis of the limb may be determined by the time spent by each cell in the process zone; this region, situated at the base of the limb bud below the apical epidermal ridge, would send signals of different durations to the cells in the limb bud, depending on when they moved out into the elongating limb bud. One remarkable consequence of the positional cuing is that if somites are transplanted from the thoracic region of a chick embryo to the brachial area, the muscles that form are those of the shoulder and upper arm rather than those of the chest wall (Butler *et al.,* 1988; Chevallier *et al.,* 1977).

Although once regarded as fanciful, the concept of diffusible morphogens has gained support. In the *Drosophila* (fruit fly) embryo, the proteins expressed by some of the homeotic genes can be detected by immunofluorescence (for review, see St. Johnston & Nusslein-Volhard, 1992), and striking gradients exist for some of the proteins at particular steps of development (figure 5.8). In the mammalian embryo, a marked gradient of gene expression in the rostrocaudal axis has been demonstrated, presumably in response to the corresponding gradient

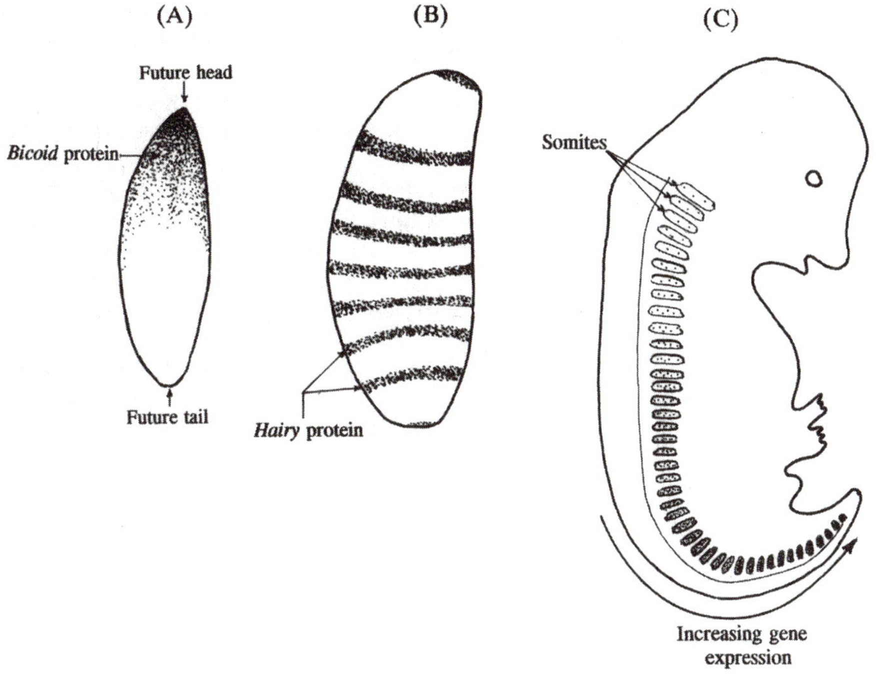

Figure 5.8 Polarization and segmentation of the embryo, under the influence of homeotic genes. (A) *Drosophila* embryo in which the protein product of the bicoid gene has been shown by immunofluorescence, shortly after fertilization. The high concentration of bicoid protein at one end of the embryo is critical for the later development of the head in this region. (B) Three hours after fertilization, another gene, hairy, is expressed in seven stripes across the embryo; these stripes serve as boundaries that eventually divide the embryo into 14 segments. Both bicoid and hairy are examples of homeotic genes. (C) A mouse embryo in which an artificially constructed gene (myosin light chain 1-chloramphenicol acetyltransferase) has been expressed in skeletal muscle within body segments. The expression is strongest at the tail end of the embryo and progressively weaker toward the head end. This difference in gene expression has been attributed to the presence of a morphogen having a concentration gradient in the rostrocaudal axis.

The drawings in A, B, and C are based on the work of Nüsslein-Volhard (1991), Langeland and Carroll (1993), and Grieshammer *et al.* (1992), respectively, and are not to the same scale.

of a morphogen (Donoghue *et al.,* 1992; Grieshammer *et al.,* 1992; see figure 5.8). In vertebrates, one possible morphogen is *retinoic acid,* a vitamin A derivative. This substance, normally found in different concentrations in the limb bud, has been shown experimentally to be associated with the establishment of all three body axes in the amphibian limb and of the anteroposterior axis in the chick limb (Summerbell & Maden, 1990). Another morphogen is FGF, which may act initially in conjunction with retinoic acid and then independently (Niswander *et al.,* 1994).

Step 5: Muscles Are Assembled

As far as the development of the limb musculature is concerned, the first step is the appearance of dorsal and ventral muscle masses in the limb bud. These blocks of tissue then split and continue developing in two stages:

Muscle masses split repeatedly.

• **Subdivision of muscle masses to form individual muscles.** After the first split, each of the newly formed muscle masses is cleaved again, and further subdivisions occur until individual muscles are established (figure 5.9). Each muscle is then assembled within a connective tissue scaffolding that defines the tendons at each end of the muscle, as well as the fibrous sheets that envelop the belly and also group the muscle fibers into bundles *(fascicles)*. Like the muscle fibers, the connective tissue is derived from mesoderm, though from a different region, the somatopleure. The somatopleural mesodermal cells destined to develop into connective tissue multiply under the influence of the PGFs (including FGF) and the proteins transcribed from the immediate-early genes. As yet, a master regulatory gene remains to be identified, but clearly some signal must commit the mesodermal cells into becoming fibroblasts.

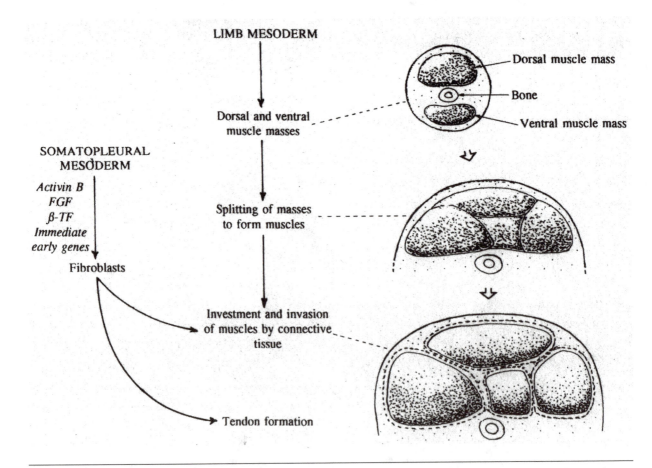

Figure 5.9 Splitting of dorsal and ventral muscle masses, formed from limb mesoderm, into separate muscles. The muscles are then surrounded and penetrated by connective tissue and attached to tendons, all derived from somatopleural mesoderm.

• **Development of muscle fiber types.** Once formed, the fibroblasts replicate and migrate into appropriate positions within and around the developing muscle, presumably in response to positional cues. In the human fetus, the first (primary) myotubes are formed at 7 weeks of gestation; their proliferation and development have been investigated in the quadriceps muscle by Draeger *et al.* (1987) and by Barbet *et al.* (1991), both of whom employed antibodies to detect the appearance of different types of myosin. According to these authors, at 8 to 10 weeks gestation, the primary myotubes contain "slow" myosin heavy chains, as well as two developmental myosin heavy chains, known as "embryonic" and "neonatal" (figure 5.10). Because the motor *axons* may have already begun to penetrate the muscle masses (Toop, 1975), the expression of the slow myosin gene may be nerve dependent. In other species, the initial fiber typing can occur

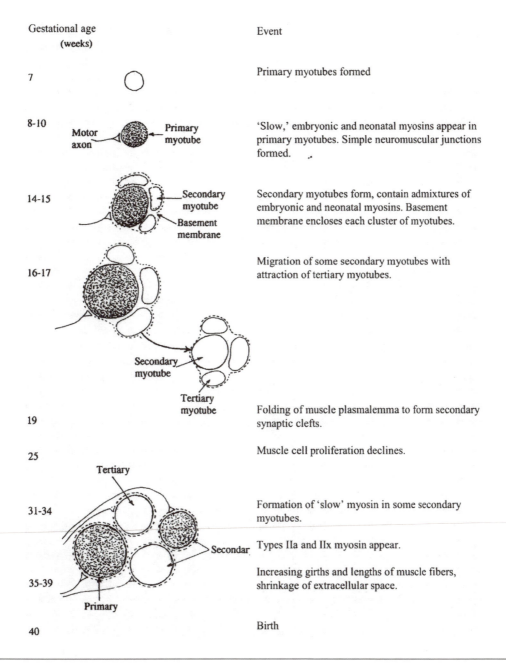

Figure 5.10 Development of muscle from primary, secondary, and tertiary myotubes. The presence of slow and fast myosins is indicated by stippled and open myotubes, respectively.

even in the absence of motor innervation (Butler *et al.,* 1982; Condon *et al.,* 1990). The process of muscle fiber type development continues thus:

-By 14 to 15 weeks gestation, the myotubes become surrounded by smaller, secondary ones that also contain admixtures of slow and fast myosin heavy chains. The formation of secondary myotubes occurs after the arrival of innervation. Both the primary and secondary myotubes lie beneath the same basement membrane so as to form a cluster, separated from other clusters by connective tissue (figure 5.10). By 16 to 17 weeks gestation, some of the secondary myotubes have moved out from under the basement membrane and have started to attract smaller, tertiary myotubes around them. By this time, the primary myotubes/myofibers are distinguishable from the others in that they are larger and contain only type I myosin heavy chain (Staron, 1997).

-The secondary myotubes/myofibers begin at this time to lose the developmental myosin heavy chain *isoforms* and begin accumulating slow myosin, or one of the mature fast forms (in humans, types IIa and IIx). The tertiary myotubes can be recognized not only by their smaller size but also by their late retention of the neonatal and embryonic myosin heavy chains. Because of their increased numbers, the myotubes are now more closely apposed and no longer give the impression of clusters. With further development, the secondary and tertiary myotubes enlarge; and by 31 to 34 weeks gestation, approximately half of the secondary myotubes have acquired slow myosin. Also at this time, myosin heavy chain types IIa and IIx, characteristic of adult human muscle, become distinguishable (Staron, 1997). Thus, the slow-twitch fibers appear to be derived from the primary and a proportion of the secondary myotubes, while the fast-twitch fibers come from the remainder of the secondary myotubes and from the tertiary population (figure 5.10).

It should be pointed out, however, that the muscle cell lineage patterns in human skeletal muscles have not been completely resolved, since there is now evidence of different "fiber types" before the arrival of the motor nerve. Ghosh and Dhoot (1998) prepared muscle cell cultures from human quadriceps at different stages of fetal development, and stained the myotubes with antibodies to fast and slow skeletal muscle myosin ATPases to see which proteins were being expressed during 6 to 18 days in culture without the influence of the innervating motor nerve. Their results showed that all cells at 10 to 18 weeks of gestation contained fast myosin heavy chains (including fetal and neonatal forms), while only 10% to 40% of these same cells stained for slow myosin heavy chain. Thus, there are at least two distinct myotube types, one of which expresses slow as well as fast myosin heavy chains, while the other expresses only fast.

The location of the primary fibers in the whole-muscle cross section exerts an influence on their ultimate myosin heavy chain expression. Primary fibers located in the muscle belly farthest from the bone have a greater chance of becoming fast-twitch fibers, while those located closest to the bone have a greater chance of becoming slow-twitch. The mechanism for this preferential localization of muscle fiber types within the muscle is not fully known, but this differentiation is known to occur in the absence of innervating neurons (Edgerton *et al.,* 1996).

Muscle development in the human fetus has also been carefully studied by Stickland (1981), who counted the numbers of myotubes and muscle fibers in sartorius muscles. From his data, it appears that there are large increases in the numbers of myotubes and muscle fibers until about 25 weeks gestational age, after which cell proliferation slows down and may even have ceased by the time of birth. Beyond 25 weeks gestational age, most of the increase in muscle cross-sectional area is due to enlargement of existing fibers. The interstitial connective tissue, initially accounting for 60% of the muscle volume, decreases to about 20% by 36 weeks gestation.

Postnatal Development of Muscle

At birth the muscle fibers are still relatively small, most no larger than 20 μm in diameter (figure 5.11). In human muscle, however, a 20-fold enlargement of the muscle takes place during childhood and puberty, assuming that the proportional contribution of muscle to the total body weight remains constant. In this section, we consider the various stimuli for muscle growth and the mechanisms involved.

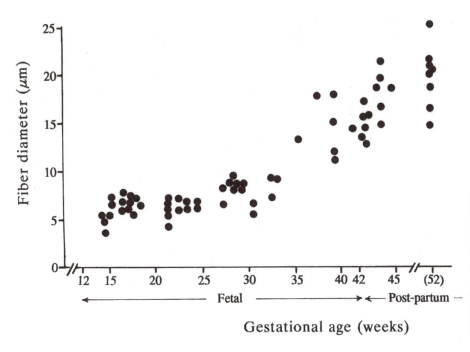

Figure 5.11 Increasing sizes of muscle fibers in human fetus and infant. The values are for fibers that have not yet differentiated into type I or type II.

Adapted, by permission, from A.S. Colling-Saltin, 1978, "Enzyme histochemistry on skeletal muscle of the human foetus," *Journal of Neurological Sciences* 39: 179.

Increase in Length and Girth of Growing Fibers

New sarcomeres are added to the ends of the growing muscle fibers.

Obviously, much of the increase in muscle size will take place in an axial direction through the formation of new *sarcomeres,* and this process has been demonstrated at the ends of growing fibers by Williams and Goldspink (1971). The nature of the stimulus for this longitudinal expansion is not known for sure, though growth hormone is certainly involved (figure 5.12). Thus, when a deficiency of growth hormone occurs, as in panhypopituitarism of childhood, there is a failure of all tissues to develop, including the somatic musculature. In experimental animals in which the pituitary glands have been ablated, it can be shown that muscle growth is restored by the injection of growth hormone (Goldberg & Goodman, 1969). Insulin is another hormone necessary for muscle growth; at the level of the muscle fiber, it has been shown to stimulate amino acid transport, enhance protein synthesis, and inhibit protein degradation (Goldberg, 1968; Fulks *et al.,* 1975). Finally, thyroid hormone and liver-derived *insulin-like growth factor*-1 (IGF-1) are necessary for normal muscle development and maturation (Adams *et al.,* 2000a, 2000b).

Local mechanical factors are also of major importance in determining muscle length and in ensuring that the number of sarcomeres remains appropriate for the distance separating the attachments of the muscle—irrespective of whether the subject is a dwarf or a giant. Tabary *et al.* (1972) have shown that if a cat soleus muscle is maintained in a shortened position, by fixing of the leg in a plaster cast, there is a 40% loss of sarcomeres within 3 weeks. Conversely, prolonged extension results in the formation of new sarcomeres. The functional advantage of an adjustable fiber

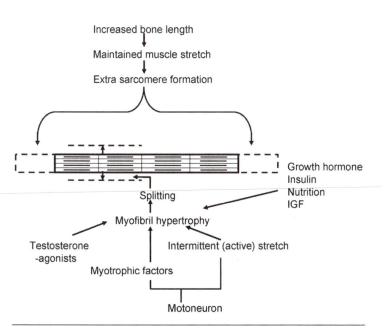

Figure 5.12 Summary of factors stimulating muscle fiber enlargement.

Reprinted from A.J. McComas, 1977, *Neuromuscular Function and Disorders* (London: Butterworth Publishing Co.), 97.

length is clear, as it allows the muscle to work at, or below, the optimal region of the length–tension curve (see chapter 11); overstretching of the muscle and damage to the sarcomeric structure are less likely to occur.

Muscle girth changes during development are partly regulated through continuing action of members of the transforming growth factor (TGF-β) superfamily; an important one of these is known as myostatin (McPherron *et al.,* 1997). Myostatin, which is expressed in developing and adult muscles, is evidently a negative regulator of muscle mass increase; in mice in which the myostatin gene is disrupted, muscles are two to three times the size of muscles of normal mice because of larger and more numerous fibers (McPherron *et al.,* 1997). It appears that as the bones grow, there is a stimulating effect on protein synthesis effected through mechanical strain on the muscle fiber, extracellular matrix, and the consequent downstream activation of gene expression via a number of mechanisms (Vandenburgh *et al.,* 1995; Goldspink, 1999). Regular load bearing is also an important component of muscle development; muscles subjected to reduced load bearing during the early rapid period of muscle development appear to have a long-lasting deficit in muscle size and in the number of myonuclei (Mozdziak *et al.,* 2000).

At the same time that the fibers are growing longitudinally, their diameters are increasing so that, in man, mean values of about 50 μm eventually will be achieved (Brooke & Engel, 1969). It is this change in muscle growth that is responsible for the increased muscle strength during development.

McComas *et al.* (1973a) measured isometric *twitch* tensions in the extensor hallucis brevis muscle in boys of different ages and found that muscle strength increased in two phases (figure 5.13). The first phase lasted until puberty, and during this time there was a gradual increase in strength such that an approximately linear relationship could be demonstrated between maximum twitch tension and age. The second phase occurred at puberty and was more dramatic in that a doubling of contractile force took place within a comparatively short time, certainly no longer than 2 years. It is probable that this spurt in muscle strength is due to the direct action of testosterone on muscle fibers; for the serum levels of this hormone are raised at puberty, and it is known that testosterone has a direct anabolic effect on muscle fibers. This anabolic effect has been demonstrated by experiments on the levator ani muscle of the rat, which is particularly hormone dependent. If an animal is castrated, marked *atrophy* of this muscle occurs; the atrophy can then be reversed by the administration of testosterone (Wainman & Shipounoff, 1941). In man, synthetic steroid substances closely related to testosterone, the anabolic steroids, have been found to increase muscle bulk and strength and have been used therapeutically. As recent Olympic Games have shown, anabolic steroids have also been taken by athletes anxious to improve their performance not only in such "heavy" field sports as shot-putting and discus throwing, but also in sprinting.

> In males, muscle strength is increased at puberty, presumably by testosterone.

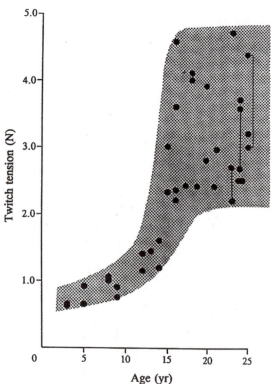

Figure 5.13 Maximal isometric twitch tensions of extensor hallucis brevis muscles in male subjects of different ages. Bilateral observations have been linked together. Note the marked increase in tension during puberty (usually 14-16 years).

Data from McComas *et al.* (1973a).

Myofibrillar Enlargement and Splitting, and the Role of Mechanical Stretch

According to MacCallum (1898) and Stickland (1981), the number of muscle fibers does not increase in a human muscle (sartorius) after birth. If this is true of all human muscles, then the increased strength at puberty can have resulted only from the synthesis of new myofibrils in existing muscle fibers. This last process has been studied in mice during the vigorous postnatal growth phase by Goldspink (1970). He found that in some fibers the numbers of myofibrils increase by as much as 15 times. On the basis of electron microscopic observations on individual myofibrils, Goldspink suggested that the myofibrils first enlarge and then split in two, with the rupture commencing in the Z-disks. Myofibril enlargement and splitting also occur during work *hypertrophy* and in recovery from the starvation-induced atrophy; the cycle may be completed in the remarkably short time of 1.5 to 2 days (Goldspink, 1965).

> **As muscles enlarge, myofibrils thicken and split.**

From observations on bodybuilding and weightlifting exercises in man (MacDougall, 1986), it is clear that muscle contractions are far more potent stimuli for muscle fiber hypertrophy if they are lengthening or isometric rather than *shortening contractions* (see chapter 11). Since the action potential is common to both types of contraction, the difference must be related to the degree of muscle stretch. In shortening contractions, all of the sarcomeres are allowed to shorten; in an isometric contraction, however, some sarcomeres will shorten, but others may be stretched, because the two ends of the muscle remain almost the same distance apart. In *lengthening contraction,* the stretching of the sarcomeres is even greater, since the ends of the muscle are separated farther. Hence stretch has two anabolic effects on muscle. Maintained stretch, as in the plaster cast experiments by Tabary and colleagues (1972), induces the formation of new sarcomeres, including fresh myofilaments, at the extremities of the fiber. Severe intermittent stretch, as in forceful isometric or lengthening contractions, results in the synthesis of new myofilaments around existing myofibrils, but there is no increase in sarcomeres.

> **Stretching of sarcomeres promotes muscle fiber growth.**

Since protein is being formed during myofibrillar enlargement, one might expect some change in the number or disposition of the ribosomes to become evident. Galavazi and Szirmai (1971) were able to demonstrate an increase in intermyofibrillar polyribosomes in electron micrographs of levator ani muscles from castrated rats treated with testosterone. Surprisingly, the increase in myofilaments occurs rather earlier than that of the polyribosomes. That the ribosomes in such animals are driven by myonuclear DNA is suggested by the rise in DNA, increased incorporation of [^{3}H]-thymidine, and proliferation of nuclei within the fibers.

Myonuclei are considered to be postmitotic in the human at birth and therefore do not produce additional DNA. In spite of this, muscle fiber hypertrophy during the growth period is associated with an increased DNA content. This is explained by the fact that satellite cells, which are adult myoblasts, are the source of DNA increase during this time. These cells, discovered by Alexander Mauro in 1961, are found in small numbers adjacent to the muscle fibers and lie beneath the *basal lamina.* In the adult, these cells are considered to be skeletal muscle stem cells (O'Brien *et al.,* 2002). When activated they proliferate, differentiate, and fuse to existing fibers. With age, and as the rapid growth of muscle fiber size subsides, the number of active satellite cells decreases (Mesires & Doumit, 2002). The satellite cells can also contribute to muscle repair and *regeneration* (see chapter 21).

It is of interest to consider what the advantage might be of myofibrillar splitting as opposed to having thicker myofibrils, because the contractile force should be no different in the two situations. One possibility is that the biochemical support systems for the myofibrils, operating through the mitochondria and sarcoplasmic reticulum, may be relatively ineffective in nourishing large cylinders of contractile material and that the myofibrils have an optimal surface:volume ratio. Figure 5.12 summarizes the factors that are known to influence muscle growth; they are both hormonal (growth hormone, insulin, and testosterone) and mechanical (muscle stretch). The ability to produce muscle atrophy by either *denervation* or starvation at any age indicates that an intact nerve supply and adequate nutrition must be considered as additional factors necessary for normal growth.

> **Splitting may improve the nutrition of the myofibrils.**

Development of Twitch Speeds and Electrical Properties of Muscle Fibers During Infancy

> **Twitch speeds of human muscle fibers alter during infancy.**

Isometric twitch studies enable the speed of muscle contraction and relaxation to be measured. In newborn kittens, all muscles have slow twitches, but in those muscles destined to become fast, a speeding up occurs during the next few weeks under the controlling influence of the *motoneuron* (Buller *et al.,* 1960a). In the human triceps surae, however, the *contraction time* is rather short during the neonatal period, but by 3 years it has reached its adult value, having increased by 50%

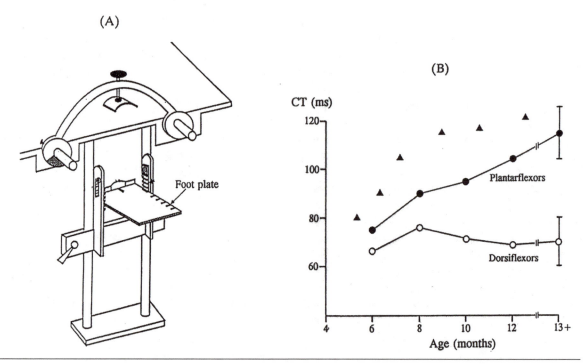

Figure 5.14   (A) Apparatus for recording isometric twitches in ankle dorsiflexor and plantarflexor muscles of infants. With the child lying supine, the knee is flexed over the end of the examining table, and the foot is strapped to the horizontal plate below. (B) Mean contraction times (CT) of dorsiflexor and plantarflexor muscles in 35 infants; vertical bars around the points for the eldest infants indicate standard deviations. The triangles show the successive values obtained for the plantarflexor muscles of a single infant.

From "Histochemical and contractile property changes during human muscle development," by G. C. B. Elder and B. A. Kakulas, 1993, *Muscle and Nerve,* 16, p. 1259. Copyright 1993 by John Wiley & Sons, Inc. Adapted with permission of John Wiley & Sons, Inc.

(Gatev *et al.,* 1977). In the human extensor hallucis brevis muscle also, the contraction time has entered the adult range by 3 years, though the relaxation phase remains slow for a rather longer time (McComas *et al.,* 1973a). Studies by Elder and Kakulas (1993) show that 6 to 12 months is a critical period in the development of leg muscles, as reflected in the twitch durations (figure 5.14); this finding suggests that the support of body weight may be an important controlling factor.

Finally, the electrical properties of the muscle fiber membranes also change during growth. In the 1-week-old mouse, the resting potentials of the fibers are still well below those of the adult. Subsequently, the potentials rise in both "fast" and "slow" muscles, with the former achieving higher values (Harris & Luff, 1970).

Applied Physiology

In this section, we will discover that muscles may sometimes fail to develop normally in the embryo and may occasionally be missing altogether. Although the mechanisms involved are not fully understood, the eventual explanations are likely to involve some of the factors that have just been considered.

Congenital Conditions

Two of the more common congenital failures of muscle development are discussed here.

• **Muscles may be missing at birth.** On rare occasions, otherwise normal infants are born lacking a muscle or a whole muscle group; the pectoralis major, with or without the other pectoral muscles, is probably the most frequent example. In such cases the deficiency is nearly always unilateral. It is likely that the abnormality results from the failure of the pectoral muscle mass to form as the cervical somites differentiate; in some cases there appears to be a genetic basis.

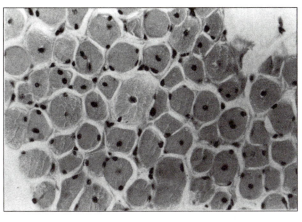

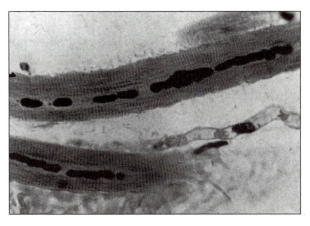

(A) (B)

Figure 5.15 Muscle fibers from an 8-year-old girl with centronuclear myopathy. (A) The muscle specimen has been cut transversely; almost every fiber has a central nucleus, and the muscle fibers are much smaller than normal (420×). (B) Two muscle fibers have been sectioned longitudinally to show rows of nuclei *(nuclear chains)* in the centers of the fibers (583×).

From B.Q. Banker, 1986, The Congenital Myopathies. In *Myology-basic and clinical,* edited by A.G. Engel and B.Q. Banker (New York: McGraw-Hill Companies), 1548. Reprinted with permission from The McGraw-Hill Companies.

• **In centronuclear myopathy, the myotubes do not mature.** Congenital myopathies are a group of disorders that can cause variable *weakness* in infancy and are due to primary abnormalities of the muscle fibers. Among these disorders is centronuclear myopathy, in which the muscles are composed of small, centrally nucleated fibers (figure 5.15). All the evidence suggests that the latter are myotubes that have failed to continue in their development, possibly because of a defect in their innervation (Elder *et al.,* 1983). Curiously, the muscle fibers inside the *muscle spindles* are unaffected and consequently remain larger than the *extrafusal fibers.*

In Thalidomide Babies and Wingless Chickens, Limb Buds Fail to Form Properly

In the 1950s, thalidomide was a commonly prescribed sleeping drug and one that apparently had few side effects. It came as a surprise, therefore, when McBride (1961) raised the possibility that thalidomide taken during early pregnancy was responsible for serious skeletal abnormalities in the fetus. The most striking of these abnormalities was the failure of the arms or legs, or both, to develop fully (amelia), so that an infant could be born with no more than stumps to which the fingers or toes were attached. X rays of the arms showed that the radius was aplastic; this observation, and the fact that the thumb was poorly developed or absent, suggested that segmental defects had occurred in the embryo. In turn, the segmental lesions were attributed to lesions of the *neural crest* or sensory neurons; either type of lesion could have deprived the limb bud of neurotrophic signals necessary for its growth and differentiation (McCredie, 1975; McBride, 1978). In key experiments, however, Strecker and Stephens (1983) inserted tantalum foil between the neural tube and somites of chick embryos to block nerve outgrowth and found that the wings still formed normally. At present, then, the neural explanation for thalidomide appears unconvincing; the possibility that thalidomide interferes with DNA transcription (Koch, 1990) does not remove the need to know how the embryonic tissues are subsequently affected.

In contrast to thalidomide embryopathy, the pathogenesis of the limb defects in genetically amelic chickens is better understood. Limbless is a mutant autosomal recessive gene that causes complete absence of the wings and legs (Prahlad *et al.,* 1979). In this condition, the exchange of tissue grafts between normal and mutant chick embryos has established that the primary defect is a failure of the apical epidermal ridge to form in the ectoderm; hence the signals necessary for the normal growth of the limb buds are missing (Carrington & Fallon, 1988).

More discrete lesions of the apical epidermal ridge can be produced by feeding retinoic acid to pregnant mice. In the embryo, the ridge cells undergo excessive cell death and cause various types of digital abnormality to appear—missing toes, split toes, or even extra toes (Sulik & Dehart, 1988). These experimental findings have considerable clinical significance, because retinoic acid has been widely used in the treatment of recalcitrant acne.

The study of muscle development has been completed, and our attention must now turn to the formation of the motoneurons and of their subsequent connections to the muscle fibers.

6

Development
of Muscle Innervation

As with muscle formation, the development of the motor innervation in the embryo can be visualized as occurring in stages:

- Passage of inductive signals from *mesoderm* to *ectoderm*
- Formation of the *neural tube*
- Neuronal proliferation and migration
- Axonal outgrowth
- Formation of the neuromuscular junction
- *Synapse* elimination and *motoneuron* death

Step 1: Inductive Signals Pass From Mesoderm to Ectoderm

The nervous system arises from ectoderm, under the influence of mesoderm.

The motoneurons, like all mammalian nerve cells, are ultimately derived from embryonic ectoderm. It will be recalled that ectoderm is itself descended from the cells making up the animal cap of the *blastula* (see figure 5.1, chapter 5).

The transition of part of the ectoderm into neural tissue is the result of a quite remarkable mechanism, first demonstrated by Spemann and Mangold (1924). In a classic experiment performed on a salamander, they removed the dorsal lip of the cleft in the blastula (blastopore) through which invagination of the mesoderm normally commences (figure 6.1). This tissue, when transplanted to a different site in a second embryo, resulted in the formation of a second primitive nervous system. Subsequent experiments, using pigmented and nonpigmented embryos as donor and host embryos, respectively, established that the nervous system was the consequence of a powerful inductive action of the mesoderm in the dorsal lip of the blastopore upon the ectoderm.

Spemann and Mangold referred to the dorsal lip as the "organizer"; and in the light of recent advances in molecular biology, it is likely that this tissue releases one or more peptide growth factors that cause the epidermal genes to be turned off and the neural genes to be turned on in the ectodermal cells (figure 6.2). Thus, during gastrulation, the gene for the *neural cell adhesion molecule (N-CAM)* is expressed and is restricted to the neural plate (Jacobsen & Rutishauser, 1986); other neural genes activated are those that code for β-*tubulin, vesicle-associated membrane protein (VAMP)*, and a nervous system-specific β-unit of Na$^+$-K$^+$ ATPase (for review, see Good *et al., 1990*). By analogy with muscle formation, it is likely that the *transcription* of those "constitutive" neural genes follows the switching on of a master neural regulatory gene.

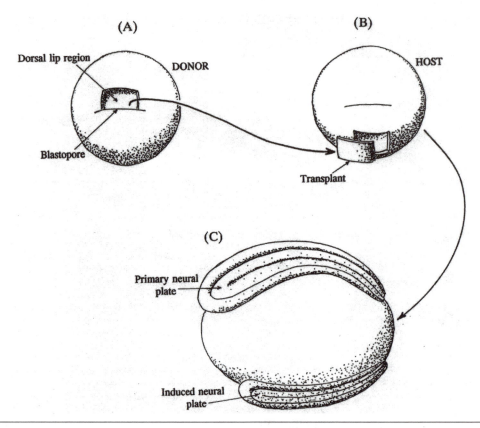

Figure 6.1 Induction of the nervous system. A neural plate can be induced in a salamander embryo by (A) removing a piece of tissue from the upper lip of the blastopore and (B) grafting it onto the ventral aspect of a second embryo. (C) A second induced plate then forms at the site of the graft. This famous experiment by Spemann and Mangold (1924) provided evidence for some kind of "organizer" in the mesoderm.

Based on Hamburger (1947).

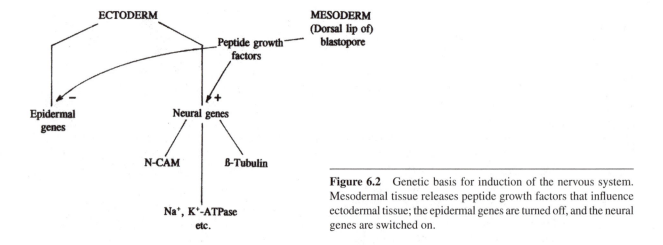

Figure 6.2 Genetic basis for induction of the nervous system. Mesodermal tissue releases peptide growth factors that influence ectodermal tissue; the epidermal genes are turned off, and the neural genes are switched on.

Step 2: The Neural Tube Forms From Thickening and Invagination of the Dorsal Ectoderm

The first visible change as a result of the inductive process is the formation of an anteroposterior neural groove in the ectoderm. As this groove becomes more pronounced, the ectoderm lying on either side of it thickens so as to form two neural plates (figure 6.3A). Because of these prominent changes in the dorsal surface, the embryo is now referred to as a *neurula*. On each side, the neural

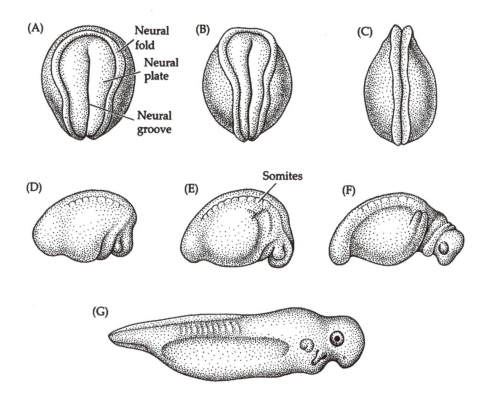

Figure 6.3 Early stages in the development of the nervous system (neurulation) in an amphibian *(Ambystoma)*. The neural folds (A) enlarge and come closer together (B, C), subsequently joining in the midline. When viewed from the side (D to G), the embryo is seen to lengthen, and the head, somite, and tail structures become more obvious.

Reprinted, by permission, from D. Purves and J.W. Lichtman, 1985, *Principles of neural development* (Ontario, Canada: Sinauer Associates, Inc.), 13.

plate develops a fold of tissue along its lateral margin; starting anteriorly, the two neural folds meet in the midline so as to form a cylinder of cells, the neural tube (figure 6.3C). It is the neural tube that will ultimately form the brain and spinal cord. The peripheral nervous system arises from a part of the neural plate that is not included in the neural tube but comes to lie dorsally and laterally to it; this is the *neural crest.* In addition to forming the spinal and autonomic ganglia, the neural crest is the origin of the *Schwann cells* in the peripheral nerves; it also gives rise to a variety of nonneural tissues.

Step 3: Nerve Cells Proliferate and Then Migrate

In this next phase, the developing nervous system rapidly increases the number of future neurons, and shortly afterward moves them to appropriate locations in the brainstem and spinal cord.

Neuroblasts multiply around the ventricular cavity.

Proliferation of Neuroblasts

The epithelium lining the cavity, or ventricle, inside the neural tube appears to have several layers. In fact, there is only a single thickness of cells, and the "pseudostratification" is due to the *nuclei* occupying different positions in relation to the ventricular cavity (figure 6.4D). The proliferation of the primitive nerve cells, or *neuroblasts,* is undertaken by cells that have their nuclei close to the ventricle. The nuclei then move away from the cavity along cytoplasmic processes that initially extend into the neural tube. As the nuclei approach their final destinations, the cytoplasmic processes lose their terminal connections.

The first motor neuroblasts are pushed outward by later ones.

Migration of Neuroblasts and Formation of Motoneuron Clusters

Initially, the entire neuroepithelium of the ventricular zone is actively proliferating, as deduced by labeling with [³H]-thymidine (Nornes & Das, 1974). The first cells to migrate, however, are the most ventral ones (thick arrows in figure 6.4A) and are motor neuroblasts. These first cells are then pushed outward by later-migrating cells, which include some that arise from a more dorsal

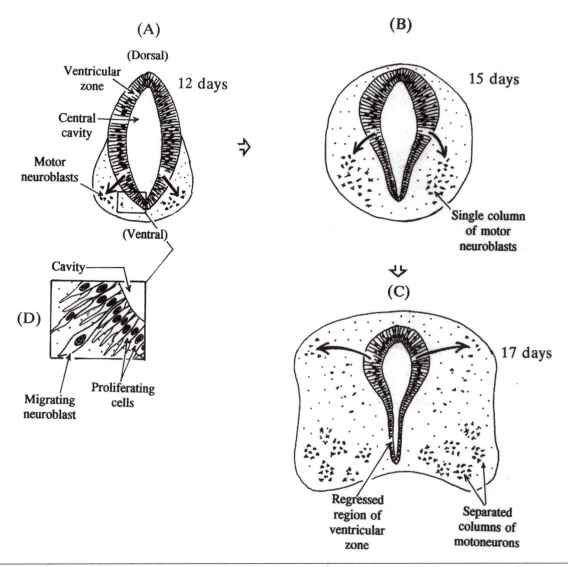

Figure 6.4   Migration of motor neuroblasts in the developing spinal cord of the fetal rat. (A) The cord is shown at 12 days of gestation, and the migrating neuroblasts (thick arrows) are seen leaving the ventral region of the ventricular zone. (B) At 15 days gestation, successively more dorsal populations of neuroblasts migrate outward, pushing the earlier cells in front of them. The ventral region of the ventricular zone is now becoming thinner as it regresses. (C) At 17 days gestation, the last cells to leave the ventricular zone are those that will form the dorsal horn; by this time the developing cord is much wider, and the motor column has broken up into six smaller ones. (D) Inset shows an expanded view of a portion of (A).

Based on Nornes and Das (1974).

region of the ventricular zone (thick arrows in figure 6.4B). The motor neuroblasts come to form one continuous column in the rostrocaudal axis of the developing spinal cord. As the neuroblasts differentiate into motoneurons, and as their *axons* reach the muscle masses in the limb buds, the motor column begins to divide.

The migration of motor neuroblasts diminishes as the ventral region of the ventricular zone begins to shrink. The process of neuroblast migration, followed by regression of the corresponding region of the ventricular zone, proceeds in a dorsal direction. These later neuroblasts will form the intermediate gray region of the spinal cord and the different laminae of the dorsal horn. The last cells to leave the ventricular zone (thick arrows in figure 6.4C) thread their way past the earlier arrivals and form the cap *(substantia gelatinosa)* on the dorsal horn.

In addition to the ventrodorsal gradient of progression just described, there is also a rostrocaudal one, such that neuroblast proliferation and migration occur rather earlier in the cervical region of the neural tube than in the lumbosacral one.

At the end of the migratory process, there are six main clusters of motoneurons for each of the limb buds. These clusters correspond to the dorsal and ventral locations of the muscle groups and also to the position of muscle groups in the proximal-distal axis of the limb bud. In general, the motoneurons that will supply dorsal muscles in the limb are situated laterally to those innervating the ventral muscles (Romanes, 1941, 1951).

The organization of motoneurons into motor pools in the spinal cord may be related to the species of *cadherin* surface molecules expressed by developing motoneurons. In the chick, 15 cadherin genes have been shown to be expressed in motor pools innervating limb muscles, and pools were distinguished by specific combinations of cadherins expressed (Guthrie, 2002; Price *et al.,* 2002).

Step 4: Axons Grow Out From the Spinal Cord Along the Extracellular Matrix

An axon starts to grow out from each motoneuron even while the cell is still in the process of migrating to its final position. The trajectories of the growing axons are probably determined by chemical guidance factors that either attract or repel them. At least four such factors have been identified as axon guidance factors in the nervous system and are known as *netrins, Slits, semaphorins,* and *ephrins.* Each factor, secreted by the target, binds to specific receptors in the *growth cone,* with different receptors to the same guidance factor dictating either attraction or repulsion. The factors most likely influence the *actin*-based motility of the growth cone via signaling pathways that influence the production and orientation of *filopodia* necessary for growth cone advance (Dickson, 2002). The axon guidance factors that are specific for motor axons have not yet been determined.

Outside the cord, the axon enters the limb bud at a time when the individual muscles have yet to form and when only the dorsal and ventral muscle masses can be distinguished. As they leave the embryonic spinal cord, the axons are grouped in bundles that, in the case of those destined for the limb buds, merge to form plexuses and are then redistributed into peripheral nerve trunks. If a segment of neural tube is transplanted from the thoracic to the brachial region, the outgrowth of axons forms a brachial plexus, indicating that the axons are affected, directly or indirectly, by their position in the anteroposterior axis of the embryo (Butler *et al.,* 1986). In the trunk and limb buds, the axons appear to follow paths that are marked out in the extracellular matrix of the trunk and limb buds by guidance factors rather different from those in the spinal cord, such as *laminin* and *fibronectin.*

Outgrowth of Axons Toward Muscles

Axons find their correct muscle destinations reliably.

In the limb buds, the peripheral nerve trunks give off main branches that enter the dorsal and ventral muscle masses and then break into smaller branches, corresponding to each of the main subdivisions of the muscle masses. The routing process that ensures that the axons from a particular motoneuron pool reach the correct muscle belly is remarkably accurate; it is also resilient in that it can compensate for different types of experimental perturbation by redirecting the axons (Lance-Jones & Landmesser, 1980a, 1980b). The mechanisms for this phenomenon are not entirely known. In *Drosophila,* muscle surface proteins *fasciclin,* semaphorin, and *connectin* act positively (fasciclin) or negatively (semaphorin, connectin) to influence development of nerve–muscle interactions by providing *trophic* signals to the growing axons (Fitzsimonds & Poo, 1998). Naturally occurring examples of successful rerouting are found in those human subjects with anomalous branching of their peripheral nerves. Thus, axons for the short abductor and opponens muscles of the thumb normally arrive in the median nerve. In some people, however, the axons cross over to join the ulnar nerve in the forearm (Martin-Gruber anastomosis) but still find their proper targets after the ulnar nerve has entered the palm. Figure 6.5 shows axonal outgrowths in the zebrafish, in which there are only three motoneurons to innervate the muscles in each ipsilateral segment of the trunk. Each axon follows a path diverging from those of the other two and supplies a particular muscle mass.

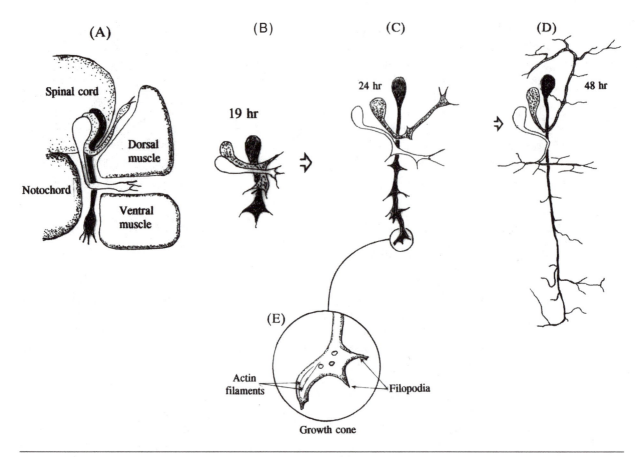

Figure 6.5 Axon outgrowth in the zebrafish. (A) A cross section made in a 19-hour-old embryo shows the three motoneurons responsible for innervating all the muscles in the trunk segment on that side. The axons have started to invade the spaces between the muscle masses. (B, C, D) Side views of the same three motoneurons illustrate the divergent paths, elongation, and branching of the respective axons. (E) An enlargement of a growth cone from one of the axon endings.

Adapted from Westerfield and Eisen (1988, pp. 20-21).

The Role of the Growth Cone

Growth cones explore the surroundings of the growing axons.

When the motor axons first arrive in the belly of the developing muscle, they are still nonmyelinated since the Schwann cells, which are derived from the neural crest, follow later. At the tip of the axon is a specialized expansion, the growth cone, that consists of a leaflet of axoplasm from which fingers (filopodia) project. The interior of the growth cone is packed with actin filaments and various types of organelle, including cisternae, vesicles, and *lysosomes* (figure 6.5E). Microscopic studies of axonal growth, either in tissue culture or in transparent regions of embryos, reveal that the growth cone has extraordinary mobility, extending and withdrawing, bending first in one direction and then in another, and all the while appearing to explore the surface of the tissue.

Fluorescence and differential interference microcopy, following the introduction of Ca^{2+}-sensitive dyes into the axoplasm, has shown that movements of the growth cone are associated with transient localized increases in Ca^{2+} concentration, which are presumably translated into sliding of the actin filaments. The first contact between a growth cone and a developing muscle fiber appears to be random, although it is always close to the entry of the nerve fibers into the muscle mass. Because the muscle fibers grow by additions to the two ends, the neuromuscular junctions remain in the approximate centers of the fibers as the latter lengthen.

These first axons are exploratory ones; they form a neuromuscular junction on each of the *myotubes* that has developed by fusion of the *myoblasts*. Subsequently, other axons arrive and establish neuromuscular junctions in the same regions of the developing fibers (Bennett & Pettigrew, 1974). Although, as already seen, the initial contacts between motor axons and developing muscle fibers are randomly located, the same sites will be preferred in later life if axons regenerate or *sprout* after a period of muscle *denervation* (see chapter 17).

Step 5: The Axons Establish Connections With the Muscle Fibers

The progress of events involved in the establishment of neuromuscular contact is summarized in figure 6.6.

Once the nerve and muscle fibers have made their first contacts, *synaptic transmission* becomes established within 1 to 2 hr (Chow & Cohen, 1983). The reason for this rapid onset is that the exploratory growth cone already contains *acetylcholine (ACh)* and can release it spontaneously. The release of ACh increases after the first nerve–muscle contact because of signals from the developing muscle fiber to the nerve terminal (Xie & Poo, 1986). The gradual increase in ACh release after initial contact is the result of an increased calcium concentration in the nerve terminal, which in turn is most likely mediated via the activity of cAMP-dependent *protein kinase* (Connor & Smith, 1994).

Differentiation of the Growth Cone by Signals From the Developing Muscle Fiber

One of the signals from the muscle fiber appears to be due to the adhesion between the fiber and the terminal; possible candidate molecules are neural cell adhesion molecule (N-CAM) and cadherin. The formation of transient *gap junctions* (Allen & Warner, 1991; Fischbach, 1972) could provide a second mechanism. Last, the myotubes could secrete molecules such as arachidonic acid and growth factors (Hall & Sanes, 1993). Two muscle factors shown to be involved in the initial establishment of the neuromuscular junction are *neurotrophin-3 (NT-3)* (Fu *et al.,* 1997) and *fibroblast growth factor* b (bFGF) (Fitzsimonds & Poo, 1998). Regardless of its nature, the effect of the molecular signal is to cause a rise in the Ca^{2+} concentration in the growth cone (Dan & Poo, 1992), an event that promotes the differentiation of the growth cone into a presynaptic terminal. The vesicle-associated proteins may play an accessory role.

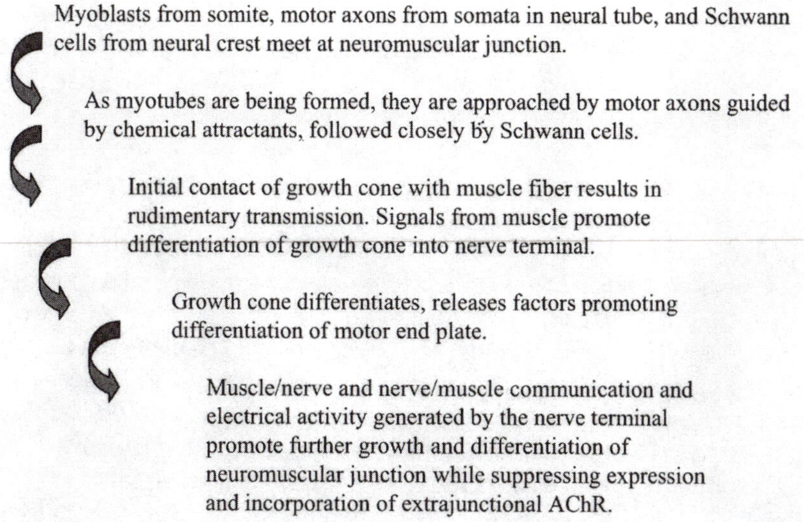

Myoblasts from somite, motor axons from somata in neural tube, and Schwann cells from neural crest meet at neuromuscular junction.

As myotubes are being formed, they are approached by motor axons guided by chemical attractants, followed closely by Schwann cells.

Initial contact of growth cone with muscle fiber results in rudimentary transmission. Signals from muscle promote differentiation of growth cone into nerve terminal.

Growth cone differentiates, releases factors promoting differentiation of motor end plate.

Muscle/nerve and nerve/muscle communication and electrical activity generated by the nerve terminal promote further growth and differentiation of neuromuscular junction while suppressing expression and incorporation of extrajunctional AChR.

Figure 6.6 Sequence of events from growth cone to establishment of a neuromuscular junction.

The motor
nerve
terminal
controls
differentia-
tion of the
synaptic
region of the
muscle fiber.

Control of the Postsynaptic Region of the Muscle Fiber by the Nerve Terminal

Just as the muscle fiber supervises the conversion of the nerve ending from a growth cone into a presynaptic terminal, so the presynaptic terminal influences the differentiation of the underlying muscle fiber. One of the best-understood changes is the clustering of *acetylcholine receptors (AChRs)* in the muscle *plasmalemma*. The clustering is made possible by the release of *agrin*, which is synthesized by the motoneuron *soma*, transported to the axon terminals, and inserted into the *basement membrane* within the *synaptic cleft* (Magill-Solc & McMahan, 1990a; Sanes & Lichtman, 1999). The agrin molecules combine with receptors in the muscle fiber plasmalemma (*muscle-specific kinases*, or *MuSKs*) and are then able to phosphorylate tyrosine in the β subunits of the AChRs (figure 3.5, chapter 3); the *phosphorylation* is evidently needed to tether the AChRs to the plasmalemma (Wallace *et al.*, 1991). In agrin-deficient knockout mice, AChR clustering and other postsynaptic specializations beneath nerve–muscle contacts do not occur (Sanes & Lichtman, 1999). A molecule with an effect similar to that of agrin is bFGF, which is also found in the synaptic basement membrane and is released by *proteolysis*.

Not only is the spatial distribution of the AChRs controlled by the nerve terminal, but the number of receptors is increased following enhanced transcription of AChR genes in the synaptic myonuclei. Two of the mediators in this effect are *calcitonin gene-related peptide (CGRP)* and *acetylcholine receptor-inducing activity (ARIA)*. Both of these mediators are produced by motoneurons, transported to nerve terminals by the axon transport machinery, and released at the nerve terminals, where they concentrate at the *basal lamina* and induce AChR synthesis (Uchida *et al.*, 1990; Loeb *et al.*, 2002). Thus, the AChR concentration changes, in mouse muscles, from a uniform distribution of approximately 1,000 receptors $\cdot$ μm^{-2} of membrane before innervation to a concentration of approximately 10,000 $\cdot$ μm^{-2} in the synaptic region and 10 $\cdot$ μm^{-2} outside the synaptic region in normally innervated adult muscles (Sanes & Lichtman, 1999).

Acetylcholine receptor-inducing activity is one of a family of at least 14 different growth and differentiation factors called *neuregulins (NRGs)*, which are *isoform* products of a single gene. Neuregulins released at nerve terminals bind to specific receptors, erbBs, which are transmembrane receptor tyrosine *kinases* related to the epidermal growth factor receptor family (Lin *et al.*, 2000). Neuregulin binding to its receptor activates a Ras/MAP kinase signaling cascade that enhances transcription of the ε subunit of the AChR in end-plate-associated nuclei while simultaneously repressing expression of the extrajunctional ε subunit (Goldman & Sapru, 1998). The result will be incorporation of "adult" AChRs at the end-plate as ε subunits replace the γ subunits—likewise, the concentration of "fetal" AChRs outside the end-plate containing the ε subunits decreases. The erbB receptors are tethered to a complex that includes the AChR, *acetylcholinesterase* (AChE), and *rapsyn* (43-kD protein) at the neuromuscular junction (Fischbach & Rosen, 1997). Acetylcholine receptor-inducing activity also plays a role in the survival and migration of Schwann cells, and erbB receptors are found in the latter. In neuromuscular junctions of erbB-deficient mice embryos, Schwann cells are absent, and junctional folding is impaired (Lin *et al.*, 2000).

Other molecules that accumulate in the specialized postsynaptic area include *spectrin* in the cytoskeleton and laminin, AChE, and *heparan sulfate proteoglycan* in the basal lamina (Edgerton *et al.*, 1996). In addition, a specialized subsynaptic *Golgi apparatus* develops, which appears to be involved in processing of the proteins specific to this region (Sanes & Lichtman, 1999).

At the same time that AChRs are accumulating at the new neuromuscular junctions, their numbers are decreasing elsewhere along the developing fibers, due to *down-regulation* of gene transcription in the nonsynaptic *myonuclei*. This reduction is due to electrical activity, in the form of *action potentials*, propagating along the fibers. This effect of electrical activity has been revealed by the ability of electrical stimulation to suppress the appearance of extrajunctional AChRs in denervated muscle fibers (Lømo & Rosenthal, 1972); conversely, different types of neuromuscular blockade in innervated fibers can induce extrajunctional AChRs.

Activity is vital to the proper development of complex neuromuscular junctions—one way in which this effect is mediated is via presynaptic presence of NRGs. When activity is suppressed in chick embryos by *curare,* presynaptic NRG expression is blocked and neuromuscular junction maturation is delayed. It appears that muscle activity, by altering the muscle concentrations of such neurotrophic substances as *brain-derived neurotrophic factor (BDNF)*, neurotrophin-3 (NT-3), and glial cell-derived neurotrophic factor (GDNF), influences the presence of NRGs in the terminal and hence neuromuscular junction architecture (Loeb *et al.,* 2002).

Development of Neuromuscular Junctions

Neuro-muscular junctions appear in human fetal muscle at 8 to 9 weeks.

In the human embryo, the neuromuscular junctions are first seen in the intercostal muscles at the age of 8.6 weeks (corresponding to a crown–rump length of 3.2 cm); their appearance in the tibialis anterior occurs rather later, at 10 weeks (4.3-cm length; Juntunen & Teravainen, 1972). Initially, the newly formed junctions have a primitive appearance, consisting of a simple (primary) cleft between the apposed nerve and locally thickened myotube membranes (Kelly & Zacks, 1969). At about 19 weeks of human fetal development, folding of the muscle membrane starts to take place, with the formation of the secondary clefts. The density of AChRs increases in the *postsynaptic membrane,* and this process is followed by the insertion of AChE into the basement membrane, usually 1 or more days after synaptic transmission has been established.

Synaptic vesicles increase in the nerve terminal.

As the neuromuscular junction continues to develop, the myonuclei migrate into its vicinity and begin to heap up on each other, increasing the complexity of the *sole-plate.* Meanwhile, in the axon terminal, increasing numbers of synaptic vesicles are seen; initially very few of these (e.g., one to four) are available for release following a single *impulse* (Robbins & Yonezawa, 1971), and the resulting *depolarization* is too small to initiate an action potential. Within a relatively short time, however, the quantal release increases and successful transmission becomes established. Some of the vesicles in the miniature axon terminal are relatively large and have dense cores; these vesicles contain CGRP and other molecules that may have special trophic functions. As these changes are taking place at the neuromuscular junction, the motor axon becomes thicker and acquires a myelin sheath from the Schwann cells that have followed the axon outward from the neural crest.

Step 6: Redundant Synapses and Motoneurons Are Eliminated

Unlike the situation in the adult, there is a phase in embryonic development when several axons come to innervate the same muscle fiber. In the mouse and rat, the surplus innervation is present at birth and is relinquished during the next 2 weeks (see "Polyneuronal Innervation"). In the human embryo, however, the results of *cholinesterase* staining suggest that the withdrawal of the motor axon terminals takes place before birth, probably between 25 weeks of gestation and term (Toop, 1975).

These important events are not the only regressive changes to take place in the motor innervation of the embryo. Thus, well before the period of synapse elimination is over, there is a marked degeneration and disappearance of motoneurons in the spinal cord; this is an example of the neuronal cell death that is found widely throughout the developing central nervous system.

Surplus motoneurons die.

The number of motoneurons in the embryonic spinal cord is influenced by the amount of muscle that has to be innervated. For example, if a limb bud is removed from a chick embryo, the number of motoneurons is reduced (Shorey, 1909). Conversely, the motoneuron population is increased if an additional limb bud is grafted onto the trunk of an amphibian embryo (Detwiler, 1920). In early studies, it was thought that the size of the "target," that is, the amount of muscle tissue, was determining the extent of neuronal proliferation in the spinal cord. However, Hamburger (1958) counted spinal motoneurons in chick embryos at different ages and found that approximately half of the cells died as part of the pattern of normal development (figure 6.7). Later studies showed that this cell death was complete before the withdrawal of *polyneuronal innervation* from the muscle fibers commenced (see "Polyneuronal Innervation," p. 80).

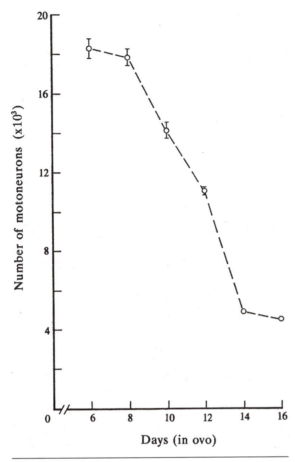

Figure 6.7 Natural motoneuron death in the brachial spinal cord of the chick embryo. The segments counted included spinal nerves 12 to 18, inclusive.

Courtesy of Dr. E. Cosmos and Dr. P. Cauwenbergs.

The probable significance of motoneuron death is that an abundance of motor axons is necessary to ensure that a sufficient number reach the muscle. Once all the muscle fibers have become innervated, initially by several axons on each fiber (polyneuronal innervation), many of the motoneurons become redundant. Indeed, studies of adult human muscle show that, despite cell death in the embryo, there is still an appreciable safety margin in muscle innervation. For example, muscle strength is not compromised in muscles that have been partially denervated as a result of chronic disease or old nerve injury if up to three-quarters of the motoneurons are lost (McComas *et al.*, 1971b).

The nature of the signal responsible for motoneuron death is presently unknown. One possibility is that the motoneurons that die are those that reach the "wrong" muscles, but this is not borne out by experiments in which axons are deliberately misrouted (Lance-Jones & Landmesser, 1980b). An alternative proposal, suggested by limb bud grafting studies, is that motoneurons compete for trophic factors produced by the muscle fibers (see chapter 18). However, there are experimental situations in which motoneuron death is not affected by the amount of muscle available for innervation (discussed next). Moreover, motoneuron death, like synapse elimination, is strongly influenced by muscle activity. If activity is blocked by the application of curare or α-bungarotoxin, motoneuron death is abolished (Laing & Prestige, 1978); on the other hand, the rate and extent of cell death are enhanced if chick embryo muscle is stimulated electrically (Oppenheim & Nunez, 1982).

Polyneuronal Innervation

For many years, it was accepted that each skeletal twitch muscle fiber was innervated by a single axon, and it was assumed that this arrangement was established as soon as embryonic muscles made their first contacts with exploratory axons. It was therefore something of a surprise when Paul Redfern, working in Thesleff's laboratory in Lund, Sweden, showed that the muscle fibers of newborn rat pups were multiply innervated (Redfern, 1970).

Redfern's crucial finding was that the sizes of *end-plate potentials* evoked in muscle fibers of the diaphragm depended on the intensity of the stimulus to the phrenic nerve, indicating that more than one axon must have supplied each neuromuscular junction (figure 6.8). It later transpired that anatomists, using the light microscope to examine specimens of

muscle in which the nerve fibers had been stained with silver, had observed multiple axons ending on muscle fibers of newborn mice many years earlier (cf. Purves & Lichtman, 1985). Redfern's important results were quickly confirmed not only by recordings of end-plate potentials but also by measurements of *twitch* and tetanic tensions. Other aspects of the multiple innervation were also described. For example, Brown *et al.* (1976) reported that, at birth, all rat soleus muscle fibers were multiply innervated; and Dennis *et al.* (1981) demonstrated that an average of three axons were present at each neuromuscular junction. Dennis *et al.* also measured the time taken for the surplus axons to be removed; they found, in rat intercostal muscle fibers, that only single motor axons remained by 13 days postpartum.

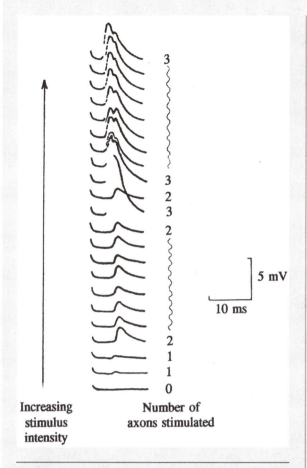

3

3
2
3
2

5 mV

10 ms

2
1
1
0

Increasing stimulus intensity

Number of axons stimulated

Figure 6.8 Microelectrode recordings from a single muscle fiber in a newborn rat, revealing the presence of multiple innervation. As the stimulus to the motor nerve is gradually increased, the end-plate responses can be seen to enlarge in three steps; this finding indicates that three axons must have been present, with different thresholds for excitation. The number of axons contributing to each response is given at the right.

Adapted, by permission, from P.A. Redfern, 1970, "Neuromuscular transmission in new-born rats," *Journal of Physiology* 209: 703.

Further insights into synapse elimination have come from repeated inspection of the same, identified, neuromuscular junctions, using the vital stain 4-Di-2-ASP to visualize motor axons and fluorescently labeled α-*bungarotoxin* to show AChRs. In sternomastoid muscles of newborn mice, AChRs begin to disappear before regressive changes are observed in the motor nerve terminals (Lichtman & Balice-Gordon, 1990; Balice-Gordon & Lichtman, 1994). At this time, there is a reduction in the sizes of the *miniature end-plate potentials,* which would be expected from the loss of receptors.

What is the mechanism for the withdrawal of the surplus axons? One clue is that the process is inhibited by paralyzing of the muscle fibers (Benoit & Changeux, 1978) and is accelerated by chronic stimulation (O'Brien *et al.,* 1978); further, the most active nerve–muscle connections promote the disappearance of the least active ones (Balice-Gordon & Lichtman, 1994). Although the answer to the question is not available at present, several explanations have been advanced (figure 6.9; see also Balice-Gordon & Lichtman, 1994). One of these is that the muscle fibers produce a limited supply of a trophic factor that can be taken up only by an active motor nerve terminal and is essential for the maintenance of that terminal (figure 6.9A; Purves & Lichtman, 1980). Support for this view comes from experiments in which muscle fibers were crushed on both sides of their neuromuscular junctions. Although the junctions were not damaged, there was a disappearance of axon terminals as the muscle fibers degenerated, followed by a reappearance when the muscle fibers began to regenerate (Rich & Lichtman, 1989).

An alternative suggestion (Changeux & Danchin, 1976) is that impulses in muscle fibers convert AChRs from a labile to a stable state and hence would tend to preserve the synaptic connections to the most active motor axons (figure 6.9B). Dan and Poo (1992) have provided evidence that, following transmission at one neuromuscular junction, there is a Ca^{2+}-mediated release of a toxin from the presynaptic ending (figure 6.9C). A Ca^{2+} mechanism has also been proposed by Vrbová and Lowrie (1989). These authors postulate that K^+, released into the interstitial spaces by muscle fiber impulses, opens Ca^{2+} *channels* in the motor nerve terminals, allowing the entry of Ca^{2+}. The Ca^{2+} then activates *proteases* that, in turn, digest the protein cytoskeleton of the axon terminal, causing the terminal to withdraw (figure 6.9D). This last suggestion is consistent with the observation of Vrbová and Lowrie that loss of terminals is prevented by the application of protease inhibitors *in vitro,* or by the lowering of extracellular $[Ca^{2+}]$, while the process is hastened by raising $[K^+]$.

There is one further aspect of synapse elimination that must also be accounted for in a satisfactory hypothesis, and this concerns the architecture of the *motor units.* When a single motor axon is stimulated repetitively, the muscle fibers belonging to that motor unit are depleted of glycogen and are then seen to be widely scattered in the muscle belly, overlapping with the territories of 20 to 30 other motor units (see chapter 12). It would appear that, during synapse elimination, there is a mechanism that discourages

(continued)

motor axons from innervating neighboring muscle fibers (Willison, 1978). Is extracellular K⁺ the key? The rise in interstitial [K⁺] would clearly be high in a region of the muscle where there were many muscle fibers supplied by the same motor axon. In keeping with the hypothesis of Vrbová and Lowrie (previously discussed), this rise could promote destruction of the majority of axon terminals until a stage was reached in which excitation of the same axon would produce changes in [K⁺] tolerated by the remaining terminals.

The process of synapse elimination, first demonstrated in muscle fibers, has since proved to be a widespread, though not universal, phenomenon in the central and autonomic nervous systems; and it is possible that the mechanisms involved are similar to those at the neuromuscular junction (cf. Purves & Lichtman, 1985).

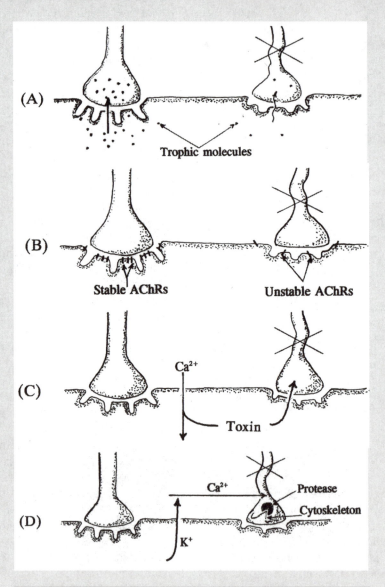

Figure 6.9 Four mechanisms proposed for the elimination of polyneuronal innervation.

A more subtle proposal, which reconciles these discordant results, is that muscle-derived trophic factors are indeed required by the embryonic motoneurons during activity; however, the limiting factor is not production of trophic factors by the muscle fibers, but rather the access to the factors by the motoneurons through their synaptic connections (Oppenheim, 1989). Thus, a motoneuron that had more synaptic connections than another would be at a competitive advantage for survival.

Such an explanation also accounts for certain observations on rats that have been subjected to crushing of their sciatic nerves (Albani *et al.,* 1988; Krishnan *et al.,* 1985). Although parts of the muscles in these animals do not become reinnervated, there are no losses of motoneurons; there is, however, increased branching of the axons and polyneuronal innervation of the muscle fibers. The question remains whether such a scheme could account for the remarkable range in motor unit sizes of normal muscles, which may vary at least 100-fold (see chapter 12).

Applied Physiology

In this section, we return to the neuromuscular clinic to examine two conditions. In the first condition, the gross structure of the spinal cord is abnormal, while in the other, there is a selective deficiency of motoneurons. Although the second disorder might be expected to be the less serious of the two, the effects in an infant can be fatal within days.

Spina Bifida

In spina bifida, the neural tube is improperly formed.

Occasionally a baby is born with a soft bulge over the lower part of the spine; the walls of the swelling are formed by the spinal meninges, and the contents consist of cerebrospinal fluid. Since the laminae and spines are missing from the lumbar vertebrae, the fluid-filled sac or meningocele is able to distend the skin and may reach a considerable size. Exploration of the sac may reveal that the spinal cord is open; that is, the nervous tissue that would normally form the roof of the central canal and become the dorsal part of the spinal cord lies to each side *(myelocele)*. In some cases the lumbosacral nerve roots enter the sac, which is then termed a *myelomeningocele*. More commonly the sac is absent, even though the spinal cord is malformed, and in these infants the neurological consequences are less severe. In patients with myelomeningoceles, however, there is usually total paralysis of muscles below the knee, loss of sensation over the feet, and lack of bowel and bladder control.

Regardless of their extents, these congenital disorders are loosely referred to as spina bifida and are usually attributed to a failure of the two neural folds to meet in the midline during embryogenesis. In fact this explanation may be incorrect, since in the human fetus, as in the chicken and the rat, the caudal part of the spinal cord forms by a different process than that seen in the rostral cord. Thus, in the early stages, there is an amorphous mass of mesenchymal and neuroectodermal tissue that subsequently forms one or more internal canals. As the cells proliferate and segregate, as part of the differentiation of the spinal cord, the multiple canals normally coalesce to form a single central canal; the latter then fuses with the canal formed rostrally by closure of the neural tube (Lemire, 1969). It is not clear why these processes may go awry in the caudal neural tube. From epidemiological studies in humans and by extension of observations made in certain mouse mutants (Park *et al.,* 1989), it is evident that a genetic factor may be responsible for an unknown proportion of cases. In addition, it is possible that some instances may be the result of environmental teratogens. *Retinoic acid* is one potential agent, as it may be used inadvertently to treat severe acne in a pregnant woman; valproic acid and folic acid antagonists have also been incriminated (Campbell *et al.,* 1986).

Spinal Muscular Atrophy

In spinal muscular atrophy, too few motoneurons are present.

In the hereditary disorder *spinal muscular atrophy (SMA)*, there is a reduced number of spinal motoneurons in the spinal cord and brainstem. The severity of the depletion in any one motoneuron pool, and the number of pools affected, vary from one patient to another. At one extreme is the infantile form, in which the resulting *weakness* affects not only the movements of the arms and legs but swallowing and breathing as well. At the other extreme is the middle-aged adult who, for the first time, notices some weakness of the limb girdle muscles, for example, in climbing the stairs or in raising the arms above the head. Regardless of the age of onset of symptoms, serial motor unit estimates (see chapter 12) indicate that the reduction in motor neurons is present at birth. The condition is evidently due to excessive neuronal death *(apoptosis)*. A spinal muscular atrophy gene localized to chromosome 5q (Melki *et al.,* 1990) has now been shown to code for an *apoptosis-inhibiting protein* (Roy *et al.,* 1995).

Specimens of muscle taken from affected infants typically show a few large fibers among expanses of much smaller, noninnervated ones. In an adult, one of the most striking findings is the presence of groups of histochemically similar fibers instead of the normal mosaic (see figure 17.7, chapter 17).

We have now finished studying the structure of skeletal muscle and its nerve supply, and also the way in which these very different types of cells develop in the embryo and come into contact with each other.

II

Putting Muscles to Work

Now that the structure of the muscle fiber and its innervation have been described, it is time to look at function. In chapter 7, we start at the level of the cell membrane by examining the molecules that act as ion channels and pumps, but we will also look at molecules that can temporarily capture or bind an ion for some particular purpose. Although the concept of muscle and nerve activity is usually associated with action potentials and muscle contraction, it must not be forgotten that there is a slower, chemical signaling that takes place between muscle and nerve. This alternative signaling requires axoplasmic transport to deliver substances from the soma to the nerve terminal, with subsequent release to the synaptic cleft and interactions with proteins on the muscle membrane. Axoplasmic transport is the subject of chapter 8. In the following chapter (chapter 9), we consider the ionic basis of the resting membrane potentials of nerve and muscle fibers, as well as the brief perturbations that form the impulses. Some of this knowledge was gleaned from the squid, because of the presence of a large axon suitable for study, but it will be seen that recordings of impulse activity can be readily made in human subjects.

Chapter 10 is largely about acetylcholine, the chemical that acts as an intermediary between the action potential in the axon and that in the muscle fiber. The extraordinary structure of the neuromuscular junction enables the release of acetylcholine from the nerve terminal, and its capture by receptors in the muscle fiber membrane, to be performed with remarkable speed and efficiency. Next comes an account of the activation of the myofilaments, with their subsequent contraction (chapter 11), and of the key role played by Ca^{2+} in this coupling process.

In chapter 12, the focus of interest shifts from molecular events to the properties of whole colonies of muscle fibers, each colony being supplied by a single motoneuron so as to form a motor unit. It will be seen that the properties of the motor units differ greatly and that each unit is specialized for the type of task in which it is called upon to participate. The study of the motor units is carried into chapter 13, which deals with their thresholds for excitation and their impulse firing patterns. Here, we will see more clearly how the structure matches the function.

The final chapter of this part of the book describes the biochemical changes that take place in the muscle fiber, as energy in the form of ATP is transformed to mechanical energy and heat as the muscle begins to contract. The replenishment of ATP is achieved through a variety of metabolic pathways that are described in chapter 14.

7

Ion Channels, Pumps, and Binding Proteins

Ions such as Na^+, K^+, and Ca^{2+} play vital roles in nerve and muscle fibers through their ability to move across cell membranes in a regulated fashion. The simplest movement, that of diffusion, results from the opening up of ion-specific channels in the membrane, in response either to a change in the transmembrane voltage or to the combination of the channel with a *ligand* such as a neurotransmitter (as in the *acetylcholine receptor*). To restore the correct distributions of ions across the membrane, the ions are pumped or otherwise transported against their respective concentration gradients by special molecules in the membrane. For Ca^{2+}, there is an additional type of molecule that binds the ions in the *cytosol* or *sarcoplasmic reticulum* (SR) and functions either as a buffer, like *parvalbumin,* or as the initiator of a chemical reaction or chain of chemical reactions, like *calmodulin.*

General P

Channels and Pumps

derstanding the structure and molecular mechanisms of the
ins over the past two decades. Much of this new information
riments in which purification of the various proteins has led
the order of nucleotide bases in a gene is known, it is but a
sequence of amino acids in the protein. This knowledge,
h parts of a molecule are in the membrane *(hydrophobic)*,
ular space. Modification of specific amino acid sequences
ion of the functions of the various regions of the protein:
activity, site of *phosphorylation,* and so on.

Ions diffuse rapidly through membrane channels; pumping is a slower process.

aqueous pores through which the appropriate ions can
lso have pores. However, there is a striking difference
bilities to transfer ions. If it were to stay open for 1 s, a
ns through the membrane, whereas in the same period
only a hundred or fewer ions. This great disparity in
lecules be very plentiful in membranes if they are to
in the cytosol. On the other hand, because they can
els need remain open for only a fraction of a second

en channels and pumps that is a consequence of the
respective *electrochemical gradient* in a channel, the ions
their concentration gradient with a pump. To make the latter possible, biological

work must be performed by the pump, which in turn requires that energy must be used. The energy for the pumps is provided by the hydrolysis of *ATP;* each pump molecule has a site for combining with ATP as well as a channel for admitting the ion.

The variety of channels and pumps can be likened to that in an orchestra. Thus, each channel or pump serves a specific role, yet coordination of these proteins permits a symphony of events while maintaining *homeostasis* with respect to ion concentration gradients. These proteins are embedded in the membranes of the muscle fiber, with certain pumps or channels gathered together in a specific "section." The Na^+-K^+ pumps are concentrated in the *sarcolemma,* as are the Na^+ and K^+ channels. The *dihydropyridine receptors (DHPR)* are concentrated in the *transverse tubules* (T-tubules), along with a variety of K^+ channels. The *ryanodine receptors* are localized in the terminal cisterns while the *sarco-endoplasmic reticulum calcium ATPase,* or *SERCAs,* are primarily confined to the longitudinal SR. The motoneuron can be considered to be the conductor or regulating entity; most of these channels respond to a change in the *membrane potential,* brought about by *activation* of the muscle fiber, and the pumps reply in response to the altered ion concentration gradients.

Selectivity of Channels

Channels select ions on the basis of charge and the sizes of their water shells.

How is it that channels can discriminate between ions so that they can pass one species but not another? The first point to make is that each of the cations is hydrated in solution; that is, the metallic atom is surrounded by a shell of water molecules. Since the hydrated molecules of different species of ion differ in size, the pores of one type of ion channel may be too narrow to pass ions of another species.

In addition, the ions are required to give up part of their water shell in order to pass through the pore, and the various species differ in their readiness to do so. Last, the ions must momentarily bind to specific amino acid residues lining the pore, and one species of ion may be attracted more strongly than another (Jan & Jan, 1989). It is now known that the pore inside an ion channel is long enough to accommodate several ions at any instant.

Na⁺ channels increase membrane excitability; K⁺ and Cl⁻ channels reduce it.

In reviewing this fascinating field, we will consider the channels and pumps for a particular ion species together before passing on to those for a different species. It will be seen that the Na^+ channels and pumps are required to give membranes, such as those of nerve and muscle fibers, the property of excitability. Thus, the opening of the Na^+ channels at successive points along the membrane enables an *action potential* to be propagated from one region of the fiber to another. K^+ channels have the opposite effect in that they reduce membrane excitability; they terminate the action potential but also ensure that the membrane does not become unstable and generate action potentials spontaneously.

Although they have been studied less extensively, Cl^- channels are also present in the membranes of excitable cells and have the same stabilizing effects as K^+ channels. *Ca^{2+} channels* have evolved to serve quite a different purpose; by allowing Ca^{2+} ions through the membrane, many chemical reactions can be initiated in the interior of the cell. These reactions are remarkably diverse and are necessary not only for much of the "housekeeping" metabolism to be found in any cell, but also for special functions such as muscle fiber contraction and the release of *acetylcholine* from the motor nerve terminal. In the case of muscle contraction, Ca^{2+} channels in the membrane of the SR result in the transfer of Ca^{2+} from the *terminal cisternae* to the *sarcoplasm.*

A Small "Gating" Current Is Produced Whenever a Channel Opens

In their classic study of the squid giant axon, Hodgkin and Huxley (1952a) were the first to suggest how ion channels might work: They did so by measuring the ionic currents that flowed when a potential difference was applied suddenly across the axon membrane and held at a specific level (see chapter 9). Hodgkin and Huxley recognized that the action potential was due to a large inward current carried by Na^+ ions; this current was able to flow because the membrane

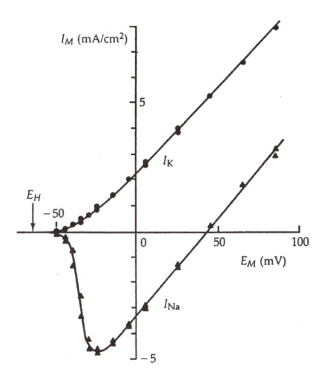

Figure 7.1 Current–voltage relationship of squid axon. Using a *voltage clamp* technique, similar to that illustrated in figure 9.4, the membrane potential is instantaneously changed from the "holding" potential (E_H) to a series of less negative, and positive values. Depolarizing the membrane in this way from a holding potential of –50 mV to –20 mV induces a large inward current following the opening of the Na$^+$ channels (I_{Na}); there is also a smaller outward current due to the K$^+$ channels opening (I_K). The net current flowing through the membrane at any instant will be the sum of the Na$^+$ and K$^+$ currents.

Adapted, by permission, from K.S. Cole and J.W. Moore, 1960, "Ionic current measurements in the squid giant axon membrane," *Journal of General Physiology* 22: 123-167.

suddenly became highly permeable to Na$^+$ ions. Following the transient Na$^+$ current was an outward current, carried by K$^+$ ions. Both currents were activated by membrane *depolarization,* as shown in the current–voltage curves in figure 7.1.

Having derived equations to describe the observed time courses of the Na$^+$ and K$^+$ currents (see figure 9.4, chapter 9), Hodgkin and Huxley (1952a) suggested that some of the mathematical terms could have physical counterparts in the structure of the respective ion channels. Thus, each K$^+$ channel might have four independent "gating" particles, all of which would have to move into a new position, under the influence of an electrical field, before the channel could open. The situation for the Na$^+$ channel was rather more complex because, in addition to the gating particles that allowed the channel to open, there was thought to be another particle responsible for "inactivating" (closing) the channel; the two types of gate were referred to as *m* and *h,* respectively.

One of the predictions from this classic work was that small charges, or *gating currents,* would momentarily flow across the membrane whenever a channel opened. Such currents were detected in Na$^+$ channels some 20 years later by Armstrong and Bezanilla (1973) and were subsequently found for other voltage-gated channels. The gating current is thought to be the consequence of reorganization of the position of a specific amino acid sequence within the channel structure. This reorganization permits ions to flow through the channel.

A simple conceptual model of a voltage-gated channel is shown in figure 7.2. In the model, there is a critically narrowed region of the aqueous pore, the *selectivity filter,* that permits only hydrated ions of a certain size to pass through. In the case of the Na$^+$ channel, the selectivity filter can allow Na$^+$ ions with a diameter of 0.1 nm to pass through but bars entry to the slightly larger K$^+$ ion (diameter 0.13 nm). The gate, which allows ions to leave the aqueous pore, is connected to a part of the channel, the sensor, that gauges the electrical field across the membrane. When the membrane potential depolarizes to the threshold level, the voltage sensor detects this and triggers the gate to open. Alternatively, the voltage sensor may be a charged particle (amino acid sequence) within the pore that actually forms the gate. For the Na$^+$ channel, the model would also include an inactivating gate (the *h* gate) at its cytoplasmic end. Na$^+$ ions would flow through the channel only when this gate, as well as the *m gate,* was open (see chapter 9).

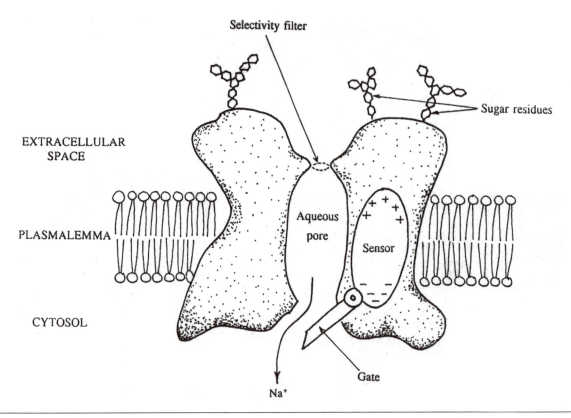

Figure 7.2 Model of an ion channel. The model has all the working parts that need to be incorporated into a functional channel, but the appearance of the channel is purely speculative.

Adapted from Hille (1992, p. 66).

The K⁺ Channel Evolved Before the Ca²⁺ and Na⁺ Channels

How did these various functions of the ion channels come about? The answer is that the cells must have found survival advantages for the channels as the different types evolved. It is probable that the K^+ channels were the first to appear in the earliest living cells and that the Ca^{2+} channels came next, following mutations of the K^+-channel gene. The last channel to appear was that for Na^+ ions, presumably by mutation of the Ca^{2+}-channel gene (Hille, 1992). One of the remarkable aspects of these events is that, once the gene evolved for a new type of channel, the nucleotide base sequence became extremely well conserved. For example, the Na^+ channels in the electroplax of an electric eel are very similar to those in a human nerve fiber. Similarly, the mammalian gene for the K^+ *delayed rectifier* channel was found only after the same gene had been cloned in the fruit fly. Indeed, yeasts, worms, and even a unicellular organism such as *Paramecium* all have K^+ and Ca^{2+} channels similar to those found in humans.

Sodium Channels

In mammalian skeletal muscle, the Na^+ channel consists of two subunits, designated α and β (figure 7.3; Barchi, 1988; Catterall, 1988). The larger α subunit has a molecular weight (MW) of 260 kD and contains substantial amounts of carbohydrate, mostly sialic acid, and lipid. The β subunit has a MW of 38 kD and also contains appreciable carbohydrate. The function

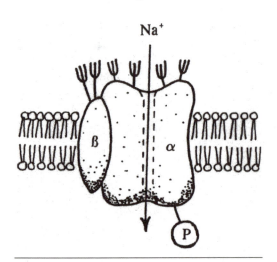

Figure 7.3 Subunit structure of the Na^+ channel in skeletal muscle. Sugar residues are shown as ψ, and phosphorylation by P.

Adapted from Catterall (1988, p. 52).

of the β subunit is not clear, since the α subunit by itself can act as a Na$^+$ channel; possibly the β subunit facilitates the insertion of the α subunit into membranes or else protects it from *proteolysis* (Auld *et al.,* 1988). In the Na$^+$ channels of the brain, each molecule contains a second β subunit, while in the electroplax organ of the eel, the β subunit is absent.

The Alpha Subunit

Using a cDNA cloning technique, the amino acid sequence of a Na$^+$ channel was first determined for the electroplax organ of the eel (Noda *et al.,* 1984) and subsequently for the α subunits of rat brain and skeletal muscle (Trimmer *et al.,* 1989). Regardless of its source, the α subunit was found to consist of approximately 2,000 amino acids; most of the latter are contained in four domains (figure 7.4, A & C), which can be recognized by the similarity of their amino acid sequences. Within each domain, there are six regions (S1 to S6) that span the cell membrane and are linked together (figure 7.4B). There is an additional membrane region, between S5 and S6, that forms the walls of the channel pore.

Which parts of the α subunit correspond to the gating mechanism of the Na$^+$ channel? There is general agreement that each of the four domains contributes one of the gating particles postulated by Hodgkin and Huxley (1952a) and that the channel becomes fully open (able to conduct ions) only when all four have been activated. The sensor would be expected to have a number of charged groups that would move as the electrical potential changed across the membrane. Examination of the transmembrane regions in each of the four domains (figure 7.4B) suggests that the S4 region is the most probable sensor, because it consists of repeated motifs in which a positively charged amino acid (usually arginine) is followed by two hydrophobic residues. If, as seems likely, the S4 region is arranged in an α-helix, then there would be a 60% rotation and a small (0.5 nm) outward movement of the helix in the membrane as the latter depolarized. This conformational change could allow the Na$^+$ ions to enter the pore.

Regardless of the detailed molecular structure of the gating mechanism, there is good experimental evidence that S4 is the transmembrane segment involved, since substitution of a single amino acid (phenylalanine for leucine) in this region produces a 25-mV shift in the membrane potential at which the Na$^+$ channel opens (Auld *et al.,* 1990).

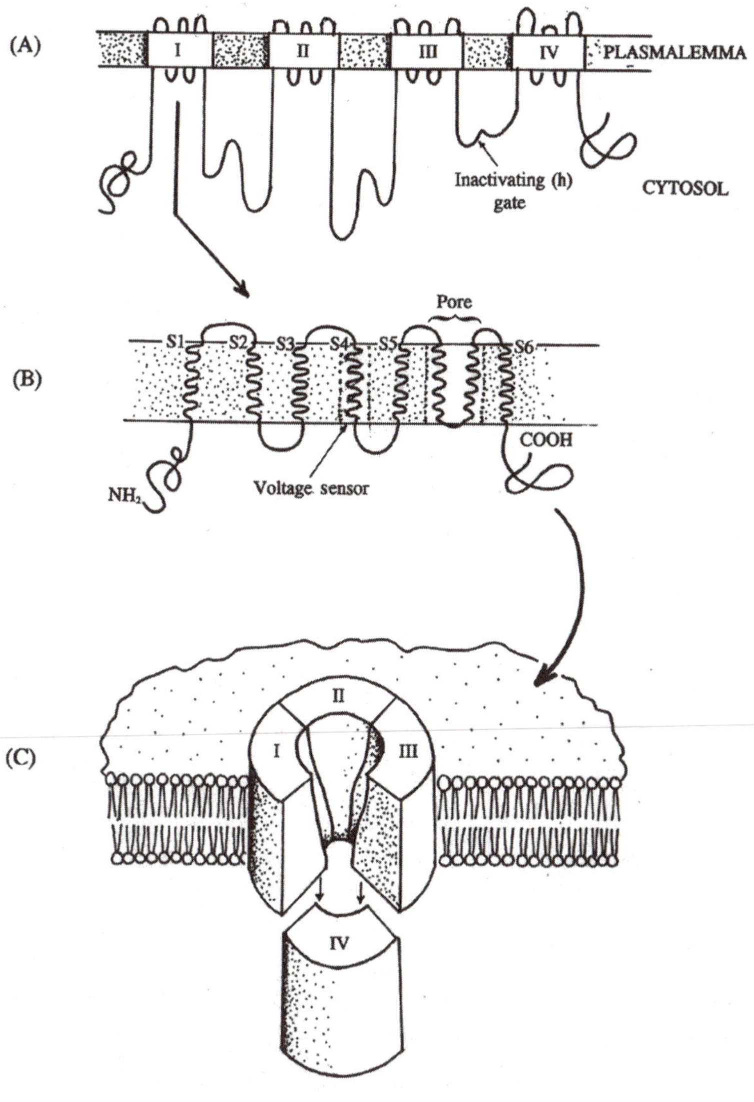

The α subunit has four domains.

The S4 region of each domain contains the voltage sensor, and the S5 and S6 regions line the channel pore.

Figure 7.4 (A) Structure of the Na$^+$ channel, showing the four domains (I to IV) in the α subunits. (B) One of the domains has been enlarged to reveal the transmembrane segments. S4 contains the voltage sensor controlling the gating mechanism, and the pore region lies between S5 and S6. (C) Three of the four domains have been assembled into a channel. Note that the basic structure of this channel is the same as those of the K$^+$ and Ca^{2+} channels.

It is not so easy, on theoretical grounds, to decide which part of the domain forms the pore, because the only clue is that the wall should be lined by hydrophilic amino acids. One experimental approach has been to investigate the function of the Na^+ channel after the amino acid sequence has been changed in different parts of the domain. This can be done using molecular biology techniques that alter the nucleotide base sequence in the cDNA clone (site-directed mutagenesis). Studies employing this technique strongly favor that part of the domain linking the S5 and S6 regions as the pore (figure 7.4B; Stevens, 1991).

Finally, there is the site of the inactivating mechanism (h gate) to consider. Armstrong *et al.* (1973) suggested that this gate was at the cytoplasmic end of the Na^+ channel, because inactivation could be abolished by the intracellular application of peptide-digesting enzymes. More precise localization has come from the use of antibodies directed against specific fragments of the channel protein; these experiments indicate that the h gate is probably formed by the polypeptide chain connecting domains III and IV (figure 7.4; Vassilev *et al.,* 1988).

The patch-clamp electrode can measure ion current through a single channel.

Special Techniques Reveal Channel Properties

With the advent of *patch clamping,* it became possible to measure the flow of ionic current through a single channel. This important technique was described by Neher and Sakmann (1976b), who devised methods for sealing the tips of glass microelectrodes against the membranes of living cells (figure 7.5A). Since the seal was made extremely tight, the background thermal noise in the circuit became sufficiently low that small ionic currents could be detected. In figure 7.5B, it can be seen that the ionic currents, in this case flowing through a Na^+ channel, start and stop abruptly; in some cases the current repeatedly flickers on and off. In the membrane of a rat muscle fiber, Sigworth and Neher (1980) found that the Na^+ channels have a mean conductance of 18 pS ($18 \times 10^{12} \; \Omega^{-1}$). In keeping with the Hodgkin-Huxley current–voltage curves, the time that a channel spends in the open state is determined by the potential across the membrane, so that nearly all channels are open after a depolarization to –50 mV (figure 7.5C; Hartshorne *et al.,* 1985).

Neurotoxins have been used to count Na^+ channels in membranes.

Distribution of Sodium Channels

The density of Na^+ channels in different types of excitable membrane can be determined by tagging the channels with radioactively labeled neurotoxins and counting the radioactivity; most studies have been made with [^{3}H]-saxitoxin or [^{3}H]-tetrodotoxin (see "Applied Physiology" at the end of this chapter). In the case of mammalian skeletal muscle, the density values have been estimated at 209 to 557 Na^+ channels $\cdot$ μm^{-2} of surface membrane (Bay & Strichartz, 1980; Ritchie & Rogart, 1977). The true density will be rather lower, however, for the estimates will have included Na^+ channels in the transverse tubules; nor is the distribution at the surface uniform, because the density in the *secondary synaptic clefts* at the neuromuscular junction is 10 times higher than in the remainder of the fiber (Beam *et al.,* 1985).

The Na^+ channels in the membranes of the motor axons are concentrated at the *nodes of Ranvier,* where the densities may be several times higher than in muscle fiber membranes. The high densities at the nodes are essential both for the rapid depolarization of the *axolemma* during excitation and for giving a sufficiently high safety factor to the action potential mechanism. This high density also contributes to a high action potential propagation velocity.

The Sodium Pump (Na^+-K^+ Pump)

Clearly if Na^+ channels are going to be periodically opened to allow a rapid movement of Na^+ into the muscle fiber, there must be a compensating mechanism that restores the ion concentration gradient to the proper level. This mechanism is the Na^+-K^+ pump.

The Purpose of the Na^+-K^+ Pump

The existence of a pump capable of extruding Na^+ was first postulated for cells in the renal tubule by Dean (1941) and was later shown to be a feature of most, if not all, living cells. In the simplest organisms, the pump, by expelling cations, prevents cells from becoming dangerously swollen

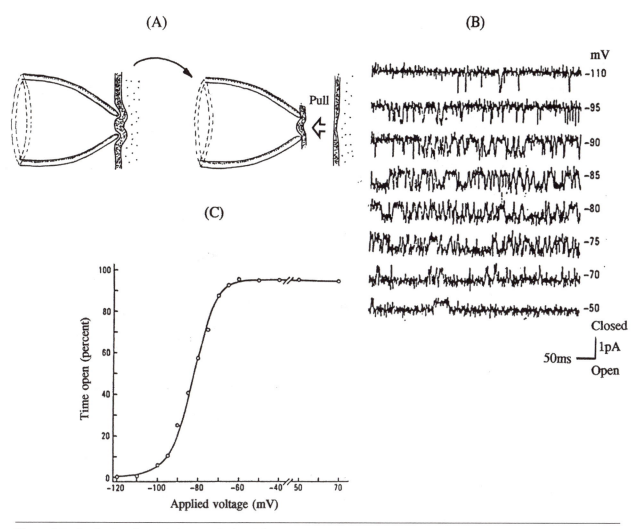

Figure 7.5 Patch-clamping technique applied to Na$^+$ channels. (A) The tip of a glass micropipette is pressed against the surface of a cell and is then withdrawn, removing a small piece of membrane. The membrane of the cell reseals spontaneously. (B) Openings of a single Na$^+$ channel inserted into an artificial membrane. As the negative potential across the membrane is reduced (depolarization), the channel opens more frequently until, at –50 mV, it is open nearly all the time. (C) Open time plotted as a function of membrane potential.

B and C: From R.P. Hartshorne et al., 1985, "Functional reconstitution of the purified brain sodium channel in planar lipid bilayers," *Proceedings of the National Academy of Sciences (USA)* 82: 241. Reprinted by permission of authors.

Ion gradients, established by the Na$^+$-K$^+$ pump, are used for many purposes.

through osmosis. In muscle and nerve cells, this pump maintains an appropriate concentration gradient for Na$^+$ and K$^+$ and contributes to generation of a charge separation across the membrane.

During evolution the presence of the pump gave important new properties to cells. By extruding Na$^+$ from the cell in exchange for K$^+$, it created an internal milieu in which the concentrations of the two cations were very different from those outside the cell. Whereas Na$^+$ is the most prevalent cation in the extracellular fluid, K$^+$ is much more concentrated in the cytosol (see chapter 9). However, the pump also creates an electrical potential across the cell membrane, because in each cycle of pump activity only two K$^+$ ions are traded for three Na$^+$ ions. Thus, with each cycle of ion transfer, there is an extra positive charge extruded, leaving the inside of the cell more negative than the outside. This charge gradient serves as a source of potential energy and is employed by cells for the transport of different types of molecules across their membranes. For example, the uptake of many sugars and amino acids by cells is driven by the electrical potential, or the concentration gradient of Na$^+$ ions across the membrane, or both; this charge gradient also runs the various cation *antiport* systems. In addition, in special cells such as nerve and muscle fibers, the electrical potential confers the

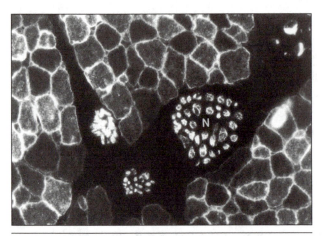

Figure 7.6 Distribution of Na$^+$-K$^+$ pump molecules (white areas) in a cross section of skeletal muscle. The specimen includes three intramuscular nerve branches, the largest of which is labeled N.

Courtesy of Dr. Douglas Fambrough.

ability to propagate action potentials from one region to another and, indirectly, to release transmitter from the motor nerve terminals and to initiate a sequence of events leading to contraction of the *myofibrils.*

Figure 7.6 shows a cross section through part of a mammalian muscle that has been exposed to a labeled antibody against the Na$^+$-K$^+$ pump. The pump molecules are seen to be associated with the surface membranes of the muscle fibers, as expected, and some fibers (type II) contain more than others. The section also includes bundles of nerve fibers; and, rather surprisingly, there is a high density of pump molecules in the myelin sheaths. Last, both the muscle fibers and the nerve fibers contain appreciable numbers of pumps in the cytosol; presumably these are molecules either undergoing degradation or awaiting incorporation into the surface membranes. Observing the presence of these pumps in the cytosol should serve as a reminder that all proteins are synthesized inside the cells, and pump and channel proteins must be inserted into the membrane to become functional. There must be a coordinated method of moving these proteins to the membrane and assuring that they are appropriately inserted.

How the Pump Works

> The α subunit has binding sites for Na$^+$, K$^+$, and ATP; there is also a β subunit.

Each Na$^+$-K$^+$ pump molecule is formed by the combination of two subunits (figure 7.7). The larger of these, the α subunit, has a MW of 112 kD, is mostly intracellular, and is folded on itself such that it spans the cell membrane from six to eight times. The Na$^+$- and K$^+$-binding sites are part of this subunit, as is the site for ATP binding. It is now known that there are several isoforms of the α subunit, and these are expressed to different extents in the various tissues of the body. The β subunit appears to exist in only one form and is a smaller molecule (35 kD) that is mainly external to the cell membrane and is glycosylated at three sites. The blocking agent, *ouabain,* combines with the α subunit, as shown in figure 7.7, and possibly with the β subunit as well.

A mechanism for the enzymatic reaction that splits ATP and redistributes Na$^+$ and K$^+$ across the membrane has been suggested by Albers (1967) and by Post *et al.* (1969). In this scheme the Na-K ATPase has two molecular conformations, E$_1$ and E$_2$ (figure 7.8). In the E$_1$ state, The ATP is split to form a phosphorylated intermediate. K$^+$ is released to the intracellular space in exchange for three Na$^+$ ions. The phosphorylated enzyme then changes back to the E$_2$ conformation. Three Na$^+$ ions are then released into the extracellular fluid in exchange for two K$^+$ ions. The initial step in the cycle, the electroneutral transition of the enzyme from the E$_2$ to E$_1$, is then repeated by release of inorganic phosphate and binding of ATP (figure 7.8). It has been estimated that each cycle takes 10 ms (Fambrough *et al.,* 1987), but the minimum time may be less, because the maximum transport capacity for a single pump molecule may be as high as 16,000 K$^+$ ions $\cdot$ min^{-1}. Under resting conditions, however, skeletal muscles only use 2% to 6% of their maximal pumping capacity (Clausen, 1986).

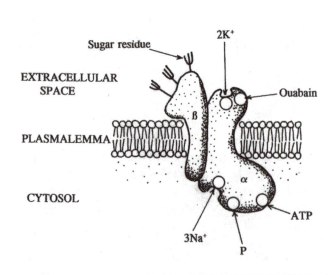

Figure 7.7 Model for Na$^+$-K$^+$ pump, showing α and β subunits, together with different binding sites.

Adapted from Horisberger *et al.* (1991, p. 567).

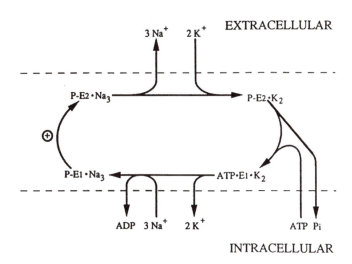

EXTRACELLULAR

INTRACELLULAR

Figure 7.8 Albers-Post model of the Na^+-K^+ pump cycle, as simplified by Horisberger *et al.* (1991, p. 567). For explanation of steps, see text.

Pumping increases after impulse activity, and pump density is regulated by use.

Regulating the Pump

A rise in Na^+-K^+ pump activity occurs immediately after stimulated or voluntary muscle contractions, and because of the electrogenic nature of the pump, the muscle fiber resting potentials may increase; this increase, in turn, causes the evoked muscle action potentials to become larger (Hicks & McComas, 1989; Galea & McComas, 1991). Other evidence of enhanced pump activity comes from measurements of the uptake of radioactive Rb^+, an ion that is treated like K^+ by the pump. In the rat, the enhancement of pump activity is greater in the soleus than in the extensor digitorum longus muscle and is apparently due to the rise in intracellular Na^+ concentration and the β-adrenergic effects of epinephrine and norepinephrine (Everts *et al.*, 1988). The elevation in pump activity affects not only the contracting fibers but also quiescent ones in their vicinity (Kuiack & McComas, 1992). The cellular mechanisms responsible for the β-adrenergic changes in pump activity involve the *cyclic AMP second messenger* system.

The density of pump sites in the surface membranes can be adjusted over a period of several hours; under physiological conditions, such modulation would presumably occur in response to altered activity levels of the muscle fibers. In cultured *myoblasts,* an increase in pumps at the fiber surface can be induced by exposure to veratridine, a drug that maintains the voltage-gated Na^+ channels in an open state. Fambrough *et al.* (1987) have postulated that the accumulation of Na^+ inside the fibers provides the necessary stimulus both in this model and in response to increased contractile activity; the same workers have shown that the higher density at the fiber surface is achieved by three mechanisms—increased synthesis and reduced turnover of pump molecules, and the transfer of pumps from the fiber interior to the surface membrane. It now appears that there may normally be an excess of α subunits in the endoplasmic reticulum and that, when additional Na^+-K^+ pumps are required, there is increased *transcription* of the gene coding for the β subunits. The newly formed β subunits combine with the α subunits in the interior of the fiber and are then conveyed to the surface of the fiber, where they are inserted into the plasmalemma (Taormino & Fambrough, 1990). The consequent changes in Na^+-K^+-pumping capacity will persist if an increase in physical activity is maintained, as has been shown in dogs (Knochel *et al.,* 1985) and in human subjects (Tibes *et al.,* 1974). An acute increase in Na^+-K^+ pump activity can also be induced by catecholamines, insulin, *insulin-like growth factor,* and *calcitonin-gene related peptide (CGRP)* (Clausen, 1996).

Longer-term changes in pump regulation, occurring over a period of several weeks, can be induced by alterations in thyroid activity. For example, pump activity may be as much as 10 times greater in muscles from rats made hyperthyroid, as opposed to those of hypothyroid animals (Kjeldsen *et al.,* 1986). *Up-regulation* of the pump is also a feature of developing muscles, at least in the first few weeks of age in rats and mice. In contrast, starvation, inactivity, and diabetes produce *down-regulation* of pump activity in experimental animals (Clausen & Everts, 1989).

Potassium Channels

Potassium channels are difficult to isolate and purify. Their molecular structure could be deduced only after the technique of chromosome walking had been used to clone the gene responsible for the Shaker mutation in the fruit fly, *Drosophila* (Papazian *et al.*, 1987; Kamb *et al.*, 1987; see "Cloning a Potassium-Channel Gene"). Previous work had suggested that the leg shaking characteristically exhibited by this mutant under ether anesthesia was due to a failure of the voltage-gated K^+ channel to cut short the action potential in the motor nerve terminals (Jan *et al.*, 1977).

Structure and Diversity of Potassium Channels

The K^+-channel gene codes for a single domain.

After cloning, it was found that the protein encoded by the K^+-channel gene contained only 616 amino acids and was therefore much smaller than the main subunits of the Na^+ and Ca^{2+} channels; rather, it corresponded to only one of the four domains of two other channels. A complete K^+ channel is evidently formed by the fusion of four of the K^+-channel proteins. Similar to the situation with the Na^+ and Ca^{2+} channels, each protein (domain) forms part of the central pore and each contains a voltage-sensitive region (S4). The four proteins may be identical, or they may differ from each other due to alternative splicing of the mRNA transcribed from the same K^+-channel gene. It is the alternative splicing that accounts for the remarkable diversity of K^+ channels, even in the same cell.

Recently, the probable structure of a K^+ channel has been determined by *X-ray crystallography* at a higher resolution (2 angstrom) than was previously possible (Morais-Cabral *et al.*, 2001). The channel contains seven main sites for K^+ along its axis, including four in the selectivity filter. The filter is just wide enough to accept K^+ ions without their shells of water. The first K^+ channel to be described was the one associated with the action potential in the squid giant axon by Hodgkin and Huxley (1952a), and it remains the best understood of the K^+ channels.

More than 30 different types of K^+ channel exist.

Over 30 different K^+ channels have been identified so far on the basis of their different kinetic and pharmacological properties. The main K^+ channel in excitable cells, such as *α-motoneurons* and striated muscle fibers, is that responsible for terminating the action potential; its kinetic properties were described by Hodgkin and Huxley (1952a). In view of the slight delay before the channel opens in response to depolarization, and because of the shape of the current–voltage relationship (figure 7.1), this channel became known as the delayed rectifier. However, since other K^+ channels also have delayed openings and have conductances that increase more slowly, the delayed rectifier channel is often referred to as the fast K^+ channel. The channel can admit Rb^+ and NH_4^+ ions, in addition to K^+, but is blocked by larger cations such as Cs^{2+} and Ba^{2+}; another very effective blocking agent is tetraethylammonium (TEA).

Another voltage-gated K^+ channel found in striated muscle fibers has the properties of an *inward rectifier*, since it opens when the membrane potential is larger than the K^+ *equilibrium potential* and allows net entry of K^+ ions into the fibers along their electrical gradients.

Most of the other K^+ channels are like the fast channels in being voltage gated, but there are some that are activated by intracellular messengers. Two of the latter are channels that open following rises in the intracellular concentration of Ca^{2+} ions, and current conducted through these channels is therefore given the symbol $I_{K(Ca)}$. One has a low conductance (SK [small] channel) and is blocked by apamin, a peptide found in bee venom; the other (BK [big] channel) has a very large conductance and is blocked by TEA. Both channels, when open, tend to stabilize the membrane by preserving a polarized state. Another K^+ channel in striated muscle fibers is normally inhibited by intracellular ATP and opens when the ATP concentration declines. Yet another K^+ channel is activated by a rising concentration of intracellular Na^+ ions.

K^+ Channels Stabilize the Membrane Potential

In all cells, the basic function of the K^+ channels is to stabilize the membrane potential at its resting level and thereby to counteract the excitatory effects of the Na^+ (and Ca^{2+}) channels. In skeletal muscle fibers, most of the K^+ channels are not in the surface membrane but in the transverse tubules, and there is another type of channel localized to the SR. In motor nerve axons, the greatest density

Cloning a Potassium-Channel Gene

The first voltage-gated ion channel for which the *DNA* could be cloned was the Na^+ channel. This was a very considerable achievement, and it took advantage of the fact that the concentration of Na^+ channels is especially high in the electroplax organ of electric eel, allowing the channels to be isolated and purified. Once part of the amino acid structure of the channel had been established, it was possible to construct the corresponding mRNA nucleotide sequence and to use the sequence to recognize the complementary DNA in the genomic library (Noda *et al.,* 1984). In comparison, the cloning of a K^+-channel gene was a much greater challenge. For one thing, K^+ channels are not found in high concentrations in nervous tissue, and for another, K^+ channels are remarkably diverse in their pharmacological properties and have few ligands that bind with them.

The solution to the cloning difficulty was to make use of one of the neurological mutations exhibited by the fruit fly, *Drosophila melanogaster.* This mutant, Shaker, was originally noticed because its legs trembled when the fly was exposed to ether. Microelectrode recordings from pupal and larval muscle fibers showed that the membranes appeared to lack the transient outward K^+ current responsible for repolarizing the membrane and terminating the action potential (figure 7.9). It was subsequently shown that the same channel was probably deficient in nerve fibers and that the broadened action potential caused excessive amounts of transmitter to be released at neuromuscular junctions, provoking repetitive firing of the muscle fibers (Jan *et al.,* 1977; Tanouye & Ferrus, 1985). The next step was to look for breakpoints in the X-chromosome of flies in which the Shaker mutation had been induced by X-irradiation, and to undertake a chromosome walk until the affected gene could be identified by DNA probes. This elegant work was undertaken independently by Mark Tanouye and by Lily and Yuh Jan and their respective colleagues. Simultaneously, both teams reported the cloning of the Shaker DNA and thereby the first isolation of a K^+-channel gene (Papazian *et al.,* 1987; Kamb *et al.,* 1987). The definitive proof that the gene did indeed code for a K^+ channel came from experiments in which synthetic *RNA,* prepared by transcribing the cloned DNA, was injected into the *Xenopus* oocyte. Within a few days, voltage-gated K^+ currents could be detected in the oocyte membrane (Timpe *et al.,* 1988).

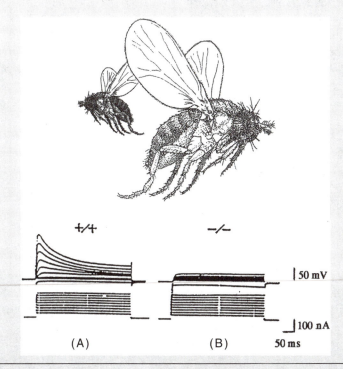

Figure 7.9 Membrane responses to depolarizing currents, injected into cells of a "wild-type" (normal) fruit fly and of a Shaker mutant. The recordings were made at 72 hr of development, corresponding to the pupal stage. In the wild-type fly (+/+), but not in the mutant (-/-), there is a fast K^+ current, shown as the upward traces. The absence of a fast K^+ current remains throughout the life of the mutant.

Adapted, by permission, from L. Salkoff, 1983, "Genetic and voltage-clamp analysis of a Drosophila potassium channel," *Cold Spring Harbor Symposia on Quantitative Biology* 48: 221-231.

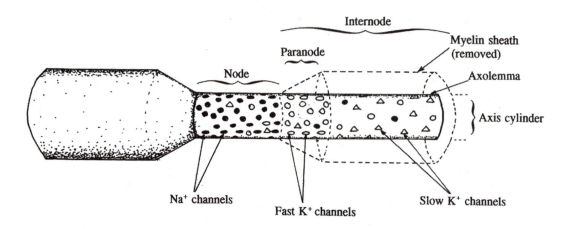

Figure 7.10 Myelinated nerve fiber, showing the distributions of Na$^+$ channels (●), fast K$^+$ channels (○), and slow K$^+$ channels (Δ) at the node, paranode, and internode regions of an axon.

Based on the data of Black *et al.* (1990).

of fast K$^+$ channels is found in the paranodal region, that is, in the axolemma lying beneath the attachments of the myelin loops on either side of the node of Ranvier (see figure 2.7, chapter 2). Presumably, the K$^+$ channels cannot be accommodated at the node itself, in view of the necessity of having a high concentration of Na$^+$ channels instead. The densities of fast K$^+$ channels in the membranes of the node and the remainder of the *internode* are only one-sixth those in the paranodal regions (Black *et al.,* 1990; figure 7.10).

The properties of the different types of K$^+$ channel in striated muscle and motor nerve fibers are summarized in table 7.1. It is worth reiterating that they all serve to maintain or restore the membrane to its resting condition. Some of the K$^+$ channels would be expected to become especially important in muscle fatigue; thus, the BK and SK channels would open as the intracellular concentration of Ca^{2+} increased, as would the K$_{ATP}$ channel when the cytosolic store of ATP in the vicinity of the membrane declined. This channel is sensitive to changes in pH and other modulating factors, such that a very small change in [ATP] in the cell could result in opening of this K$^+$ channel (Renaud, 2002). Rising concentrations of Na$^+$ would open another type of K$^+$ channel. If the membrane potential of the muscle fibers was higher than the K$^+$ equilibrium potential, under the influence of the electrogenic Na$^+$-K$^+$ pump, K$^+$ would be attracted from the interstitial spaces back into the muscle fibers. The inward rectifier K$^+$ channel would also have such an effect.

Calcium Channels and Pumps

We have seen that Na$^+$ channels confer excitability on cell membranes and that K$^+$ channels have an opposite effect by stabilizing the membrane potential. In contrast to both, Ca^{2+} channels, which are found in all living cells, are required to link electrical or chemical signals at the surface membrane to molecular events within the cell. This linkage is made possible by the entry of Ca^{2+} ions into the interior of the cell; the Ca^{2+} ions are then seized by special binding proteins, and this combination, in turn, activates *protein kinases* that phosphorylate certain key molecules.

Ca^{2+} Is a Second Messenger

If the depolarization of the surface membrane constitutes the initial instruction to a cell, then Ca^{2+} ions can be considered second messengers. Among other actions, they are responsible for initiating the contractile process in the muscle fibers and for releasing acetylcholine from the motor nerve terminals. The signaling action of Ca^{2+} is made possible by the ability of the resting cell to maintain the internal concentration of this ion at a very low level, of the order of 10^{-7} M. Consequently,

Table 7.1　Potassium Channels in Striated Muscle[a]

Name(s)	Gating mechanism	Conductance	Blocking agents	Actions
Fast (delayed outward rectifier)	Depolarization	17-64 pS	TEA, Ba^{2+}, Cs^+	Terminates action potential; contributes to refractory period
Inward (anomalous rectifier)	Membrane potentials negative to E_K	5-28 pS	TEA, Ba^{2+}, Cs^+	Stabilizes resting membrane; allows entry of K^+ into fibers during fatigue
Sarcoplasmic reticulum channel	Depolarization of SR	150 pS	TEA	?
Calcium-activated (small [SK], intermediate [IK], big [BK])	Increased intracellular [Ca^{2+}]	6-250 pS	Apamin, charybdotoxin	Stabilizes membrane and diminishes Ca^{2+} entry; increases K^+ entry into fibers during fatigue (?)
ATP sensitive [K_{ATP}]	Decreased intracellular [ATP] and pH	20-90 pS	Tolbutamide, 4-aminopyridine	Stabilizes membrane and increases K^+ entry into fibers during fatigue (?)
Na^+ activated	Rise in intracellular [Na^+] > 20 mM	220 pS	TEA, 4-aminopyridine	Stabilizes membrane and increases K^+ entry into fibers during fatigue (?)

[a]See Castle et al. (1989), Rudy (1988). ? = unknown action.

Several different mechanisms reduce Ca^{2+} in the cytosol.

even a relatively small influx of Ca^{2+} will change the internal concentration appreciably. During a muscle contraction, the level rises above 10^{-6} M. Assuming a cell 5 cm in length and 50 μm in diameter, this concentration change could be achieved by the transfer of 5.3×10^7 ions. However, this calculation seriously underestimates the amount of Ca^{2+} released into the *cytoplasm* due to a high level of buffering within the cytoplasm.

Calcium is important to the muscle fiber for another reason. If the plasmalemma is torn as a result of injury or disease, Ca^{2+} ions will flow into the fiber down the electrochemical gradient and disrupt both the structure and the function of the cytoplasmic contents. The disruption is due to the arrest of ATP production by the mitochondria and to activation of proteolytic enzymes; these enzymes, by degrading the intermediate filament proteins, weaken the scaffolding of the fiber. More information on this process is presented in chapter 21.

How are the low resting levels of Ca^{2+} in the cytoplasm achieved? The answer is that some Ca^{2+} ions are removed from solution by binding to special proteins, and other Ca^{2+} ions are pumped or attracted across the various membranes of the cell. These membranes include the surface membrane, the membrane enclosing the endoplasmic reticulum (and SR in the case of the muscle fiber), and the inner membrane lining the mitochondria (figure 7.11).

At the cell surface (figure 7.12), the Ca^{2+} ions are transferred out by a pump that is fueled by ATP. The surface membrane also contains another transporter, which exchanges Na^+ ions for Ca^{2+} ions; the energy for this exchange comes from the Na^+ concentration gradient established across the cell membrane by the Na^+-K^+ pump. Since three Na^+ ions are admitted for each (divalent) Ca^{2+} ion, there is a net gain of one positive charge by the cell interior.

In the inner membrane of the mitochondrion, there is another type of Na^+-Ca^{2+} exchanger, which expels one Ca^{2+} ion for two Na^+ ions and is therefore electrically neutral (figure 7.12). There is also a "uniporter" molecule that allows Ca^{2+} to move into the mitochondrion down the large electrical gradient (–180 mV) across the inner membrane. This Ca^{2+} removal mechanism is extremely powerful in most cells, because the foldings of the mitochondrial inner membrane provide a large surface

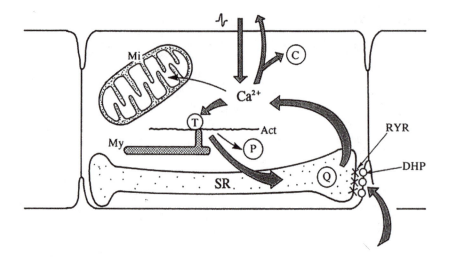

Figure 7.11 Movements of Ca^{2+} into and within the muscle fiber. Act = *actin;* C = calmodulin; DHP = dihydropyridine receptor; Mi = mitochondrion; My = *myosin;* P = parvalbumin; Q = calsequestrin; RYR = ryanodine receptor (channel); SR = sarcoplasmic reticulum; T = troponin C.

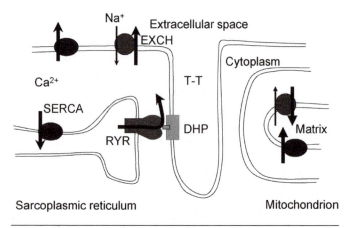

Figure 7.12 Calcium transport systems in three types of muscle fiber membrane: the cell surface, the sarcoplasmic reticulum (SR), and the inner membrane of the mitochondrion. Thick arrows represent Ca^{2+} movement. Ovals represent Ca^{2+} ATPase; circles represent antiport exchangers.

for Ca^{2+} removal. Finally, if the cytosolic Ca^{2+} level rises to very high levels, for example, after injury to the cell surface, the mitochondria can admit phosphate ions and precipitate the Ca^{2+} as crystals of calcium phosphate (hydroxyapatite). This precipitation mechanism has also been proposed to account for trapping of Ca^{2+} within the SR during fatigue (Kabbara & Allen, 1999).

The third type of membrane that removes Ca^{2+} from the cytosol is that of the endoplasmic reticulum, much of which has become specialized in the muscle fiber to form the SR. Approximately 90% of the SR membrane is occupied by a Ca^{2+} ATPase; this enzyme acts swiftly and sensitively in response to rises in intracellular Ca^{2+} concentration (figure 7.12). This Ca^{2+} ATPase is referred to as SERCA.

The relative activities of the different Ca^{2+}-transporting mechanisms within the muscle fiber depend on the concentration of Ca^{2+} in the cytosol. In resting muscle, the amount of Ca^{2+} removed from the cytosol by the SERCA is about double that which is extruded across the cell surface, mainly by the Na^{+}-Ca^{2+} exchanger. During muscle contraction, the Ca^{2+} pump in the SR becomes even more important and now expels up to 90% of the transported ions. If the intracellular Ca^{2+} concentration rises still higher (10^{-5} M), as when the fiber is damaged by injury, exercise, or disease, the uniporter in the mitochondrion becomes increasingly active and removes a substantial proportion of the ions (see Carafoli & Penniston, 1985 for review).

There are different types of Ca^{2+} channel.

Calcium channels are classified as the T (tiny or transient), L (large or long-lasting), and N (neither T nor L, or neuronal). The best understood of these is a type of L-channel found in high density in the transverse tubular membranes of skeletal muscle fibers; this is the DHP channel, so named because it is bound, and blocked, by dihydropyridine. A fourth Ca^{2+} channel, discovered more recently in skeletal muscle fibers, is the ryanodine channel; this last channel is localized to the SR and is probably connected to the DHP channel in the transverse tubules (see chapter 11).

The Dihydropyridine Channel

The purified DHP channel is composed of five polypeptide chains (figure 7.13; Catterall, 1988; Hofmann *et al.,* 1990). The α_1 subunit is the central component of the complex and has a MW of 165 kD. It strongly resembles the Na^{+} channel in having four domains, each of which has at least six regions that span the membrane. By analogy with the Na^{+} channel, it seems probable that the positively

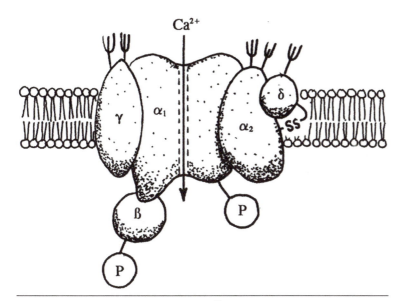

Figure 7.13 Subunit structure of the Ca^{2+} channel. Sugar residues and phosphorylation sites shown as ψ and P, respectively.

Adapted from Catterall (1988, p. 52).

charged, hydrophilic S4 region acts as the voltage sensor and that the segment linking the S5 and S6 regions forms part of the pore for Ca^{2+} ions. The Ca^{2+} channel also differs in containing more subunits; in addition to the α_1 subunit, there are α_2, β, and γ subunits (MW, 135 kD, 55 kD, and 28 kD, respectively; Hofmann *et al.,* 1990). Unlike the α_1 subunit, the α_2 and γ subunits are heavily glycosylated. It is probable that the β subunit serves the same function as the β subunit of the Na^+ channel; Catterall (1988) has suggested that it is attached to the α_1 subunit (figure 7.13). The Ca^{2+} channel can be phosphorylated on both the α_1 and β subunits, and this increases its functional activity significantly.

Like the Na^+ and K^+ channels, the Ca^{2+} channels are activated by membrane depolarization; and it is possible, too, that a part of the Ca^{2+}-channel complex helps to inactivate the channel during depolarization, as in the case of the Na^+ channels. However, at least in the case of cardiac muscle, a more important source of inactivation is the rise in intracellular Ca^{2+} concentration that follows channel opening.

Calcium channels can be opened by depolarization and by intracellular mechanisms.

The Ca^{2+} channel can also be opened by signals transmitted from the cell interior. For example, epinephrine binds to a surface receptor and activates adenyl cyclase, which produces cyclic adenosine monophosphate *(cAMP)*. Increased cytoplasmic cAMP concentration activates a protein kinase that then phosphorylates the DHP channel and causes it to open.

In skeletal muscle, the DHP channel has a peak conductance of approximately 20 pS but is much slower to activate and inactivate than channels of the same type in smooth and cardiac muscle. It also differs in that, unlike the other channels, it does not appear to be inhibited by a rise in intracellular Ca^{2+} concentration.

Dihydropyridine channels can be blocked by divalent cations and by drugs.

Dihydropyridine channels are blocked by divalent metallic ions such as Ni^{2+}, Cd^{2+}, Co^{2+}, and Mn^{2+}. In contrast, the channel is fully permeant to Sr^{2+} and Ba^{2+} ions because of their smaller sizes. The DHP channels are also blocked by a number of drugs that are used in medicine to control heart arrhythmias and to relax smooth muscle. These drugs fall into three distinct classes, each of which acts at a separate site on the Ca^{2+} channel; these drugs are the dihydropyridines (e.g., nifedipine and nimodipine), the phenylalkylamines (e.g., verapamil and D-600), and the benzothiazepines (e.g., diltiazem).

The ryanodine receptor releases Ca^{2+} ions from the SR.

Embedded in the membrane of the terminal cisternae, but in close approximation to the DHPR, the ryanodine receptor (RYR) acts as a very effective Ca^{2+} channel. Its function is to release Ca^{2+} ions rapidly into the cytosol, where they will combine with troponin C and trigger the contractile mechanism of the muscle fiber. The channel is composed of four large subunits, each of which has a MW of 564 kD. It is probably coupled mechanically to the DHP channel in the transverse tubules (see chapter 11), and it can be locked in the open state by ryanodine. It has been demonstrated that calcium-induced calcium release is an important mechanism for opening the RYR in cardiac muscle, and may be important in those RYRs of skeletal muscle that are not coupled to DHP receptors. Under normal circumstances, the channel is inactivated by phosphorylation induced by a protein kinase (Wang & Best, 1992).

Calcium-Binding Proteins

The Ca^{2+}-binding proteins are necessary for the signaling actions of Ca^{2+} ions in the cell interior. Thus, the combination of Ca^{2+} with a binding protein initiates a conformational change in the latter and, in turn, triggers a chemical reaction (Hiraoki & Vogel, 1987). There are several known binding proteins, and four of these are especially important in skeletal muscle fibers: calmodulin, parvalbumin, *troponin,* and *calsequestrin.*

The combination of Ca^{2+} with calmodulin activates enzymes.

Calmodulin is a Ca^{2+}-binding protein with a MW of 16.8 kD and contains four Ca^{2+}-binding domains. Each domain contains 12 amino acid residues that form a "pocket" for a Ca^{2+} ion; the four pockets are separated from each other by amino acid α-helices (figure 7.14). Once Ca^{2+} is bound by calmodulin, the latter molecules are able to attach themselves to a wide variety of enzymes in the cytosol and thereby activate them. It is not yet clear how the Ca^{2+}:calmodulin complex is able to select the appropriate target enzyme among the many other types present. One of the enzymes activated is the Ca^{2+} ATPase in the surface membrane; by increasing the extrusion of Ca^{2+} from the cell, the enzyme completes a negative feedback loop across the membrane. Other calmodulin-activated enzymes in muscle fibers include *adenylate cyclase,* cyclic nucleotide phosphodiesterase, a multifunctional calmodulin-dependent protein kinase, phosphorylase kinase, myosin light chain kinase, and calmodulin-dependent phosphoprotein phosphatase (England, 1986). Certain other enzymes, such as the mitochondrial dehydrogenases and protein kinase C, can be directly activated by Ca^{2+} ions without the intervention of calmodulin (figure 7.15).

Parvalbumin reduces $[Ca^{2+}]$ in fast-twitch fibers.

A second Ca^{2+}-binding protein found in skeletal muscle is parvalbumin, which has a MW of 12 kD and contains two Ca^{2+}-binding pockets in each molecule. Parvalbumin is found in appreciable amounts in amphibian twitch fibers and fast-twitch skeletal muscle fibers of small mammals. Parvalbumin functions as a buffer for Ca^{2+} in the cytoplasm. When Ca^{2+} concentration rises in response to activation, parvalbumin will bind Ca^{2+}, accelerating the decline in $[Ca^{2+}]$. This permits the rapid relaxation that is characteristic of fast-twitch skeletal muscle. The slowing of relaxation in a prolonged *tetanus* is thought to be due to saturation of parvalbumin with Ca^{2+}, as once all of the binding sites are occupied, the decrease in $[Ca^{2+}]$ is slowed. Consequently, parvalbumin-deficient mice (gene knockout) have slower relaxation than wild-type mice, and the rate of relaxation does not change with prolongation of *tetanic stimulation* (Raymackers *et al.,* 2000).

Ca^{2+}–troponin interactions allow cross-bridge engagement.

Unlike parvalbumin, a third Ca^{2+}-binding protein in skeletal muscle is essential, being necessary for the contraction process to take place. This is troponin, a globular protein (MW 80 kD) composed of three nonidentical subunits, T, I, and C. Troponin C has a MW of 18 kD and acts as the calcium receptor; like calmodulin, it has four pockets for attaching Ca^{2+} ions, and these are separated by α-helices. Troponin T interacts with *tropomyosin.* The latter protein, to which troponin is attached, is a long, rodlike molecule running alongside the actin filaments (see figure 11.10, chapter 11). In the absence of Ca^{2+} but in the presence of troponin T and troponin I, tropomyosin prevents the myosin *cross-bridges* from attaching themselves to the actin filaments. However, the conformational change in the troponin molecule, following the combination of Ca^{2+} with troponin C, lifts the tropomyosin rod away from the actin filaments and allows the myosin cross-bridges to engage.

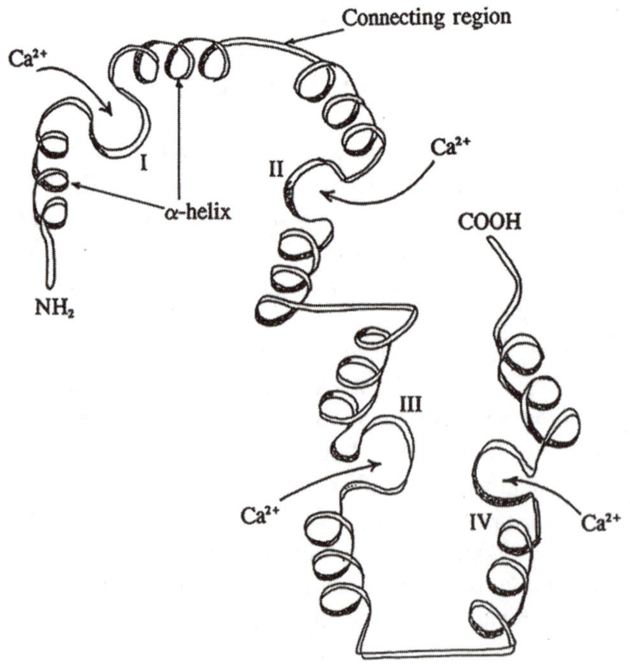

Figure 7.14 Structure of calmodulin. The molecule consists of a single chain of 148 amino acids. Within the chain are four Ca^{2+}-binding domains (I to IV), each of which is flanked by an α-helix connected to a similar helix of the adjacent domain. The structure of troponin C is very similar.

Adapted from England (1986, p. 376).

The fourth binding protein, calsequestrin, is also extremely important in the skeletal muscle fiber and is responsible for collecting Ca^{2+} ions in the SR when the fiber is not contracting. It is found in highest concentration in the terminal cisternae and is attached to that part of the membrane facing the transverse tubules. Calsequestrin has a MW of approximately 40 to 50 kD. The Ca^{2+} binding of calsequestrin is impressive; MacLennan and Wong (1971) have estimated that a single molecule can capture 970 nmol of Ca^{2+}. Calsequestrin is therefore

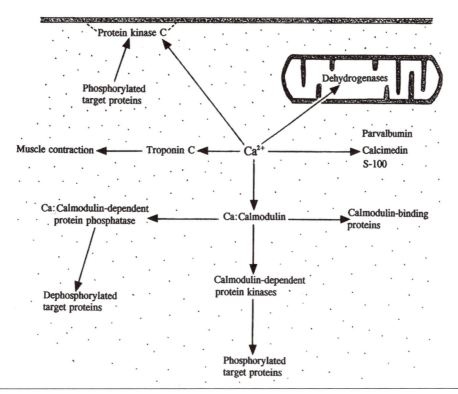

Figure 7.15 Actions of Ca^{2+}, alone or with its different binding proteins, within the muscle fiber.

Adapted from England (1986, p. 375).

**Calseques-
trin binds
Ca^{2+} in the
SR.**

considerably more powerful than the other Ca^{2+}-binding proteins in the skeletal muscle fiber. The free $[Ca^{2+}]$ in the SR is about 1.5×10^{-3} M. However, since the volume of the SR (and in particular, the volume of the terminal cisternae) is very small, the $[Ca^{2+}]$ would fall considerably during repeated activations. Dissociation of Ca^{2+} from calsequestrin prevents this possible decline in $[Ca^{2+}]$.

Table 7.2 summarizes the main features of the major Ca^{2+}-binding proteins in the skeletal muscle fiber.

Calcium Pumps

The surface membrane of the skeletal muscle fiber, like that of other vertebrate and invertebrate cells, has a pump that expels Ca^{2+} from the interior of the fiber (Schatzmann, 1989). Because of the essential role of Ca^{2+} ions in the contractile process, the muscle fibers have a second type of pump in the membrane of the SR (figure 7.12). This pump transfers Ca^{2+} from the cytosol into the lumen of the SR, thereby terminating the activation of the contractile filaments.

Table 7.2   Major Ca^{2+}-Binding Proteins of Skeletal Muscle

Protein	MW (kD)	Location	Function
Calmodulin	16.8	Cytoplasm	Ubiquitous Ca^{2+}-dependent modulator of protein kinase and other enzymes
Parvalbumin	12	Cytoplasm (mainly amphibian and fast-twitch muscle fibers of small mammals)	Accelerates relaxation by decreasing $[Ca^{2+}]$
Troponin C	18	Actin filaments	Triggers muscle contraction
Calsequestrin	40-50	Interior of SR	Sequesters Ca^{2+} within terminal cisternae of SR

Not surprisingly, Ca^{2+} pumps in the surface and SR membranes of the muscle fiber are similar. The following is a summary of the key features of these pumps:

• The two types of Ca^{2+} pump have extensive similarities in their molecular structures, and each consists of two identical polypeptide chains. The surface membrane pump (SMP) has a MW of 140 kD and is rather larger than the SR pump (SERCA; MW, 110 kD). Part of the reason for the greater size of the SMP is that the molecule requires a Ca^{2+}:calmodulin-binding site in order for the pump to be activated. In contrast, the SERCA responds to Ca^{2+} ions directly. Both pumps move Ca^{2+} ions against large concentration gradients (approximately 1:10,000), and both are stimulated by the presence of Mg^{2+} ions.

• As expected of such ionic pumps, there is a higher affinity for Ca^{2+} ions in the cytosol than in the interstitial fluid or in the lumen of the SR. As with other pumps, the energy comes from the hydrolysis of ATP; for each molecule of ATP, two Ca^{2+} ions are transported across the membrane. It now appears that the pump moves H^+ ions in the opposite direction. The exchange is not electrically neutral, however, since only one H^+ ion is returned for each divalent Ca^{2+} ion; like the Na^+-K^+ pump, the Ca^{2+} pump increases the negativity of the cytosol bordering the membrane. It is probable that the phosphorylation of the pump, using ATP, takes place at an aspartate residue in a large region extending into the cytosol. The resulting perturbation is transmitted to parts of the molecule within the membrane, and Ca^{2+} ions are then displaced (Inesi & Kirtley, 1990).

• As in the Na^+-K^+ pump, the membrane portion of the Ca^{2+} pump is arrayed in several helices that traverse the membrane and are linked by amino acid chains. Indeed, there appear to be 10 helices (Clarke *et al.,* 1989). Among four of the transmembrane segments are six regions that contain *glutamate* or other acidic amino acid residues and that have been identified as the sites of calcium binding; it is likely that the four helices form a pore, rather like that inside an ion channel. The crystal structure of SERCA1 has recently been resolved (Toyoshima *et al.,* 2000). Compared with an ion channel, however, the maximum rate at which ions can be transferred by the pump is quite low. Thus, whereas 10×10^8 ions can flow through a single cation channel in 1 s, a Ca^{2+} pump can deliver only 20 ions in the same time. The SR compensates for this low pumping capacity by having the Ca^{2+} pump molecules packed tightly together in the membrane of the terminal cisternae. Furthermore, ion channels flicker open for a fraction of a second whereas ion pumps can be continuously active.

• The SERCA2a isoform of the SR Ca^{2+} pump is the form that is present in cardiac and slow-twitch skeletal muscle. This SERCA isoform is inhibited by *phospholamban,* which is a peptide that co-localizes with SERCA2a in the SR membrane. Modulation of SERCA2a activity occurs by phosphorylation of phospholamban, which releases this inhibition and increases Ca^{2+} pump activity. Sarcolipin, which has some homology with phospholamban, co-localizes with SERCA1a, the isoform of the Ca^{2+} pump in fast-twitch skeletal muscle. Sarcolipin increases the maximum rate of transport by SERCA1a, but it is not yet known whether or not this activity can be acutely modulated.

Anion Channels

Although the more dramatic events of cell life are mediated by cations, the surface membrane also contains channels for anions. In the case of skeletal muscle fibers, the most abundant anion in the extracellular fluid, Cl^-, evidently has more channels available to it than does K^+. Thus, the *permeability* of the resting membrane is several times higher to Cl^- than to K^+ (Hutter & Noble, 1960).

Plentiful Cl^- channels stabilize membranes.

The Cl^- channels resemble the K^+ ones in that they tend to stabilize the membrane potential. In other tissues, Cl^- channels are involved in cell volume regulation, signal transduction, and transport across epithelial surfaces. Although most channels are ligand gated or voltage gated, the Cl^- channel associated with cystic fibrosis appears to have a different mechanism. In skeletal muscle, it is the voltage-gated channel that is especially important; abnormal functioning of this channel, as the result of a genetic mutation, leads to hyperexcitability of the muscle fiber plasmalemma with spontaneous contraction (*myotonia;* see "The Myotonic Goat").

The structures of the Cl^- channels have been determined using molecular biology techniques. The first channel to be cloned was that of the electric organ of the electric ray *Torpedo marmorata* (Jentsch *et al.*, 1990); this, in turn, led to cloning of the skeletal muscle channel in the same laboratory (Steinmeyer *et al.*, 1991). The skeletal muscle channel (ClC-1) probably contains four subunits (Steinmeyer *et al.*, 1994), each with a MW of 100 kD. The subunits are unlike those of the cation channels in having a greater number of transmembrane domains (12), with a further domain in the cytoplasm.

It is likely that the Cl^--channel density in the muscle fiber membrane is relatively high, since the current admitted by each channel is small (1 pS); for the same reason it has not been possible to carry out patch clamping. Nevertheless, it is known that channel opening is greatest at membrane potentials slightly higher than the normal resting potential and that it decreases when the membrane depolarizes. The Cl^- channel can be blocked by Zn^{2+} and by a number of aromatic monocarboxylic acids.

The Myotonic Goat

The first identification of a disorder caused by an abnormal ion channel was made in a strain of goats. These animals, introduced into the southern United States during the 1880s under mysterious circumstances, attracted attention by their remarkable behavior when startled. Instead of running away, their legs became so stiff that the goats were unable to move and often toppled over. The stiffness could last for up to 1 min, after which the muscles gradually relaxed and the animals were able to move about once more. Some engine drivers, aware of this phenomenon, delighted in observing the effects of the train whistle as they approached a field containing these animals. An amusing anecdote was given by Mayberry (see McComas, 1977):

> I've heard a story about a new hired man who, a few days before a barbeque, was given a scatter gun and told to go into the back pasture and kill one of the goats. He was advised that the goats were very shy and that he should be very careful not to apprise them of his presence until ready to shoot. After crawling up cautiously, he picked out a nice fat kid, took careful aim, and fired. Goats dropped in every direction, some thirty animals collapsing simultaneously. Aghast, and without waiting for the resurrection, he ran back to the house. "I don't know how it happened," he panted. "I only fired once but I killed every damn one of them goats."

Why did the muscles of these goats become stiff?

The first investigation of the myotonic goat was made by Brown and Harvey (1939). They recorded from affected muscles with needle electrodes and found that the stiffness was associated with rapid discharges of action potentials. Further, the electrical activity could still be elicited after nerve section and after curarization. This last observation excluded the possibility that the discharges could have arisen at the *end-plate* following spontaneous release of acetylcholine. Rather, it appeared that there was abnormal excitability of the muscle fibers themselves. The story of these fascinating animals was taken up some 30 years later by Bryant and his colleagues in Cincinnati. By injecting current through an intracellular microelectrode, Bryant found that the resistances of the muscle fiber membranes were much higher in myotonic animals than in normal ones (Bryant, 1969). The increased resistance was later shown to be due to a reduced Cl^- permeability of the membrane; further, normal fibers became hyperexcitable if placed in a chloride-free bathing medium.

Bryant later took some of the myotonic goats to Cambridge, England, and carried out further studies with Richard Adrian. They showed that the repetitive discharges in the muscle fibers arose from unusually large *negative after-potentials* that followed the spike potentials (figure 7.16). The after-potentials and the repetitive activity were abolished by disrupting the necks of the transverse tubules with glycerol. The researchers reasoned that the negative after-potentials were due to depolarization of the muscle fibers by the accumulation of K^+ ions in the transverse tubules following *impulse* activity. In a normal fiber, the after-potentials were much smaller because the high Cl^- permeability of the surface membrane acted as a stabilizing influence; that is, the membrane

(continued)

potential of the fiber would be intermediate between the equilibrium potentials for Cl⁻ (at the surface membrane) and for K⁺ (at the tubular membrane). One can appreciate this point by referring to the Goldman-Hodgkin-Katz equation (see chapter 9).

It was through the study of the myotonic goat that Lipicky and Bryant (1973) were able to identify a similar Cl⁻ impermeability in the muscles of patients with the genetic disorder, myotonia congenita.

There is a very interesting and similar disorder of mice, caused by a mutant gene, in which the animals are almost totally paralyzed. This condition, named muscular dysgenesis, is considered in chapter 11.

 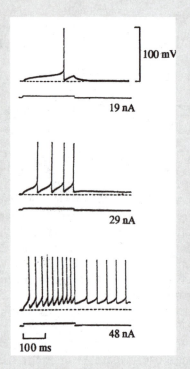

Figure 7.16 Responses of a muscle fiber from a myotonic goat to depolarizing pulses of current injected through an intracellular microelectrode (shown below each action potential tracing). *(top)* A slow depolarization to threshold is observed, with a typical single action potential following (note negative after-potential). *(middle and bottom)* Unlike the responses of a normal fiber, there are repetitive discharges of action potentials as the stimulus intensity increases; this pattern is characteristic of myotonia.

Adapted, by permission, from R.H. Adrian and S.H. Bryant, 1974, "On the repetitive discharge in myotonic muscle fibres," *Journal of Physiology* 240: 508.

Applied Physiology

In this section, there is rather a lot of ground to cover, because, as we shall see, the Na⁺ channels can be affected by a number of agents, and these and other channels can be at fault in various hereditary disorders.

Na²⁺ Channels Are Blocked by Local Anesthetics and Biological Toxins

It has long been known that cocaine, when applied to peripheral nerves or mucous membranes, can cause anesthesia and paralysis. Derivatives of cocaine, such as procaine and xylocaine, have been employed as local anesthetics for many years and produce their effects by blocking the Na⁺ channels. Powerful though these agents are, they are much less effective than some of the naturally occurring toxins, all of which may cause death. *Tetrodotoxin (TTX)* is secreted by the Japanese puffer fish. Its poisonous effects were well known to the Chinese, who referred to it in the first herbal pharmacopoeia, *Pen-T'so* (written during the life of the Emperor Shen Nung, 2838-2698 b.c.). One of the victims of TTX poisoning was the circumnavigator, Captain James Cook, who fortunately survived and was able to provide a fascinating account of the paralysis and loss of sensation that had overtaken him (Cook, 1777).

Saxitoxin (STX), another Na^+-channel blocker, is produced by the marine alga, *Gonyalaux catanella* (Kao, 1966). When plentiful in the sea, the algae form a "red tide" and infest shellfish; the latter, when eaten, induce paralysis and can prove fatal. Batrachatoxin also causes severe paralysis and is secreted in the skin of a small South American frog. Other toxins interfere with Na^+ channels by keeping them open and can be equally deadly. Thus, a sting from the tail of a scorpion causes local swelling and a sharp, burning sensation, followed by sweating and salivation. Numbness and muscle twitching develop and are associated with restlessness and increasing respiratory distress. There is pain in the abdomen and chest, together with vomiting. Finally, in severe cases, convulsions and loss of consciousness lead to death.

Aconitine is another Na^+-channel-opening poison. The alkaloid is contained in the monkshood, a blue-flowered wild plant found in Europe and North America. Ingestion causes tingling and numbness in the mouth, throat, and skin. Restlessness, incoordination, vomiting, diarrhea, and convulsions also occur. Rather similar symptoms result from swallowing veratridine, one of the alkaloids produced by the false hellebore plant.

Episodes of Paralysis Can Occur in a Rare Familial Disorder

The condition familial hyperkalemic periodic paralysis (also known as adynamia episodica hereditaria or Gamstorp's syndrome) is one in which those affected are prone to attacks of muscle weakness. The disorder is inherited by an autosomal dominant gene and therefore affects both sexes. The attacks of paralysis tend to occur during daytime and can be provoked by resting after heavy exercise, by fasting, or, diagnostically, by ingesting K^+. The weakness may be preceded by muscle stiffness, due to spontaneous firing of action potentials by the muscle fibers (myotonia). Associated with the weakness is a rise in the K^+ concentration in the plasma, such that values of 6 to 7 mM are reached; the attacks characteristically last from 1 to 3 hr and do not progress to complete paralysis of the affected muscles. In one patient with this disorder, Brooks (1969) was able to record resting membrane potentials during an attack and found that the muscle fibers were severely depolarized (figure 7.17); further, they could not respond to stimuli applied through the intracellular microelectrode.

Subsequently, in a specimen of excised intercostal muscle from another patient, Lehmann-Horn *et al.* (1987) found that the defect in the muscle fibers was due to the presence of an abnormal Na^+

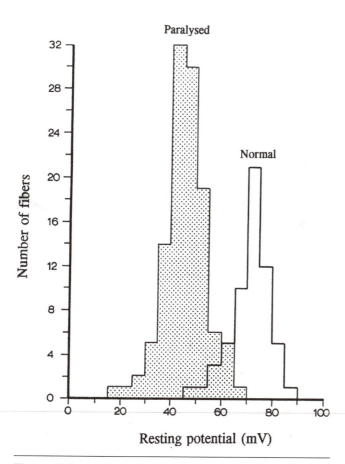

Figure 7.17 Measurements of resting membrane potentials in a patient with the hyperkalemic type of familial periodic paralysis. In between attacks of paralysis, the values were normal (open histogram); but during an episode of weakness, depolarization occurred (shaded histogram), and the fibers became inexcitable on direct stimulation.

Adapted, by permission, from J.E. Brooks, 1969, "Hyperkalemic periodic paralysis. Intracellular electromyographic studies," *Archives of Neurology* 20: 13-18. Copyright 1969, American Medical Association.

channel that opened in response to muscle depolarization, or to small elevations in K^+ concentration, but differed in failing to inactivate; that is, the channel stayed open. More recently, very tight linkage has been found in a large pedigree between the susceptibility to the episodic weakness and the Na^+-channel gene on human chromosome 17 (Fontaine *et al.,* 1990). This linkage indicates that the abnormal Na^+ channel has the molecular defect in this condition; recent work has shown that there is an amino acid substitution in that part of the protein joining the S5 and S6 segments (figure 7.4). This and other types of Na^+-channel disorder have been reviewed by Rüdel *et al.* (1993).

Myotonic Muscular Dystrophy: A Na$^+$-K$^+$ Pump Defect?

So far there are no clinical disorders that are definitely known to arise from defective Na^+-K^+ pumping. One possibility, however, is the myotonic form of muscular dystrophy (dystrophia myotonica; Steinert's disease). This is a hereditary disorder, transmitted by an autosomal dominant gene, in which there is degeneration of muscle fibers and also a characteristic stiffness of the muscles. After making a fist, for example, patients may require several seconds before they can straighten their fingers again. As in familial hyperkalemic periodic paralysis (see the preceding section), the stiffness, or myotonia, is due to the spontaneous firing of action potentials by the muscle fibers. What makes this disease especially fascinating is the great variety of associated clinical defects—frontal baldness, hyperostosis of the skull, cataracts, ovarian and testicular *atrophy,* excessive sleepiness, and insulin-resistant diabetes. Even the central nervous system is affected, for patients may undergo progressive mental deterioration and eventually become demented.

The gene for this condition is found on chromosome 19 and codes for a serine-threonine protein kinase (Fu *et al.,* 1992). Several observations suggest that one of the consequences of an abnormality in this enzyme is impaired Na^+-K^+ pumping. Among the various clues are the severalfold reduction in Na^+-K^+-pumping activity in skeletal muscle biopsies, the increased intracellular Na^+ concentration in resting muscle fibers, and the inability of the muscle fibers to potentiate their action potentials promptly during contractions (Hicks & McComas, 1989; a Na^+-K^+ pump mechanism, see Fenton *et al.,* 1991).

Abnormal Cl$^-$ Channels Also Cause Muscle Stiffness

As already noted, myotonia is the clinical term used to describe an involuntary stiffness of muscles, caused by hyperexcitability of muscle fibers. The membrane defect can be due to prolonged opening of Na^+ channels (in familial hyperkalemic periodic paralysis) or, possibly, to defective Na^+-K^+ pump activation, allowing K^+ to accumulate in the interstitial spaces and to depolarize the muscle fibers (myotonic muscular dystrophy). In myotonia congenita, however, there is a third cause for muscle stiffness—the muscle fiber membranes have a reduced Cl^- permeability. An abnormality of Cl^- channels was first detected in a strain of goats with hereditary myotonia. These curious animals, all descended from a small flock in Tennessee, are prone to develop rigidity of their legs and to topple over when startled (see "The Myotonic Goat"). Bryant (1969) showed that the membrane resistance in myotonic goat muscle fibers at rest was considerably increased; this elevation was later shown to be due to the absence of the chloride component of the membrane conductance (Bryant & Morales-Aguilera, 1971).

How could a Cl^- defect in the membrane cause myotonia? It will be recalled that in a resting mammalian muscle fiber, there are probably more open Cl^- channels than K^+ ones, because the membrane conductance is several times greater for Cl^- than for K^+. Further, both types of channel are responsible for stabilizing the membrane potential by counteracting small increases in Na^+ permeability, which might otherwise become regenerative. This stabilizing effect of K^+ and Cl^- channels is evident from the inclusion of terms for permeability and concentration ratio for both these ions in the Goldman-Hodgkin-Katz equation for the resting membrane potential (see chapter 9). In a later study of goat myotonia, Adrian and Bryant (1974) pointed out that if Cl^- permeability is reduced in the surface membrane of the muscle fiber, then the accumulation of K^+ ions in the transverse tubules during contractile activity will exert a powerful depolarizing effect on the fiber. This depolarization, in turn, will activate the Na^+ channels and cause impulses to be initiated spontaneously.

Following the recognition of a Cl⁻-channel defect in the myotonic goat, Lipicky and Bryant (1973) were able to demonstrate the same phenomenon in four of six patients with myotonia congenita. Studies in molecular biology have provided the final proof that a defective chloride channel is responsible for myotonia congenita. An abnormal gene on chromosome 7 has been found. In two families with the autosomal recessive form of the condition (Becker's disease), this gene coded for a Cl⁻ channel (CLCN1) with a phenylalanine-to-cysteine substitution in one of the transmembrane domains (Koch *et al.*, 1992). A different mutation of the same gene accounts for those patients with autosomal dominant inheritance (Thomsen's disease; Steinmeyer *et al.*, 1994). In neither type of myotonia congenita is there any degeneration of the muscle fibers so that the disorder cannot be classified as a muscular dystrophy. If anything, the muscles are usually well developed, presumably because the prolonged contractions cause muscle fiber *hypertrophy*. An interesting aspect of the disorder is that the stiffness is reduced if repeated voluntary contractions are made; this is the warm-up phenomenon.

Although a large amount of material has been covered in this chapter, it is essential for a proper understanding of the ionic basis of the resting and action potentials (chapter 9) and for biochemical events underlying *excitation-contraction coupling* (chapter 11). Before embarking on these topics, however, there is one more aspect of the motoneuron to examine. *Axoplasmic transport* is a process that allows communication between the *soma*, the axon, and the nerve terminal. Axoplasmic transport will be considered in the next chapter.

8

Axoplasmic Transport

The *motoneuron* axon functions as a communication pathway between the spinal cord and muscle. Rapid signaling to the periphery is required if the muscle is to contract in a coordinated fashion and in quick response to proprioceptive and sensory input. This rapid signaling is achieved by the nerve *impulse;* the ionic mechanisms underlying this event are considered in the next chapter. Equally important, however, is a slower transport system that enables organelles, proteins, and peptides to be sent in both directions between the cell body of the motoneuron and the muscle fibers that it innervates. This transportation system is essential for the maintenance of the normal structure and cellular metabolism of both the muscle fiber and the motoneuron; the integrity of the *Schwann cells* and of the *axis cylinder* is also dependent on chemical signals transmitted by this system. These controlling or *"trophic"* factors are described in more detail in chapter 18.

Of concern in this chapter is the nature of the *axoplasmic transport* that makes this chemical transport/signaling system possible. It will be shown that a fast transport system operates in both directions between the motoneuron *soma* and muscle fiber; there is also a slow transport system that runs in a distal (toward the muscle fiber) direction only. These transport systems serve several needs; they carry enzymes and other special proteins to the nerve terminal, together with newly formed mitochondria and membrane-bound vesicles. In addition, aging mitochondria are returned to the soma for degradation or repair. Finally, there is a very slow movement of *microtubules* and *neurofilaments* down the axon.

Confirmation and Categorization of Axoplasmic Transport

Constricting the nerve causes local swelling.

It has been recognized from as early as the middle of the 19th century that the soma is essential for maintenance of the integrity of the axon. Considering that essentially all protein synthesis is accomplished in the soma, it could only be assumed that there was some manner of transportation available to deliver these whole proteins throughout the axon and to the nerve terminal. However, it was not until the middle of the 20th century that substantial progress was made in understanding axoplasmic transport, and only recently have we acquired a reasonable understanding of how this is achieved. The earliest direct observation of evidence for axoplasmic transport came from a series of experiments by Paul Weiss and his colleagues in the 1940s (Weiss, 1944; Weiss & Davis, 1943; Weiss & Hiscoe, 1948). The results of these and later experiments have been summarized by Weiss (1969).

Weiss and colleagues applied a gentle constriction to the sciatic nerve of an animal by slipping a short sleeve of artery over the proximal stub of a cut nerve. As the arterial cuff gradually constricted, due to smooth muscle contraction, the nerve fibers underwent compression to varying extents, and the chronic effects of this compression were studied up to several months. It was found that the segment of nerve abutting the proximal end of the constriction was swollen, mainly on account of distortions and distensions of the axons themselves. With the light microscope, the axis cylinders

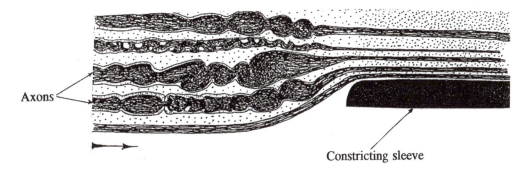

Axons

Constricting sleeve

Figure 8.1 Drawings of four axons showing characteristic deformations following chronic nerve constriction. Weiss and Hiscoe (1948) described the deformations, from below upward, as "telescoping, ballooning, corkscrewing, and beading."

From Weiss and Hiscoe (1948, p. 330).

appeared to be either beaded, corkscrewed, ballooned, or telescoped (figure 8.1). Within and beyond the constriction, the axis cylinders were narrowed.

To Weiss and his colleagues, the appearances of the axons suggested that there had been a "damming up" of cellular contents at the site of constriction. Weiss and his colleagues observed that these swellings and contortions could not simply be due to bulk transfer of *cytoplasm* from the cell body to the axons. Such a process would require a proximal-distal pressure gradient and would result in swelling all along the proximal stump of the axon, with greater expansion at the proximal end. A complete understanding of these observations has not yet been reached, but the research that has stemmed from the need to understand how the cellular substances could accumulate at the constriction has resulted in considerable advance in our understanding of axoplasmic transport.

Content and Categorization of Axoplasmic Transport

Once axoplasmic transport was demonstrated, it was important to determine what was actually transported and how this was done. These experiments led to the categorization of transport according to the speed with which this was achieved.

Axoplasmic transport has been studied by injection of radiolabeled amino acids into the dorsal root ganglion or ventral horn. These amino acids are rapidly incorporated into proteins and subsequently appear in the axons. Evaluation of the radiolabeled substances in the axons reveals what has been transported.

Newly synthesized proteins, organelles, and other material are transported.

Included among the transported proteins are the enzymes *acetylcholinesterase (AChE)* and *choline acetyltransferase.* In addition, Miledi and Slater (1970) have shown that a fast-traveling factor(s) is necessary for maintenance of the axon terminal at the neuromuscular junction; this would be considered to be a trophic factor (see chapter 18). Among a large number of transported proteins as yet unidentified are those that supervise the metabolic machinery of the axon, Schwann cells, and muscle fibers (see chapter 18). Of the particulate matter transported down the axon, the most prominent fraction is the mitochondria, and these can be observed in motion with phase-contrast and interference microscopes. It has been suggested that mitochondria are transported to the axon terminal to supply energy needed for the synthesis of ACh and that their aging components are continually replaced. The damaged components would be conveyed back to the soma by fast retrograde axoplasmic transport. Other transported particles include various kinds of vesicles, *smooth endoplasmic reticulum,* and plasma membrane. Other studies have shown that lipids and polysaccharides are also rapidly conveyed down the axon.

The Categorization of Axoplasmic Transport

Axoplasmic transport can be classified as either fast or slow, with subcategories of these designations (see table 8.1). Within the subcategories, there appears to be a fairly broad range of values for rate of transport, and it has been suggested that maximal rates of transport ($\sim$400 mm $\cdot$ day^{-1})

Table 8.1 Characteristics of Fast Versus Slow Axoplasmic Transport

	Fast			Slow	
Subtype	I	II	III	SCb	SCa
Velocity (mm · day^{-1})	240-410	34-68	4-8	2-4	0.1-2
Transported material	Organelles, mitochondria, vesicles			Cytoplasm, enzymes, microtubules, neuro-filaments, actin filaments	
Transporting structure	Microtubules			Neurofilaments, microtubules, actin filaments	
Driving mechanism	Protein motors (kinesin—anterograde; dynein—retrograde)			Extension of microtubules and filaments	
Blocking agents	Inhibitors of oxidative metabolism (dinitrophe-nol, sodium cyanide, anoxia) and colchicine			Inhibitors of protein synthesis (cyclohexi-mide) and colchicine	

may be dependent on axon length. Considering that the longest axons in humans can be greater than 1,000 mm, at 1 mm · day^{-1} it would require years for the transported substance to complete the journey! Transport of substances from the soma to the end of the axon is called *anterograde transport,* and the reverse is *retrograde* transport. The categorization into speeds of transport also permits designation of substances carried in a given category.

Slow Axoplasmic Transport

Slow axoplasmic transport has two components.

At present, two components of slow transport are recognized, and both are largely associated with movements of the three main cytoskeletal proteins—*actin, tubulin,* and neurofilament protein. The slower fraction, slow component a (SCa), travels at 0.1 to 1.1 mm · day^{-1} and contains mainly tubulin and neurofilament protein. The faster phase, slow component b (SCb), moves at 2 to 8 mm · day^{-1}; it contains actin and tubulin, together with cytoplasmic enzymes, *clathrin, spectrin,* and *calmodulin* (Lasek *et al.,* 1984).

Exactly how the cytoskeletal proteins move down the axon is not clear. It would appear that for much of the time the microtubules, neurofilaments, and actin filaments are stationary, possibly because of *phosphorylation* (see chapter 2). When the microtubules and filaments begin to move again, they may do so either as an interconnected lattice or as separate elements. Further, it is probable that molecules of the cytoskeletal proteins may be added from the axoplasm to the formed structures or else removed from the latter (for review, see Nixon & Sihag, 1991; Vallee & Bloom, 1991).

Fast Axoplasmic Transport

Fast axoplasmic transport has been studied with radioactive isotopes.

After the clear demonstration of axoplasmic transport with the nerve compression experiments described earlier, attempts were made to explore the phenomenon more fully by labeling specific molecules (usually amino acids) with radioactive isotopes and then studying their passage down the axon. The first experiments involved radioactive phosphorus, and then labeled carbon in glucose and amino acids was used; these methods have been superseded following the availability of tritiated amino acids. A commonly used strategy is to inject [^{3}H]-leucine into the ventral gray matter of the lumbar region of the spinal cord; the amino acid is then taken up by the cell bodies of neighboring motoneurons and rapidly incorporated into newly synthesized proteins. At a given time after injection, the sciatic nerve is excised and cut into segments of uniform length. The amount of radioactivity in each segment is then measured with a scintillation counter and expressed as a function of distance from the spinal cord. This approach provides a snapshot of the position of radioactive protein in the cord at any given time. Measurement conducted over a variety of times provides an indication of the velocity of transport.

An example of fast axoplasmic transport is given in figure 8.2, the lower curve of which shows the distribution of radioactivity in the sciatic nerve of the cat 6 hr after injection of [^{3}H]-leucine

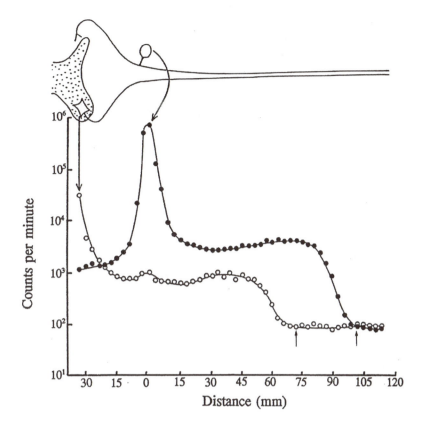

Figure 8.2 Anterograde transport of radioactive leucine in sensory nerve fibers (●) and motor fibers (○) of cat sciatic nerve, measured 6 hr after injection of labeled material into dorsal root ganglion and ventral horn, respectively. See text for description of technique and results.

From "Characteristics of the fast transport system in mammalian nerve fibers," by S. Ochs and M. Ranish, 1969, *Journal of Neurobiology,* 1, p. 258. Copyright 1970 by John Wiley & Sons, Inc. Adapted with permission of John Wiley & Sons, Inc.

into the L7 segment of the (ventral) cord (Ochs & Ranish, 1969). Included for comparison, and displayed in the upper curve, are the results for sensory nerve fibers following an injection given at approximately the same time into the L7 dorsal root ganglion on the opposite side. The peak of radioactivity in the sensory fiber curve corresponds to isotope remaining in the ganglion; similarly, the high value at the left of the motor fiber curve denotes the activity left in, and around, the cell bodies of the motoneurons. The arrows at the bottoms of the two curves indicate the advancing fronts of the labeled axoplasm; the spatial disparity corresponds to the extra length of the motor pathway due to the inclusion of the L7 *ventral root.* Studies such as these have demonstrated that there is a faster transport of substances in nerves with a maximum velocity, in mammals, of about 410 mm · day^{-1}.

Another technique for measuring rate of transport is to produce a temporary arrest of axoplasmic transport by cooling a short length of nerve; the transported material then accumulates in, and proximal to, the cooled region. Upon rewarming of the nerve, the movement of the material is resumed, and its velocity can be determined. This "stop-transport" method has been used by Brimijoin (1975) to study the passage of the enzyme dopamine-β-hydroxylase in rabbit sciatic nerve; a transport rate of 300 ± 17 mm · day^{-1} was found.

Components of Fast Axoplasmic Transport

Fast axoplasmic transport has three components.

The presence of fast axoplasmic transport is a property of all the motor and sensory axons that have been tested to date, and it has been demonstrated that the velocity of the transport is independent of fiber diameter. Isotope studies suggest that the fast-transported material travels at three different rates, approximately 240 to 410 mm · day^{-1}, 34 to 68 mm · day^{-1}, and 4 to 8 mm · day^{-1}. However, the absolute values vary between species and also between different nerves within the same species (Grafstein & Forman, 1980). Impulse activity has no effect on the velocity of particle movement. However, it may increase the amount of material leaving the cell soma (e.g., Lux *et al.,* 1970) and possibly also that appearing in the *synaptic cleft* (Musick & Hubbard, 1972). Finally, it should be noted that the *dendrites* also exhibit fast and slow axoplasmic transport.

Mechanism of Fast Axoplasmic Transport

How is the material conveyed along the axon by fast axoplasmic transport? In the case of slow transport, it was seen that entire neurofilaments and microtubules moved, either as relatively short segments or as extended linear structures. For fast transport, however, the microtubules serve as tracks along which *molecular motors* carry the particulate matter. It is probable that in any living cell there are 50 or so distinct molecular motors that run on microtubules (and actin filaments) within the cytoplasm; additional motors move along strands of *DNA* and messenger *RNA* (Spudich, 1994). The picture that emerges is not unlike a complicated model railway layout, in which engines are constantly pulling wagons in different directions along crisscrossing tracks. Of these motors, *kinesin* is especially important in axoplasmic transport and was originally identified in the squid giant axon and in chick and calf brains (Brady, 1985; Vale *et al.,* 1985a).

Energy for Axoplasmic Transport

Fast axoplasmic transport requires a local supply of ATP.

It has been suggested that the *Golgi apparatus* of the cell body acts as a "gate" controlling the delivery of the material into the axon (Ochs, 1972). As far as the transport mechanism itself is concerned, a number of observations are relevant. First, axoplasmic transport is able to proceed in both directions at a normal rate even when the axon has been removed from the body; this finding indicates that the transport system obtains its energy from local sources in the axon (or the Schwann cell).

Other experiments show that fast transport is dependent on *ATP* supplied by oxidative metabolism. For example, transport can be blocked if a small segment of nerve is made anoxic. Alternatively, blocking can be produced by interfering with oxidative metabolism, by means of agents such as sodium cyanide or dinitrophenol. Another requirement for axoplasmic transport is a critical concentration of Ca^{2+} in the interstitial fluid.

Kinesin's Role in Fast Axoplasmic Transport

Kinesin is a motor molecule involved in fast axoplasmic transport.

Kinesin's role as a motor was demonstrated by its capacity to move coated latex beads along an *in vitro* array of microtubules (Vale *et al.,* 1985b). For kinesin to transport an organelle *in vivo,* however, requires the additional presence of one or more accessory factors in the cytoplasm (Sheetz *et al.,* 1989; figure 8.3). Since kinesin moves particles only toward the distal (+) end of a microtubule, it is obviously the main motor for anterograde transport. Analysis of the purified kinesin protein reveals that it has a molecular weight of 380 kD and is composed of two pairs of polypeptides. Within each pair there is a *globular head* connected to a feathered tail by a long stalk. The globular head is the motor domain and contains tubulin-binding properties as well as ATPase activity. The oppo-

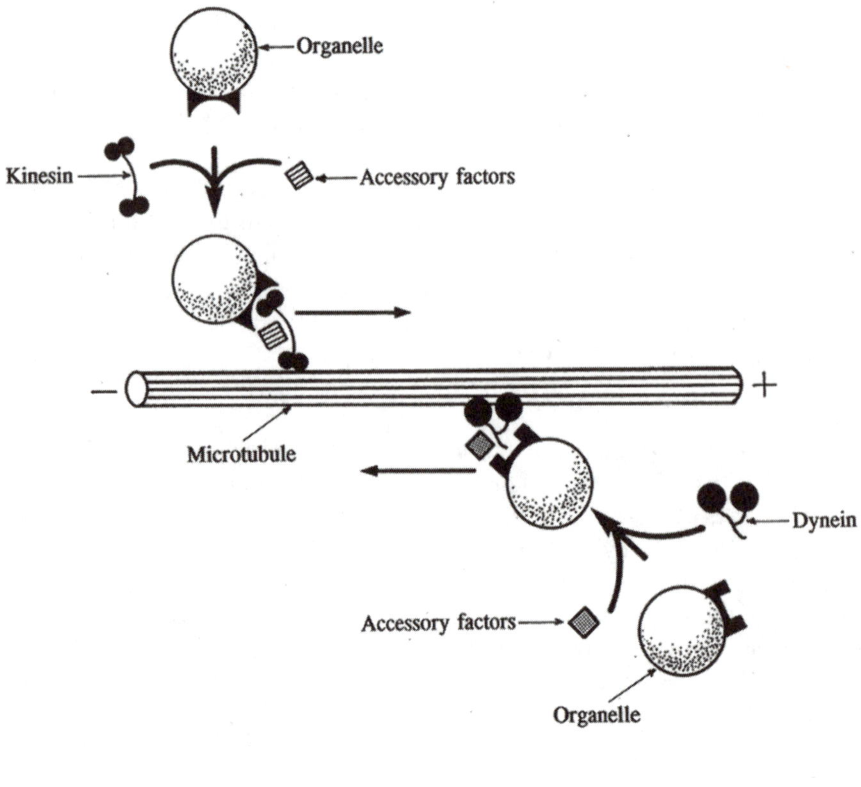

Figure 8.3 Assembly of "organelle translocation complexes" by combination of organelles with molecular motors (kinesin or *dynein*) and appropriate accessory factors.

From M.P. Sheetz. E.R. Steuer, and T.A. Schroer, 1989, "The mechanism and regulation of fast axonal transport," *Trends in Neurosciences* 12: 476. Copyright 1989 by Elsevier Trends Journals. Adapted with permission.

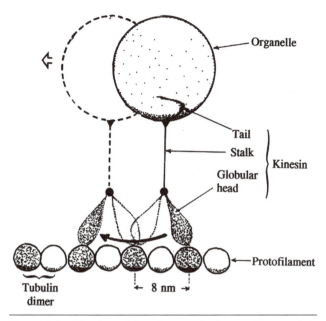

Figure 8.4 Kinesin walking along a tubulin protofilament. It is probable that, in order to step, kinesin alters its shape in the same way that the myosin molecule does (see figure 11.9, chapter 11). The kinesin head would flex on the stalk once ATP had been inserted into the molecule and hydrolyzed. Note that one globular head retains contact with the protofilaments while the other one (stippled) steps past and engages the next tubulin dimer.

site end contains the cargo-binding site. Hydrolysis of ATP provides the energy for the transport function accomplished by these molecular motors.

Until recently it was not known whether the kinesin molecules remained attached to the microtubules or to the transported organelle. In the first case, the organelle would be passed along the microtubule from one kinesin molecule to another; in the second situation, a kinesin molecule would keep hold of an organelle and would carry it along the microtubular track. Svoboda *et al.* (1993) have now shown the second possibility to be the correct one. With an extraordinarily sensitive technique, combining optical "tweezers" with laser interferometry, these authors were able to study individual kinesin molecules. They found that each kinesin molecule moves along one of the protofilaments of a microtubule in a series of steps (see figure 2.4, chapter 2). The length of one of the steps is 8 nm, which is exactly that required to span a tubulin dimer in the protofilament (figure 8.4). By keeping one of its heads attached while the other head steps past, the kinesin molecule is prevented from straying onto adjacent protofilaments (Berliner *et al.,* 1995) or from losing contact with the protofilament. An animation of kinesin transport is available on the website of Ron Milligan at www.scripps.edu/Milligan. Click on "Research," then choose the MPEG or Quick-Time version of the kinesin movie.

Retrograde Transport

So far no consideration has been given to the axoplasmic transport that occurs in a retrograde direction, that is, from the muscle fiber to the motoneuron. Evidence for such a function has come from several sources. For example, Lubinska and Niemierko (1970) found that AChE is transported proximally as well as distally; the rate of the retrograde transport is about one-half that of anterograde transport. Other workers have observed particles, probably mitochondria, being carried in axons away from the periphery. Presumably these are damaged organelles being transported to the soma for degradation or reconstitution.

Transported material reaches motoneuron cell bodies.

Proof that some of the retrograde-transported material arrives in the cell bodies of the motoneurons has been obtained by Glatt and Honegger (1973). These workers injected albumin coupled with the fluorescent dye, *Evans blue,* into triceps muscles of the rat forelimb and were able to detect dye in the ventral horns some 12 hr later. In the study by Kristensson and Olsson (1971), *HRP* was used as a marker instead. This substance had the advantage that it could produce an electron-dense reaction product, and with the electron microscope, its intracellular location was seen to be in small cytoplasmic granules around the *nucleus.* Just as the anterograde axoplasmic transport largely mediates the trophic control of axon, Schwann cell, and muscle fiber by the soma, so the retrograde transport exerts a trophic influence on the motoneuron soma by the periphery. Interruption of this trophic influence, as in cutting the axon, sets in motion the phenomenon of chromatolysis in the motoneuron soma and attempts at *regeneration* in the proximal nerve stump (see chapter 18).

Dynein is the molecular motor for retrograde transport.

Dynein Is Another Molecular Motor

While anterograde transport is made possible by the protein motor, kinesin, retrograde transport depends on another motor, dynein (Vallee *et al.,* 1989). The latter molecule was originally identified as that responsible for the movements of cilia and flagella in other types of cells. It was differenti-

Microtubule protofilament

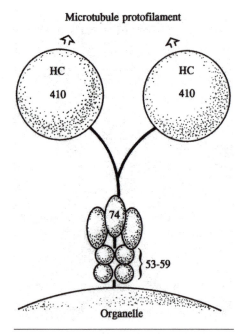

Figure 8.5 Model of dynein. The two heavy chains (HCs) interact with the microtubule protofilaments in the same way as kinesin, and the smaller subunits grip the organelle. Numbers refer to molecular weights (in kD) of subunits.

Adapted from Vallee *et al.* (1989, p. 69).

ated from other microtubule-associated proteins by the ability to split ATP, a property shared with kinesin (and also myosin; see chapter 11). Proof of dynein's capacity to serve as a protein motor came from experiments in which a glass coverslip, coated with dynein, was able to propel microtubules in the presence of ATP (Paschal *et al.*, 1987). In contrast to kinesin, dynein is attracted to the slow-growing (–) end of the microtubule, which points to the motoneuron soma. Dynein is a larger protein than kinesin, being composed of two heads, each with its own stalk, and seven smaller subunits; the complete molecule has a molecular weight of 1,200 kD (see figure 8.5). At present it is not known if dynein and kinesin molecules can run along the same microtubule, carrying their respective organelles and other particles in opposite directions.

Linker Molecules Join Motor to Cargo

There must be some specific mechanism for the molecular motors, kinesin and dynein, to identify and bind their specific cargoes. In the case of kinesin, there are a variety of isoforms, and these isoforms could have different structures in their cargo-binding domains. In the case of dynein, different light chain and intermediate chain composition could confer variable binding properties. In addition, there are thought to be specific linking or scaffolding proteins that bind to the cargo and to the molecular motor. One such linking protein has been given the name "Sunday driver" or Syd (Almenar-Queralt & Goldstein, 2001).

Applied Physiology

In this visit to the neuromuscular clinic, we encounter a second type of peripheral neuropathy. Whereas Charcot-Marie-Tooth disease, discussed in chapter 2, is a hereditary disorder, the second type is "iatrogenic." This Greek-derived word means "caused by the physician" and sounds ominous; however, in most patients the benefits of the drug, which causes the neuropathy, outweigh the neurological complications.

Vincristine and *vinblastine* are two alkaloids produced by the periwinkle plant *(Vinca rosea)*, both of which are useful in the treatment of cancer, especially acute lymphoblastic leukemia, Hodgkin's disease, multiple myeloma, and lung cancer. Of the two compounds, vinblastine was the first to be used, in 1958, but vincristine was then found to be the more powerful and became the preferred therapeutic agent. Unfortunately, both drugs produce a peripheral neuropathy as a side effect. By binding with the tubulin in the axis cylinders, they prevent microtubules from assembling and hence remove the track required for fast axoplasmic transport (Green *et al.*, 1977). There is also an accumulation of neurofilaments in the peripheral nerve fibers (Shelanski & Wísniewskí, 1969). These complications are unavoidable in that it is the combination of the vinca alkaloids with tubulin that also interferes with the formation of the mitotic spindle and, by preventing cancer cells from dividing, produces a beneficial therapeutic result. Through the reduction in axoplasmic transport, the motor and sensory neurons are no longer able to maintain the structural integrity of their respective axons, which then become unexcitable and degenerate.

In patients undergoing chemotherapy with vincristine, the first symptoms usually appear by 2 weeks and consist of numbness and tingling in the fingers and toes. Sensory loss is difficult to demonstrate at this stage, although an early rise in the thresholds of skin receptors can be demonstrated in experimental animals (Leon & McComas, 1984). Weakness appears later and is initially recognized by clumsiness of the fingers with some degree of wrist and foot drop; muscle cramps may also be prominent (Casey *et al.*, 1973). An example of a severe vincristine-induced neuropathy is shown in figure 8.6. The patient, a 34-year-old plumber with embryonal carcinoma of the testis,

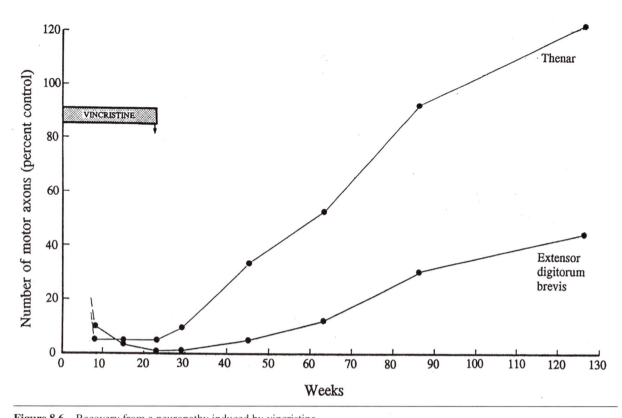

Figure 8.6 Recovery from a neuropathy induced by vincristine.

Adapted, by permission, from A.J. McComas, 1977, *Neuromuscular Function and Disorders* (London: Butterworth Publishing Co.), 277.

was treated with surgery and multiple chemotherapy. When seen 8 weeks later in the *EMG* clinic, he had severe weakness with some wasting of distal limb muscles; there was also considerable numbness and diminished sensation. An EMG at this time revealed very few intact *motor units* in the intrinsic muscles of the hand and foot. Figure 8.6 shows that after vincristine was stopped, there was a slow recovery of motor nerve fiber function over the next 2 years, though this was still incomplete for the foot muscle studied.

In the next chapter, we will really begin to see nerve and muscle fibers in action, starting with the membrane mechanisms that are responsible for the phenomenon of excitability. We are entering the world of the neurophysiologist.

9

Resting and Action Potentials

In this chapter, we will consider the nature of the electrical potential that exists across the surface membranes of nerve and muscle fibers, as well as how this potential may suddenly change so as to form an *impulse* that propagates along the membrane. An understanding of the *membrane potential* requires knowledge of the chemical composition of the fiber and its fluid environment and also an understanding of the properties of the membrane that envelops each cell.

Resting Membrane Potential

It has been known for more than a century that a difference in electrical potential exists between the inside of a cell and its external fluid environment. It is possible to determine this potential difference directly by inserting a microelectrode through the cell membrane into the interior of the fiber and connecting the other terminal of the voltage-measuring device to a reference electrode in the fluid bathing the cell (figure 9.1A). This arrangement allows detection of the voltage difference across the membrane. The difference can be amplified (DC amplifier) and displayed (CRT or voltmeter in figure 9.1A). A computer data collection system can be used to digitize the observed values and quantify the difference. In the case of inactive mammalian muscle, the inside of the cell has been found to be about 85 mV negative with respect to the outside; this potential difference is termed the resting membrane potential (figure 9.1B). A similar potential is present across the membrane of a nerve fiber.

K^+ Ions Are Concentrated Inside the Cell, Na^+ and Cl^- Outside

An understanding of the resting membrane potential requires knowledge of the ions present on the two sides of the membrane. Inside the muscle fiber, the most common cation is K^+; the anions are supplied by PO_4^{3-}, SO_4^{2-}, and HCO_3^-, together with amino acids, polypeptides, and proteins. Outside the cell, the interstitial fluid has a composition similar to that of plasma; Na^+ and Cl^- are the dominant cation and anion, respectively, and smaller amounts of K^+, Ca^{2+}, Mg^{2+}, HCO_3^-, and PO_4^{3-} are also present.

The concentrations of the various ions in mammalian skeletal muscle *cytoplasm* and in its extracellular fluid are given in table 9.1. These values represent the total cellular content of the various ions, including the content of the various organelles and the ions bound to various constituents of the cell. However, it is the free ion concentration that is relevant with respect to membrane potentials. Where the free concentration differs considerably from the total value, the free concentration is presented in parentheses. However, it should be pointed out that differences in ion distribution across organelle membranes can result in subcellular membrane potentials; mitochondria and *sarcoplasmic reticulum* would be expected to have such a membrane potential.

Ion-sensitive microelectrodes have been used to measure the ion concentrations in muscle cytoplasm. A microelectrode is manufactured from a special glass known to be selective for ions of a

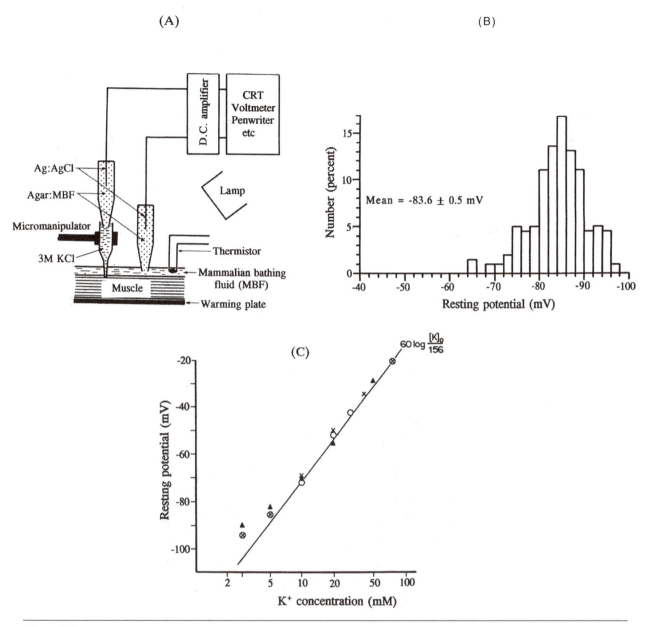

Figure 9.1 (A) Experimental setup for measuring muscle fiber membrane potentials. Observations can be made not only on an excised muscle, as shown here, but on fibers in tissue or organ culture and, of course, on muscles *in vivo*. (B) Resting potentials in biceps brachii and quadriceps of three healthy human subjects, recorded *in vivo*. (C) Resting potentials of human intercostal fibers; the muscles were excised and subjected to different concentrations of K^+ in the bathing fluid. The straight line is a plot of the K^+ *equilibrium potential* for an internal K^+ concentration of 156 mM, at 37° C. The small deviation between the observed and predicted values for low K^+ concentration is due to the Na^+ permeability of the membrane.

B: Data from McComas et al. 1968; C: Reprinted, by permission, from H.P. Ludin, 1970, "Microelectrode study of dystrophic human skeletal muscle," *European Neurology* 3: 117.

particular species; alternatively, an ordinary glass microelectrode can be coated on the inside with an ion-selective resin. When an ion-sensitive electrode is dipped into a solution, the activity difference between the outside and inside of the electrode generates a potential at the electrode tip, which is proportional to the activity of the selected ion in the sampled solution. When inserted into a muscle fiber, however, the ion-sensitive electrode will also record the electrical potential across the cell membrane. One must make a correction by subtracting the resting membrane potential as measured with a standard microelectrode or with the untreated second channel of a double-barreled electrode.

Table 9.1   Ionic Composition of Mammalian Skeletal Muscle Fibers and of the Interstitial Fluid

	Interstitial fluid (mM)	Intracellular fluid (mM)
Sodium	145	10
Potassium	4	160
Calcium	5	2 (.00007)
Magnesium	2	26 (1)
Total cations	156	198
Chloride	114	3
Bicarbonate	31	10
Phosphate	2	100 (3-30)
Sulfate	1	20
Organic acids	7	–
Proteins	1	65
Total anions	156	198

Note. Values in parentheses represent free concentration values; otherwise these values represent total content of these entities.

From M. H. Maxwell and C. R. Kleeman, 1968, *Clinical disorders of fluid and electrolyte metabolism* (New York: McGraw-Hill Companies), 28. Reprinted with permission of The McGraw-Hill Companies.

In practice, the reliability of the determination of ion concentration is hampered by the appreciable junction potentials that occur among dissimilar solutions in the microelectrode, bridge, and reference electrodes. Mean values obtained by the ion-sensitive electrode method range from approximately 120 to 180 mM for $[K^+]_i$ and are around 10 mM for $[Na^+]_i$; slow-twitch muscle fibers tend to have a lower $[K^+]_i$ and a higher $[Na^+]_i$ than fast-twitch fibers (cf. Charlton *et al.*, 1981; Fong *et al.*, 1986; Juel, 1986).

Membrane Permeability

A muscle or nerve membrane cannot completely prevent movement of ions across the membrane. Typically, there will be a slow leak of ions across the membrane when the cell is in its resting state. This ion transfer may occur through nonspecific elements in the membrane, or may rely on random brief openings of ion-specific channels in the membrane. The ease with which any ion can move across the membrane is referred to as membrane *permeability*. This movement of ions through the membrane is passive and typically occurs down the concentration gradient. However, since ions carry a charge, the membrane potential can influence this transfer of ions. In the case of K^+ ions, the high concentration inside will result in a tendency for this ion to cross the membrane from inside to outside. Since the membrane potential at rest is about –85 mV, this electrical potential will oppose the tendency of K^+ to diffuse outward.

There are three reasons why the compositions of the intracellular and extracellular fluids are so different:

The resting membrane is much more permeable to K^+ and Cl^- than to Na^+.

- First, a large proportion of the organic anions in the interior of the fiber are part of the cell structure and are therefore unable to migrate. These molecules have a net negative charge, so they contribute to an electrical potential across the membrane.

- Second, the resting membrane is semipermeable, allowing K^+ and Cl^- ions to diffuse through more easily than Na^+. The internal anions exert an electrostatic attraction for cations that can be satisfied only by a high internal concentration of K^+, because Na^+ ions have difficulty in crossing the resting membrane. The external Na^+ ions will, however, be effective in balancing the anions of the extracellular fluid, the most prevalent of which is Cl^-. This unequal distribution of K^+ and Cl^- ions is termed a Donnan equilibrium, such that

$$\frac{[K]_o}{[K]_i} = \frac{[Cl]_i}{[Cl]_o} \tag{1}$$

where the brackets signify the concentration of the ion inside (i) or outside (o) the cell.

- The third factor that contributes to distribution of specific ions across the membrane is the Na^+-K^+ pump. This contribution will be discussed later.

The Special Role of Potassium

The resting potential is mainly due to the $[K^+]$ distribution across the membrane.

The potential difference across the membrane arises because the K^+ ions, although attracted into the cell by the electrical force generated by the internal anions, are required to move uphill against their concentration gradient. Hence the internal K^+ concentration does not quite balance the anions present within the cell; the slight excess of the latter is responsible for the internal negativity of the resting membrane potential. The size of this electrical potential is related to the concentrations of K^+ on the two sides of the membrane by the Nernst equation, in which

$$E_K = \frac{RT}{F} \log_e \frac{[K]_o}{[K]_i} \qquad (2)$$

where E_K is the K^+ equilibrium potential, R is the universal gas constant, T is the absolute temperature, and F is the Faraday constant. The situation can be described in a different way by stating that the equilibrium potential is the transmembrane potential that would theoretically balance the charge separation resulting from a concentration gradient for a given ion. If the membrane potential was equal to the equilibrium potential for a given ion, then making the membrane freely permeable to that ion should result in no net change in the concentration gradient. The charge separation associated with the membrane potential would be sufficient to counteract the concentration gradient. For a mammalian muscle at 37° C, we can simplify the equation for equilibrium potential (for K^+) by inserting values of the various constants and transforming the logarithms so that

$$E_K = 61.5 \log_{10} \frac{[K]_o}{[K]_i} \qquad (3)$$

Assuming values of 4.5 and 160 mM, respectively, for external and internal concentrations of K^+, E_K would be –95 mV, which is slightly more negative than the value actually observed for the resting membrane potential. This calculation shows that it would require an electrical potential equal to –95 mV to maintain this concentration difference across the membrane if the membrane was freely permeable to K^+.

Adding Extracellular K^+

The membrane potential can be altered by changing $[K^+]_o$.

An obvious way to test the postulated ionic basis of the resting membrane potential would be to observe the effect of altering $[K^+]_o$, the concentration of K^+ in the fluid surrounding the muscle fiber. In keeping with the Nernst equation, an increase in the external K^+ concentration causes the resting potential to fall (depolarization), while a reduction in concentration induces a rise in potential (hyperpolarization; figure 9.1C). This experiment demonstrates that the membrane potential remains close to the equilibrium potential for K^+. This occurs because the membrane is more permeable to K^+ than to Na^+.

Equilibrium potentials can also be calculated for both Cl^- and Na^+ ions. It is found that the equilibrium potential for Cl^- has a value and polarity very similar to the K^+ equilibrium potential. Further, because the membrane is relatively permeable to Cl^-, one would expect the resting potential to be affected by changes in external concentration of the ion. Hodgkin and Horowicz (1959) were able to show that this was so but that the effect was short-lived because diffusion of Cl^-, K^+, and water subsequently took place until the previous concentration gradient for Cl^- was restored. In effect, the Cl^- ion had been redistributed so as to form a new Donnan equilibrium with K^+. In the case of Na^+, the calculated equilibrium potential has a polarity opposite to the resting membrane potential. The discrepancy indicates that Na^+ ions cannot be contributing significantly toward the resting potential and is consistent with the low permeability of the membrane to Na^+ described previously. It follows that an alteration of the external Na^+ concentration should have little effect on the membrane potential, and experimentally this can be shown to be so.

Tubules Have High K^+ Permeability

So far the membrane of the resting muscle fiber has been discussed as if it were a homogeneous structure, with the permeability of one region similar to that of another. It turns out that this is not so. One of the first indications of such a variation came from the experiments of Hodgkin and Horowicz (1959), already mentioned, in which the ionic composition of the solution bathing single frog muscle fibers was changed abruptly. These authors discovered that whereas alterations in Cl^- concentration produced an immediate alteration in membrane potential, the effects of K^+ were slower, particularly when the external concentration was being reduced. Hodgkin and Horowicz suggested that the reason for the slowness of the K^+ effect was that the membrane permeability for this ion was restricted to the T-tubules and that diffusion into, and out of, these channels would be retarded because of their narrowness.

Eisenberg and Gage (1969) put this hypothesis to direct test by measuring the K^+ and Cl^- permeabilities of muscle fibers after the necks of the T-tubules had been disconnected from the surface plasmalemma by treatment with glycerol. They found that the Cl^- permeability was restricted to the surface but that the K^+ permeability was shared by the surface membrane and the T-system, the permeability of the latter being twice as high as that of the former.

The Goldman-Hodgkin-Katz Equation

The theoretical basis for these observations can also be stated using the constant field concept, originally developed by Goldman (1943) and subsequently explored by Hodgkin and Katz (1949). This concept, which links together ion concentration, membrane permeability, and membrane potential, can be expressed as an equation (the Goldman-Hodgkin-Katz, or GHK, equation). It takes the form

The membrane potential can be predicted from the GHK equation.

$$E_m = \frac{RT}{F} \log_e \frac{P_K[K]_o + P_{Na}[Na]_o + P_{Cl}[Cl]_i}{P_K[K]_i + P_{Na}[Na]_i + P_{Cl}[Cl]_o} \qquad (4)$$

where P denotes the permeability of the membrane for an ion species and E_m is the resting potential. Since at rest P_{Na} is very low, and the effects of changes in Cl^- concentration are transient, the behavior of the resting membrane will be that of a K^+ electrode, and the equation will simplify to equations (2) and (3). Suppose, however, that the permeability of the membrane to Na^+ were to increase as a result of electrical or chemical stimulation. Under these circumstances, the potential would change toward the equilibrium potential for Na^+. If the membrane permeability to Na^+ increased even further, so that the K^+ and Cl^- permeabilities were comparatively small, the membrane potential would approximate to E_{Na}, the Na^+ equilibrium potential. In fact, this is precisely what happens at the summit of the *action potential,* which will be considered later.

Membrane Potential Simulation

The properties of the membrane can be simulated by electrical components.

Another way of analyzing the behavior of the membranes is to depict it as an electrical analogue (figure 9.2). The various equilibrium potentials can be represented as batteries, with the polarity of the Na^+ battery opposite to those of the K^+ and Cl^- cells. The permeability of the membrane for an ion may be symbolized as an electrical resistance because resistance is a measure of the ease, or difficulty, with which that ion will traverse the membrane. A highly permeable membrane would therefore be shown as having a low electrical resistance. The arrow over each resistance in figure 9.2 emphasizes that the resistances may vary depending on whether the membrane is at rest or

OUTSIDE (extracellular fluid)

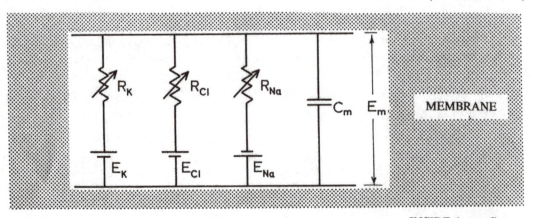

INSIDE (cytosol)

Figure 9.2 Electrical analogue of membrane. E_m = membrane potential; C_m = membrane capacity. E_K, E_{Cl}, and E_{Na} are ionic equilibrium potentials, and R_K, R_{Cl}, and R_{Na} are the corresponding membrane resistances.

Reprinted, by permission, from A.J. McComas, 1977, *Neuromuscular Function and Disorders* (London: Butterworth Publishing Co.), 18.

participating in impulse activity. From the figure it is obvious that if any one resistance becomes very small (permeability very high) in comparison with the others, the potential appearing across the membrane will be that of the corresponding ionic equilibrium potential.

One may determine the net resistance of the membrane to the various ions experimentally by measuring the change in potential produced by passing current through the membrane from an intracellular stimulating electrode. According to Ohm's law, the fiber resistance will be equal to the induced potential divided by stimulating current. In figure 9.3A, the smallest of the three currents, 2.5×10^{-8} A, produced a depolarization of 12 mV (trace a), which would give a value for the resistance of 0.48 MΩ. The resistance measured in this way is the "input" resistance of the fiber; it depends on the resistance of the fiber interior as well as on that of the membrane. One may determine the contribution of the various species of ion toward the membrane resistance by replacing each ion, in turn, with an impermeant substance. For example, by replacing Cl⁻ in the external bathing solution with methylsulfate, Hutter and Noble (1960) were able to show that the resting membranes of frog muscle fibers were twice as permeable to Cl⁻ as to K⁺.

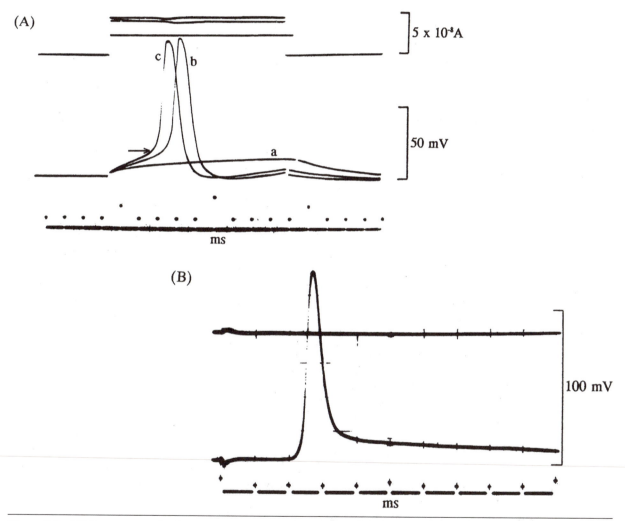

Figure 9.3 (A) Responses evoked in a normal human muscle fiber *in situ,* using intracellular stimulation through a Wheatstone bridge circuit. At the top are shown three rectangular current pulses of different intensities; the middle section displays the corresponding changes in membrane potential. Response a, produced by the smallest current, is subthreshold; but in b and c, the depolarizations are large enough to reach the firing level (arrow) for an action potential. (B) Response of a mouse tibialis anterior muscle fiber *in situ* to stimulation of the sciatic nerve. The action potential reached the recording point 2 ms after the stimulus had been delivered (artifact at left of trace) and had an overshoot of 39 mV; the resting potential was –80 mV. Note the *negative afterpotential* following the spike.

Adapted, by permission, from A.J. McComas et al., 1968, "Electrical properties of muscle fibre membranes in man," *Journal of Neurology, Neurosurgery, and Psychiatry* 31:437.

Membrane capacitance is largely due to the T-tubules.

It should be noted that the electrical analogue shown in figure 9.2 includes the symbol for *capacitance* (C_m). This is required because the lipid membrane of the fiber acts as a dielectric between two conducting media, namely the interstitial fluid and the *cytosol*. Each time a depolarizing current is made to flow across the membrane, the capacitor has to be discharged; hence the change in membrane potential is never instantaneous but takes place exponentially, as in figure 9.3A for the smallest current (response a). Similarly, there is an exponential decline in potential as the stimulus is terminated, caused by the recharging of the membrane capacitance. The time constant of the membrane, τ_m, is a measure of the capacity of the membrane, being equal to the product of the capacity and the membrane resistance. It can be determined experimentally by measuring the time taken for the membrane potential to reach 63.2% of its final value following the onset of the rectangular pulse of current. The capacitance of the mammalian muscle fiber membrane is 4 to 5 $\mu F \cdot cm^{-2}$ of surface membrane (Lipicky *et al.,* 1971) and is much larger than that estimated for nerve fiber membrane (about 1 $\mu F \cdot cm^{-2}$). This increased capacitance of the muscle fiber is due to the substantial invaginations of membrane that form the transverse tubular system (T-system). Convincing proof of this supposition came from the experiments of Gage and Eisenberg (1969b) in which glycerol was used to disrupt the connections of the *T-tubules* to the surface *plasmalemma;* the membrane capacitance was decreased by a factor of almost 3.

Rectification and Negative After-Potentials

The T-tubules are responsible for rectification and for negative after-potentials.

The slow diffusion of K^+ out of the T-tubules is responsible for two other effects. The first of these is the high resistance displayed by the muscle fiber to the movement of K^+ ions outward through the fiber membrane, as when an intracellular stimulating electrode is used to pass current. By discriminating between inward and outward K^+ currents, the fiber is said to act as a rectifier. The second effect of the T-tubules is that, during impulse activity, there is an efflux of K^+ from the muscle fiber into the T-system. Since this K^+ is slow to diffuse away, the concentration gradient for K^+ will be reduced across the tubular membrane, and so the K^+ equilibrium potential, E_K, will be correspondingly smaller (less negative). This, in turn, will cause the membrane potential to fall provided that the permeability of the fiber to K^+ is significantly larger than that to Cl^-. The condition of a high K^+ permeability occurs during and immediately after the spike component of the action potential. Because of the accumulation of K^+ in the tubules, the membrane potential will remain partially depolarized for a few milliseconds, producing a negative after-potential. This potential is more obvious in frog muscle fibers than in mammalian ones (see, however, figure 9.3B), and it is enhanced by cooling or by repetitive impulse activity (because there is a greater accumulation of K^+). Sometimes the negative after-potential becomes so large that it triggers another impulse.

These observations are also relevant to a consideration of the disease phenomenon of peripheral myotonia, in which the fibers are hyperexcitable, discharging long trains of impulses spontaneously and after transient mechanical or electrical stimulation of the fiber membrane. This uncontrolled hyperexcitability results in spontaneous and sustained muscle contractions. It has been shown that in the myotonic fiber the Cl^- permeability is greatly reduced (see chapter 7); hence K^+ in the T-tubules is likely to be effective in reducing the resting potential and thereby bringing the membrane to the threshold for firing spontaneous impulses. A similar situation would be predicted to occur if all the Cl^- were removed from a solution bathing a normal muscle fiber; such fibers do indeed become hyperexcitable and exhibit myotonic-like behavior (Bryant & Morales-Aguilera, 1971).

The Role of the Na$^+$-K$^+$ Pump

The Na$^+$-K$^+$ pump expels Na$^+$ ions that have leaked into the fiber.

Until now the membrane has been treated as if it were an entirely passive structure obeying physico-chemical laws. Suppose, however, that the muscle fiber is allowed to become anoxic or is treated with a chemical, such as cyanide or iodoacetate, that interferes with oxidative metabolism. Under either of these conditions, the resting membrane potential will slowly fall to a low level despite the large differences in ionic concentrations that existed across the membrane at the start of the experiment. The explanation for the slow depolarization involves the small permeability of the membrane to Na^+, noted earlier, and also the loss of the mechanism by which the cell normally counteracts the effects of the Na^+ permeability. These two factors may now be considered in more detail.

Since there is a small permeability of the membrane to Na^+ in the resting condition, Na^+ ions will leak into the muscle fiber down their concentration gradient and under the influence of the strong electrostatic attraction of the internal negativity of the fiber. As they accumulate in the cell, the Na^+ ions will tend to depolarize the membrane because their positive charges will make the fiber interior less negative. This depolarizing tendency is normally balanced by a pump in the cell membrane that returns Na^+ to the interstitial fluid in exchange for K^+. Since the pump consumes *ATP* as its energy source, it can also be regarded as an enzyme. This enzyme is the Na^+-K^+ ATPase (see chapter 7).

For each cycle of the pump, three Na^+ ions are expelled from the fiber and only two K^+ ions are admitted; thus, the pump not only maintains the normal chemical composition of the fiber but also contributes to the internal negativity. This "electrogenic" action of the pump, in turn, will promote the net inward transfer of K^+ ions into the fiber and will thereby maintain the K^+ concentration gradient.

How much potential does the Na^+-K^+ pump generate? Under resting, steady state conditions, the contribution of the pump can be calculated if the ionic concentrations on either side of the membrane are known, together with the relative permeabilities of the membrane for Na^+ and K^+. Thus, according to Mullins and Noda (1963),

$$E_m = \frac{RT}{F} \ln \frac{r[K^+]_o + b[Na^+]_o}{r[K^+]_i + b[Na^+]_i} \tag{5}$$

where E_m is the resting potential, r is 1.5 and is the Na^+:K^+ ion exchange ratio for the pump, and b is the ratio of membrane permeabilities for Na^+ and K^+. This value of E_m, which corresponds to the observed resting potential as measured with a microelectrode, will be rather higher than that calculated by the GHK equation using the same values of ion concentration and permeability. The difference in E_m will be the voltage contributed by the electrogenic Na^+-K^+ pump.

It is now appropriate to substitute real values of ion concentration and resting potential into equation (5) so that the permeability factor, b, can be calculated. With reference to table 9.1, $[K^+]_o = 4$ mM, $[Na^+]_o = 145$ mM, $[K^+]_i = 160$ mM, $[Na^+]_i = 10$ mM, and E_m (the observed resting potential) -85 mV for mammalian muscle of mixed fiber type. At 37° C, the terms

$$\frac{RT}{F} \ln$$

simplify to $61.5 \log_{10}$, so that equation (5) becomes

$$-85 = 61.5 \log_{10} \frac{1.5[4] + b[145]}{1.5[160] + b[10]} \tag{6}$$

and b, the ratio of the membrane permeabilities for Na^+ and K^+, will be 0.03. This value for b can now be inserted into the GHK equation (4). Under steady state conditions (in which the inward ion fluxes are equal to the outward ones), Cl^- can be ignored for the reason given earlier. For the same values of $[Na^+]$ and $[K^+]$ used in equation (6), the calculated resting potential would be -79.4 mV. The difference between the observed and calculated values of E_m is then -5.6 mV; this is the voltage contributed by the electrogenic Na^+-K^+ pump to the resting membrane potential.

During muscle contraction, however, or following the application of depolarizing drugs, the pump may produce as much as -30 mV of internal negativity (Creese *et al.*, 1987; Hicks & McComas, 1989). This is so because the pump activity will increase when intracellular $[Na^+]$ increases. During exercise the electrogenic contribution of the pump maintains the excitability of the muscle fiber membranes and prevents the onset of K^+-induced depolarization block (see chapter 15).

The Action Potential

The ability of their membranes to develop transient changes in potential, which can be transmitted from one point to another, confers upon nerve and muscle fibers the special property of excitability. The propagated change of membrane potential is the action potential, or impulse.

The Shape of the Action Potential

In order to record the full amplitude of the impulse, it is necessary to measure, during excitation, the potential that exists between an electrode in the cytosol of the fiber and an external electrode. In figure 9.3A, the same electrode was used simultaneously to stimulate and record from a human muscle fiber, and a Wheatstone bridge circuit was employed to nullify the stimulus artifact. It can be seen that the smallest rectangular pulse of current produced only a small transient depolarization of the membrane (A). In contrast, each of the two larger pulses evoked impulses that included reversals of membrane potential (B, C), such that the inside of the fiber became 13 mV positive with respect to the outside for about 1 ms.

One of the advantages of the intracellular stimulation technique is that it enables measurement of the critical amount of membrane depolarization required to trigger the action potential mechanism. From figure 9.3A it can be seen that the 12-mV depolarization induced by the weakest stimulus was insufficient, whereas depolarizations of 14 mV, produced by the two larger stimuli, were adequate. In one case, the onset of the action potential from the evoked depolarization has been indicated by an arrow. Figure 9.3B shows, for comparison, the action potential set up in a muscle fiber following electrical stimulation of its motor axon; the transient reversal of membrane potential and the negative after-potential following the spike are both clearly shown.

The Mechanism of the Action Potential

Like the resting membrane potential, the action potential has an ionic basis, depending on the permeability of the membrane to the ions on its inner and outer surfaces, as well as upon the concentrations of those ions. Many years ago it was realized that Na^+ ions had an important role, because nerves and muscles bathed in Na^+-free solutions lost their excitability.

The key to our present understanding of the ionic mechanisms involved in the action potential is the voltage clamp experiments of Hodgkin and Huxley (1952a, 1952b). These authors chose to study the giant axon of the squid because its large diameter, from 0.5 to 1.0 mm, permitted them to insert both a stimulating and a recording electrode into the axoplasm; the stimulating and recording circuits were completed by two external electrodes in the artificial seawater bathing the fiber (figure 9.4A; see also "The Squid and Its Giant Axon"). Their technique was to rap-

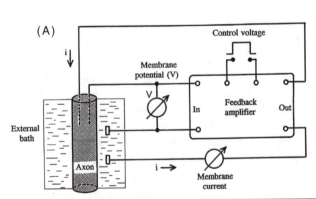

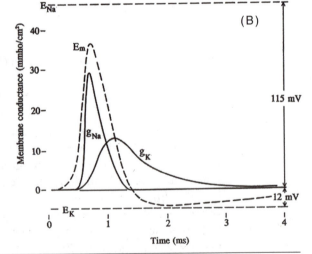

Figure 9.4   (A) Experimental arrangement used in the voltage clamp studies of Hodgkin and Huxley (1952b, 1952a) on the squid giant axon. The potential between an intracellular and an extracellular electrode is measured (V) and compared with a controlling voltage; any difference is automatically eliminated by the passage of a current (i) through another pair of electrodes situated on either side of the membrane. (B) Time courses of the ionic permeability changes during an action potential as determined by the voltage clamp experiment. The permeabilities of the membrane to Na^+ and K^+ are indicated by the corresponding conductances (g); E_m is the action potential recorded when the clamp was no longer applied. (Since the recordings were made from an axon, in which the T-system is not present, there is no negative after-potential; the hyperpolarization recorded instead is a consequence of the fact that the potassium equilibrium potential, E_K, may be larger than the resting membrane potential if the fiber is not in an optimal condition.)

A: From B. Katz, 1966, *Nerve, muscle and synapse* (New York: McGraw-Hill Companies), 82. Adapted with permission from The McGraw-Hill Companies.
B: Adapted, by permission, from A.L. Hodgkin and A.F. Huxley, 1952, "A quantitative description of membrane current and its application to conduction and excitation in nerve," *The Physiological Society* 117: 530.

idly change the potential across the membrane to any desired level by passing current between the stimulating electrodes. By using a feedback circuit from the recording electrodes, they were able to keep the membrane potential steady at its new level. The amount of current necessary to clamp the membrane in this way was equal to, and opposite in sign to, the ionic current that would have normally been flowing through the membrane. Hodgkin and Huxley then determined the contribution of Na$^+$ ions to the total current by substituting *choline,* an impermeant cation, in the bathing solution. They were able to show that as the membrane potential was reduced, there was an initial flow of Na$^+$ ions into the cell, and that this was followed by a flow of K$^+$ ions in the opposite (outward) direction.

The Squid and Its Giant Axon

It was the eminent British zoologist J. Z. Young (1936) who first drew attention to the possible usefulness of squid (figure 9.5A) studies of the nerve impulse. These creatures, like certain other invertebrates, have a few unusually large axons that are essential for the rapid transmission of impulses in predatory or escape reactions. Figure 9.5D shows a longitudinal section through *Loligo pealei,* the

common squid of the coast of North America. The animal is able to swim by undulating its tail fins but, if necessary, can move more rapidly by jet propulsion. In the latter mode, seawater is first drawn into the cavity of the mantle through the siphon and is then expelled rapidly through the same aperture by a sudden contraction of the mantle muscles. The synchronous timing of the contraction depends on

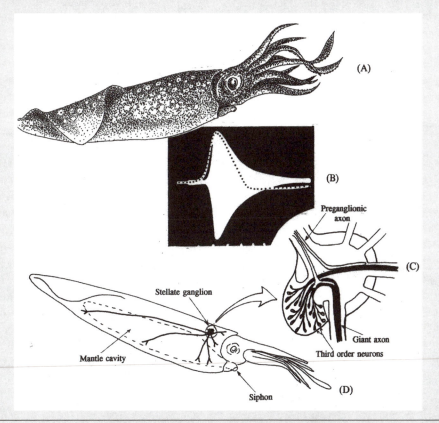

Figure 9.5   The common squid and its giant axons. (A) An example of *Loligo pealei;* (D) drawing is based on the supposition that one of these animals has been cut longitudinally so as to reveal the innervation of the muscles that supply the mantle cavity. (C) An enlargement of the stellate ganglion, showing that each giant axon is formed by the fusion of many smaller fibers belonging to third-order neurons. (B) Change in membrane conductance (white envelope), measured by a bridge circuit during the passage of an action potential (dotted line). White dots at bottom are milliseconds.

Reprinted, by permission, from K.S. Cole and H.J. Curtis, 1939, "Electric impedance of the squid giant axon during activity," *Journal of General Physiology* 22: 655. Copyright permission of the Rockefeller University Press.

(continued)

the rapid transmission of impulses from the stellate ganglion along the giant axons on each side (figure 9.5D). Each axon is a true neural syncytium, formed by the fusion of several hundred smaller axons as indicated in figure 9.5C. Whereas the largest vertebrate axons have diameters of 20 μm or so, which include the *myelin* sheaths, the squid giant axons are 0.5 to 1.0 mm in width.

Using the squid while working at Woods Hole on Cape Cod, Curtis and Cole (1938) applied surface electrodes and an oscillating current to demonstrate the enormous increase in membrane permeability (conductance) during the action potential (figure 9.5B). The usefulness of the squid axon was quickly appreciated by Alan Hodgkin during a visit to Woods Hole in 1938 and was later exploited with Andrew Huxley in a series of brilliant experiments at the Marine Biological Station, Plymouth, United Kingdom. These experiments started in 1939, while Huxley was still an undergraduate student, and were interrupted by the Second World War. Hodgkin's account of this

work (Hodgkin, 1977) describes the first insertion of an electrode down the inside of the axon and the significant early finding that the action potential exceeded the resting potential.

After the war, Hodgkin and Huxley adopted the idea of Cole and Marmont for clamping the membrane of the squid axon at any given potential with a feedback amplifier and measuring the current required (figure 9.4A). They were then able to plot the current–voltage relationship for the membrane and to determine its conductance. The results were fitted by a series of differential equations, which Huxley solved numerically with a hand-operated calculating machine. It is said that not a single error was made in the many thousand steps in the calculations! This pioneering work put the ionic basis of the resting and action potentials on a firm quantitative basis and, for the first time, encouraged ideas as to how ion channels might open and close. For their work on the squid giant axon, Hodgkin and Huxley were awarded the Nobel Prize in 1963, sharing it with John Eccles.

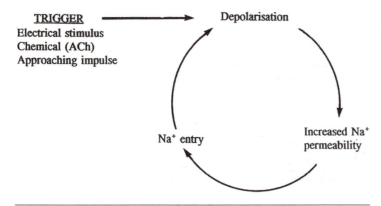

Figure 9.6   The regenerative nature of the increase in Na$^+$ permeability of the membrane.

Reprinted, by permission, from A.J. McComas, 1977, *Neuromuscular Function and Disorders* (London: Butterworth Publishing Co.), 23.

It is obvious that the membrane permeability must have changed during the action potential, since under resting conditions the membrane is largely impermeable to Na$^+$. However, during the action potential, not only does the Na$^+$ permeability increase, but the increase becomes self-regenerative. The explanation for this phenomenon is that as the membrane becomes permeable to Na$^+$, these ions will diffuse into the fiber down their concentration gradient and also under the attraction of the internal negativity of the cell. However, the entry of Na$^+$ ions will depolarize the membrane further because, being positively charged, the ions make the inside of the cell less negative. In turn, the further depolarization causes the membrane to become even more permeable to Na$^+$, and the cycle is repeated (figure 9.6; see also figure 7.3 in chapter 7).

K$^+$ Permeability of the Membrane

The action potential ends by a rise in the K$^+$ permeability of the membrane.

From a consideration of the electrical analogue of the membrane (figure 9.2) and of the GHK field equation (see p. 122), it is apparent that a membrane that is much more permeable to Na$^+$ than to K$^+$ or Cl$^-$ should have a potential across it equal to the Na$^+$ equilibrium potential. The subsequent return of the membrane potential to its resting value is due to two additional properties of the membrane; first, the Na$^+$ permeability mechanism is soon switched off (inactivation), and second, the membrane increases its permeability to K$^+$ as that to Na$^+$ declines. The time courses of these changes are shown in figure 9.4B, in which the Na$^+$ and K$^+$ permeabilities have been expressed as conductances. It is the increase in K^+ permeability that is responsible for the delayed outward current recorded in the voltage clamp experiment. In the normal unclamped fiber, the increased K^+ permeability serves to restore the membrane potential to its resting level.

At the end of an action potential, the fiber will have gained Na^+ and lost some K^+. The quantities are so small, however, that a motor nerve axon or muscle fiber can conduct several thousand action potentials without difficulty. In a thin muscle fiber or axon, the situation is rather different, for appreciable changes in internal Na^+ concentration occur after fewer impulses. During and following periods of impulse activity, the changes in ionic composition of the fiber are corrected by means of the Na^+-K^+ pump; the pump actively expels Na^+ and has been described in chapter 7.

Experiments with squid axon sheaths confirm the ionic basis of the membrane potentials.

Following its formulation, the ionic hypothesis of the resting and action potentials was verified in a strikingly elegant series of experiments by Baker *et al.* (1962b, 1962a). These authors were able to extrude the axoplasm from a squid giant axon and to fill the membrane left behind with solutions of any desired composition. By altering the constituents of the extracellular fluid as well, they could alter the resting potential at will. For example, if the external concentration of K^+ was higher than the internal one, the resting membrane potential reversed its sign and became positive inside the fiber, in agreement with predictions from the Nernst equation.

Special Membrane Channels

During the action potential, Na^+ ions move rapidly through special membrane channels.

Once the changes in Na^+ and K^+ permeability during the action potential had been recognized, a major problem was to understand the nature of the underlying molecular events in the membrane. Initially it was thought that the Na^+ and K^+ ions might be transported by "carrier" molecules that would combine with the respective ions, transfer them across the membrane, and then release them. This concept was challenged when estimates were made of the densities of the different types of channels in excitable membranes.

In the case of the Na^+ channel, use was made of two biological toxins that combine with the channel and inactivate it, *tetrodotoxin (TTX)* and saxitoxin (*STX;* see chapter 7). By using very dilute solutions of labeled TTX or STX, it has been possible to estimate the density of the Na^+ channels in excitable membranes. It appears that the channels are remarkably sparse in muscle fibers and nonmyelinated nerve fibers; for example, in lobster axons, there are probably fewer than 13 Na^+ channels $\cdot$ μm^{-2} of membrane (Moore *et al.*, 1967), while for rat diaphragm, the corresponding figure is about 21 channels (Colquhoun *et al.*, 1974). Such low densities imply that during the impulse, Na^+ ions must flow through each channel at a very high rate, probably in the region of several million ions per second (Keynes *et al.*, 1971).

This flow is far too high to be accounted for by the activation of carrier molecules, and it was necessary to postulate instead that Na^+ ions enter through special water-filled pores in the membrane. Hodgkin and Huxley (1952a, 1952b) suggested that each channel would have two components; one part would sense the potential difference across the cell membrane and would control access to the second part, now known to be the critically narrowed pore. Hodgkin and Huxley also predicted that the conformational change in the sensing part of the channel would give rise to a small current, the *gating current.*

Patch Clamp and Molecular Biology

Ion channels are studied by patch clamping and by molecular biology techniques.

Further insights into the nature of ion channels came from the introduction of *patch clamping* by Neher and Sakmann (1976a), both of whom were awarded the Nobel Prize in 1991. These authors reported that it was possible to record the currents flowing through single ion channels once a small piece of cell membrane had been tightly sealed to the tip of a glass microelectrode. Then abrupt onsets and terminations of the currents were attributed to the openings and closings of the ion channels; the frequency and durations of the opening could be controlled by applying different voltages across the patch of membrane or by exposing the latter to selected drugs or neurotoxins.

The latest approach to the understanding of ion channels has been to deduce their three-dimensional structures and amino acid sequences using molecular biology techniques. This approach was first applied successfully to the Na^+ channels of the electroplax organ of the electric eel (Noda *et al.*, 1984) and subsequently extended to Na^+ channels in skeletal muscle and rat brain, as well

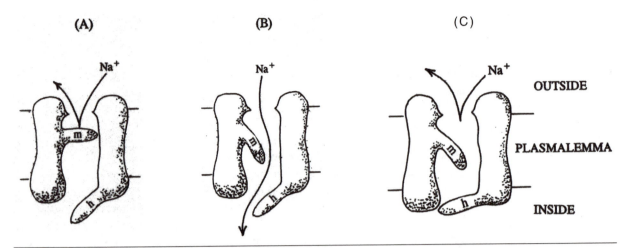

Figure 9.7   Gating mechanisms of the Na$^+$ channel. (A) At rest, the m gate is shut, barring Na$^+$ ions from the channel. (B) Depolarization of the membrane opens the m gate, allowing Na$^+$ ions to pass through the central pore of the channel. (C) After a short interval (1-2 ms), the same depolarization inactivates the channel by closing the h gate.

as to K$^+$ and Ca^{2+} channels (see chapter 7). It is now known that each Na$^+$ channel consists of four identical domains and that each of the domains contributes to the formation of the central pore. In turn, each domain is composed of six regions that span the membrane; one of these regions acts as the gating mechanism (see p. 88), and another forms part of the pore (see figure 7.4, chapter 7). In such a model, the channel will open and admit Na$^+$ ions when gating movements have taken place in all four domains. These domains collectively form the m or activation gate. However, the same depolarization that opens the m gate and allows Na$^+$ ions through the channel will also initiate closure of the channel. This closing, or inactivation, is brought about by a second type of gate, the h gate. Thus, Na$^+$ ions can flow across the membrane only when the m and h gates are both open (figure 9.7).

Inability of sufficient Na$^+$ channels to open causes refractoriness.

The Refractory Period

The Na$^+$ permeability system has added significance in that it underlies the phenomena of the *refractory period* and *accommodation*. The refractory period of a nerve or muscle fiber is the time following an impulse during which there is a depression of membrane excitability. At first the fiber is absolutely refractory and cannot respond with an impulse to a test stimulus, no matter how large the stimulus is. This phase is succeeded by the *relative refractory period* in which the fiber is able to fire an action potential provided that the stimulus intensity has been increased beyond its initial threshold value. In humans, the *absolute refractory periods* of muscle fibers vary from 2.2 to 4.6 ms (Farmer *et al.,* 1960), and that for the largest motor axons is about 0.5 ms. The explanation of the absolute refractory period is that the Na$^+$ channels can no longer function because the h gates have closed; that is, inactivation has taken place. However, in the relative refractory period, some, but not all, of the Na$^+$ channels have recovered and are available for participation in a further action potential. Sometimes the repolarization of an action potential overshoots the resting potential, producing an *after-hyperpolarization;* this occurs when the K$^+$ permeability is still increased so that the membrane potential moves closer to E$_K$, the potassium equilibrium potential. This phenomenon is another factor contributing to relative refractoriness.

A membrane is said to display accommodation if it fails to fire an action potential in response to a slowly rising current pulse, even though a rapidly increasing pulse is effective. In terms of the Na$^+$ permeability mechanism, one can envisage that during the slowly rising pulse, some channels are already inactivating while others have yet to be opened; in addition, the K$^+$ channels will already be opening. During the rapidly rising pulse, more Na$^+$ channels will be opened initially and, through the regenerative action of the resulting depolarization, will be more effective in causing others to open.

Action Potential Propagation

So far only the events that take place locally in the nerve or muscle fiber membrane have been considered. The functional significance of the action potential mechanism is that, once an impulse has been initiated, excitatory changes automatically take place in neighboring regions of the membrane and so cause the impulse to travel the length of the fiber. This spread of excitation depends on the difference in potential that exists between the region of membrane at which the action potential is momentarily located and more distant regions that are still in the resting state. At the crest of the action potential, the interior of the fiber is some 30 mV positive with respect to the exterior, while farther along, the fiber will exhibit the normal resting membrane potential such that the inside of the fiber is some 85 mV negative to the outside. Because of this difference in potential, current will flow between the two regions of membrane. The resting membrane is said to act as a source of current for a sink at the site of the action potential. This current comes from the discharge of the membrane capacity in the resting membrane. It is usual to state the direction of a current by indicating the movement of the positive charges; in this situation, cations will flow inward through the membrane at the sink. As the resting (source) membrane discharges, the potential will fall to the critical depolarization necessary for local impulse generation; in this way, the excitatory disturbance is transmitted along the fiber.

If a segment of nerve or muscle membrane is depolarized to threshold, an action potential will occur, and it will propagate to the end of the nerve or muscle fiber. The strength of the signal (electrical, chemical, or mechanical) that causes the initial depolarization will not affect the action potential, but will determine only whether or not an action potential, all-or-none response, will occur. The situation can be likened to lighting a string of firecrackers with a lighter that has a variable flame height. As long as the fuse to the firecrackers gets ignited, the firecrackers will go off. This response is independent of the size of the flame that initiated the event.

The term "all-or-none response" should not be construed to indicate that the action potential will always have the same shape (amplitude and duration). Minor alterations in ion concentration can modify the shape of the action potential, but under a given set of circumstances the action potential will either occur or not occur.

Saltatory Conduction

In nerve fibers, a modification of structure enables the action potential to be conducted much more efficiently than in muscle fibers. This modification is the incorporation of a special lipid insulating layer around the nerve fiber membrane; the extra coat is the myelin sheath and is formed by the *Schwann cells* (see chapter 2). At regular intervals, the myelin sheath is interrupted by the *nodes of Ranvier* at which the nerve membrane, the *axolemma,* is exposed to the extracellular fluid and *gap substance.* The effect of the myelin sheath is to prevent current from leaking across the axolemma; that is, R_e in figure 9.8 becomes very large in relation to the other resistances in the circuit. Thus,

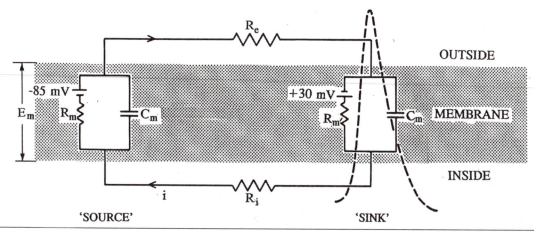

Figure 9.8 Electrical analogue of the membrane during the passage of an action potential.

Reprinted, by permission, from A.J. McComas, 1977, *Neuromuscular Function and Disorders* (London: Butterworth Publishing Co.), 25.

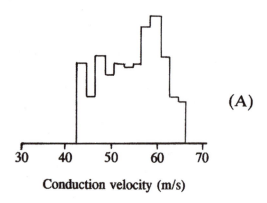

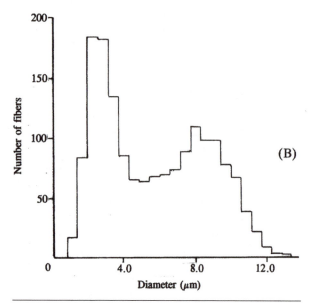

Figure 9.9 (A) Spectrum of impulse conduction velocities in motor axons of the median nerve in a single subject, as derived by mathematical decomposition. (B) Diameters of myelinated sensory and motor axons in the recurrent branch of the median nerve to the thenar muscles.

A: Adapted, by permission, from K.L. Cummins, L.J. Dorfman, and D.H. Perkel, 1979, "Nerve fiber conduction-velocity distributions. II. Estimation based on two compound action potentials," *Electroencephalography and Clinical Neurophysiology* 46: 653; B: Adapted, by permission, from R.G. Lee et al., 1975, "Analysis of motor conduction velocity in the human median nerve by computer simulation of compound muscle action potentials," *Electrocephalography Clinical Neurophysiology* 39: 230.

as the action potential advances, only the axolemma at the nodes of Ranvier will be able to depolarize to the levels necessary for impulse initiation. In this way, the action potential jumps from node to node, skipping the intervening myelinated segments of axon.

This form of propagation is termed *saltatory conduction,* and it enables the impulse to travel much faster than in nonmyelinated structures. Saltatory conduction has the additional advantage of restricting the ionic exchange during the impulse to the nodes of Ranvier. As a result, the fiber expends less energy in pumping out Na^+ in exchange for K^+. In addition, nerve conduction velocity is much faster in a myelinated nerve than a nonmyelinated nerve of the same diameter. The reason for this faster conduction is usually given as "because the action potential jumps from node to node," but this explanation is incomplete. This faster passive spread of current may relate to a reduced capacitance of the myelinated membrane, which would reduce the time constant (τ) for discharging the membrane. Not all mammalian nerve fibers are myelinated; in fact, both cutaneous and muscle nerves contain more nonmyelinated fibers than myelinated ones. The nonmyelinated fibers (C-fibers) are thin, most less than 1 μm in diameter, and they are arranged in small colonies wrapped by individual Schwann cells. However, as far as muscle is concerned, all the nerve axons supplying the *extrafusal muscle fibers* have myelin sheaths. In humans, the largest of these axons have diameters of about 13 μm, including the thicknesses of the myelin sheaths (see figure 9.9B); thicker axons are found in some of the lower mammals.

Applied Physiology

In this section, we will see how impulse conduction can be studied in human nerves and will look at the effects of damage to the myelin sheath on conduction. Some of the methods for measuring impulse conduction in muscle fibers will also be described.

Impulse Conduction in Human Nerves

One can readily determine impulse velocity in the motor axons by successively stimulating a nerve at two points and measuring the time elapsing before the muscle fibers are excited on each occasion. The difference between these times is then divided into the extra distance that the impulse had to travel when the stimulating electrodes were applied to the nerve farthest from the muscle (figure 9.10). The impulse conduction velocities are higher in the upper limb than in the lower limb, about 50 to 65 m · s^{-1} for the median and ulnar nerves in the forearm and 40 to 55 m · s^{-1} for the tibial and peroneal nerves below the knee. These values are the velocities in the fastest-conducting axons only, since the time measurements are made from the earliest responses in the muscle.

To measure the impulse velocities in the slowest-conducting fibers requires a special technique such as the use of a second stimulus, given at a different point over the nerve and at varying time

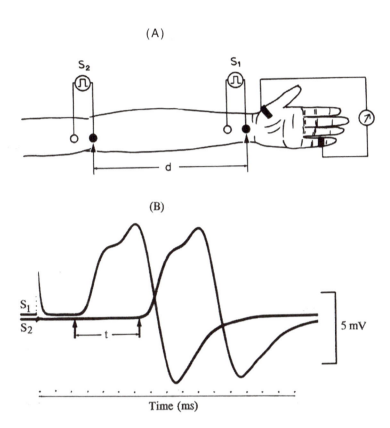

(A)

(B)

Time (ms)

Figure 9.10 Determination of the maximum impulse conduction velocities in motor axons of the median nerve. (A) The nerve is maximally stimulated separately at two sites by the electrode pairs S_1 (wrist) and S_2 (elbow). The corresponding muscle responses are recorded by a stigmatic electrode over the thenar muscles and an indifferent electrode on the little finger. A ground electrode is placed on the dorsum of the hand. (B) The distance, d, separating the cathodal electrodes (ˉ) at the wrist and elbow, respectively, is divided by the difference, t, between the latencies of the two evoked muscle potentials; the result is the impulse velocity of the fastest-conducting motor axons. In this example, d = 25 cm and t = 4.2 ms, giving a conduction velocity of approximately $60 \text{ m} \cdot \text{s}^{-1}$.

B: Reprinted, by permission, from A.J. McComas, 1977, *Neuromuscular Function and Disorders* (London: Butterworth Publishing Co.), 25.

intervals, to produce a series of collisions between the two volleys of nerve impulses (Hopf, 1962, 1963). Other techniques for measuring the full range of axonal impulse conduction velocities are based on F-wave latencies (Doherty *et al.*, 1994) and mathematical decomposition of compound action potentials (Cummins *et al.*, 1979; Lee *et al.*, 1975; figure 9.9A). Since impulse velocities are proportional to nerve fiber diameters, the velocity measurements can be used to calculate the distribution of axon sizes in the motor nerve trunk. For example, in cat peripheral nerves, the conduction velocity, in meters per second, is equal to 6.0 times the myelinated fiber diameter, in micrometers (Hursch, 1939). In the baboon, the corresponding factor is 5.3 (McLeod & Wray, 1967).

Demyelinating Lesions

Reduced conduction velocities are found in patients with demyelinating lesions.

Measurements of impulse conduction velocity are invaluable in the *EMG* clinic, where they can be used to detect injury, degeneration, or inflammation of nerve fibers. Each of these agents interferes with impulse conduction by destroying part, or all, of the myelin sheaths covering the *axis cylinders*. The successive stages of myelin destruction are illustrated schematically in figure 9.11 (top), with partially demyelinated motor axons in a human muscle biopsy shown below.

The effect of myelin loss *(demyelination)* is to make impulse conduction less efficient (see the following section), and an example of this phenomenon in the EMG clinic is given in figure 16.8A (chapter 16). This figure may be compared with the normal results in figure 9.10; the former shows the greatly delayed response in the small muscles of the hand following stimulation of the median nerve at the elbow.

Experimental Diphtheritic Neuropathy

Internodal conduction is slowed in experimental diphtheritic neuropathy.

Figure 9.12 illustrates the effect of demyelination in more detail and is taken from the work of Rasminsky and Sears (1972). These authors devised a special electrophysiological technique that enabled them to record action potential currents at successive points along the outside of single *ventral root* fibers (figure 9.12A). In the normal animal, the latency of the recorded current increased

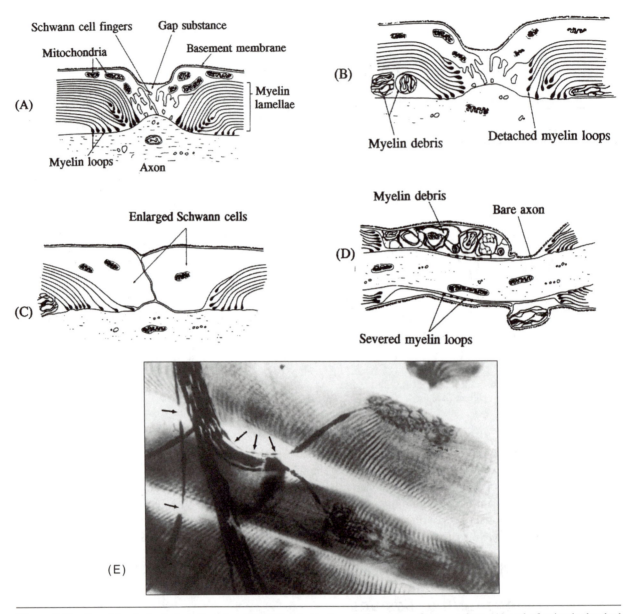

Figure 9.11   Top: Morphological studies of nodes of Ranvier in the sciatic nerves of a normal rat (A) and of animals that had received intraneural injections of diphtheria toxin. (B) An early change, detectable at 3 days, is disorganization of the myelin loops; (C) by 7 days the loops have become detached from the axolemma at the node, and swollen Schwann cells take their places. (D) At 10 days, much more of the axolemma has become denuded of myelin; myelin debris can be seen in the cytoplasm of the Schwann cells. (E) Demyelination in human nerve fibers. The preparation, a "squash," shows parts of three muscle fibers and a bundle of motor axons, most of which have various degrees of demyelination (arrows). The neuromuscular junction in the uppermost fiber is clearly depicted.

A-D: Reprinted from *Brain,* vol. 92, G. Allt and J.B. Cavanagh, Ultrastructural changes in the region of the node of ranvier in the rat caused by diptheria toxin, pgs. 459-468, Copyright 1969, with permission from Elsevier; E: Courtesy of Dr. D. F. Harriman.

in definite steps along the axon, with each step corresponding to the location of a node (figure 9.12, B & C). These findings were, of course, to be expected if saltatory conduction was occurring. In nerve fibers treated with diphtheria toxin, discrete increments of latency were also observed with increasing distance along the axon. In contrast to the findings in the normal animal, in which the conduction time from one node to the next (approximately 1 mm) was about 20 μs, values as large as 300 μs were observed in the demyelinated preparation (figure 9.12D).

Why is conduction slowed in these poorly myelinated axons? The obvious explanation depends on the fact that an affected axon becomes a less efficient cable. The loss of myelin insulation reduces the electrical resistance between the inside and the outside of the axon, and the thinner layer of

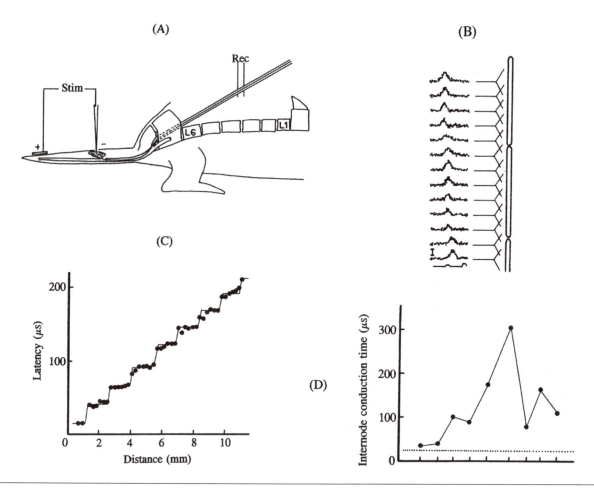

Figure 9.12 Studies of impulse propagation in the normal rat and in an animal treated with diphtheria toxin. (A) The stimulating electrodes (Stim) are shown applied to peripheral nerve fibers in the tail. The recording electrodes (Rec) consist of two fine wires mounted 400 to 600 μm apart, which are moved to different positions along the course of a ventral root fiber. (B) The external currents recorded during passage of an impulse, together with the positions of the recording electrodes in relation to the nodes of Ranvier. (C) The increments in the latency of the action current as the electrodes are moved; note their regularity. Each step corresponds to a node of Ranvier. (D) The results for an axon with partial demyelination. Notice the large and irregular values for *internode* conduction time compared with the average values for a normal axon (......).

Adapted, by permission, from M. Rasminsky and T.A. Sears, 1972, "Internodal conduction in undissected demyelinated nerve fibres," *Journal of Physiology* 227: 323-350.

myelin remaining causes the capacitance to increase. The decreased resistance will allow a larger part of the outwardly flowing action potential current to leak through the internodal region of the axolemma, and the increased capacitance will store more of the charge carried by the current. In both instances, current is expended in the internode instead of being concentrated at the next node. Hence the current must flow for a longer time before the critical depolarization is reached at the node and allows an impulse to be initiated.

What happens if an axon becomes completely demyelinated? In the large-diameter axons, impulse conduction is usually blocked, as is evident in figure 16.8A (chapter 16). In smaller-diameter axons, the density of Na^+ channels in the internodal axon membrane, although much smaller than that at the node, may sometimes be sufficient to allow impulse conduction—at a greatly reduced rate, however (Bostock & Sears, 1978; Black *et al.*, 1990).

Action potentials propagate more slowly in muscle fibers than in motor axons.

Muscle Fiber Conduction Velocity

Although not used in the EMG clinic, impulse conduction velocity measurements are easily made in skeletal muscle fibers. For example, the muscle can be stimulated directly or through its nerve supply, and the arrival of the action potentials can be timed at different points along the fibers

(A)

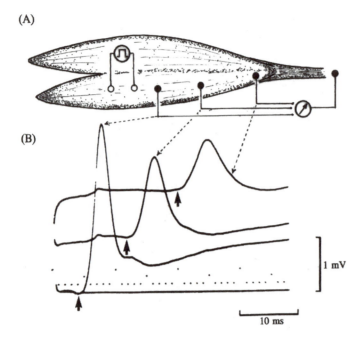

(B)

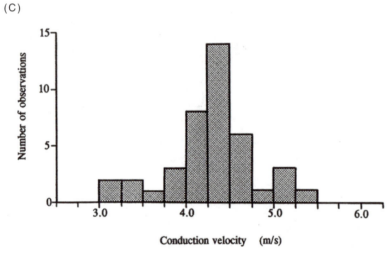

(C)

(Buchthal *et al.,* 1955; Kereshi *et al.,* 1983; figure 9.13). Alternatively, the subject is asked to make a small voluntary contraction, and the passage of an action potential down a muscle fiber can be detected at successive leadoff surfaces in a special recording needle (Stålberg, 1966). Finally, the same *motor unit* potential can be recognized at different sites in a muscle during weak effort (Nishizono *et al.,* 1979). All these methods indicate that there is a range of velocities, two of the mean values for the brachial biceps muscle being 4.02 ± 0.13 m · s^{-1} (Buchthal *et al.,* 1955) and 3.37 ± 0.67 m · s^{-1} (Stålberg, 1966). In a large human muscle such as the tibialis anterior, the fibers with the highest impulse conduction velocities belong to the motor units that have the fastest *twitches* and develop the greatest forces (Andreassen & Arendt-Neilsen, 1987). In *Duchenne muscular dystrophy,* in which abnormally large and small muscle fibers are present, one would expect the range of conduction velocities to be increased. Surprisingly, no abnormality was found by Buchthal *et al.* (1960), although previous work from the same laboratory had indicated that the atrophied fibers of denervated muscle may have 50% to 75% reductions in velocity (Buchthal & Rosenfalck, 1958).

In the next chapter, we will see how an action potential in a motor axon can provoke a similar signal in a muscle fiber. The task is a formidable one, for the nerve fiber is many times thinner than the muscle fiber and cannot provide enough current to excite the muscle fiber directly. In evolution the problem has been solved by the introduction of a chemical amplifier; the chemical is *acetylcholine.*

Figure 9.13 Measurement of impulse conduction velocities in muscle fibers of human biceps brachii. (A) Arrangement of stimulating and recording electrodes on skin overlying muscle. (B) Muscle compound action potential *(M-wave)* recorded at different distances from stimulating electrodes. Large arrows indicate response onsets of fastest-conducting muscle fibers. (C) Values for fastest-conducting fibers in 41 muscles.

10

Neuromuscular Transmission

The structure of the neuromuscular junction was described in chapter 3 (see figure 3.1), and an account of the ionic mechanisms underlying the resting and *action potentials* was given in chapter 9. Using this framework and additional experimental observations on the junction itself, we can describe how excitation spreads from the motor axon to the muscle fiber. In this excitation sequence (see figure 10.1), the *impulse* invades the motor nerve terminal and allows Ca^{2+} ions to enter; this triggers the release of *acetylcholine (ACh)* from the synaptic vesicles into the *synaptic cleft*. The ACh must be released quickly and only for a brief time to maintain the ratio of one *motoneuron* action potential to one muscle fiber action potential. The ACh diffuses across the synaptic cleft and combines with receptors in the muscle fiber membrane; this combination alters the *permeability* of the membrane, causing it to depolarize and fire an impulse that is propagated along its entire length. The presence of *acetylcholinesterase* in the synaptic cleft assures that ACh works for only a brief period of time before it is hydrolyzed to *choline* and acetate.

Acetylcholine Release

The role of ACh is to act as a chemical intermediary in the transfer of excitation from nerve to muscle. The need for such an intermediary is clear, for it serves to amplify the small impulse current in the motor nerve terminal into one that is sufficiently large to trigger an impulse in the muscle fiber. Processes in the nerve terminal are responsible for synthesizing the ACh and packaging it into vesicles, for releasing ACh by co-coordinated docking and fusion of the vesicles with

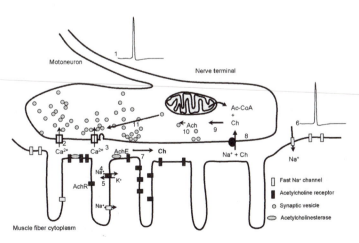

Figure 10.1 Sequence of neuromuscular transmission. (1) action potential invades nerve terminal; (2) Ca^{2+} channels open, local increase in $[Ca^{2+}]$; (3) docked vesicles fuse with nerve terminal membrane; (4) ACh diffuses and into secondary clefts, binding with acetylcholine receptor (AChR), and opening the cation-specific pore in this receptor; (5) Na^+ and K^+ pass through the pore, depolarizing the *end-plate;* (6) when threshold is reached, voltage-gated Na^+ channels open, allowing additional Na^+ to enter the muscle cell; a propagated action potential is initiated at the edge of the end-plate region; (7) acetylcholinesterase (AChE) hydrolyzes ACh to acetate and choline; (8) choline is taken up into the nerve terminal via choline transporter; (9) *choline acetyltransferase* synthesizes ACh by combining choline with acetyl CoA from the mitochondria; (10) ACh is taken up by vesicles via vesicle acetylcholine transporter; (11) filled vesicles move to ready position near active zone.

the nerve terminal membrane, and for recycling the ACh for subsequent release. Processes at the muscle end-plate are responsible for detecting the released ACh and transforming that signal into a propagated action potential. Each of the steps in the transmission process is described in this section in more detail; the account is particularly satisfying because the structural basis of most of the steps is understood. However, there are still some details to be worked out.

Synthesis and Packaging of Acetylcholine

Acetyl-choline is synthesized in the nerve ending by the acetylation of choline.

The synthesis of ACh is undertaken by the enzyme choline acetyltransferase, which, as its name implies, transfers an acetyl group from *acetyl-coenzyme A (acetyl-CoA)* to choline. The acetyl-CoA is produced in the mitochondria of the nerve terminal while the choline acetyltransferase will have been carried to the terminal from the motoneuron *soma* (Jablecki & Brimijoin, 1974). The axon terminal obtains its supply of choline by uptake from the extracellular fluid; some of the choline is derived from the plasma and the rest from the local hydrolysis of ACh by the enzyme acetylcho-linesterase (AChE) in the *basement membrane*. The choline uptake is mediated by a specialized transporter and can be inhibited by the substance *hemicholinium (HC-3)*. The most widely held view is that the synthesis of ACh occurs in the axoplasm and that the transmitter is then taken up and concentrated by the synaptic vesicles, again using a specialized transporter molecule, *vesicular ACh transporter*. The rate of synthesis is geared to the amount of impulse activity in the terminal, but a small amount of synthesis continues even in the resting state. Inhibition of choline transport across the *axolemma* by HC-3 results in decreased amounts of ACh in the synaptic vesicles. Since the amount of ACh per vesicle is reasonably constant, each vesicle content is referred to as a "quantum." In this case there is decreased *quantal size*.

Acetylcholine Is Packaged Into Synaptic Vesicles for Release

Experiments using repetitive nerve stimulation have indicated that the ACh can be visualized as existing in three compartments in the nerve terminal. The first of these comprises the ACh that is readily available for release; electron micrographs suggest that this corresponds to the synaptic vesicles found bordering the synaptic region of the axolemma (figure 3.1, chapter 3). Some of the vesicles are seen to have already docked with the axolemma, and others are in close proximity to it. The latter vesicles appear to be attached to filaments and are localized opposite the *secondary synaptic clefts*. The second compartment is the large remaining population of vesicles found more centrally within the terminal; these form the main store of ACh. Finally, there is an appreciable amount of ACh within the *cytoplasm* of the terminal; some of this ACh enters the synaptic cleft in the resting state and is responsible for the electrical "noise" that can be recorded at the end-plate. If the depolarizing action of this ACh is blocked by curare (see later), then the end-plate membrane hyperpolarizes. Isotope studies show that the cytoplasmic ACh exchanges freely with ACh in the synaptic vesicles. As already stated, ACh can be synthesized quickly and in surprisingly large amounts by the terminal during continuous impulse activity; it appears that the newly formed transmitter is largely directed into the "readily available" compartment.

Further information about the vesicular ACh has come from studies of the electric organ of the electric ray, *Torpedo*. This structure is especially rich in ACh and is a good source of vesicles; in these vesicles the concentration of ACh is as high as 0.5 M. It has been established that the interiors of the vesicles are acidic and that it is the H^+ gradient across the vesicle membrane that is used to energize the transfer of ACh into the vesicle (Anderson *et al.*, 1982). The transport of H^+ into the vesicles is driven by *ATP*, and Cl^- is cotransported for electroneutrality (Van der Kloot, 2003). Although ACh is the main component of the vesicles, the vesicles also contain an appreciable amount of ATP and a proteoglycan. The ATP is evidently released with ACh into the synaptic cleft (Silinsky & Hubbard, 1973), where the concentration can increase from 6 nM to 20 nM; a small part of the increase in ATP comes from the muscle fiber (Santos *et al.*, 2003). By combining with a purinergic receptor in the nerve terminal membrane and activating a *protein kinase* C signaling pathway, the ATP in the synaptic cleft reduces the spontaneous release of ACh (Galkin *et al.*, 2001). This helps to preserve presynaptic ACh stores and facilitate repolarization of the *postsynaptic membrane*.

At each
neuro-
muscular
junction,
20 to 300
vesicles are
emptied.

Quantal Size

How many synaptic vesicles release ACh after a single impulse invades the nerve terminal? A comparison of the sizes of *miniature end-plate potentials* (*MEPPs;* see chapter 3) with those of action potential-induced *end-plate potentials* in curarized nerve-muscle preparations would indicate that at least 100 vesicles are emptied at each neuromuscular junction. Another estimate, obtained by a different technique that avoided the use of curare, is that 200 to 300 quanta (vesicles) of ACh are released per impulse (Hubbard & Wilson, 1973). However, both these values, which were obtained in small laboratory animals, are considerably higher than those estimated by Slater *et al.* (1992) in biopsy specimens of human quadriceps muscles. In this preparation, the estimated mean number of ACh quanta at 37° C was 28 and was similar to the corresponding value obtained for human intercostal muscle (41 quanta; Engel *et al.,* 1990). Since there are many active zones in a given neuromuscular junction, only a few vesicles need to be emptied at each of them, even if individual zones respond intermittently to successive impulses (see p. 142).

The amount of ACh released from the nerve terminal with a single action potential is substantially more than is necessary to trigger an action potential in the muscle fiber membrane, and the excess is referred to as the safety margin. During repetitive stimulation, the recycling process (see later) is accelerated, and individual vesicles are moved to the axolemma for release before being completely filled with ACh; because of the safety margin, however, there is still enough ACh released to depolarize the end-plate to threshold.

The Mechanism of Transmitter Release

Impulses
travel to the
endings of
the motor
nerve twigs.

How is the ACh released from the synaptic vesicles? The first step, obviously, is the transmission of the action potential into the distal reaches of the motor axon. Katz and Miledi (1965) were able to show that the impulse travels all the way to the synapse by recording a propagated action potential with an extracellular microelectrode positioned close to the axon terminal. As expected on theoretical grounds, the conduction velocity of the impulse is much lower in the fine terminal branches of the axon than in the main trunk. In the frog, the velocity in the terminals is $0.3 \text{ m} \cdot \text{s}^{-1}$, as opposed to about $20 \text{ m} \cdot \text{s}^{-1}$ for motor axons within a peripheral nerve of the same species.

Voltage-Gated Ca^{2+} Channels

When the impulse invades the axon terminal, it depolarizes the membrane in the usual way, through the opening of Na^+ channels (see chapter 9). This *depolarization* then activates voltage-gated Ca^{2+} channels, allowing Ca^{2+} ions to enter the axon terminal down their concentration gradient; this inward Ca^{2+} current starts rather slowly and lasts longer than the Na^+ current.

The particles
in the active
zone are
the Ca^{2+}
channels.

Acetylcholine release occurs by a Ca^{2+}-regulated *exocytosis*. The entry of Ca^{2+} into the nerve terminal is essential for the release of ACh. Thus, neuromuscular transmission in *in vitro* preparations will cease if Ca^{2+} is omitted from the bathing fluid or if the concentration of Mg^{2+} is raised; Mg^{2+} ions appear to compete for the same binding sites as Ca^{2+} and in this way can block the effect of the latter. The extent of the rise in Ca^{2+} concentration within the vertebrate motor nerve terminal is not known because of the difficulty in examining such a small structure. In the giant synapse of the squid, however, the axon terminal may measure $50 \text{ }\mu\text{m} \times 1,000 \text{ }\mu\text{m}$ and is large enough to permit the introduction of one or more recording electrodes as well as the injection of Ca^{2+}-sensitive dyes such as Fura-2 and arsenazo III (Miledi & Parker, 1981; Charlton *et al.,* 1982). It has been found that an impulse invading the axon terminal causes a 10-fold rise in Ca^{2+} concentration in the region of axoplasm adjacent to the *plasmalemma.* Freeze-fracture studies of the axolemma overlying the synaptic cleft reveal multiple arrays of particles, each array consisting of two sets of double rows (figure 10.2, top). These structures comprise the active zones, for it is in their immediate vicinity that the ACh is discharged into the synaptic cleft; it is now known that the particles are the Ca^{2+} channels themselves (Pumplin *et al.,* 1981) and that the channels are of the P/Q type (Urbano *et al.,* 2002).

Docked Vesicles Fuse With the Axolemma

The events between the entry of Ca^{2+} and the release of ACh are still not fully understood, but several unique proteins are involved. There are two steps that vesicles must achieve for successful ACh release: docking and fusion. The first of these is completed prior to the arrival of the action

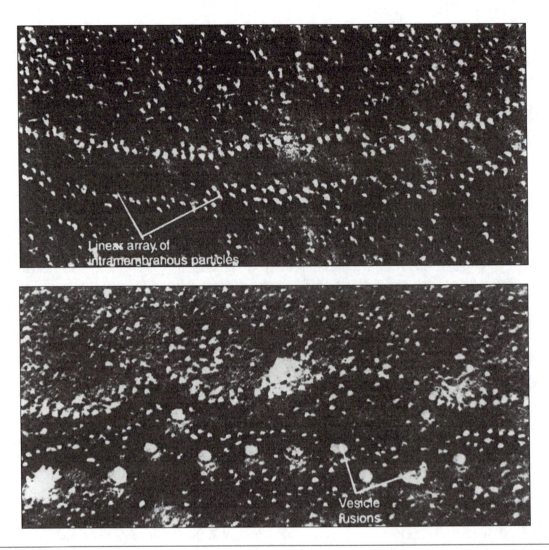

Figure 10.2   Top: Freeze-fracture electron micrograph of the axolemma of the motor nerve terminal. The two double rows of intramembranous particles identify an active zone and are thought to be Ca^{2+} channels. Bottom: Following an impulse, synaptic vesicles dock and fuse with the axolemma close to the Ca^{2+} channels before emptying. The fusion can be seen as the pale circular structures in the electron micrograph.

Reproduced from *The Journal of Cell Biology*, 1979, 81: 281, by copyright permission of The Rockefeller University Press.

potential at the nerve terminal. Several vesicles are in position at the axolemma, associated with the active zones. There are two types of specialized proteins involved in docking of the vesicle or fusion of the vesicular membrane with axolemma: vesicle membrane proteins and axon membrane proteins. These proteins associate to form what is known as the *SNARE complex*. In the second step of exocytosis, the vesicle membrane fuses with the axolemma, and ACh is released into the synaptic cleft. Since Ca^{2+} is necessary for transmitter release, it is likely that this ion triggers fusion, but it may also facilitate movement of additional vesicles into docking position. The steepness of the curve relating postsynaptic response to Ca^{2+} concentration suggests that from three to four Ca^{2+} ions cooperate in the emptying of a single vesicle (Dodge, Jr. & Rahamimoff, 1967). A number of proteins have been identified as likely to be involved in vesicle transactions (Bennett & Scheller, 1993; Warren, 1993).

Of these proteins, *synaptobrevin* and *syntaxin* probably mediate vesicle docking. Synaptobrevin, also known as the *vesicle-associated membrane protein (VAMP)*, is attached to the walls of the synaptic vesicles. Syntaxin is linked to the axolemma. In the presence of SNAPs (synaptosome-

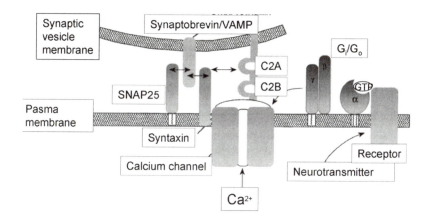

Figure 10.3 Model of protein binding at the active zone of the nerve terminal membrane. Vesicle membrane proteins synaptobrevin and *synaptotagmin* interact with *SNAP-25*, syntaxin, and the Ca^{2+} channel. When the Ca^{2+} channel opens, the local increase in $[Ca^{2+}]$ triggers fusion of the vesicle membrane with the neurolemma.

Reprinted, with permission, from the *Annual Review of Cell and Developmental Biology*, Volume 16 ©2000 by Annual Reviews. www.annualreviews.org.

associated proteins), synaptobrevin and syntaxin are able to interact (Söllner *et al.,* 1993), a process that would thereby allow docking to occur (see figure 10.3). In molecular terms, synaptobrevin and syntaxin are similar in that each has a single membrane-spanning domain with most of the protein extending into the cytoplasm (Kutay *et al.,* 1993).

In relation to the fusion step, an important experimental result was obtained by Bennett *et al.* (1992). By blocking synaptotagmin-acceptor sites with an engineered peptide fragment, these authors showed that synaptotagmin was essential for transmitter release, though not for vesicle docking. They therefore proposed that the combination of this protein in the wall of the vesicle, with its acceptor on the axolemma, enables membrane fusion to take place. It is possible that the acceptor is syntaxin (Bennett *et al.,* 1992). Since synaptotagmin binds Ca^{2+}, the fusion process is likely the Ca^{2+}-dependent step in the release of ACh. The sites of fusion appear to be just to the side of the active zones in the nerve terminal, since the technique of freeze-fracture electron microscopy (Heuser *et al.,* 1974; figure 10.2, bottom) reveals a series of depressions in this region immediately after impulse activity. These depressions are likely the inverted (inside out) vesicles, now fused with the axolemma, having released their ACh.

A Period of Facilitation Follows the Discharge of Acetylcholine

Most of the *synaptic delay* between the excitation in the nerve and its subsequent appearance in the muscle fiber is due to the Ca^{2+}-mediated release of transmitter. For the frog neuromuscular junction this delay is of the order of 0.5 ms, but at the mammalian junction the value is smaller (about 0.2 ms; Eccles & Liley, 1959; Hubbard & Schmidt, 1963). Experiments with double or multiple stimuli have shown that for the first 100 ms or so following an impulse, the axon terminal is in a hyperexcitable state. Thus, although some 30 to 100 vesicles will have been emptied of ACh, the transmitter remaining is more available for release by a second impulse. This *facilitation* is mainly due to residual Ca^{2+} in the nerve terminal following the first impulse; this Ca^{2+} is added to that admitted by the second impulse, to produce a larger than normal Ca^{2+} concentration and hence a greater release of ACh (Katz & Miledi, 1968).

The period of facilitation is succeeded by a depression of synaptic transmission, which typically lasts for several seconds. During this time, a second stimulus releases a smaller than normal number of ACh quanta, due to delays in the movement of synaptic vesicles to the active zones from more remote parts of the axon terminal. A detailed analysis of the postexcitation changes in the nerve terminal has been made by Magleby and Zengel (1982).

Recycling Nerve Terminal Membrane

One interesting question concerns the fate of the vesicles after they have discharged their ACh into the synaptic cleft. If the membranes of the vesicles fused with the axolemma, as electron micrographs indicate, the surface area of the axolemma would be increased to such an extent that some later mechanism would have to be available for removing the excess membrane. By using extracellular markers, Heuser and Reese (1973) have shown that recycling takes place; the presynaptic

For a short time after any action potential, the nerve terminal is more sensitive to subsequent activation.

Membrane is removed from the nerve terminal to form new synaptic vesicles.

membrane is reformed into vesicles toward the periphery of the synapse *(endocytosis)*. In the frog neuromuscular junction, endocytosis can be detected within 1 s of ACh release and is maximal at 30 s; the process is almost over by 90 s (Miller & Heuser, 1984). This process not only maintains the membrane area of the nerve terminal, but also permits recycling of the specialized vesicular membrane proteins.

The Saturating Disk

Each active zone and its opposing AChRs constitute a "saturating disk."

The active zones releasing ACh from the motor nerve terminal respond differently from one impulse to the next. It is therefore appropriate to consider the neuromuscular junction as an array of *saturating disks*, in which each part of the postsynaptic membrane is able to respond maximally to a corresponding region of the motor nerve terminal (Fertuck & Salpeter, 1976; Engel, 1987).

In each disk, the AChRs in a *junctional fold* are close to an active zone in the nerve terminal. Since the density of AChRs ($10,000\text{-}20,000 \cdot \mu m^{-2}$) is much greater than that of the AChE molecules in the basement membrane ($2,000\text{-}3,000 \cdot \mu m^{-2}$), most of the ACh will combine with its receptor. Further, a quantal packet of ACh needs to diffuse only 0.3 μm along the top, or down the side, of a synaptic fold in order to saturate the receptors. The small sizes of the saturating disks prevent them from overlapping and ensure that only a small fraction of the AChRs are saturated when a number of ACh quanta are released by a nerve impulse. Because successive impulses will activate different active zones and hence different saturating disks, the AChRs are prevented from becoming desensitized by continued exposure to ACh.

Postsynaptic Events

Acetylcholine combines with its receptor.

When ACh is released from a micropipette near an end-plate, the electrical response of the muscle fiber can be detected almost instantaneously. It seems, therefore, that once ACh is discharged from the synaptic vesicles into the synaptic cleft, it diffuses very quickly to combine with receptors in the muscle fiber membrane. That the ACh-binding sites are on the outer surface of the muscle fiber membrane was shown by early experiments in which ACh was released iontophoretically from a micropipette. When the tip of the pipette was inside the fiber, no response to injected ACh could be detected, whereas external application of the substance produced a rapid fall in muscle fiber membrane potential.

Channel Properties of the Acetylcholine Receptor

Combination of ACh with its receptor increases permeability to Na^+ and K^+.

The AChRs have already been described (chapter 3). The receptors are attached to the muscle fiber plasmalemma where they line the crests of the junctional folds and the upper parts of the secondary synaptic clefts (figure 3.1C, chapter 3). The AChRs, which can be seen with high-powered electron microscopy, each consist of five subunits, two of which (the α subunits) carry the ACh-binding sites (see figure 3.4, chapter 3). Whenever two molecules of ACh combine with the α subunits, the cation-selective channel in the center of the receptor opens sufficiently to allow K^+ and Na^+ ions to flow through. The ions travel in opposite directions, moving down their respective *electrochemical gradients;* thus, K^+ ions leave the fiber while Na^+ ions enter the fiber. The membrane potential will tend to fall to a value between the respective *equilibrium potentials* for Na^+ and K^+ (see chapter 9). The resulting depolarization is termed the end-plate potential (EPP) and, if fully developed, would bring the membrane potential from its resting value of approximately –85 mV to about –15 mV (Fatt & Katz, 1951). The extent of this depolarization, which can be seen only in the presence of an agent that blocks the voltage-gated Na^+ channels, reflects the safety margin referred to earlier. The blocker μ-*conotoxin* can be used to block muscle voltage-gated Na^+ channels, permitting full development of the EPP.

Acetylcholine receptor channels open for 1 ms or longer.

One way to examine the full time course of the EPP is to block a proportion of the AChRs with curare so that the EPP becomes too small to trigger an action potential. The EPP is then seen to have a sharply rising onset and an exponential decay lasting several milliseconds. In view of the time needed to discharge the *capacitance* of the muscle plasmalemma (see chapter 9), the current flowing at the end-plate is briefer than the EPP. Since this current is the aggregate of those of the AChR channels, it is important to determine the periods of time that individual ACh channels are open. This task can be accomplished by patch-clamp recording, and the AChR was the first ion channel to be studied by this technique (Neher & Sakmann, 1976b). It was found that individual channels opened for periods of time that could be as short as 1 ms or as long as 50 ms. However, these were extrajunctional AChR, induced by denervation, and the measurements were conducted at 8° C with a –80-mV holding potential. Apparently junctional AChRs will open for shorter durations; increased temperatures and decreased membrane potential will further shorten the open time.

The Muscle Fiber Action Potential

The EPP triggers an action potential in the muscle fiber.

Under normal circumstances, the rising EPP is overtaken by an action potential, which arises in the membrane immediately adjacent to the AChRs and is due to the opening of voltage-gated Na^+ channels; these channels are especially plentiful in the end-plate region, and large numbers line the deeper regions of the secondary synaptic clefts (Flucher & Daniels, 1989). Once initiated, the action potential propagates in both directions toward the ends of the muscle fiber. It has been calculated that during the EPP, each of the activated AChRs allows 17,000 Na^+ ions to flow into the fiber and a smaller number of K^+ ions to leave it.

Analysis of end-plate "noise" indicates that the opening of one ACh channel depolarizes the membrane by 0.3 μV. Since the fully developed EPP would be approximately 70 mV (see the preceding section), it follows that more than 200,000 AChRs must be activated following a single impulse in the motor nerve terminal. At human neuromuscular junctions, where relatively few ACh quanta are released by a single impulse, the EPP is smaller and probably not much larger than the threshold potential for action potential generation; that is, the safety margin is smaller.

Slater *et al.* (1992) have suggested that the Na^+ channels in the deeper parts of the secondary synaptic clefts provide an extra safety margin by acting as an amplifying system.

Acetylcholinesterase terminates synaptic transmission.

The enzyme AChE is one of the proteins in the basement membrane and is therefore positioned in the synaptic cleft between the ACh release sites in the axolemma and the AChRs in the muscle fiber plasmalemma. By virtue of its location, the AChE will capture a fraction of the ACh discharged into the synaptic cleft, before the transmitter can reach the AChRs. However, because there are many more ACh molecules released than there are AChE molecules to catch them, sufficient ACh nevertheless reaches the receptors. After opening the pores in the receptors, the ACh molecules dissociate and diffuse toward the basement membrane where they are hydrolyzed by AChE. The hydrolysis of ACh brings the synaptic transmission process to a conclusion. The end-plate membrane will repolarize to the resting level in preparation for the next bombardment of ACh.

Applied Physiology

In view of the complex sequence of events involved in neuromuscular transmission, it is hardly surprising that there are many ways in which the process can be disrupted. In this section, we will study first a number of drugs, including some that are used in the operating theater; then a bacterium that can produce a fatal paralysis; and last, two immunological disorders.

Drugs Affecting the Neuromuscular Junction

Neuromuscular transmission can be blocked to produce muscle paralysis during surgery.

During surgery it is usually important to have the muscles of the patient relaxed so that the trachea can be intubated by the anesthetist and so that, at the site of the operation, the muscles can be more easily incised and pulled to one side. The favored drug for producing relaxation is suxamethonium. It is a quaternary ammonium compound that blocks neuromuscular transmission by combining with AChRs and producing a sustained depolarization. As the depolarizing action commences, the muscle fibers begin to *twitch* spontaneously, indicating that their membrane potentials have fallen to the critical level for action potential initiation. As depolarization proceeds, the muscle fibers become refractory to motor nerve excitation, and paralysis ensues. In comparison with that of the natural transmitter, ACh, the action of the drug is prolonged, and this enables it to be used as an effective muscle relaxant during surgery. Suxamethonium can be administered by intravenous drip in the form of its chloride *(succinylcholine)*. Decamethonium is a similar type of compound.

Tubocurarine is a different type of drug and is less frequently used. It is an alkaloid that can be prepared from crude extracts of the plant toxin, curare. By competing for the same postsynaptic receptors as ACh, tubocurarine diminishes the action of the transmitter. Since the drug does not produce depolarization of the muscle fiber membrane itself, it is classified as a nondepolarizing blocking agent. Experimentally, microelectrode recordings from single muscle fibers reveal progressive reductions in the sizes of the EPPs as they continue to fall below the level required for impulse initiation. In humans, the action of tubocurarine usually lasts about 30 min. Gallamine is another drug with a similar type of action.

Anticholinesterase drugs have opposite effects; they potentiate neuromuscular transmission by preventing the hydrolysis of ACh by the AChEs attached to the basement membrane. Consequently, ACh released from a nerve terminal exerts a larger and more prolonged depolarization of the muscle fiber. Examples of anticholinesterases used in medicine are neostigmine, pyridostigmine bromide, and edrophonium chloride.

Organophosphate nerve agents used in weapons of chemical warfare (examples are tabun, sarin, and soman) are anticholinesterases. They are so powerful that even very small doses cause ACh to accumulate at neuromuscular junctions; this accumulation, in turn, desensitizes the AChRs and results in paralysis and death.

Botulism

The toxins produced by the bacterium, *Clostridium botulinum,* remain the most powerful yet known to man; as little as 0.5 μmol of toxin A is fatal. The bacterium is anaerobic and is found in soil and in animal feces. Poisoning is fortunately uncommon and usually results from the improper canning or bottling of food; if heat sterilization has been insufficient, the spores persist in the food and then germinate in the anaerobic environment.

• **Very small amounts of botulinum neurotoxin can produce a fatal paralysis.** Symptoms of botulinum poisoning appear within 48 hr of ingestion of the toxin; initially they consist of double vision (diplopia) and unsteadiness on standing. Subsequently, the lower cranial nerves are affected, producing paralysis of speech and swallowing. Unless treatment is available, death occurs from respiratory paralysis within a few days. Until the end, the patient is fully conscious and is without any disorder of sensation.

• **The paralysis is due to the inability of impulses to release ACh.** *Electromyography* reveals small evoked muscle responses following nerve stimulation, although the muscle can be shown to respond normally to direct stimulation. Repeating the nerve stimulation at low rates (e.g., 2 Hz) causes further decrement in the muscle responses, whereas rapid stimulation (at 50 Hz) produces potentiation; this pattern of behavior resembles that found in the *Lambert-Eaton Myasthenic syndrome* (see p. 149). Together, these findings indicate that the paralysis results from a failure of neuromuscular transmission, and this has been confirmed by microelectrode investigations in frog muscle. Thus, the finding of spontaneously occurring MEPPs indicates that transmitter is available within the axon terminal and that there is no loss of postsynaptic responsiveness. However, although the nerve impulse still invades the axon terminal during the phase of paralysis, none of the available transmitter can be released (Harris & Miledi, 1971).

• **The various botulin neurotoxins attack synaptic vesicle proteins.** There are seven types of botulinum neurotoxins, each of which can block neuromuscular transmission. It turns out that each type is a zinc-containing endopeptidase, capable of degrading one of the proteins necessary for the release of ACh from the nerve terminal. As an example, Schiavo *et al.* (1992) have shown that botulinum B neurotoxin splits the protein synaptobrevin (as does *tetanus toxin*), while Blasi *et al.* (1993) found that neurotoxin A cleaves SNAP-25. Not only do observations of this type explain the actions of this deadly family of neurotoxins, but they also demonstrate that the targeted proteins are essential for either vesicle docking or membrane fusion.

• **Botulinum neurotoxin (Botox) can eliminate frown lines.** As remarkable as it seems, this deadly toxin is now used to remove unwanted wrinkles from the brows of individuals who would prefer not to show these manifestations of age. Very small doses, repeated at 3-month intervals, paralyze a fraction of the subcutaneous muscle fibers, preventing their tonic contraction. The same drug is extremely useful in reducing spasticity and the spontaneous twitching of muscles around the eye and in the side of the face that affects some individuals.

Myasthenia Gravis

The diseases that primarily affect the neuromuscular junction are relatively infrequent but are nevertheless well understood as a result of studies employing electron microscopy, immunology, and microelectrode recordings. In the least common conditions, all of which are genetic in nature, the defects range from faulty ACh packaging (familial myasthenia gravis) to the impaired synthesis of AChRs and AChE.

An Immunological Disease

The neuro-muscular junction may be subjected to immuno-logical attack.

Two other transmission disorders are immunologically determined; in *myasthenia gravis,* the *antigen* is the AChR, while in the Lambert-Eaton myasthenic syndrome, it appears to be the voltage-gated Ca^{2+} channels in the motor nerve terminal.

The 17th-century physician Thomas Willis may have been the first to describe a patient with myasthenia. In 1672, Willis wrote:

> Nevertheless those labouring with a want of spirits, will use these spirits for local motions as well as they can; in the morning they are able to walk firmly, to fling their arms about hither and thither or take up any heavy thing; before noon the stock of spirits being spent, which had flowed into the muscles, they are scarcely able to move hand or foot. At this time I have under my charge a prudent and honest woman who for many years has been subject to this sort of spurious palsy, not only in her members, but also in her tongue. She for some time can speak hastily or eagerly, she is then not able to speak a word, but becomes suddenly as mute as a fish, nor can she recover the use of her voice for an hour or two.

Myasthenia usually affects ocular muscles first.

Myasthenia attacks about one in 10,000 persons, mainly young women and older men. The presenting symptom is often drooping of an eyelid or diplopia, and is most noticeable in the late afternoon and evening. In more severe cases, there is difficulty in swallowing, talking, and chewing, and the arms and legs also fatigue easily. In the most seriously affected patients, even the respiratory muscles are involved, and artificial ventilation may be needed.

Diagnosis of Myasthenia

Diagnosis is confirmed by pharma-cological, immuno-logical, and electrical testing.

The intravenous injection of a small dose of edrophonium chloride can improve muscle strength within a minute, as in the woman shown in figure 10.4, who was then able to open her eyes and smile. Edrophonium blocks AChE, allowing ACh to linger in the synaptic cleft and to combine repeatedly with AChRs. Since myasthenia is an autoimmune disorder, it is not surprising that antibodies to the

Figure 10.4 Facial appearance of patient with myasthenia gravis before (left) and approximately 1 min after (right) an injection of an anticholinesterase drug. Notice how the drooping of the eyelids (ptosis) disappears.

Courtesy of Dr. H. S. Barrows.

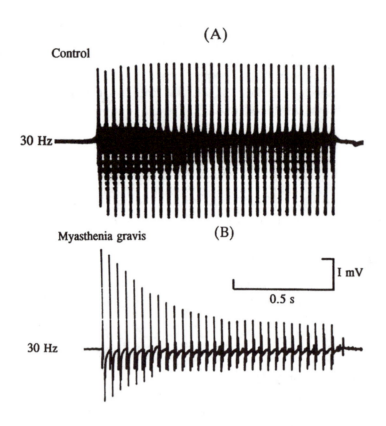

(A)

Control

30 Hz

Myasthenia gravis   **(B)**

I mV

0.5 s

30 Hz

AChR can be detected in the plasma of 90% of patients with the disorder; in general, the more ill the patient, the higher is the antibody titer.

Electrical testing is based on the observation that repeated excitations cause fluctuations in the efficacy of neuromuscular transmission. For example, if the motor nerve is stimulated repetitively, the evoked muscle responses decline, as in figure 10.5. A more accurate and less painful electrical test is single-fiber *EMG,* devised by Ekstedt and Stålberg (1967). The principle of the test is to record with a needle electrode from two muscle fibers belonging to the same *motor unit.* In normal subjects, the time interval between the respective fiber action potentials fluctuates slightly as the person carries out a weak contraction; in myasthenic patients, however, the fluctuations, or "jitter," are more pronounced, as in figure 10.6.

Figure 10.5 (A) Normal response (absence of decrement) in the extensor digitorum brevis muscle of a healthy subject; the peroneal nerve was stimulated at 30 Hz. (B) Decrementing response in the corresponding muscle of a 35-year-old woman with generalized myasthenia gravis.

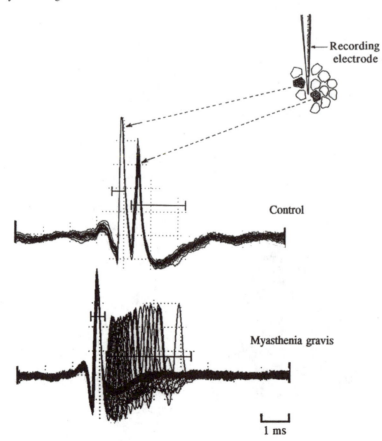

Recording electrode

Control

Myasthenia gravis

1 ms

Figure 10.6 Superimposed recordings made from two muscle fibers belonging to the same motor unit in a healthy subject and in a patient with myasthenia gravis. Notice how the interval between each pair of impulses fluctuates markedly in the patient (jitter phenomenon) but not in the healthy control subject. These recordings are examples of single-fiber EMG.

Experimental Induction of Myasthenia

Myasthenia can be induced in experimental animals.

While attempting to prepare antibodies to the AChR, Patrick and Lindstrom (1973) were puzzled to find that rabbits injected with receptor from the electric organs of the electric eel became weak and floppy. Vanda Lennon, who had earlier suggested that the AChR might act as an antigen in myasthenia (Lennon & Carnegie, 1971), recognized that the injected animals had become myasthenic and was later able to show that the disease takes place in two stages (Lennon *et al.*, 1976). In the first stage, thymus-derived lymphocytes (T-cells) become sensitized to AChR at 4 to 5 days after inoculation of AChR. The sensitized T-cells then react with the animal's own AChRs and release pharmacologically active lymphokines; these increase vascular permeability and cause masses of *inflammatory cells* to accumulate at the neuromuscular junction (figure 10.7). Destruction of the synapses then occurs and is responsible for the first episode of weakness; during this phase, the axon terminals become separated from the postsynaptic membranes.

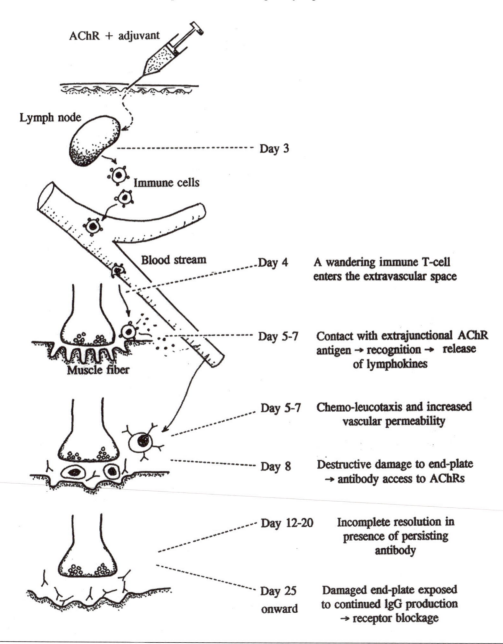

Figure 10.7   The postulated mode of production of a myasthenia gravis-like illness in animals following the injection of AChR and adjuvant.

Adapted from Lennon *et al.* (1976, p. 295).

After the acute inflammation response subsides, the neuromuscular junctions are partially repaired but now assume a simpler structure with fewer secondary clefts. Further degeneration of the synapses takes place following the appearance of circulating antibody to the AChR; this stage is responsible for the second phase of muscle weakness. Electron microscope studies have revealed a strong similarity between the structures of the neuromuscular junctions in this chronic phase of experimental myasthenia and in that of the human disease (Engel *et al.,* 1976).

Acetylcholine receptor antibodies have several myasthenic effects.

How do the AChR antibodies produce their effects in myasthenia? It is now clear that there is more than one type of AChR antibody but that most types bind to the α subunit (Tzartos *et al.,* 1983), though not usually at the ACh-binding site. A more serious effect of the receptor antibody is to cause *complement* fixation, which, in turn, results in destruction of the synaptic folds. Thus, not only are AChRs lost as pieces of membrane are shed into the synaptic cleft, but there is less membrane available for the insertion of newly synthesized receptors (Engel, 1987). A further action of the antibodies is that, after combination with receptors, they hasten the internalization and degradation of the receptors (Stanley & Drachman, 1978).

Neuromuscular Transmission in Myasthenia

The EPPs are diminished in myasthenia.

In myasthenia, neuromuscular transmission is jeopardized for two reasons. First, there are too few receptors for the ACh to combine with, and second, the widened *primary synaptic cleft* encourages the diffusion of ACh away from the receptors.

Since the ACh-AChR combinations are diminished, following an impulse in the nerve terminal, the EPP develops more slowly than usual and may not reach the threshold for generating an action potential. Such impulse failures can be seen in figure 10.8, from a study in which a microelectrode was used to record from the end-plate region of a human intercostal muscle fiber. On those occasions when the threshold for excitation is reached, the timing of the action potential varies considerably; this variation is the basis for the increased neuromuscular jitter, which can be recorded with a single-fiber EMG electrode (see figure 10.6). As would be expected, the deficiency of AChRs also reduces the sizes of the MEPPs (Elmqvist *et al.,* 1964). It must be emphasized, however, that there is nothing wrong with the synthesis and release of ACh from the motor nerve terminal in this form of myasthenia; indeed, measurements with gas chromatography and mass spectrometry indicate that the amount of ACh may be increased in myasthenia (Ito *et al.,* 1976).

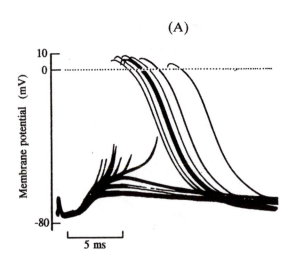

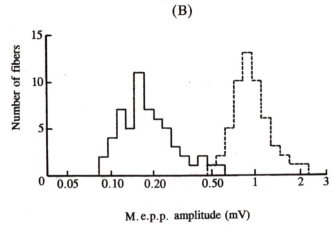

Figure 10.8   (A) Intracellular recordings from an intercostal muscle fiber of a patient with generalized myasthenia gravis. When the nerve was stimulated at 5 Hz, the evoked responses from the fibers often consisted of subthreshold end-plate potentials instead of action potentials. (B) Amplitudes of miniature end-plate potentials (MEPPs) in intercostal muscle fibers of myasthenic patients (solid line) and control subjects (dashed line).

Adapted, by permission, from D. Elmqvist et al., 1964, "An electrophysiological investigation of neuromuscular transmission of myasthenia gravis," *Journal of Physiology* 174: 417-434.

The Lambert-Eaton Myasthenic Syndrome

The eponym for the next disorder, Lambert-Eaton myasthenic syndrome (LEMS), is well justified. Lee Eaton was a senior neurologist at the Mayo Clinic who had under his care a group of patients with an unusual form of muscle weakness. Ed Lambert not only recognized the unusual nature of the disorder but continued for many years as the most distinguished electromyographer in North America.

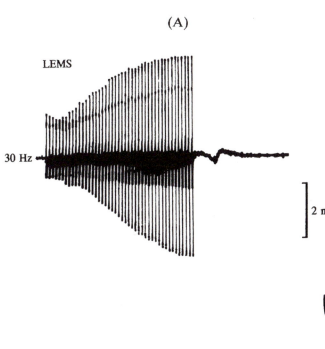

(A)

Eaton and Lambert (1957) pointed out that the myasthenic-like weakness was often associated with a malignancy, usually small cell carcinoma of the lung. Further, the condition could be differentiated from myasthenia gravis by the response of the muscles to high-frequency motor nerve stimulation. In myasthenia, the muscle response is of normal amplitude initially and then declines, whereas in LEMS, the response is at first small and then potentiates markedly as stimulation is continued (figure 10.9); potentiation of the evoked responses can also be achieved by voluntary contraction.

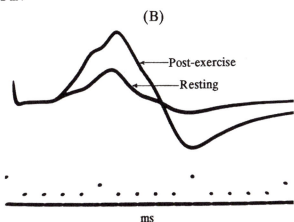

(B)

Figure 10.9 (A) Incrementing intramuscular EMG during repetitive nerve stimulation at 30 Hz in a patient with LEMS. The progressive enlargement of the electrical responses should be compared with the situation in the healthy subject of figure 10.5A. (B) A simple, and less painful, test for LEMS is to contrast the muscle EMG responses to single shocks delivered before, and immediately after, a brief voluntary contraction. The enlargement is obvious in a LEMS patient, as in the example here. This enlargement is not seen in control (healthy) subjects.

In LEMS, insufficient ACh is released from the nerve endings.

The fact that the muscle responses grow larger as stimulation continues strongly suggests a presynaptic disorder, and microelectrode studies on intercostal muscle biopsies have confirmed this prediction (Lambert & Elmqvist, 1971; see also Hofmann *et al.,* 1967). The MEPPs are found to be of normal size; this finding indicates that the synaptic vesicles contain normal amounts of ACh and that the AChRs are available and responsive. Direct measurements of ACh contents in LEMS muscle also yield normal results (Molenaar *et al.,* 1982). When the motor nerve is stimulated, however, too few quanta are released to depolarize the muscle fiber membrane to the threshold for generating an action potential. The quantal release can be enhanced by raising the Ca^{2+} concentration in the fluid bathing the muscle fibers or by adding guanidine. All of these findings, then, point to defective release of synaptic vesicles in LEMS.

The possibility that the voltage-gated Ca^{2+} channels might be abnormal in the motor nerve terminal has been investigated by subjecting this membrane to freeze-fracture. The parallel rows of particles, which constitute the active zones for ACh release (figure 10.2, top), are seen to be markedly decreased, although other randomly disposed particles are normal in number (Fukunaga *et al.,* 1982).

Presynaptic Ca^{2+} channels are attacked by antibodies raised against tumor cells.

Convincing evidence for an autoimmune basis for LEMS is the ability to induce the electrophysiological features of the condition in mice by injecting them with immunoglobulin from affected patients (Lang *et al.,* 1983; Vincent *et al.,* 1989). The animals develop the same type of disruption of the active zones in the motor nerve membrane as do LEMS patients. Supporting evidence favoring an autoimmune etiology in LEMS comes from the finding that removing immunoglobulin from patients by the technique of plasmapheresis is often beneficial.

The most plausible explanation for the etiology and pathogenesis of LEMS is that the voltage-gated Ca^{2+} channels in the motor nerve become the targets of antibodies. In patients with cancer, the antibodies are probably raised against voltage-gated Ca^{2+} channels in the tumor cells and then react with similar channels in the motor nerve endings. In the minority of LEMS patients without tumors, the origin of the antibodies is unknown. The combination of antibodies with Ca^{2+} channels causes the latter to become cross-linked, destroying the normal structure of the active zones and preventing the influx of Ca^{2+} during impulse invasion of the motor nerve terminal.

The last step in muscle activation is, of course, the contraction itself and the production of force or movement. In the next chapter, we will learn that muscle contraction, like neuromuscular transmission, is a complex process involving a number of steps.

11

Muscle Contraction

Even the simplest unicellular creatures move about. Further, individual cells in multicellular organisms may change their sizes and shapes, depending on their functions and stages of growth. These movements involve different types of protein filament: *actin, myosin,* intermediate filaments, and *microtubules.* A different type of movement, considered in chapter 8, is the mechanism responsible for *axoplasmic transport.* However, the most specialized type of movement, and that most completely studied and understood, is the contraction and relaxation of skeletal muscle fibers. It will be seen that, like axoplasmic transport, the muscle movements are brought about by *molecular motors* interacting with filaments; in this case, the motors are myosin molecules, and the filaments are actin. The various ways in which skeletal muscle is studied are described in "Studying Skeletal Muscle Contractions" (see below). Each of these techniques has advantages and limitations. These need to be kept in mind when one is evaluating the various findings associated with muscle contraction.

Sliding Filament Theory of Muscle Contraction

Shortening and lengthening of skeletal muscle is achieved by sliding of filaments.

A major contribution to our knowledge of the way in which skeletal muscle fibers contract was deduced by two quite different types of microscopic study in the early 1950s, and in each case a Huxley was involved. Andrew Huxley, of the voltage clamp experiments (see chapter 9), made his observations with an interference microscope of his own design, while Hugh Huxley (no relation) used an electron microscope to examine what were, in those days, exquisitely thin sections of muscle.

Studying Skeletal Muscle Contractions

The contractile response of skeletal muscle has been studied in different ways. These ways can be called *in vivo, in situ,* and *in vitro.* Each way of studying skeletal muscle has advantages and disadvantages. Figure 11.1 illustrates the different ways in which contractile responses can be studied.

In Vivo Study of Skeletal Muscle

When skeletal muscle is studied *in vivo,* it is studied within the living body. This approach can include whole body exercise, voluntary contrac-

tion of muscle groups, or stimulation of individual muscles or *motor units.* Measurements that are typically obtained during *in vivo* study of skeletal muscle include muscle force (or more likely, joint moment), work or power, *EMG,* temperature, whole body or regional oxygen uptake, and so on. Force measurements can include the time courses of the *twitch* and the *force–frequency relationship,* maximal and submaximal voluntary torque, and torque due to reflex *activation* of a muscle. Knowledge of metabolic events during *in vivo* study of muscle

(continued)

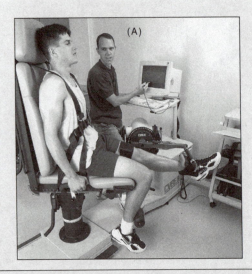
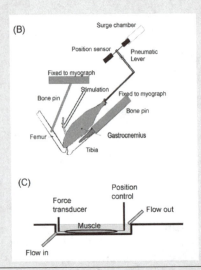

Figure 11.1 Muscle is studied (A) *in vivo,* (B) *in situ,* and (C) *in vitro.* Each approach has advantages and disadvantages.

can be obtained from biopsies or from noninvasive *magnetic resonance spectroscopy.* Ultrasound and *magnetic resonance imaging* can be used to quantify muscle size and changes in *fascicle* and tendon length during contractions.

The ultimate goal of research on muscle is to understand how it works in the body, so *in vivo* research is a vital approach. The disadvantage of *in vivo* research is that the muscle under study is operating in a complex system, with the experimenter having very little control of several circumstances that can influence the measurement. For example, the extracellular fluid contains hormones, neurotransmitters, ions, and proteins that make it unique and difficult to duplicate. This complexity makes interpretation of results difficult.

In Situ Study of Skeletal Muscle

When a muscle is studied *in situ,* it is also studied within a living body, but this term is usually reserved for a situation in which there is some partial isolation. Muscle groups, individual muscles, and motor units have been studied in this way. Generally, a muscle tendon is isolated and connected directly to a force or displacement transducer, or both, for direct measurement of work, force, or power, while the origin of the muscle is maintained in its natural circumstance with respect to attachment, blood flow, and innervation. Additional measurements that are often combined with this approach include EMG, temperature, blood flow, and oxygen uptake. Biopsies can be taken, or the whole muscle can be frozen for biochemical or histochemical analysis. Magnetic resonance spectroscopy has been used to evaluate energetics and metabolism of *in situ* muscle preparations. Occasionally, researchers

have probed the extracellular concentration of various substances by using ion-sensitive electrodes inserted into the muscle or by microdialysis. More commonly, exchange of substances with the muscle can be studied by measuring blood flow and the arterio-venous concentration differences. Sonomicrometry (micrometric ultrasound) has been used to quantify muscle fascicle length changes during isometric and dynamic contractions.

In Vitro Study of Skeletal Muscle

In vitro means in glass (or in a test tube). Essentially, *in vitro* study of skeletal muscle involves study of the muscle or some part of the muscle when it has been completely removed from the body. *In vitro* studies have been used with whole muscles and with single (intact or skinned) fibers, *myofibrils,* muscle homogenates, filaments, and individual molecules. Measurements have been made of force, displacement, *membrane potential,* intracellular ion concentration, and enzymatic activity. The advantage of *in vitro* study is that the investigator has full control of the environment (ionic composition, hormones, temperature, etc.); further, a reductionist approach enables individual molecules to be studied. The disadvantage is that the tissue under study may not behave in the same way that it would *in vivo.*

Combined Approach

It is sometimes more revealing to combine the different types of study. For example, exercise or stimulation of muscles *in vivo* can be followed by excision of part of the muscle (i.e., biopsy) for subsequent study *in vitro;* the latter may involve muscle homogenates or skinned fibers with evaluation of substrate concentrations or enzyme activities.

With the electron microscope, it was seen that, within a muscle fiber, each myofibril was composed of many *myofilaments* and that the latter were of two types, thick and thin. In cross sections of muscle fibers, each thick filament was seen to be surrounded by a hexagonal array of thin filaments (figure 11.2, top); in longitudinal sections the thick filaments of a myofibril were found to be in register with each other (figure 11.2, bottom). For a long time it had been known that the contractile proteins of muscle were myosin and actin, and next it was shown that these proteins corresponded to the thick and thin filaments, respectively. This step was achieved by dissolving the myosin in KCl solution and demonstrating that the thick filaments were no longer visible with electron microscopy.

At the same time that Hugh Huxley was undertaking these important studies, Andrew Huxley was using the interference microscope to examine the muscle striations of living frog muscle fibers during contraction and relaxation (an *in vitro* preparation; see "Studying Skeletal Muscle Contractions," and see also chapter 1 for a description of the *sarcomere* and the corresponding bands and regions within the sarcomere). He observed that during contraction the light *I-band* became shorter while the dark *A-band* remained the same length; within the A-band, however, the pale *H-zone* narrowed and could disappear completely. Quite independently, the Huxleys proposed that their respective findings could be explained by a sliding movement of the actin and myosin filaments past each other (Huxley & Niedergerke, 1954; Huxley & Hanson, 1954). This sliding filament theory is now generally accepted but is not without opposition (Pollack, 1995).

Figure 11.3 shows that the I-band is the region of fiber where only actin filaments are present; the A-band corresponds to the position of the myosin filaments. In the relaxed state at a long length (figure 11.3D), although there is some overlap between the actin and myosin filaments, opposing actin filaments within each sarcomere are separated from each other by a gap where only myosin filaments are observed. This gap between the ends of the opposing actin filaments is responsible for a pale region (H-zone) at the center of the A-band.

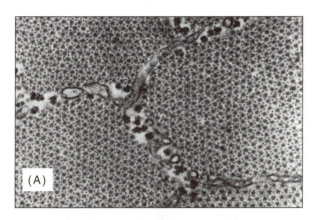

Figure 11.2 A: Lattice of actin and myosin filaments. Six actin (thin) filaments surround each myosin (thick) filament in a hexagonal array, as shown diagrammatically in figure 11.3F. The electron micrograph shows part of a frog sartorius myofiber in cross section. B: Thin, longitudinal sections through four myofibrils showing overlapping thick and thin filaments. See also figure 1.9 in chapter 1 for identification of features.

Reprinted, by permission, from H.E. Huxley, 1972, Molecular basis of contraction in cross-striated muscles. In *The structure and function of muscle*, edited by G.H. Bourne (New York: Academic Press), 316.

When the muscle shortens, actively or passively (figure 11.3E), the opposing actin filaments within each sarcomere slide along the intervening myosin filament. As the ends of opposing thin filaments approach each other, they cause the H-zone to become narrower; similarly, as more of each actin filament is drawn into the space between the myosin filaments, the I-band becomes shorter, as does the sarcomere. Since the myosin filaments do not alter their shape, the length of the A-band stays unchanged. It was the apparently constant lengths of the filaments that led to the sliding filament theory of muscle contraction.

The function of the Z-line, or Z-disk, in addition to providing an anchor for proteins associated with the *costamere* (see chapter 1), is to tether the actin filaments to the ends of the sarcomere. The *M-line* in the center of the A-band corresponds to links between the myosin filaments; both types of structure maintain the orderly geometrical relationship of the filaments to one another (see chapter 1). The structural organization of the filaments is aided by *nebulin* and *titin* (see figure 1.11).

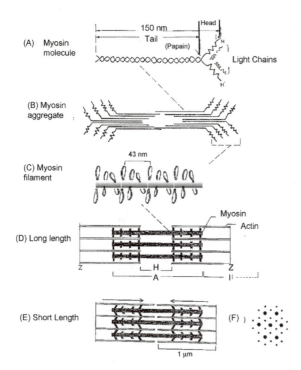

Figure 11.3 The myosin molecule and its arrangement in the thick filaments of the myofibrils. (A) A single molecule, consisting of a double-helical rod terminating in two *globular heads,* each of which has two light chains attached. The site of enzymatic cleavage, with papain, is shown. (B) In solution, myosin molecules spontaneously aggregate to form filaments with heads toward both ends. (C) In a thick filament, the two globular heads of a myosin molecule project to form a *cross-bridge.* The next bridge is separated by 14.3 nm and is rotated 60° relative to the previous one (see text). (D) The overlap of actin and myosin filaments in a myofibril is maximized at optimal length, and the various refractive bands that are created are labeled. (E) Contraction is produced by actin filaments sliding over the myosin filament, causing approximation of the *Z-lines (disks)* and narrowing of the H-region and I-band. (F) A cross section through a portion of a myofibril to display the hexagonal disposition of the actin filaments around the myosin filaments. The diagram can be compared with the electron micrograph in figure 11.2 (top).

The Cross-Bridge Theory of Skeletal Muscle

The sliding mechanism is produced by cross-bridges.

The next step in understanding the mechanism of muscle contraction was the demonstration, using electron micrographs at a high magnification, of small projections from the myosin filaments (H. E. Huxley, 1958; see figure 11.4). It was proposed that these projections could momentarily attach themselves to the actin filaments, forming cross-bridges, and propel the thin filaments to new positions. The projections are approximately 13 nm long and lie in six rows along the myosin filament; each row can engage one of the actin filaments in the hexagonal array surrounding the myosin filament. As shown in figure 11.3C, the projections are arranged in pairs separated by 180°; the next pair is always 14.3 nm away, with its axis rotated through 60°.

Single-Molecule Experiments

Further insights into the mechanism of cross-bridge action have come from studies of the alga *Nitella;* the advantage of using this preparation is that the cells are relatively long and contain parallel bundles of actin filaments. If the cells are cut open and spread out, the *cytoplasm* can be washed away, leaving the bundles of actin filaments in place. Sheetz and Spudich (1983)

Figure 11.4 Myosin cross-bridges. A: Two thin (actin) filaments separate adjacent thick (myosin) filaments. (Inspection of figure 11.2, top, and figure 11.3F shows how this appearance arises.) The small projections, perpendicular to the overlapping filaments, are the myosin heads. B: The projections are seen more clearly.

Reprinted, by permission, from H.E. Huxley, 1972, *Molecular basis of contraction in cross-striated muscles* (Orlando, Florida: Academic Press), 320.

Isolated cross-bridges can be made to work *in vitro*.

applied tiny plastic beads coated with myosin to the treated *Nitella* cells and observed that, in the presence of *ATP*, the beads were moved along the actin filaments, presumably through the action of the myosin cross-bridges. An extension of this type of experiment was to replace *Nitella* with an artificial membrane to which longitudinally oriented actin filaments had been attached (Spudich *et al.*, 1985). In subsequent experiments from the same laboratory, the researchers eliminated the need for beads by coating a glass surface with myosin filaments, applying ATP, and observing the movement of fluorescently labeled actin filaments under the microscope. All three types of experiment (*in vitro* motility assay) indicated that the cross-bridge movements occur with a constant velocity and that the direction of movement is determined by the polarity of the actin filaments (see p. 158).

Recent advances in nanotechnology have permitted the study of the interaction of individual myosin cross-bridges with actin (Finer *et al.*, 1994). Dual laser beams hold a styrene bead sparsely coated with myosin in place, and direct interaction of one myosin molecule with an actin filament can be studied (Simmons *et al.*, 1996). In a single "working stroke," a cross-bridge moves an actin filament through 5 to 10 nm. It is thought that each cross-bridge cycle results in hydrolysis of one molecule of ATP (see H.E. Huxley, 1990), but some investigators have suggested that multiple cycles are performed for each ATP hydrolyzed (Yamaguchi *et al.*, 1996; Yanagida *et al.*, 2000).

Independent Force Generators

Contractile force depends on the number of cross-bridge interactions.

One of the predictions of the cross-bridge theory was that each cross-bridge would act as an independent force generator and that the force developed in a contraction would therefore depend on the number of simultaneous interactions between myosin heads and the actin filaments within each half sarcomere. Experimentally, the number of cross-bridge–actin interactions can be varied by stretching the muscle fiber and so altering the amount of overlap between the actin and myosin filaments. This type of experiment is easily carried out *in situ* by recording the isometric force developed by a muscle when it is set at a variety of lengths (figure 11.5). It can be seen that the developed force increases as the muscle is tested at longer lengths, up to a maximum, and then, with further stretching, starts to decline.

A disadvantage of the experiment illustrated in figure 11.5 is that the contractions are "fixed-end" contractions. Although the length of the muscle is held constant, the muscle fibers are allowed to shorten while the tendon and aponeurosis are stretched in proportion to the force exerted by the muscle. Therefore, the actual length of the sarcomeres in these contractions cannot be easily known. This is partly overcome here by measurement of the fascicle length. The values for passive force can be matched to values for peak force at slightly shorter fascicle lengths. Sarcomere length is still not known.

To overcome this problem, and to determine the length dependence of force with respect to sarcomere length, Gordon *et al.* (1966b) devised an elegant experiment in which they examined the length dependence of force generation in single muscle fibers of the frog. These authors fixed two small pieces of gold leaf to the fiber and used these to interrupt two beams of light passing from cathode ray tubes to two photocells (figure 11.6). The output from the photocells was fed into a control circuit that operated a servomotor attached to one end of the muscle fiber. The other end of the fiber was attached to a strain gauge. By using negative feedback, the system enabled the length of the fiber segment between the gold-leaf markers to be kept at a preset value (constant average sarcomere length) while the active force during *tetanic stimulation* was measured.

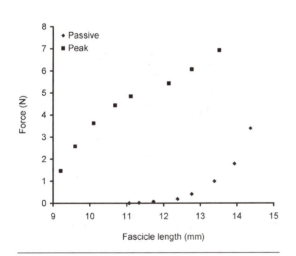

Figure 11.5 Effect of altering the lengths of muscles on passive and peak force (torque if measured *in vivo* without implanted force transducers), produced during brief tetanic stimulation. The fascicle lengths decrease during the contraction, so the passive force prior to a contraction is shown at a length just slightly longer than the length at the peak of the contraction.

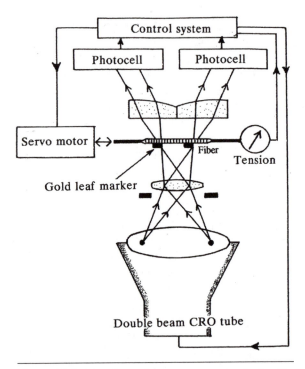

Figure 11.6 Experimental system to examine the length–tension relationship in short lengths of single muscle fibers. See text.

Adapted, by permission, from A.M. Gordon, A.F. Huxley, and F.J. Julian, 1966, "Tension development in highly stretched vertebrate muscle fibres," *Journal of Physiology* 184: 149.

The investigators observed that, over a certain range (traces 1 and 2 in figure 11.7), the tension developed is indeed proportional to the degree of overlap between the actin and myosin filaments and hence to the number of available cross-bridges on the latter. The sarcomere length corresponding to the greatest overlap and allowing the highest active tension is termed the optimum length. In other experiments, Gordon and colleagues stretched the muscle fiber so far that there was no overlap between the myosin and actin filaments. Because the myosin heads were unable to reach the actin filaments, no active tension could be developed (beyond 3.65 μm in figure 11.7). This is exactly the situation that is thought to occur in cardiac failure associated with ventricular distension.

Whether excessive lengthening can ever take place in normal skeletal muscles is doubtful, for the permissible ranges of joint movement are probably too small. In diseased muscles, however, the possibility of hyperextension seems more likely, since the partial replacement of muscle fibers by relatively inelastic fibrous tissue could well allow surviving segments of fibers to be stretched excessively. Similarly, immobilization of a muscle at a short length results in reduction of the number of sarcomeres in series. Under these circumstances, passive tension rises to a high level to oppose overextension.

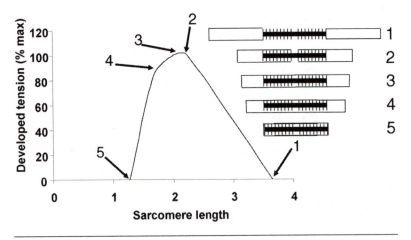

Figure 11.7 Length–tension results of experiment depicted in figure 11.6. At the right is shown a sarcomere representing the numbered positions on the *force–length relationship*. The amount of overlap of the actin and myosin filaments at different sarcomere lengths is the primary factor determining the active force at positions 1 to 3. At shorter sarcomere lengths, thin filaments from opposite ends of the sarcomere interfere with each other. When the sarcomere length decreases to the length of the thick filament, developed force is zero.

Adapted, by permission, from A.M. Gordon, A.F. Huxley, and F.J. Julian, 1966, "The variation in isometric tension with sarcomere length in vertebrate muscle fibres," *Journal of Physiology* 184: 170-192.

The Key Contractile Proteins: Myosin and Actin

The cross-bridge theory requires that the thick and thin filaments interact in a manner that generates tension and allows sliding of the filaments past one another. The mechanism of this interaction is better understood with a close look at the molecules that make up these filaments.

Myosin

Much more has been learned about the myosin molecule through the use of high-powered electron microscopy and *X-ray diffraction*. The complete myosin molecule is a structure some 150 nm long, consisting of two globular heads and a single tail (figure 11.8A). The heads can be snapped apart from the tail by the proteolytic enzyme papain. The relatively long tail is formed by two α-helices that coil round each other; the tails of several hundred myosin molecules associate together to form a single myosin filament. This aggregation occurs spontaneously and can be demonstrated in myosin precipitated from solution. The structure of the tail also ensures that, within each filament, half the myosin molecules face away from the other half such that the middle part of the filament is devoid of cross-bridges (figure 11.3B).

Myosin Head Structure

The cross-bridges are the globular heads of the myosin molecules.

The two globular heads of each myosin molecule are often referred to as S1 fragments (figure 11.8A). Each head consists of a heavy chain of about 2,000 amino acids and has two light chains attached to it. The globular heads, and the necks to which they are attached, form the cross-bridges. Biochemical studies have identified three components of the globular head, with molecular weight of 50 kD, 25 kD, and 20 kD, respectively (figure 11.8B). The two light chains are attached to the 20-kD moiety, while the 50-kD segment contains a pocket for binding and hydrolyzing ATP, as well as sites for attachment to actin (cf. Vibert & Cohen, 1988). Rayment *et al.* (1993b) used computerized image analysis to study the results of X-ray crystallography and electron microscopy of myosin. The major discovery was a cleft through the 50-kD segment, dividing the latter into upper and lower domains (figure 11.8B). The width of the cleft is controlled by the interactions of ATP with the pocket; the opening of the cleft weakens the binding of the 50-kD segment to actin, while the closure of the cleft strengthens the binding. The ATP pocket can also open, tilting the myosin head and thereby altering the angle of the cross-bridge (discussed next).

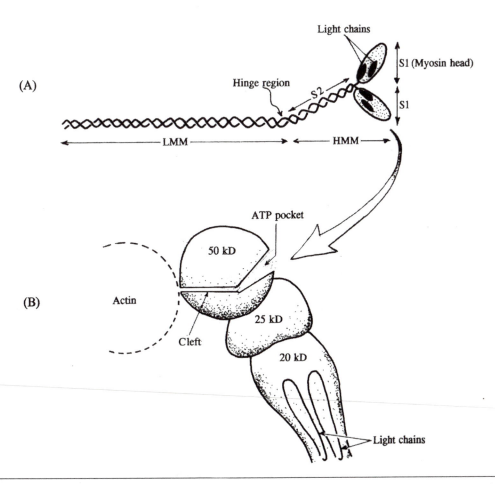

Figure 11.8   (A) Different components of the myosin molecule. Various proteolytic enzymes cleave the molecule into heavy mero-myosin (HMM) and light meromyosin (LMM). The point of cleavage is flexible in the intact molecule (hinge region) and serves to bring the cross-bridge to the surface of the myosin filament. The HMM moiety comprises the remainder of the α-helical rod (S2 fragment) and the two globular heads (S1 fragments), to which the light chains are attached. The two globular heads form a cross-bridge. (B) Enlargement of one globular myosin head to show the three components (segments). The 50-kD segment is divided into upper and lower domains by a cleft, the size of which is regulated by the interaction of ATP with its "pocket."

Based on Vibert and Cohen (1988) and Rayment *et al.* (1993a).

The Light Chains of Myosin

Myosin
light chains
regulate
the myosin
heads.

As noted previously, each myosin head contains two light chains, with different isoforms, and each of these has a molecular weight of approximately 20 kD. The two light chains appear to be attached to the neck region of the myosin head (figure 11.8A) and are termed regulatory and essential, respectively; they give the cross-bridge structural stability.

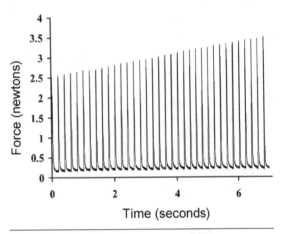

Figure 11.9 Active force during sequential twitch contractions of rat medial gastrocnemius muscle stimulated at $5 \cdot s^{-1}$.

Reprinted, by permission, from B.R. MacIntosh and D.E. Rassier, 2002, "What is fatigue?" *Canadian Journal of Applied Physiology* 27: 42-55.

Removal of the regulatory light chains of myosin increases Ca^{2+} *sensitivity* of the contractile proteins (Metzger & Moss, 1992), suggesting that the presence of the regulatory light chain decreases Ca^{2+} sensitivity. When Ca^{2+} concentration increases during muscle activation, some of the Ca^{2+} will bind to *calmodulin*, and this will activate *myosin light chain* kinase, resulting in *phosphorylation* of the regulatory light chains. It is thought that phosphorylation of the regulatory light chain causes the cross-bridge to swing out from a position close to the thick filament to a position that places the actin-binding portion of the myosin head in close proximity to actin (Levine *et al.*, 1996). Phosphorylation of the regulatory light chains is thought to contribute to *activity-dependent potentiation (staircase* and *posttetanic potentiation)* as illustrated in figure 11.9 and discussed later. Not only do the essential light chains influence the maximal velocity of shortening, but their removal reduces the force per cross-bridge by half (VanBuren *et al.*, 1994).

Thin Filaments

In chapters 1 and 2, actin filaments were encountered as part of the cytoskeletons of muscle and *motoneurons,* while in the present chapter, the actin filaments, through their attachments with the myosin heads, are contractile proteins. In the present section, we will learn a little more about their structure and about the special proteins, troponin and *tropomyosin,* that are associated with actin in the thin filaments of the muscle fiber.

The polarity
of actin
filaments
dictates the
direction
of force
generation
by cross-
bridges.

• **Actin filaments have a polarity.** With the electron microscope, an actin filament is seen as a thread about 1 μm long and 5 to 10 nm wide. The filament contains about 400 actin molecules, and each of these, in turn, is made up of 375 amino acids. Although the actin molecules are nearly spherical, they have a distinct polarity, and they line up facing in the same direction. Each filament is a double-helical strand of individual actin molecules (figure 11.10). This union of the actin molecules is brought about by polymerization, a process that requires the presence of a nucleotide (either ADP or ATP). An actin filament can be "decorated" by exposing it to myosin heads, which then appear as arrowhead-like structures along the filament. Since the actin molecules in each filament face the same way, the arrowheads behave similarly, pointing to the minus end of the filament. This interaction between actin and myosin demonstrates that the S1 segment of myosin can bind to actin.

Tropomyosin
and
troponin are
associated
with actin
filaments.

• **Additional thin filament proteins.** There are two important regulatory proteins that occur at regular intervals along the thin filaments: tropomyosin and troponin (see figure 11.10). Tropomyosin is rather like the tail of the myosin molecule. It consists of a double helix that lies on the surface of the actin filament and stiffens it; each tropomyosin molecule spans seven of the actin molecules. Troponin is a complex of three polypeptides, one of which (troponin T) attaches it to tropomyosin. Troponin I, in contrast, binds to actin and indirectly prevents the latter from interacting with the myosin head. During a muscle contraction, this inhibitory effect of troponin I is overcome by a sudden rise in cytosolic $[Ca^{2+}]$; each troponin C molecule binds four Ca^{2+} ions (see chapter 7) and,

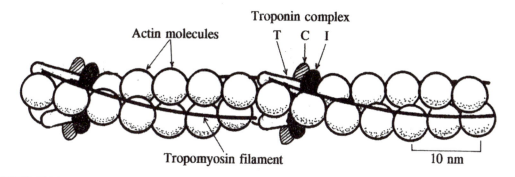

Figure 11.10   Schematic drawing of part of an actin filament showing the relationship between successive pairs of actin molecules, filaments of tropomyosin that span seven actin monomers, and the three types of troponin, regulating each tropomyosin. Another depiction of a (nonmuscle) actin filament is given in figure 2.4A.

as it does so, undergoes a conformational change that lifts the tropomyosin molecule away from the myosin-binding sites on the thin filament. This movement of tropomyosin away from the sites on the actin filament to which the myosin heads can become attached allows cross-bridge interaction to proceed.

The Role of Adenosine Triphosphate

The myosin head hydrolyzes ATP and tilts to produce the power stroke.

Since the original description of the sliding filament mechanism, evidence has gradually accumulated to show how the myosin cross-bridge might undergo repositioning to give a *power stroke* to the actin filament and how this movement could be related to the hydrolysis of ATP. It was long known that the combination of ATP with the myosin head was necessary to free the latter from the actin filament. Thus, if ATP is unavailable or incapable of being hydrolyzed in the fiber, the two types of filament remain locked together in the *rigor* state, as after death (rigor mortis). Lymn and Taylor (1971) proposed that the hydrolysis of ATP to *ADP* and Pi led to the formation of an intermediate complex with myosin (the myosin products complex). A further feature of this model was that the binding of actin to myosin took place in two stages, being weak initially and then strong. Another concept, suggested by several authors (e.g., Huxley & Simmons, 1971), was that the movement of the cross-bridge might arise from rotation of the globular myosin head. With the improved definition of the myosin head by Rayment *et al.* (1993b), it has been possible to provide a more complete account of the probable structure and function of the myosin head. This account, which incorporates several of the previous ideas, has as its main feature the opening and closing of the recently discovered cleft in the 50-kD segment of the myosin head. The cleft is associated with the actin-binding region of the myosin head. A lever arm action of the cross-bridge is thought to accomplish a translation of the thin filament by about 5 nm.

The sequence of steps involved in a cross-bridge cycle is illustrated in figure 11.11. Prior to activation (Ca^{2+} release), each myosin S1 segment is in a dissociated state, with ATP bound in the pocket (figure 11.11B). Partial hydrolysis of ATP to ADP·Pi causes the head to move to a "cocked" position, ready to bind to actin (figure 11.11C). When the actin site for myosin binding has been exposed by Ca^{2+} binding to *troponin* C, the following sequence occurs:

1. Myosin binds with actin in a weak-binding state (figure 11.11D).
2. Actin stimulates the complete hydrolysis of ATP to ADP, resulting in the release of Pi and transition to a strong-binding state corresponding to lever arm movement of the myosin head (the power stroke, figure 11.11E).
3. ADP is now released (figure 11.11A); and with replacement by ATP, the S1 segment dissociates from actin. ATP enters the pocket with the terminal γ phosphate first and with the adenosine ring still protruding from the pocket. This incomplete entry is sufficient to open the narrow cleft between the upper and lower domains of the 50-kD myosin segment, weakening its binding to actin and permitting the S1 segment to dissociate (figure 11.11B).

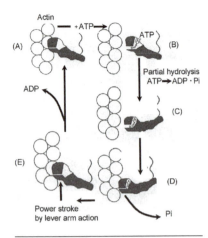

Figure 11.11 Hypothetical scheme of cross-bridge cycling proposed by Rayment *et al.* (1993a); the dark structure is the cross-bridge. See text.

Reprinted with permission from I. Rayment et al., 1993a, "Structure of the actin-myosin complex and its implications for muscle contraction," *Science* 261, 58-65. Copyright 1993 by AAAS.

Absence of ATP (as in death) prevents release of the cross-bridge from actin and stalls the cycle in the strong-bound state. This is referred to as rigor. A quick-time movie of the cross-bridge cycle is available at www.scripps.edu/milligan/projects.html.

The duration of a cross-bridge cycle varies, depending on temperature, *myosin isoform,* and velocity of shortening. The cycle could last as long as 50 ms; the myosin head remains attached to the actin filament in the strong-bound state for less than 50% of this time, and at maximal velocity of shortening, this duration can be less than 1 ms. The power stroke of fast-twitch myosin is a very brief event, probably lasting just 2 to 10 ms for an isometric contraction. In extraordinarily delicate experiments, in which single myosin molecules and actin filaments were allowed to interact, it was possible to estimate the forces and movements developed by individual myosin cross-bridges (Finer *et al.,* 1994). The movements, measured without any opposing force, are 5 to 11 nm. In some cases, this is rather larger than the movements predicted by Rayment *et al.* (1993a; see earlier discussion). When feedback is applied with a laser system, so as to prevent movement of the actin filament, a single myosin head generates a force of 3 to 4 pN, and the time spent in the strong-binding position is quite long.

Excitation-Contraction Coupling

Calcium is the second messenger that triggers muscle contraction.

As previously noted, the signal for a contraction to begin is a sudden rise in the concentration of Ca^{2+} ions in the vicinity of the myosin and actin filaments. In this way Ca^{2+} is a *second messenger,* acting as an intermediary between the *action potential* and the contractile process. The cellular mechanism by which the Ca^{2+} concentration is increased is part of the process called *excitation-contraction coupling* and takes place in two steps:

1. *Depolarization* of the *T-tubules* and activation of the dihydropyridine receptors
2. Diffusion of Ca^{2+} ions from the *sarcoplasmic reticulum* into the myoplasm through specialized channels called *ryanodine receptors* (figure 11.12)

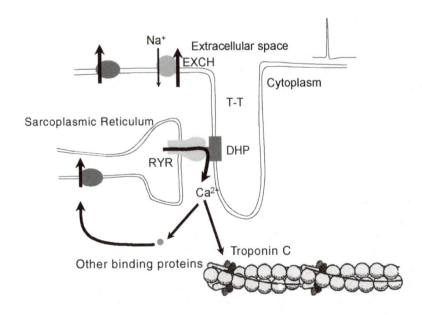

Figure 11.12 Excitation-contraction coupling. The action potential propagates along the surface *plasmalemma* and down the transverse tubule (T-T). Ca^{2+} is released from the sarcoplasmic reticulum via interaction of the *dihydropyridine receptor (DHPR)* with the ryanodine receptor (RYR), and diffuses throughout the myoplasm. Ca^{2+} is quickly sequestered by various binding proteins, including troponin C, SERCA, and calmodulin among others. Myoplasmic free Ca^{2+} concentration is restored to the resting level by transport primarily back into the sarcoplasmic reticulum, but also across the *sarcolemma.*

The Transverse Tubules

The detailed structure of the transverse tubular system (T-tubules) has been determined with the electron microscope. The tubules are situated at regular intervals along the muscle fiber and run inward toward its center, meeting with each other to form a continuous structure (see Peachey, 1965 and figure 1.13 in chapter 1). In the frog, the tubules are situated in the vicinity of the Z-disk and thus occur once per sarcomere. In mammals, however, the T-tubules are to be found at the junctions of the A- and I-bands, giving two tubular arrays per sarcomere. Electron micrographs have revealed that the tubules open onto the surface of the muscle fiber so that the tubular membrane becomes continuous with the sarcolemma. The existence of such openings has been confirmed by the ability of relatively large molecules to pass into the T-system from solutions bathing the muscle fiber. The substances chosen for this type of experiment have included ferritin (Huxley, 1964) and fluorescent dyes (Endo, 1966), as well as albumin, thorium dioxide, and *horseradish peroxidase.*

The key role of the T-tubule system in excitation-contraction coupling was shown by two contrasting experimental approaches. First, A. F. Huxley and Taylor (1958) applied weak stimulating currents at different points along the surface of frog muscle fibers. They found that local contractions of myofibrils occurred only when the electrode tip was over the I-band, in the center of which (at the Z-disk) the T-tubules are situated. In the second type of experiment, Gage and Eisenberg (1969a) demonstrated that action potentials were unable to elicit twitches in muscle fibers treated with glycerol. In these experiments, glycerol was added to the physiological bathing fluid and penetrated the muscle fibers. When the fibers were returned to a glycerol-free solution, the T-tubules swelled and broke off from the surface plasmalemma, preventing transmission of activation into the muscle fiber.

The speed with which the signal to contract travels into the fiber has been measured by González-Serratos (1971). The technique used was to embed single frog semitendinosus fibers in gelatin-Ringer and to compress them longitudinally to increase the diameter of the fibers; the myofibrils were photographed with a high-speed ciné camera; in shortening, they became straight instead of wavy. It was found that the evoked potential caused the superficial myofibrils to contract first and the central ones last. At 20° C the velocity of inward spread of activation was 7 cm $\cdot$ s^{-1}.

By themselves, the measurements of inward activation velocity and of its change with temperature do not enable the nature of the inward spread to be determined. In theory, two such mechanisms are possible. One is a passive electrotonic spread of depolarization along the T-system, and the other is a mechanism in which the tubules propagate action potentials into the center of the fiber. One way of deciding between these possibilities has been to abolish any propagated action potentials with tetrodotoxin and to depolarize the surface plasmalemma using an intracellular stimulating electrode. In this way, only the effects of electrotonic spread are observed. Adrian *et al.* (1969) found that when brief stimuli were used, only the superficial myofibrils would contract when the fiber membrane had been depolarized to 0 mV; larger surface depolarizations were necessary to make the central myofibrils shorten. In an ingenious extension of their work, the same authors imposed an artificial action potential on the membrane by using a voltage clamp circuit driven by an action potential in another fiber. It appears that the reversal of membrane polarity during the spike is just sufficient to activate the central myofibrils by electrotonic spread along the T-system. If, however, the same type of experiment is carried out on toad muscle at 20° C, it is found that the surface action potential produces only about 30% of the normal twitch tension, the remaining 70% presumably requiring a propagated *impulse* in the T-tubules (Bastian & Nakajima, 1974).

Other indirect evidence for a T-impulse has come from the work of Costantin (1970), who investigated the responses of single frog muscle fibers to focal stimulation through a microelectrode. Through reduction of the concentration of Na$^+$ in the solution bathing the fiber, the initiation of an action potential in the surface plasmalemma was prevented. Costantin found that the central and superficial myofibrils had very similar thresholds for contraction, suggesting that an active mechanism was present and still effective in the T-system even though it was absent at the fiber surface. This conclusion was strengthened by the results of experiments involving tetrodotoxin, in which the superficial myofibrils now had lower thresholds than the central ones and were the

The T-tubules conduct impulses into the interior of the fiber.

first to shorten. Under these last conditions, only passive electrotonic depolarization of the fiber interior could have taken place.

The most convincing evidence for an inwardly propagated action potential would come from recording it. Although there have been several attempts to do this (Natori, 1975; Strickholm, 1974), there is doubt as to the significance of the recorded potentials.

Ca^{2+} Ions Are Needed for Contractions to Take Place

It has been known that Ca^{2+} ions play an important role in muscle contraction ever since Ringer (1883) showed that the frog heart stops beating if this ion is omitted from the bathing solution. Mines (1913) was able to demonstrate that the quiescent heart still generates action potentials, indicating that Ca^{2+} ions must be required for some step beyond the excitation of the fibers. Similarly, in the case of skeletal muscle, Frank (1958) showed that frog muscle fibers will not develop *contractures* when placed in K^+-rich solutions unless Ca^{2+} is also present. Further evidence for the important role of Ca^{2+} is that injections of this ion into a muscle fiber through a micropipette evoke local contractions; Na^+, K^+, and Mg^{2+} ions are ineffective (Heilbrunn & Wiercinski, 1947). Another experiment has been to desheath a muscle fiber by removing the plasmalemma and then to induce a contraction by applying calcium ions directly to myofibrils with a micropipette (the Natori preparation; see Podolsky, 1964). This approach, as well as slight modifications to it, has been used to investigate the relationship between Ca^{2+} concentration and force (see "The Force–pCa^{2+} Relationship").

The Force–pCa^{2+} Relationship

The skinned fiber preparation has been successfully used to investigate the relationship between Ca^{2+} concentration and force in skeletal muscle fibers. In this approach, the membrane of the fiber is physically removed, or chemicals like triton or glycerol are used to make the membrane permeable. Under these circumstances, the fiber can be immersed in various solutions that simulate the intracellular environment. Different concentrations of Ca^{2+} can be used in these solutions, and force can be measured at each Ca^{2+} concentration, usually expressed as the negative log of the $[Ca^{2+}]$ or pCa^{2+}.

Figure 11.13 illustrates a force–Ca^{2+} relationship and demonstrates that this relationship can be modulated by several factors: temperature, pH, regulatory light chain phosphorylation, inorganic phosphate, length, myofilament lattice spacing, and *caffeine*. The primary mechanism by which Ca^{2+} sensitivity is altered in skeletal muscle is by changes in cross-bridge kinetics; and most of the modulators just listed operate by altering the rate at which cross-bridges enter the strongly bound state. This rate increases when the actin and myosin filaments are brought into close proximity or the cross-bridges swing out away from the thick filament backbone, closer to the actin fila-

ments (MacIntosh, 2003). This is associated with increased Ca^{2+} sensitivity.

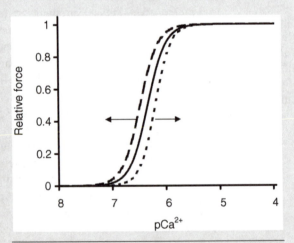

Figure 11.13 The sensitivity of the myofilaments to Ca^{2+} is expressed as force developed at any $[Ca^{2+}]$. Ca^{2+} concentration is expressed as the negative log of $[Ca^{2+}]$, pCa^{2+}. This kind of graph is typically created by measurement of force in a skinned fiber preparation exposed to a variety of Ca^{2+} concentrations. The solid line represents measurements obtained under control conditions, while the dashed lines represent conditions with increased (left shift) or decreased (right shift) Ca^{2+} sensitivity.

Ca²⁺ is
released into
the cytosol
from the
sarcoplasmic
reticulum.

Triggered Calcium Release

The Ca^{2+} ions, which are necessary for contraction to proceed, are released from the sarcoplasmic reticulum; the onset and duration of the efflux can be measured with Ca^{2+}-sensitive dyes. In the first such study, Ashley and Ridgway (1970, 1968) injected large, single muscle fibers of barnacles with aequorin, a protein that luminesces in the presence of Ca^{2+} ions. Subsequent measurements with faster-responding Ca^{2+} dyes have permitted a better understanding of how the Ca^{2+} transient relates to force of contraction in skeletal muscle. As an example, the twitch contractions and corresponding Ca^{2+} transients are illustrated in figure 11.14 for different fiber types of toadfish muscle fibers at 16° C. These measurements were obtained using *furaptra*, a very fast-responding fluorescent dye. Clearly Ca^{2+} concentration increases rapidly to a peak value, then decreases with a slower time course. The tension response is much slower. Relaxation begins at a time when free $[Ca^{2+}]$ has already decreased substantially.

Relaxation involves discontinuation of cross-bridge cycling. When Ca^{2+} dissociates from troponin, the cross-bridges in the area of control will complete the current cycle, then maintain a state of dissociation. It is not clear what limits the rate of relaxation in normal circumstances, the rate of fall in $[Ca^{2+}]$, dissociation of Ca^{2+} from troponin, or cross-bridge cycling.

The duration of the Ca^{2+} transient resulting from a single stimulus is dependent on the rate and duration of Ca^{2+} release from the *terminal cisternae* and the binding and pumping capacity of cellular entities. The time course of opening of individual ryanodine receptors (see later) can be evaluated by studying Ca^{2+} sparks, but this probably results in overestimation of this duration because a rising $[Ca^{2+}]$ is thought to promote closure of the Ca^{2+} release channel. The rate of rise of $[Ca^{2+}]$ in the vicinity of the ryanodine receptors would be faster if more channels were active. The duration of opening is also dependent on the type of muscle fiber that is studied, the temperature under which the study is conducted, and probably additional factors. Twitch fibers of frogs and fast-twitch fibers of small rodents contain *parvalbumin*, which accelerates the decrease in free cytoplasmic Ca^{2+} concentration. The binding and pumping properties of the sarcoplasmic reticulum Ca^{2+} ATPase are also very important.

The Ca^{2+} transients have also been measured when, instead of a single stimulus, a series of shocks are given to the muscle fiber. The Ca^{2+} concentration is seen to rise to a higher level and then to plateau (see figure 15.8, chapter 15). The rise in Ca^{2+} concentration in the cytosol following excitation is impressive; the resting concentration is only 10^{-7} M, but following successive stimuli it can rise to 10^{-5} M—a 100-fold change.

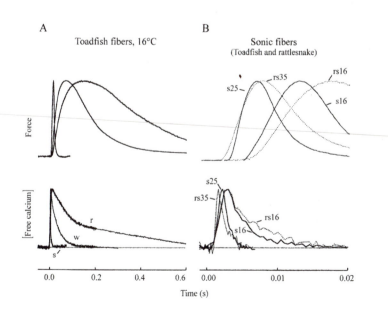

Figure 11.14 Transient rise and fall in Ca^{2+} concentration within the *cytosol* of muscle fibers is shown with corresponding force transients. These traces represent particularly fast muscles, so the rise and fall of $[Ca^{2+}]$ is very fast. Note that the rise of $[Ca^{2+}]$ precedes the development of force and that the speed of the Ca^{2+} transient corresponds with the speed of the muscle (fiber type: slow-twitch, r; fast-twitch, w; and superfast swim bladder muscle, s). Temperature also affects the speed of the Ca^{2+} transient, as shown in B for rattlesnake tail shaker muscle (rs) and toadfish swim bladder muscle (s) at 16° and 36° C.

Reprinted, by permission, from L.C. Rome et al., 1996, "The whistle and the rattle: The design of sound producing muscles," *Proceedings of the National Academy of Sciences of the United States of America* 93: 8095-8100.

The Ca^{2+} release mechanism involves two types of Ca^{2+} channel.

The Role of Calcium Channels

The action potential traveling down the T-tubules activates one type of Ca^{2+} channel, which interacts with another to release Ca^{2+} from the sarcoplasmic reticulum (SR). The first type is termed the dihydropyridine receptor (DHPR) because it binds dihydropyridine; it is an example of an L (large or long-lasting) Ca^{2+} channel and is described more fully in chapter 7. This type of channel is found in high concentration in the membrane of the T-tubules. According to Fosset *et al.* (1983), the density of DHPRs found in the T-tubules of skeletal muscle is from 50 to 100 times greater than that in any other tissue. Unlike L-channels in other membranes, however, the function of the skeletal muscle DHPR is not to admit Ca^{2+} ions, because vertebrate muscle fibers can still contract if this ion is omitted from the bathing fluid (Armstrong *et al.,* 1972). Rather, the DHPR appears to act primarily as a *voltage sensor* that transmits a signal to a second type of Ca^{2+} channel, found in the membrane of the SR. This second channel is the ryanodine receptor (RYR); it is a huge protein, consisting of four subunits, each of which has a molecular weight of 564 kD (see chapter 7). There are at least three isoforms of RYR: RYR1 is the form that is present in skeletal muscle; RYR2 is in cardiac muscle; and RYR3 is present in many cell types throughout the body.

Low-power electron micrographs of muscle fibers show that the T-tubules and SR have a special relationship; a T-tubule and the SR terminal cistern on either side form a *triad* (figure 11.15). Higher-power electron micrographs can identify the *DHP channels* in the T-tubule and the ryanodine channels in the SR. The two types of channel are then seen to have a precise relationship to each other; thus, the DHP channels occur in groups of four, and each of these *tetrads* faces every second ryanodine channel (figure 11.16). Even in some of the earliest electron microscopic studies of the triads, it was possible to identify additional structures, the *junctional feet* (Franzini-Armstrong, 1970), which span the 15-nm gap separating the T-tubular and SR membranes and are now known to compose part of the ryanodine channel. Three-dimensional reconstruction from multiple electron micrographs has revealed further details of the ryanodine channel (Wagenknecht *et al.,* 1989). There appears to be a central pore running through that part of the molecule that lies in the SR membrane. This pore is connected to four radial outlets in the foot part of the molecule, and these, in turn, open into the gap between the SR and T-tubule (see figure 11.17B). The central pore and the four radial outlets presumably provide the pathway for Ca^{2+} ions to exit from the SR into the cytosol.

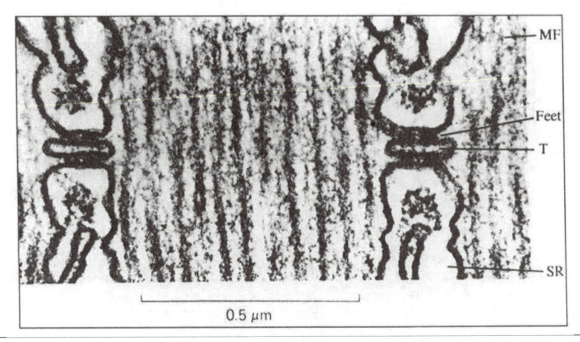

Figure 11.15 Longitudinal electron micrograph through part of a toadfish muscle fiber showing two triads. Note the flattening of the T-tubules (T) and the junctional feet linking the tubules to the adjacent SR membranes (SR). MF = myofibril.

Reprinted, by permission, from C. Franzini-Armstrong, 1980, "Structure of sarcoplasmic reticulum," *Federation Proceedings* 39: 2404.

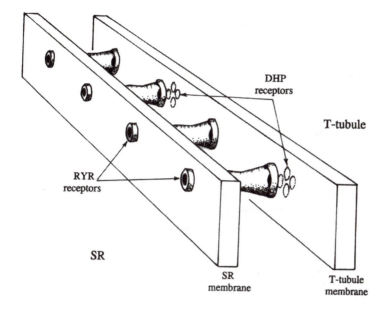

Figure 11.16 Schematic diagram of alternating arrangement of RYR and DHP receptors. The DHP receptors occur in tetrads (groups of four) and face every second RYR receptor.

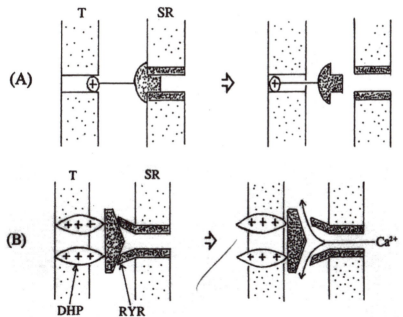

Figure 11.17 Probable connections between Ca^{2+} channels in the T-tubular and SR membranes: (A) is the mechanical "plunger" system proposed by Chandler *et al.* (1976), while (B) incorporates subsequent knowledge of the molecular structure of the ryanodine channel (RYR).

Adapted from Ríos *et al.* (1991, p. 129).

The DHP and ryanodine channels are connected mechanically.

How can a signal get from the DHP channel to the ryanodine channel? Well before the structures of the two channels were known, Chandler *et al.* (1976) suggested that they were physically coupled to each other. In this model the DHP channel, after activation, would be able to unplug the ryanodine channel by means of a connecting rod (figure 11.17A). The new information concerning the RYR, previously described, would be quite consistent with such a model. For example, the top of the ryanodine channel foot could be attached to the DHP channel; the conformational change in the DHP channel, brought about by a voltage signal in the T-tubule, could then lift the top off the ryanodine channel and allow Ca^{2+} to flow through the four radial outlets (Ríos *et al.*, 1991).

Chemical links between the DHP and ryanodine channels have not been excluded.

The "plunger" mechanism proposed by Chandler *et al.* (1976) may not be the only means by which the DHP channel can activate the ryanodine channel. Additional mechanisms of channel activation may involve Ca^{2+} ions and *inositol triphosphate* (IP_3). Such a role for Ca^{2+} would be consistent with the Ca^{2+} transients recorded in the cytosol following excitation; also, Ca^{2+} release from the SR does appear to be positively affected by the Ca^{2+} level in the cytosol. It is generally thought that a small increase in $[Ca^{2+}]$ will cause Ca^{2+}-induced Ca^{2+} release, but a substantial increase in $[Ca^{2+}]$ in the triad region will cause closure of these channels.

The evidence favoring IP_3 is that concentration of this messenger increases rapidly in the cytosol after stimulation of the muscle fiber (Nosek *et al.,* 1990) and that, even in very low concentrations, it can release Ca^{2+} ions from the SR. One obvious role for Ca^{2+} or IP_3 would be to turn on the alternate ryanodine channels that are not apposed by the DHP tetrads.

Inactivation of Contraction

Ca^{2+} ions are pumped into the SR from the cytosol.

Muscle contractions must be terminated at the appropriate moment, and the muscle fiber achieves this by pumping Ca^{2+} ions from the cytosol back into the SR. The average Ca^{2+} concentration during repetitive stimulation will be the result of a balance between Ca^{2+} release, with each activation, and Ca^{2+} removal by the Ca^{2+} ATPase and the Ca^{2+}-buffering molecules. Once repetitive activation is terminated, $[Ca^{2+}]$ will fall rapidly.

Since the pump has to move Ca^{2+} ions against a large concentration gradient, it requires ATP to provide the necessary energy. With each cycle of the pump, two Ca^{2+} ions are transported into the SR in return for two K^+ ions; this unequal exchange of positive charges causes the cytosol near the SR to become increasingly negative. On gaining entry to the SR, the Ca^{2+} ions are detached from the pump and taken up by the Ca^{2+}-binding protein *calsequestrin*. Calsequestrin has a high capacity for binding Ca^{2+}, so very large quantities of Ca^{2+} can be stored in a relatively small volume. Since individual pumps have a low capacity for transporting ions (e.g., $20 \cdot s^{-1}$), it is necessary for the SR to have a high density of pumps in its membrane in order to quickly bring down the Ca^{2+} concentration in the cytosol. Freeze-fracture electron micrographs reveal rounded particles, almost certainly the Ca^{2+} pumps, that are packed tightly against each other in the SR membrane. Further details of the Ca^{2+} pump were considered in chapter 7.

The Contractile Response

The elementary contractile event is the twitch contraction.

When the free $[Ca^{2+}]$ increases in response to a stimulus and Ca^{2+} binds to troponin, the cross-bridge cycle is initiated in all activated segments of the thin filaments. This is the contractile response of the muscle. The intensity and time course of the contractile response will depend on the number and frequency of stimulations delivered to the muscle fiber or motor unit.

The Twitch Contraction

The smallest contractile response that can be elicited is the twitch. The delay between the detection of the action potential and the first indication of contractile activity (shortening or force development) is referred to as the electromechanical delay. This delay occurs because of the time needed for conduction of activation into the T-tubules, release of Ca^{2+} into the cytoplasm, binding of Ca^{2+} to troponin, and initiation of cross-bridge cycling. The T-tubule action potential results in brief opening of the RYR channels and flow of Ca^{2+} from the terminal cisternae into the cytoplasm. The released Ca^{2+} binds to troponin C, permitting cross-bridge cycling. Each troponin molecule will bind up to two Ca^{2+}, and the amount bound will depend on the magnitude and duration of the Ca^{2+} transient. This magnitude and duration can be influenced by several factors, including fiber type, temperature, species, and parvalbumin concentration. The proportion of troponin that is saturated with Ca^{2+} will determine the number of cross-bridges that can participate in force development. Ca^{2+} is then removed from the cytosol by the pumping action of the Ca^{2+} ATPase in the SR; the cross-bridges stop cycling; and the muscle relaxes.

The rise and fall of tension in the twitch contraction has been used to characterize the elementary contractile capabilities of a muscle (or motor unit or muscle fiber). Figure 11.18 illustrates a twitch contraction of the rat medial gastrocnemius muscle at 37° C. This figure illustrates the measurement of developed tension, *contraction time,* peak rate of tension development, peak rate of relaxation, and *half-relaxation time.* It has recently been demonstrated that developed tension at long lengths should be estimated by subtraction of the passive force expected at the fascicle length reached at the peak of the contraction (MacIntosh and MacNaughton, 2005).

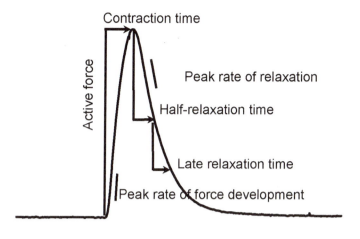

Figure 11.18 A twitch contraction of the rat medial gastrocnemius muscle, showing the measured parameters: active force, contraction time, half-relaxation time, peak rates of force development and relaxation. The late relaxation time is more representative of an exponential half-time for decay of force than is the traditional half-relaxation time.

Tetanic Contractions

Summation of contraction results from repetitive stimulation.

When a second stimulation and subsequent stimulations are applied before complete relaxation occurs, *summation* results, and the contractile response is referred to as a tetanic contraction. This is illustrated in figure 11.19. Peak force in a tetanic contraction depends on the frequency of activation and the number of activations. When the stimulation is of sufficient duration to achieve a plateau, at each of several frequencies, a force–frequency relationship can be obtained (see figure 11.20). Characteristic features of the force–frequency relationship include the following: *fusion frequency,* peak force, and the twitch:*tetanus* ratio. Fusion frequency is the lowest frequency for which oscillations in force are not evident. Peak force is the highest force obtained, and the twitch: tetanus ratio is the ratio of the developed force of a twitch contraction to peak force. These characteristic properties are different between fiber types within a species, and between species for a given fiber type. Furthermore, these properties are dependent on local circumstances within a given muscle, such as prior contractile activity, temperature, and extracellular fluid composition.

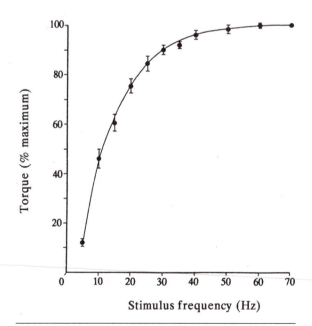

Figure 11.20 Force–frequency curve for human plantarflexor muscles. As the stimulus frequency to the tibial nerve is raised, the force developed by the muscle (expressed here as torque) becomes larger. However, beyond a certain optimal frequency, in this case 60 Hz, no further increase in force can be generated.

Reprinted, by permission, from D.G. Sale et al., 1982, "Extent of motor unit activation during effort," *Journal of Applied Physiology* 51: 1637.

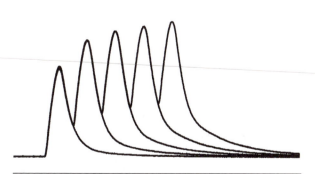

Figure 11.19 Active force for twitch and brief tetanic contractions of the medial gastrocnemius muscle, obtained by stimulation of the rat sciatic nerve at 40 Hz with one to five pulses. When the interpulse interval is decreased, the increment in force increases, resulting in greater summation.

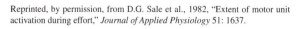

Activity-Dependent Potentiation

A further property of contracting muscles is that after the tetanus has been stopped, the twitch response to a single shock, given during the next few minutes, will be enlarged; the same phenomenon can be demonstrated following a voluntary contraction (see figure 11.21), particularly in fast-twitch muscles. This behavior has been termed posttetanic potentiation after electrically elicited contractions and postactivation potentiation after voluntary contractions. Such activity-dependent potentiation has been attributed to phosphorylation of the regulatory light chains on the myosin head (Moore & Stull, 1984), but another possible mechanism is elevation of Ca^{2+} in the cytosol for a short while after the end of the tetanus or voluntary activation (Allen *et al.*, 1989). Another phenomenon with a similar underlying basis is the positive staircase, which was originally observed in the frog gastrocnemius muscle but has also been demonstrated in fast-twitch mammalian muscles including the human adductor pollicis muscle (Slomic *et al.*, 1968). It consists of a progressive enlargement of isometric twitches when stimuli are given at a low repetition rate, such as 5 Hz, as shown in figure 11.10. Repetitive activation can also result in enhanced force of brief submaximal tetanic contractions (MacIntosh & Willis, 2000).

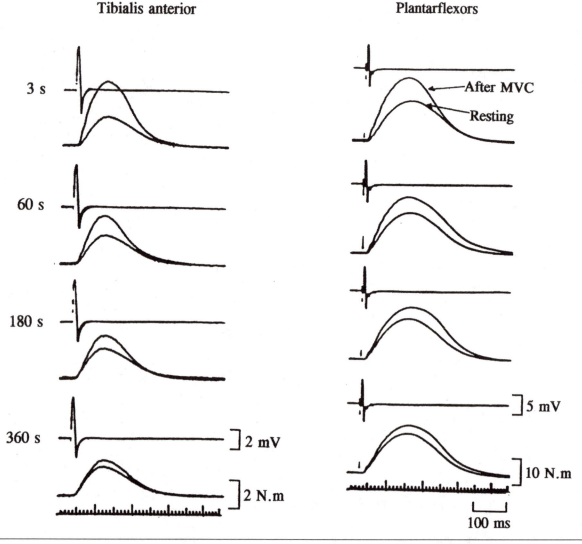

Figure 11.21   Twitch potentiation at different times after a brief maximal voluntary contraction (MVC). In each pair of traces, the upper is the *M-wave* (compound muscle action potential), and the lower shows two twitches, the smaller of which is the control (resting) twitch and the larger is the potentiated one (after MVC). Note that potentiation is greater for the fast-twitch tibialis anterior than for the slow-twitch plantarflexors and that it is not associated with any change in the M-wave.

Adapted, by permission, from A.A. Vandervoort, J. Quinlan, and A.J. McComas, 1983, "Twitch potentiation after voluntary contraction," *Experimental Neurology* 81: 145.

Chemicals and Drugs Can Modify the Contractile Response

The duration of a twitch contraction is prolonged by cooling the muscle fiber or by applying a variety of chemical agents (see Sandow, 1965 for review). Lyotropic anions such as bromide, nitrate, and methylsulfate seem to act by reducing the mechanical threshold of the fiber. Thus, the twitch is prolonged because the contractile machinery is switched on with less membrane depolarization during the action potential. Divalent metal ions such as Zn^{2+} and uranyl can augment the twitch tension by a factor of two to three times; they do so by increasing the duration of the action potential and hence of the calcium transient. Caffeine in low concentrations (0.5 to 4 mM for rat muscle) will enhance Ca^{2+} release and increase Ca^{2+} sensitivity (see figure 11.13). At higher caffeine concentrations a contracture will occur. Caffeine produces this effect by spontaneous release of Ca^{2+} from the SR and inhibition of its subsequent reuptake by the Ca^{2+} pump.

In contrast, the magnitude of the twitch response can be diminished by reducing the availability of Ca^{2+} ions to the myofibrils. One method to achieve this is to bind the Ca^{2+} as an organic salt by injecting potassium citrate or potassium oxalate into the fiber. Another technique is to immerse the fiber in hypertonic saline; although the fiber shrinks, the T-tubules and the *lateral sacs* become dilated. This behavior would be expected of the T-system, because this is really an extracellular space that will draw water from the fiber by osmosis; the SR appears to behave as an intermediate compartment between the intra- and extracellular fluids. Evidently, the end result of the distension of the triads is to reduce the release of Ca^{2+} from the lateral sacs (Ashley & Ridgway, 1970). The effect of epinephrine is to speed up the isometric twitch; in theory, the tension developed should be rather smaller, but Marsden and Meadows (1970) observed little change in the human soleus muscle. These authors interpret their findings in terms of an action of epinephrine on β-receptors on the muscle fiber membrane.

Dantrolene sodium also reduces the magnitude of the twitch contraction and does so by interfering with the release of Ca^{2+} from the SR (Desmedt & Hainaut, 1977). Patch-clamp experiments have demonstrated a direct effect of this drug on SR Ca^{2+} channels (Suarez-Isla *et al.*, 1986). Dantrolene is an important drug in the treatment of patients with *malignant hyperthermia* (see p. 173).

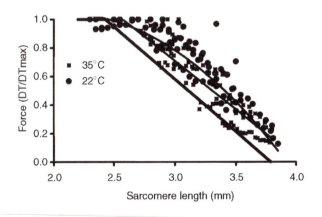

Figure 11.22 The descending limb of the force–length relationship for mouse extensor digitorum longus muscle fiber bundles *in vitro*. The thick solid line represents the theoretical values, based on filament lengths. Observed developed force is shifted to the right, probably due to the "fixed-end" nature of these contractions. Sarcomere length was measured at the beginning of the contraction but would be expected to decrease during the contractions (similar to what is seen in the fascicles in figure 11.5). Temperature has no effect on the force–length relationship.

From D.E. Rassier, 2002, "Sarcomere length-dependence of activity-dependent twitch potentiation in mouse skeletal muscle," *BioMed Central-Physiology* 2: 19. By permission of D.E. Rassier.

The Length Dependence of Force

When a muscle is maximally activated, the isometric force that is developed depends on the length at which the muscle is held (Rassier *et al.*, 1999). At very short lengths, active force is small. The length–tension relationship has an ascending limb, a plateau, and a descending limb (from short to long length). When the muscle is tested at progressively longer lengths, the active force increases to a maximum, which corresponds with the sarcomere length at which myofilament overlap is optimal. Beyond optimal length, isometric active force decreases, but not in the linear manner that would be predicted by the experiments of Gordon *et al.* (1966b) described earlier. The length–tension relation observed for small bundles of muscle fibers is illustrated in figure 11.22. The contractions with which this relation was obtained were fixed-end tetanic contractions.

There is an incomplete understanding of the mechanisms responsible for less than maximum force along the ascending limb of the length–tension relation. Possible explanations include misalignment of the thin filaments, due to overlap of thin filaments from opposite

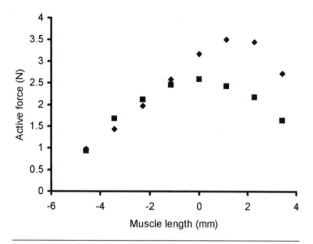

Figure 11.23 The force–length relationship varies with the level of activation. Here, brief tetanic contractions of the rat medial gastrocnemius muscle are shown (2 pulses at 200 Hz, squares; or 10 pulses at 50 Hz, diamonds). Muscle length is relative to the length at which the double-pulse stimulation gave the greatest active force. Note that the largest 50-Hz contraction occurred at a longer length. This is typical of length-dependent activation. When activation is submaximal, length has an effect on Ca^{2+} sensitivity (see figure 11.13), such that force continues to increase at longer lengths despite the decrease in filament overlap.

ends of the sarcomere; and at very short lengths, the thick filaments abut the Z-disk.

These considerations relate to the special circumstance when activation is maximal. The length dependence of force is somewhat different when activation is submaximal, particularly when sarcomere length is greater than optimal length. At long sarcomere lengths, Ca^{2+} sensitivity increases, due to increasing proximity of the myofilaments (see "The Force–pCa^{2+} Relationship"). This increased sensitivity at long lengths results in a shift in the apparent plateau in the relationship between length and active force. Peak force occurs at a longer length for twitch contractions and contractions elicited with low-frequency (submaximal) stimulation (see figure 11.23).

It has been known for quite some time that the length dependence of active force with submaximal activation does not have the same relationship as with maximal activation (Rack & Westbury, 1969). Surprisingly, active force continues to increase beyond the length associated with maximal myofilament overlap, reaching a peak at a muscle length that is longer than the length at which maximal activation yields the strongest contraction. It has been confirmed that changes in Ca^{2+} sensitivity account for this length dependence of activation (Endo, 1973). An increase in Ca^{2+} sensitivity means that greater force is obtained at a given $[Ca^{2+}]$ than would otherwise be predicted. This effect is able to compensate for the decreasing number of cross-bridges available for interaction.

The mechanism of length-dependent activation is associated with the geometry of the myofilaments. As muscle length is increased, the actin filaments are positioned closer to the myosin filaments. This proximity of the myofilaments increases the probability of cross-bridge interaction and permits more of the available cross-bridges to be in the strong-binding state than would be the case when inter-myofilament spacing was greater. This effect of proximity of the myofilaments is illustrated in figure 11.24.

In vivo contractions depend on joint angle.

In general, muscles in the human body are thought to operate along the ascending limb and at the plateau of the length–tension relationship. In addition to the sarcomere length dependence of force, the torque generated by a muscle depends on the moment arm, which depends in turn on the location of attachment of the tendons to the bones and the angle of the bones about the joint of concern.

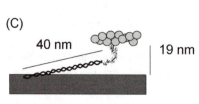

Dynamic Contractions

So far in this chapter, consideration has been given only to isometric or fixed-end contractions. In reality, muscles are used in a variety of ways. They can shorten or elongate during activation, or they can remain at a

Figure 11.24 Calcium sensitivity in skeletal muscle is altered primarily by proximity of the S1 head of myosin with the actin filament. (A) At long sarcomere lengths, the myofilaments are brought close together, because of the constant-volume nature of a muscle fiber. Ca^{2+} sensitivity is high under this condition. (B) At short sarcomere lengths, the filaments are farther apart and Ca^{2+} sensitivity is decreased. Several factors can increase Ca^{2+} sensitivity under this condition by (C) causing the myosin head to swing out toward the thin filament. These factors include increased pH, decreased temperature, and increased myosin light chain phosphorylation.

Reprinted, by permission, from B.R. MacIntosh, 2003, "Role of calcium sensitivity modulation in skeletal muscle performance," *News in Physiological Sciences* 18: 222-225.

fixed length (isometric contraction). More often than not, muscles are used with a combination of these forms of contraction.

When an object is lifted and then held in a fixed position, the relevant contraction would be shorten-isometric. In many ballistic movements, muscles undergo stretch and then shortening, or a stretch-shorten combination (also known as a stretch-shorten cycle). This is true for the knee extensors in jumping movements and for the shoulder flexors in throwing movements. Even during walking, the gluteus maximus is active in a stretching contraction during the later stages of the swing phase, and this is followed by a *shortening contraction* during the stance phase.

In the past, the terms "concentric" and "eccentric" have often been used to describe shortening or *lengthening contractions,* respectively; but Faulkner (2003) has recently indicated that these terms are not appropriate. For this reason, the terms shortening and lengthening contractions will be used in this book.

The Force–Velocity Relationship

Consider a situation in which progressively larger loads need to be lifted. It is a well-known observation that with maximal effort the speed of the lift becomes increasingly slower as the load is made larger. In the laboratory, this phenomenon may be investigated by stimulating the muscle repetitively (tetanically) and measuring the velocity of muscle shortening for different loads (figure 11.25A). This type of experiment represents a special circumstance in which the load that the muscle shortens against remains constant for each contraction. These contractions are referred to as *isotonic.* One can see that not only is the velocity of shortening reduced when a relatively large load is applied, but there is a latent period before the object is moved at all. During this latent period, the muscle is contracting isometrically until sufficient tension has been developed to equal the load. As shown in figure 11.25A, not only is the rate of shortening reduced for heavy loads, but also the amount of shortening is decreased. An experiment of this type enables the force–velocity properties of muscle to be plotted (see figure 11.25B). It is unlikely that true isotonic contractions occur *in vivo,* because the moment arm changes during excursions of the limbs. This means that with a constant torque, muscle force must be changing.

A point to be emphasized is that irrespective of the circumstances of the contraction—shortening, isometric, or lengthening—the interactions between the myosin heads and the actin filaments remain the same, in that the cross-bridges engage the actin filaments and attempt to slide them along. In a shortening contraction with a small load, the sliding movement allows the myosin filaments to become completely overlapped by actin. In an isometric contraction, the amount of filament overlap depends on the length at which the muscle is held prior to activation. During activation the cross-bridges repeatedly make and break connections with the actin filaments, producing tension equal to the external load, but no movement of the actin filaments occurs. In a lengthening contraction, the cross-bridges

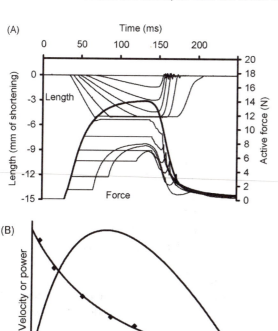

Figure 11.25 (A) A series of isotonic contractions of the rat medial gastrocnemius muscle *in situ* is shown. All contractions begin at the same length, and load is regulated by pressure in a cylinder through which the lever pushes a piston during the tetanic muscle contraction. Redevelopment of force occurs when the lever reaches the limit of excursion (for lower pressures). Force rises isometrically until the force exerted by the muscle overcomes the load imposed; then force remains constant while shortening occurs at a velocity that is greater when the load is smaller. (B) The force–velocity and power–velocity relationships are shown. Individual points represent values obtained from isotonic contractions similar to those in A, while the line is the line of best fit of the Hill equation.

collectively generate less tension than the external stretching force applied to the muscle, and the opposing actin filaments in the sarcomeres are pulled away from each other.

Power Output of Muscle

By definition, the power developed by a muscle in a contraction is equal to the product of the load and the velocity with which it is moved. Thus:

$$\text{power} = \text{load} \times \text{velocity}$$

Inspection of the *force–velocity relationship* (figure 11.25B) indicates that power will be zero both when the load on the muscle is zero and when the load is so heavy that it cannot be moved at all (velocity = zero); at some intermediate load, the power will be greatest. Hill (1938) employed a hyperbola to describe the relationship between force and velocity in isolated muscles. However, although this is satisfactory over a broad range of values, when the load exceeds 80% of isometric force, the velocity of shortening is less than the predicted value. This discrepancy is evident in figure 11.25B. In view of this discrepancy, Edman and colleagues (1988) have described the force–velocity relationship with a double hyperbola.

The double hyperbola can be explained if one assumes that each cross-bridge has the potential to engage an actin filament in two strong-binding positions. In an isometric contraction, cross-bridges are unable to make the transition to the second position, but with shortening at less than 80% of maximal isometric force, this transition is permitted.

The ability to perform work at a high rate or generate power is an important aspect of human movement. This is true for athletic pursuits such as jumping, kicking, throwing, and sprinting as well as for activities of daily living. In each instance, the force–velocity properties of individual muscles set the upper limits of performance. In turn, the shape and magnitude of the force–velocity relation are dictated by three parameters: maximal isometric force, maximal velocity, and the degree of curvature. Maximum isometric force is dictated by the physiological cross-sectional area of the muscle, and maximal velocity is a function of fiber type and muscle length (number of sarcomeres in series). The degree of curvature depends on temperature and also on fiber type, since slow-twitch fibers have a greater degree of curvature in their force–velocity relationship than fast-twitch fibers.

Applied Physiology

In this section, we consider the curious condition of muscle contracture, in which the fibers remain contracting even though they are no longer firing impulses. We will discover that the most serious of these disorders is due to abnormal Ca^{2+} fluxes and can cause death during anesthesia. There is a strain of mouse, however, in which a Ca^{2+} defect results not in contracture, but in paralysis.

Spontaneous Muscle Contracture

A contracture differs from a contraction in that the activation of the myofibrils is prolonged in the absence of action potential activity on the plasmalemma. One method of producing a contracture is to immerse the fiber in a solution rich in K^+ ions. By reducing the K^+ *equilibrium potential,* these ions maintain the membrane in a depolarized state and thereby cause a lasting release of Ca^{2+} ions from the SR. Another technique is to add caffeine to the bathing solution; the nature of this drug action was described earlier in the chapter. Contracture may also occur as part of a disease process; in *McArdle's disease* (*myophosphorylase* deficiency), the muscle fibers remain in a contracted state and swell after physical exertion, at a time when impulse activity in the membrane has ceased (see chapter 14).

It should be noted that the term "contracture" is used here in a physiological sense. Unfortunately, the term is also applied by clinicians to permanent shortening of muscle produced by excessive growth of fibrous tissue within the muscle belly and possibly by resorption of sarcomeres as well (Tabary *et al.,* 1972). One of the best examples of this type of contracture is the severe shortening of the calf muscles that affects boys with advanced *Duchenne muscular dystrophy* (see chapter 1). In such cases, it is best to refer to the phenomenon as a "pathological" contracture to avoid confusion with the "physiological" type considered in this section.

In rigor there is a special state of contraction that is independent of action potential activity. In this condition, unlike a contracture, the fiber is depleted of ATP. Because of this deficiency, the cross-bridges on the myosin molecules cannot be detached from the actin filaments, and the myofibrils are unable to relax. A rigor state supervenes after death, once the ATP in the fiber has been consumed by the various autolytic reactions that take place during the early degenerative processes.

Malignant Hyperthermia

A hereditary disorder, malignant hyperthermia, can cause sudden anesthetic deaths.

Denborough and Lovell (1960) discovered an Australian family in which no less than 10 members had died unexpectedly after having been given a general anesthetic. The manner of the deaths was alarming. Within minutes of starting an anesthetic, the patient would develop a rapid pulse, become cyanosed, and have increasing stiffness of his or her skeletal muscles. Since the body temperature could rise as high as 43° C (109° F), the condition was termed malignant hyperthermia (MH). The Australian cases were identical to other reported instances of unexpected anesthetic deaths, and it was clear that the condition was a hereditary one, transmitted by an autosomal dominant gene.

The incidence of the MH gene in the population is much higher than the incidence of MH reactions during surgery (1:15,000 in children), for many patients with the susceptibility can tolerate one or more general anesthetics without complications. The anesthetic agents that have been incriminated are halothane, ether, ethyl chloride, methoxyflurane, trichloroethylene, ethylene, and cyclopropane. Another triggering agent is *succinylcholine,* a drug that depolarizes skeletal muscle fibers by combining with the ACh receptors and is commonly employed as a muscle relaxant. With the increased awareness of MH, the routine monitoring of body temperature during surgery, and the availability of dantrolene (an excitation-contraction blocking agent, see p. 169) and of body-cooling apparatus, the incidence of anesthetic deaths from MH has declined from 80% of events to 7%.

Impaired Calcium Handling

Malignant hyperthermia is due to abnormal calcium release.

It was suspected that MH might be due to abnormal handling of Ca^{2+} inside the muscle fibers. For one thing, muscle contraction is controlled by the release of Ca^{2+} from the SR into the cytosol, and were the latter to be prolonged for any reason, the muscle fibers would stay contracted as they do in MH. At the same time, Ca^{2+} would stimulate a number of muscle enzymes, including phosphorylase kinase; the latter would increase the breakdown of glycogen in the muscle fibers, producing excessive amounts of lactate and heat—two other features of the MH reaction. Suggestive evidence for a Ca^{2+}-based mechanism is that muscle fibers from MH-susceptible patients are unusually sensitive to caffeine, a compound that increases Ca^{2+} release from the SR and can cause both potentiation of the twitch and muscle contractures. Indeed, the most reliable test of genetic susceptibility until now has been to expose a biopsy specimen of muscle to a bathing fluid containing halothane and caffeine, and to find the concentrations of drug that, singly and in combination, produce contractures of the fibers (Kalow *et al.,* 1970).

Faulty Ryanodine Receptors

In pigs and in some human MH subjects, the RYR is the culprit.

The first clue to the location of the abnormal gene came from studies on strains of pigs susceptible to MH. These pigs were bred for their heavy muscling and leanness, but it was also noted that occasionally during stress (e.g., during transport or prior to slaughter) animals might die spontaneously and also develop large, pale, soft areas in their muscles. Genetic studies showed tight linkage between the genes for MH and for the enzyme glucose phosphate isomerase (GPI). In humans, the GPI gene occurs on the long arm of chromosome 19 and is close to the gene for the RYR. The next step was to show, in MH-susceptible pigs, that the RYR was the abnormal protein due to the substitution of a single amino acid (cystine for arginine; Fujii *et al.,* 1991). This small change in protein structure is apparently sufficient to prevent the ryanodine channel from closing normally (Fill *et al.,* 1990).

Although the RYR is the culprit in all breeds of pig susceptible to MH, the situation in humans is more complicated. In a minority of patients, the RYR is indeed abnormal, but in others the MH trait is due to some other deficit in the cell or is part of another recognized neuromuscular disorder, such as central core disease or Duchenne muscular dystrophy. Evidently, any lesion that can produce abnormally large or prolonged Ca^{2+} levels in the muscle fiber cytosol is capable of initiating an MH reaction (MacLennan & Phillips, 1992).

Mice With Muscular Dysgenesis

There is a very interesting disorder of mice, caused by a mutant gene, in which the animals are almost totally paralyzed. In this condition, named muscular dysgenesis, physiological recordings show that despite the absence of contractile activity, action potentials propagate normally along the *myotubes* (Powell & Fambrough, 1973). This dissociation of events strongly suggests that the defect is due to a disorder of excitation-contraction coupling. Furthermore, in affected animals, the triads are disorganized in their microscopic structure; there is a 5- to 10-fold reduction in DHP-binding sites (Pinçon-Raymond *et al.,* 1985); and the slow Ca^{2+} current, normally mediated by the DHP channel, is absent (Beam *et al.,* 1986). This combination of findings points to the DHP channel as the primary abnormality in muscular dysgenesis; and indeed, injection of an expression plasmid containing the cDNA sequence for the DHP channel can restore excitation-contraction coupling in the myotubes (Tanabe *et al.,* 1988). It is not yet known, however, how the affected myotubes can be rescued by coculture with normal spinal cord (Rieger *et al.,* 1987). Contractions are also restored if dysgenic *myoblasts* are cultured with normal *fibroblasts* (Courbin *et al.,* 1989); in this case it is probable that the two types of cell fuse together and that the fibroblasts provide the genetic code for the DHPRs needed by the myoblasts.

In the last three chapters, we have seen how impulses, discharged from motoneurons in the spinal cord, result in contraction of the muscle fibers. In the next chapter we will find that the colonies of muscle fibers, controlled by single motoneurons, differ from each other in various ways, including the speeds of their contractions.

12

Motor Units

So far the muscle and nerve fibers have been treated as single cells. It is time to consider the anatomical and physiological relationships between these two types of cells and then to study how the body makes use of the neuromuscular system. As Sherrington (1929) appreciated, all the varied reflex and voluntary contractions of a muscle are achieved by different combinations of active *motor units*. This term, motor unit, describes a single *motoneuron* and the many muscle fibers to which its *axon* runs (figure 12.1). The motor unit can be considered the elementary functional unit of our regulated motor responses. Expressed differently, all movements are planned by the central nervous system in terms of motor units rather than individual muscle fibers. Some of the more important aspects of motor unit function can now be discussed.

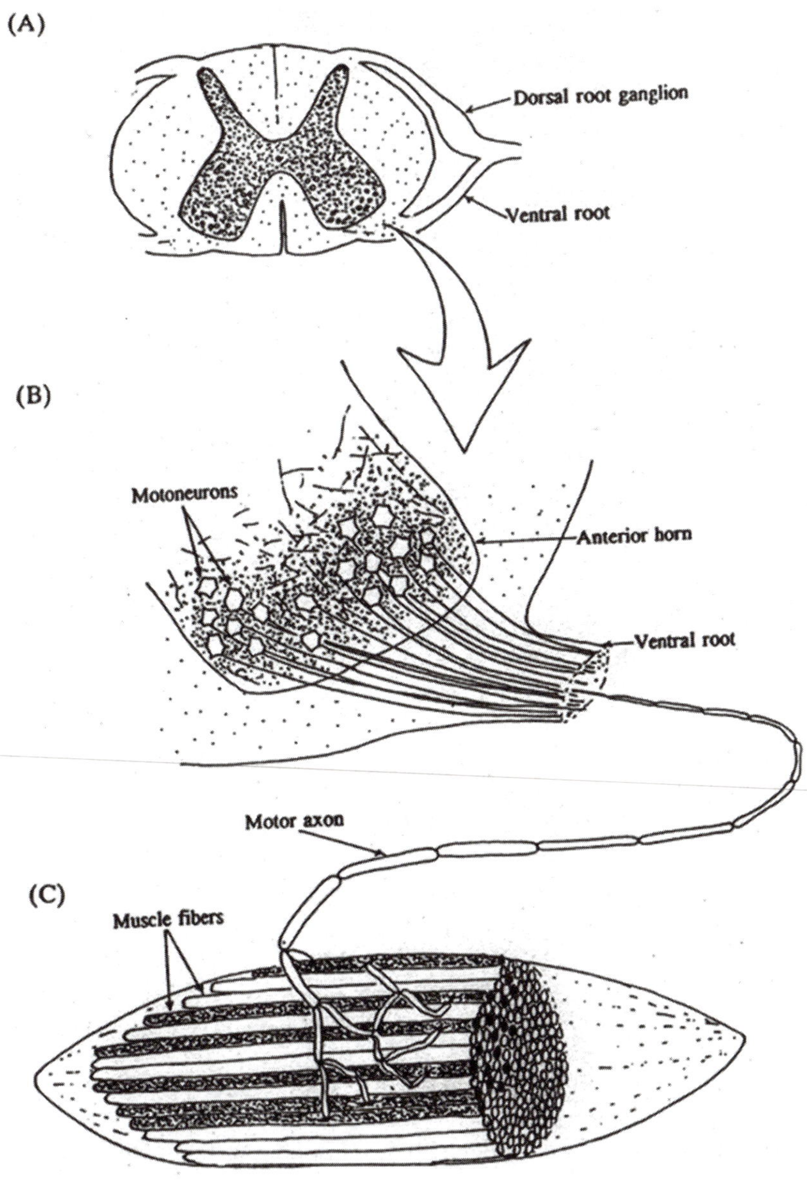

(A)

Dorsal root ganglion

Ventral root

(B)

Motoneurons

Anterior horn

Ventral root

Motor axon

(C)

Muscle fibers

Figure 12.1 The anatomy of a motor unit. (A) Cross section of the spinal cord to show gray and white matter, with dorsal and ventral nerve roots. (B) Enlargement of anterior (ventral) horn with motoneurons, one of which is shown sending its axon through the *ventral root* to the muscle (below). (C) The muscle fibers supplied by the branches of the single axon are darkened and can be seen on the surface of the muscle belly and where the belly has been transected. This population of fibers, together with the motor axon and motoneuron, constitutes a single motor unit. Most muscles have a hundred or more such units.

Organization of Motor Units

One of the problems that has repeatedly interested anatomists concerns the number of motor units in individual muscles. In this chapter, we will examine how these numbers can be determined in animal and even human muscles. Of equal interest, perhaps, are the sizes of the motor units, that is, the numbers of muscle fibers supplied by single motoneurons through branching of their axons. In addition, there is the architecture of the motor units to consider, by which is meant the way that the muscle fibers of any one motor unit are distributed within the muscle belly. Last, the type of motor unit, which affects the speed, strength, and fatigue characteristics of the movement in which the unit is recruited, will be considered.

Estimating Motor Unit Numbers

Some motor nerve fibers innervate the muscle spindles.

To find the numbers of motor units in individual muscles, the usual approach in animals has been to cut dorsal roots distal to their ganglia (i.e., dorsal rhizotomy) and to allow time for degeneration of the severed sensory nerve fibers. If the nerve to a muscle is then examined, the surviving nerve fibers are those that left the cord in the ventral roots and are motor in function. Until the important study of Leksell (1945), it was not realized that the smallest motor axons in the ventral roots, the gamma (γ)-axons, are those passing from fusimotor neurons to the small muscle fibers inside spindles (see chapter 4). In the first studies, the numbers of motor units estimated were therefore too high, in most cases by 20% to 40% (see, for example, Eccles & Sherrington, 1930). Even in these early studies, however, it was clear that there were considerable differences in the numbers of motor units between muscles within the same animal.

In animals, motoneurons can be counted after retrograde labeling.

The numbers of motor units can also be determined in animals by the *retrograde* labeling of motoneurons. The substance originally used for this purpose was *horseradish peroxidase (HRP)*; when applied to a muscle belly as a conjugate with wheat germ agglutin, HRP is taken up by the motor nerve terminals and transported in the axons to the motoneuron cell bodies. After 24 hr or longer, depending on the length of the motor axons, the spinal cord is fixed, removed, and sectioned serially. The neurons are stained with tetramethyl benzidine, which reacts with HRP, and motoneurons specific to the injected muscle can then be recognized under the light microscope and counted (Mesulam, 1982). Since HRP may sometimes spread from the injected muscle to neighboring ones, more reliable results may be obtained by cutting the motor nerve and placing the proximal stump in a solution containing HRP. As an alternative to HRP, new types of tracers have been developed; these include fluorescent markers such as the dextran-amines (Glover *et al.,* 1986). The latter allow double- and even triple-labeling of motoneurons so that motoneurons identified as to their innervating muscle can be categorized according to other biochemical and ultrastructural properties.

In human muscles, anatomical estimates of motor units have depended on comparisons with pathological material.

For obvious reasons, it has not been possible to determine numbers of motor units in human muscles with the same precision as in the animal studies. The best that can be done in humans is to count the number of large myelinated fibers in a muscle nerve and to assume that a certain proportion of these are alpha (α) motor and that the remainder are sensory fibers from spindles and tendon organs. The uncertainty introduced by this assumption is evident from the fact that the proportion of large fibers that are motor varies between 40% and 70% in nerves of different muscles in the cat hindlimb (Boyd & Davey, 1968). In the study of Feinstein *et al.* (1955), 60% of the large-diameter fibers were regarded as α-motor. This value was chosen on the basis of comparison with observations of the muscle nerves of a single patient who had died from poliomyelitis; no motor axons were thought likely to have been left. Leaving aside the questionable validity of this assumption, there is no doubt that Feinstein and colleagues (1955) have provided the best anatomical data for the numbers and sizes of human motor units.

Estimates of Motor Unit Numbers in Various Human Muscles

Small muscles of the eyes and face have large numbers of small motor units.

Table 12.1, which includes most of the results obtained by Feinstein *et al.* (1955), emphasizes the considerable differences in motor unit numbers between limb muscles. Note that the intrinsic muscles of the foot and hand (lumbrical, first dorsal interosseous) have fewer motor units than the

Table 12.1 Comparison of Numbers and Sizes of Motor Units in Various Human Muscles

Muscle	No. of motor units	Muscle fibers/unit
External rectus[a]	2,970	9
Platysma[a]	1,096	25
First lumbrical[a]	96	108
First dorsal interosseous[a]	119	340
Thenar group[b]	203	—
Brachioradialis[a]	333	>410
Tibialis anterior[a]	445	562
Gastrocnemius (medial)[a]	579	1,934
Cricothyroid[c]	112	166
Masseter[d]	1,452	640
Temporalis[d]	1,331	936
Rectus lateralis[e]	4,150	5
Flexor digiti minimi[f]	130	108

[a]Feinstein et al. (1955). [b]Lee et al. (1975). [c]Faaborg-Andersen (1957), from table 1 of Enoka (1995). [d]Carlsoo (1958), from table 1 of Enoka (1995). [e]Torre (1953), from table 1 of Enoka (1995). [f]Santo Neto et al. (1998).

larger muscles (brachioradialis, tibialis anterior, and medial gastrocnemius) situated nearer the trunk. However, because the motor units in the distal muscles are relatively small, these muscles have greater numbers of units in relation to their masses. Table 12.1 also draws attention to the remarkable properties of the external eye muscles (external rectus) and the superficial muscles of the face and neck (platysma), which contain large numbers of very small units.

Although the anatomical data are valuable, axon counting is tedious and depends on the availability of post mortem tissue. A very different approach enables motor unit numbers to be estimated during life and can be performed rapidly; this electrophysiological method is based on the comparison of an average motor unit potential with the response evoked from the whole muscle (see "Estimation of Human Motor Units").

Table 12.2 gives some of the results obtained by this method. The computer-based version produces rather lower estimates than the original "manual" technique, partly because corrections can be made for "alternation" (see "Estimation of Human Motor Units"). How well do the numbers of units estimated by the computer-based method compare with the anatomical determinations? The sparse anatomical data are not particularly reliable, for reasons given previously; nevertheless, comparison of tables 12.1 and 12.2 shows that there is fair agreement for the tibialis anterior muscle and a better match for the median-innervated thenar muscle group. Since the median nerve normally supplies two of the thenar muscles completely (abductor pollicis brevis and opponens pollicis) and part of a third muscle (flexor pollicis brevis), it would appear that each of the intrinsic muscles of the hand has about 100 motor units. This conclusion would be consistent with Feinstein et al.'s values of 119 units for the first lumbrical (table 12.1). A surprising result is that the biceps brachii, a large muscle, appears to have a number of motor units similar to that in one of the small muscles in the hand. Further, if the values for humans are compared with those of smaller mammals (table 12.3), it would seem that in the course of evolution the motoneuron populations supplying the small muscles of the hand increased, in relative terms, much more than those innervating the larger limb muscles. Presumably this trend allowed the extraordinary manual dexterity of human beings to develop.

Physiological estimates of human motor units agree with anatomically derived numbers.

Table 12.2 Numbers of Motor Units in Control Subjects Below the Age of 60 Years, Estimated by Automated and Manual Incremental Methods[a]

Muscle	Automated method [mean ± SD (n)]	Manual method [mean ± SD (n)]
Thenar	230 ± 90 (90)	342 ± 89 (115)
Hypothenar	411 ± 174 (41)	390 ± 94 (109)
Biceps brachii	109 ± 43 (80)	—
Extensor digitorum brevis	143 ± 73 (86)	210 ± 65 (151)
Tibialis anterior	256 ± 107 (22)	—
Soleus	—	957 ± 254 (41)
Vastus lateralis	224 ± 112 (24)	—

[a]For information about the methods of estimation, see "Estimation of Human Motor Units."

Table 12.3   Number of Motor Units in Muscles of Small Animals

	Mouse	Author	Rat	Author	Cat	Author
Extensor digitorum longus	20	Bateson & Parry (1983)	40	Close (1967)	130	Boyd & Davey (1968)
Tibialis anterior	223	Harris & Wilson (1971)	—	—	200	Boyd & Davey (1968)
Soleus	22	Lewis & Parry (1979)	30	Close (1967)	155	Boyd & Davey (1968)
Lumbrical	—	—	11	Betz *et al.* (1979)	—	—
Gastrocnemius	—	—	41 (lat)	Gillespie *et al.* (1986)	280 (med)	Boyd & Davey (1968)
			103 (med)	Arasaki *et al.* (1997)		
			134 (med)	Vanden Noven *et al.* (1994)		

lat = lateral head; med = medial head.

Motor unit populations vary among subjects.

Variation of Motor Unit Sizes Among Individuals and Within Muscles

A further surprise from motor unit estimations in humans is that there is so much variation in the results for the same muscle between healthy individuals of the same sex, size, and age. Part of this variation is undoubtedly due to methodological factors, especially in the sampling of motor units of different sizes. However, it is quite clear that some individuals have significantly more motor units than others, and this phenomenon tends to be generalized throughout the body musculature. It is not yet known whether this is a genetic endowment and, if so, whether it is due to greater proliferation of motoneurons during embryogenesis or to reduced cell death (see chapter 6). Evidence of variation among individual muscles has also come from animal experiments using HRP to label motoneurons; for example, Wasserschaff (1990) obtained a coefficient of variation of 21% for the mouse tibialis anterior muscle.

Bulky muscles contain the largest motor units.

Various methods have been employed to determine the sizes of motor units, that is, the number of muscle fibers innervated by individual motoneurons. For a given muscle, the average size of the motor units may be estimated simply by dividing the total number of muscle fibers by the number of motoneurons. It has been found, for example, that a relatively large muscle, the human medial gastrocnemius, has an average of about 2,000 fibers in each unit, whereas the much smaller external rectus muscle has only 9 (Feinstein *et al.,* 1955; table 12.1).

Even in the same limb muscle, motor units may vary 10 to 100 times in their sizes.

Electrophysiological studies show that, even within the same muscle of an individual animal or human subject, the sizes of motor units vary so greatly that the estimation of mean size has limited significance. For example, Burke (1967) stimulated single motoneurons and measured the tensions produced in the cat triceps surae muscle. He found that the *twitch* tensions ranged from 2 to 950 mN and noted that the scatter of results was much greater in the gastrocnemius muscle than in the soleus (see figure 12.8). Although these results indicate that the motor unit sizes vary considerably, they do not permit the absolute numbers of muscle fibers to be determined. These determinations may be performed using a technique developed independently by Edström and Kugelberg (1968) and by Brandstater and Lambert (1969). In this method, a single motor axon is stimulated repetitively, and the muscle is then excised and stained with periodic acid-Schiff (PAS) reagent for glycogen. The glycogen-depleted fibers are those that have been stimulated and therefore correspond to a single motor unit (figure 12.2A). In the rat tibialis anterior muscle, the sizes of the motor units are within the 50- to 200-fiber range (Brandstater & Lambert, 1973; Edström & Kugelberg, 1968). As might be expected, the comparatively large tibialis anterior

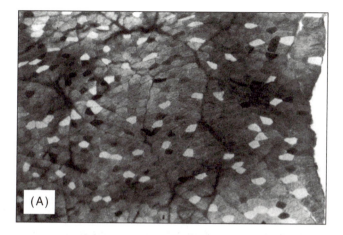

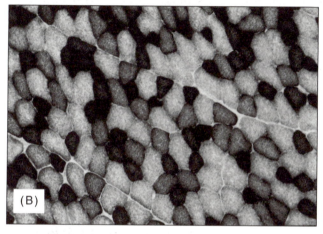

Figure 12.2 (A) Part of a motor unit in the tibialis anterior muscle of a rat. A single motor axon was stimulated repetitively; the activated muscle fibers became depleted of glycogen and failed to stain with PAS, appearing white in the photograph. (B) Transverse section of tibialis anterior muscle from a 3-month-old mouse, stained for succinic dehydrogenase activity. Three types of fiber can easily be differentiated on the basis of their staining reactions; the darkest fibers are type I and the lightest are type IIB. Magnifications: A, 150×; B, 320×.

A: Courtesy of Dr. M.E. Branstater; B: Courtesy of Dr. Ethel Cosmos.

In human muscles, the relative sizes of motor units can be determined by comparing their electrical and mechanical responses.

and extensor digitorum longus muscles of the cat have large units, the range observed by Mayer and Doyle (1970) being 43 to 1,099 fibers. In the cat triceps surae complex, the sizes of the motor units in the soleus range from less than 50 to more than 400 fibers, while in the medial gastrocnemius, an average unit contains between 400 and 800 fibers (Burke *et al.*, 1974; Burke & Tsairis, 1973).

Of course, this technique also has its limitations. Most muscles do not contain fibers that extend throughout the whole length of the muscle proper. Therefore, a sample of the muscle cross section will contain only a representative portion of the muscle's entire *complement* of muscle fibers. Even in relatively small but architecturally complex muscles of small mammals, this phenomenon is one for which it is difficult to correct—in transverse sections taken some distance apart, it is difficult to recognize the same fibers and therefore to know which depleted fibers are the same as those on the previous section. Monti and colleagues (2001) reported, for example, that in the cat tibialis anterior, motor unit territories spanned from 56% to 62% of the muscle length and that fibers within the motor unit were often distributed at different positions along the longitudinal axis of the muscle.

Estimation of Motor Unit Properties in Humans

In humans, it is not possible to determine the absolute sizes of motor units by the glycogen-depletion technique. However, the relative sizes of motor units can be assessed by measuring the sizes of their electrical and mechanical responses when motor axons are excited by threshold stimuli (McComas *et al.*, 1971a; Garnett *et al.*, 1979). This can be accomplished via *percutaneous* stimulation, intramuscular stimulation, or intraneural microstimulation.

In one of the earliest investigations of human motor unit properties, Sica and McComas (1971) used graded percutaneous electrical stimulation to demonstrate that the twitch tensions of 122 single units in the extensor hallucis brevis (EHB) varied from 20 to 140 mN (mean 54 ± 22 mN). Using intramuscular stimulation, Garnett *et al.* (1979) reported that, in the medial gastrocnemius, the range of motor unit tension was 15 to 2,000 mN. A problem with this approach is that individual axons branch within the nerve well before they reach the muscle and some of these measurements may have included just part of a motor unit. A recent stimulation method has been to excite motor axons by fine tungsten wires inserted through the skin and into the nerve trunk (intraneural microstimulation technique; Westling *et al.*, 1990). With this technique, the forces developed by thenar motor units, following stimulation of median nerve axons above the elbow, range from 3 to 34 mN (Thomas *et al.*, 1990b). First dorsal interosseous motor units range from 1 to 430 mN (Elek *et al.*, 1992; Young & Mayer, 1981), and EDL/EHL/EDB units from 6.3 to 78.1 mN (Macefield *et al.*, 1996) with use of this technique.

Twitch tensions of human muscles have also been studied by the technique of *spike-triggered averaging* during voluntary contractions (see figure 13.3, chapter 13). With this approach, single motor units in the first dorsal interosseous muscle of the hand are found to have a 100-fold range in twitch tension (Milner-Brown *et al.*, 1973b). Other results for this and different muscles are given in table 12.4.

One of the problems with the spike-triggered averaging technique is that, even during the weakest contractions, the frequency of motor unit discharge (8-12 Hz; Monster & Chan, 1977) is still sufficiently high to cause the twitches to become partially fused. This fusion, in turn, would be expected to reduce the amplitudes of the twitches by approximately half and to shorten the *contraction times* to an even greater extent (Thomas *et al.*, 1990a). The fact that such reductions are not observed, when the results of spike-triggered averaging and motor axon stimulation are compared for the same muscle, may indicate that other mechanisms may counteract this tendency. These mechanisms might include synchronization of the measured motor unit with that of another unit, potentiation and compliance effects, and nonlinear force summations (Thomas *et al.*, 1990a; Kossev *et al.*, 1994).

Rather less satisfactory is to measure the sizes of the motor unit potentials recorded with large surface electrodes. Although the size of each potential will be proportional to the number of muscle fibers in that unit, the potential will also be influenced by the diameters of the fibers and their proximity to the stigmatic recording electrode. Figure 12.3 should be viewed with these qualifications in mind; it suggests that there is a greater than 30-fold variation in motor unit size in the extensor digitorum brevis muscle but that the number of large units is small.

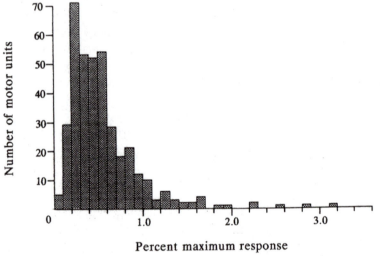

Figure 12.3 The sizes of 380 motor unit potentials in the extensor digitorum brevis muscles of 41 control subjects below the age of 60. Each value has been expressed as a percentage of the maximum response in order to compensate for variations in electrode placement and in the thickness of the tissues overlying the muscle.

Reprinted, by permission, from A.J. McComas et al., 1971, "Electrophysiological estimation of the number of motor units within a human muscle," *Journal of Neurology, Neurosurgery, and Psychiatry* 34: 128.

How Are Muscle Fibers Within the Motor Unit Distributed in the Muscle?

It was once thought that the muscle fibers supplied by a single axon were distributed in clusters, or subunits, within the muscle belly. The validity of this concept was questioned by Ekstedt (1964) on the basis of recordings with a special electrode that had small recording surfaces 180° apart (the "Janus" electrode). He observed that when a motor unit fired, usually only one of the two surfaces was next to an active fiber; had subunits been present, the electrode should have been surrounded by discharging fibers. It is now recognized that the subunit arrangement is true of motor units only in partially denervated muscles, in which *collateral reinnervation* of muscle fibers has taken place (see chapter 17).

The correct architecture of the motor unit has been revealed by the glycogen-depletion method. Within a given volume of muscle, the fibers in a single motor unit are arranged randomly among those of other units, with only a few fibers in the same unit being contiguous (figure 12.2A, Brandstater & Lambert, 1969; Edström & Kugelberg, 1968).

Glycogen-depletion studies also show that fibers belonging to the same motor unit can be distributed over a surprisingly large volume of muscle, for example, up to a quarter of the cross-sectional area in the tibialis anterior and up to three-quarters of the soleus in both cat and rat (Edström & Kugelberg, 1968; Bodine-Fowler *et al.,* 1990; Monti *et al.,* 2001; Burke & Tsairis, 1973). The motor unit types also differ in average territory in the order *FF > FI > FR > S* (Monti *et al.,* 2001). When seen in muscle cross sections, the motor unit territories tend to be elliptical rather than circular and to contain regions in which the fibers may be present in high densities or else altogether lacking, as in the FR (fast contracting, fatigue resistant) unit in figure 12.4; the uneven distributions have been attributed to the presence or absence of major axonal branches (Bodine-Fowler *et al.,* 1990). In figure 12.5, the fiber distributions in muscle cross sections have been displayed graphically, and their unevenness is reflected in the varying heights of the peaks. Although the fibers of a single motor unit tend to be randomly disposed within a dense region, remarkably few fibers touch (Kelly & Schotland, 1972; Willison, 1978), with the implication that the elimination of *polyneuronal innervation* during embryogenesis involves a mechanism that discriminates against the clustering of fibers belonging to the same unit (see chapter 6).

Investigation of Human Motor Units Using Needle Electrodes

In humans, the information available on motor unit territories has come from electrophysiological recordings, using either a multilead electrode (Buchthal *et al.,* 1957) or a scanning electrode that is slowly drawn through the muscle belly. In the scanning technique, a second electrode is maintained in a constant position and is used to ensure that the potentials picked up by the scanning electrode come from the same motor unit (Stålberg & Antoni, 1980). Both techniques show that the territories of individual human motor units have cross-sectional diameters of 5 to 10 mm, with 10 to 25 units overlapping with each other. Even greater degrees of overlap, involving at least 40 to 50 motor units, have been found by Burke and Tsairis (1973) in the cat medial gastrocnemius muscle.

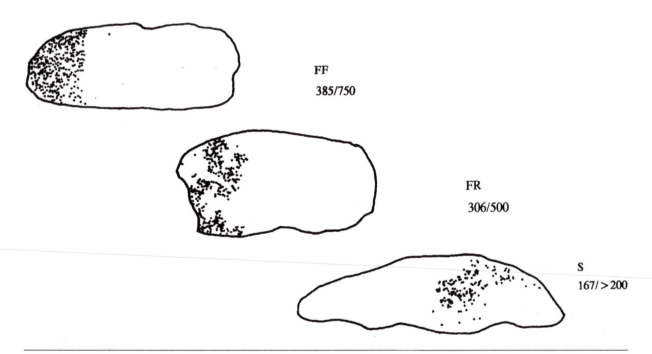

FF
385/750

FR
306/500

S
167/ > 200

Figure 12.4   Territories of three different motor units in the cat medial gastrocnemius muscle, as determined by the glycogen-depletion method. Since in this muscle the fibers do not run the full length of the belly, only about half of the respective fiber populations can be seen in a single cross section. For each of the three units shown, the number of depleted fibers in the section/estimated total number of fibers in the unit is given at the right, together with the designations of the motor units. (The designations FF, FR, and S are discussed later in this chapter.)

Adapted, by permission, from R.E. Burke and P. Tsairis, 1973, "Anatomy and innervation ratios in motor units of cat gastrocnemius," *Journal of Physiology* 234: 749-765.

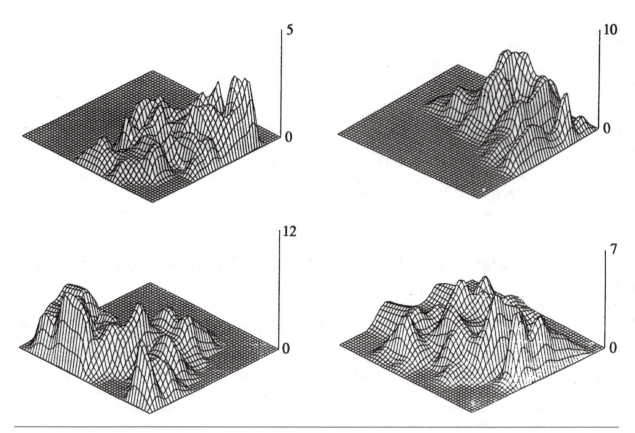

Figure 12.5   Distributions of muscle fibers within four single motor unit territories of the rat tibialis anterior. The muscle cross sections have been stained and divided into "quadrants," based on the territorial boundary of a single motor unit; the number of fibers in each quadrant has been counted and displayed on the vertical axis.

From "Spatial distribution of muscle fibers within the territory of a motor unit," by S. Bodine-Fowler et al., 1990, *Muscle and Nerve*, 13, p. 1138. Copyright 1990 by John Wiley & Sons, Inc. Reprinted with permission of John Wiley & Sons, Inc.

Finally, it is important to recognize not only that motor units may extend over a considerable part of the muscle cross-sectional area but also that, in long muscles, they may include muscle fibers in different regions of the longitudinal axis. For example, the hamstring muscles of the thigh are composed of two or more serial arrays of muscle fibers separated from each other by fibrous *inscriptions* (Loeb *et al.,* 1987). A single motor axon may innervate fibers belonging to different arrays (Manzano & McComas, 1988; Wines & Hall-Craggs, 1986).

Physiological and Biochemical Properties of Motor Units

Pale fibers are most plentiful in superficial parts of muscle bellies and fiber fascicles.

Motor units differ from each other not only in size but also in the biochemical and physiological properties of their muscle fibers. Ranvier (1873) provided evidence of differences, observing in a comparative study of several species that some muscles contracted more slowly than others and had a more pronounced red coloration. The red color of the muscle is due to the increased amounts of the oxygen-binding pigment myoglobin in the muscle fibers. These fibers also have a rich capillary network. Denny-Brown (1929) noted that the slowly contracting red muscles are those that are used continuously in the maintenance of posture. In contrast, the fast-contracting pale muscles are employed intermittently, being reserved for movements of a nonrepetitive nature.

Denny-Brown (1929) pointed out that some mammalian muscles contain both pale and red parts; for example, the deeper region of the cat tibialis anterior is red, but the superficial part is pale. In other muscles, and this is particularly true of humans, red and pale muscle fibers are generally intermingled so as to form a mosaic. Within *fascicles* there is a tendency for pale fibers to form a ring under the perimysial *connective tissue* border, with a layer of dark fibers immediately underneath (Pernuš & Erzen, 1991).

With the development of methods for staining muscles histochemically, further differences between muscle fibers have emerged. Some fibers are found to stain strongly for glycolytic enzymes such as phosphorylase and α-glycerophosphate dehydrogenase. Other fibers, because of their numerous mitochondria, are rich in oxidative enzymes such as malate and succinate dehydrogenase, which are associated with these organelles (figure 12.2B). Some of the differences between muscle fibers will now be considered in more detail.

Differences Among Motor Units in Twitch Contraction Time

Twitch contraction times vary among mammalian motor units.

Following the lead of Henneman and his colleagues (McPhedran *et al.*, 1965; Wuerker *et al.*, 1965), it has become routine to record the twitch and tetanic forces developed by single motor units in mammalian muscles. The techniques employed have been either to divide ventral root filaments until stimulation yields all-or-nothing contractile and electromyographic responses, or to excite single motoneurons through intracellular microelectrodes (Burke *et al.*, 1973). Although it is not possible to review the results for the various species examined, it can be said that, for most muscles, there are considerable ranges in the times taken for motor units to contract and relax. One conventional measure is the contraction time of a twitch, also known as the time to peak tension; this is the interval elapsing between the onset of tension and the peak value (see figure 11.18, chapter 11). In the cat triceps surae, the contraction times are as short as 20 ms in the medial gastrocnemius and as long as 130 ms in the soleus (Burke *et al.*, 1974). In the corresponding muscles of smaller mammals, such as the rat and mouse, the motor unit contraction times also display considerable variations but are shifted toward briefer values. Thus, the smaller the animal, the shorter the possible duration of contraction. In all species investigated, the soleus has the motor units with the longest contraction times, while other hindlimb muscles, such as the extensor digitorum longus and plantaris, are mainly composed of motor units with fast twitches.

Human motor units also differ in their contraction times.

In human muscles, some motor units have contraction times as short as 20 ms but other contraction times as long as 140 ms. An example of the spectrum of contraction times is given in figure 12.6 for the extensor hallucis brevis muscle, and table 12.4 contains values for other human muscles. Note that the masseter has mostly fast-twitch motor units, even though the histochemical profile of this specialized muscle does not reflect a predominance of "fast" muscle fibers (Butler-Browne *et al.*, 1988; see table 12.6). In the extensor digitorum communis and medial gastrocnemius, there are fairly even mixtures of fast- and slow-twitch units (Garnett *et al.*, 1979; Monster & Chan, 1977); in all of these muscles,

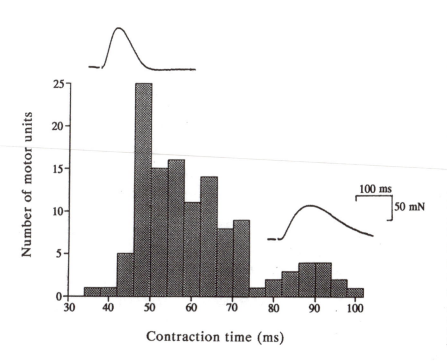

Figure 12.6 Contraction times of 122 motor units in human extensor hallucis brevis muscles. (The contraction time is the interval between the onset of the twitch and the peak tension.) Examples of fast and slow motor unit twitches are shown above the histogram.

Adapted, by permission, from R.E.P. Sica and A.J. McComas, 1971, "Fast and slow twitch units in human muscle," *Journal of Neurology, Neurosurgery and Psychiatry* 34:118.

Table 12.4 Contractile Properties of Human Motor Units

Muscle	CT[a] (ms) Mean ± SD	CT[a] (ms) Range	Force[b] (mN) Mean ± SD	Force[b] (mN) Range	Method[c]	Authors
Masseter	57 ± 14	24-91	58 ± 58	1-329	STA	Yemm (1977)
	35 ± 10	20-72	—	20-763	STA	Nordstrom & Miles (1990)
Temporalis	49 ± 11	30-76	76 ± 59	3-297	STA	Yemm (1977)
Abductor pollicis brevis	74 ± 20	37-102	28 ± 25	2-163	STA	Thomas et al. (1987a)
Thenar	50 ± 9	35-80	11 ± 8	3-34	INS	Thomas et al. (1990b)
	49.5 ± 11.5	28-72	8.8 ± 7.4	1.5-33.2	SS	Doherty & Brown (1997)
Extensor carpi radialis	38 ± 13	10-921	3.5 ± 7.6	0.13-16.57	STA	Schmied et al. (1997)
Tibialis anterior	46 ± 14	20-86	25.5 ± 21.5	0.7-171	STA	Van Cutsem et al. (1997)
Toe extensors (EDL/EHL/EDB)	75 ± 4	50-107	20 ± 4	6.3-78.1	INS	Macefield et al. (1996)
First dorsal interosseous	56 ± 13	30-100	15 ± 18	1-122	STA	Milner-Brown et al. (1973b)
	—	32-122	—	2-294	STA	Stephens & Usherwood (1977)
	65 ± 18	34-140	34 ± 47	2-423	IMS	Young & Mayer (1981)
	68	40-99	52	3-215	STA	Thomas et al. (1986)
	—	—	—	1-100	STA	Dengler et al. (1988)
	—	36-140	39	6.3-3,500	STA	Vogt et al. (1990)
	63 ± 15	10-110	16 ± 18.7	1-137	IMS	Elek et al. (1992)
	47 ± 13	20-90	18 ± 20	0.2-105	STA	Kossev et al. (1994)
Flexor carpi radialis	—	—	—	0.2-98	STA	Calancie & Bawa (1985)
Extensor digitorum communis	—	40-70	—	5-49	STA	Monster & Chan (1977)
Extensor hallucis brevis	63	35-96	54 ± 22	20-140	SS	Sica & McComas (1971)
Medial gastrocnemius	—	40-110	200 ± 304	15-2,000	IMS	Garnett et al. (1979)

[a]CT = contraction time (i.e., the time elapsing between the start of the contraction and the moment of peak tension during an isometrically recorded twitch). [b]Published values expressed in grams were converted to mN. [c]STA = spike-triggered averaging; INS = intraneural stimulation; IMS = intramuscular stimulation; SS = surface stimulation.

the ranges of contraction times are similar to those of the very much smaller muscles in the hand and foot (see table 12.4).

It is obvious that the shape of the whole-muscle twitch will depend on the proportions of motor units with different contraction times. Figure 12.7 contrasts the twitches of four limb muscles in the same individual, recorded with a piezoelectric device pressed over the muscle bellies. It can be seen that in the facial muscles the units are nearly all of the fast-twitch type, because the mean contraction time is only 43 ms; in contrast, the relatively long contraction time in the human calf muscles indicates the presence of a substantial slow-twitch motor unit population. Last, in the small muscles of the hand and foot, the contraction times are intermediate between those of the calf and facial muscles, indicating similar numbers of fast- and slow-twitch units (see also table 12.5). Of

Contraction times of whole muscles reflect motor unit composition.

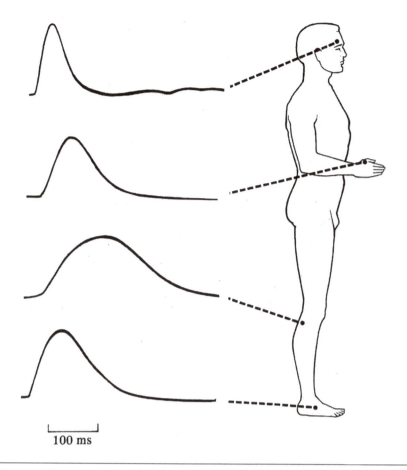

Figure 12.7 Isometric contractions recorded from four human muscles using a stiff piezoelectric bender that was pressed into the muscle bellies. Note the marked variation in twitch duration.

Reprinted, by permission, from A.J. McComas and H.C. Thomas, 1968, "Fast and slow twitch muscles in man," *Journal of the Neurological Science* 7: 304.

Table 12.5   Varying Twitch Speeds of Human Muscles (Selected Values)

Muscle	CT[a] (ms) (Mean ± SD)	Authors
Frontalis/orbicularis oculi	43 ± 4.3	McComas & Thomas (1968)
Abductor digiti minimi	63	Burke *et al.* (1973)
Adductor pollicis	65 ± 7	Slomic *et al.* (1968)
First dorsal interosseous	65 ± 3.3	McComas & Thomas (1968)
Vastus lateralis	66 ± 4	Round *et al.* (1984)
Biceps brachii	66 ± 8.9	Bellemare *et al.* (1983)
Extensor digitorum brevis	67 ± 3.3	McComas & Thomas (1968)
Tibialis anterior	81 ± 7.4	Marsh *et al.* (1981)
Diaphragm	88 ± 7	Faulkner *et al.* (1979)
Triceps surae	112 ± 11.1	Sale *et al.* (1982a)
Lateral gastrocnemius	118 ± 6.5	McComas & Thomas (1968)
Soleus	120	Buller *et al.* (1959)

[a]CT = contraction time (i.e., the time elapsing between the start of the contraction and the moment of peak tension).

course, the size of the various motor units (i.e., number of fibers innervated by the motoneuron) in the muscle will also have an impact on the whole-muscle contractile response. Since fast motor units contain more fibers than slow units, a muscle with equal numbers of fast and slow units will have a faster twitch than one with equal cross-sectional areas of fast and slow muscle fibers.

Histochemical Differences Among Fiber "Types"

Different types of muscle fiber can be recognized with enzyme histochemistry.

Many different staining methods have been devised to study the enzyme activity of muscle fibers (see, for example, Dubowitz, 1985), but the most useful one for many purposes is that for *myosin ATPase* (Engel, 1962). The fibers that stain strongly following preincubation at pH 10.4 are referred to as type II, while the type I fibers show low ATPase activity following this treatment. Brooke and Kaiser (1970) have divided the type II fibers into A, B, and C subtypes on the basis of differences in staining that can be brought out by preincubating the tissue at various pH values (4.3 and 4.6). Brooke and Kaiser (1974) considered the IIC fibers to be precursors with the ability to develop into type IIA or IIB fibers. We now realize that IIC fibers, like the IC fibers, contain mixtures of type I and type II myosin ATPase enzymes. Individual fibers expressing more than one *myosin isoform* are called *hybrid fibers*. The myosin ATPase activities of the fibers are related to their rates of shortening, as measured from *force–velocity* studies (see chapter 11). The reason is that the velocity with which the *actin* and *myosin* filaments can slide over each other depends in part on the speed with which the heads of the myosin molecules can break down *ATP*.

Histochemical studies show consistent differences between some human muscles.

The myosin ATPase staining reaction has been employed extensively in the study of normal and diseased muscles and has proven a useful technique for detecting *denervation*. Prior to the 1970s, however, little was known concerning the distributions of fiber types among normal human muscles. In an autopsy study of 36 human muscles, Johnson *et al.* (1973) showed that in most muscles there is a considerable variation in the proportions of type I and type II fibers. Exceptions include the soleus and adductor pollicis muscles, which are largely composed of type I fibers, and the orbicularis oculi muscle, in which type II fibers predominate. Results from some of the reports of type I and type II fiber proportions in human muscles are listed in table 12.6. Other histochemical studies, also included in table 12.6, make use of the fact that some of the muscle fibers are specially equipped for non-oxidative metabolism. These fibers are able to produce ATP by breaking glycogen down to glucose and thence to *lactic acid;* they contain correspondingly large amounts of glycogen phosphorylase, α-glycerophosphate dehydrogenase, and lactate dehydrogenase that can be detected histochemically. Other fibers have numerous mitochondria and therefore react strongly in histochemical tests for products of the various enzymes associated with these organelles, for example, succinate dehydrogenase and malate dehydrogenase.

Table 12.6   Type I Fiber Compositions of Selected Human Muscles

Muscle	% type I	Authors	Muscle	% type I	Authors
Orbicularis oculi	15	Johnson *et al.* (1973)	Quadriceps	52	Round *et al.* (1984)
Triceps brachii	50	Johnson *et al.* (1973)	First dorsal interosseous	57	Johnson *et al.* (1973)
Biceps brachii	38 42	Brooke & Kaiser (1970) Johnson *et al.* (1973)	Masseter	60-70	Butler-Browne *et al.* (1988)
EDB	45	Johnson *et al.* (1973)	Abductor pollicis brevis	63	Round *et al.* (1984)
Vastus lateralis	46	Green *et al.* (1981)	Tibialis anterior	73	Johnson *et al.* (1973)
Gastrocnemius (lateral)	49	Green *et al.* (1981)	Adductor pollicis	80	Johnson *et al.* (1973)
Diaphragm	50	Bellemare *et al.* (1986)	Soleus	80	Gollnick *et al.* (1974c)

Immunohistochemical Differences Among Fiber Types

A more recent refinement of fiber typing is the use of immunohistochemistry to identify different molecular forms *(isoforms)* of key muscle proteins. The protein most extensively studied in this way has been myosin itself, which consists of two heavy chains and four light chains (see chapter 11). The heavy chains determine the rate of *cross-bridge* reactions with the actin filaments and hence the speed of muscle shortening (Reiser *et al.,* 1985; Bottinelli *et al.,* 1991). The myosin heavy chains also control the pH sensitivity of the ATP-splitting reaction and are therefore responsible for the depth of histochemical staining at the various pHs in the Brooke and Kaiser (1970) procedure. It is now established that there are at least four major heavy chain isoforms and eight light chain isoforms in adult mammalian muscle (Staron & Pette, 1990). The four myosin heavy chain isoforms of importance in limb muscles are I, IIa, IIx, and IIb (small letters are used here to designate the myosin heavy chain, while capital letters designate the fiber type). In human muscles, fibers containing exclusively or almost exclusively one myosin heavy chain are known as type I (containing I heavy chain), type IIA (containing heavy chain IIa), and type IIB (containing heavy chain IIx). It was originally thought that human type IIB fibers contained type IIb myosin heavy chain; it now appears that the protein found in type IIB fibers in humans is closer in identity to the IIx than to the IIb heavy chain found in smaller mammals—thus the human IIB fiber should really be called IIX (Smerdu *et al.,* 1994)! The IIX fiber is also known as the IID fiber in the literature. Human leg muscles also contain a large proportion of IIAB (IIAX) fibers, such that the fibers contain both IIa and IIx myosin heavy chains (Stephenson, 2001). Type IIC fibers are those containing mixtures of type I and IIa myosin heavy chains (more IIa than I), while type IC fibers contain mixtures of type I and type IIa myosin heavy chains (more I than IIa).

In addition, developing muscle has its own unique isoforms, and so far embryonic, fetal, and perinatal types have been described (e.g., Draeger *et al.,* 1987). During muscle development, there are transitions from one of the early isoforms to another and eventually to the appropriate adult isoforms. At any given stage, however, there may be more than one isoform expressed in the same muscle fiber; in later life this situation is quite common and is especially evident when muscles are undergoing a change of usage as the result of training programs or a period of disuse (see chapter 20). Table 12.7, adapted from Staron and Johnson (1993) and Pette and Staron (2000), summarizes the myosin heavy and light chain composition of human muscle fibers and is based on the use of monoclonal antibodies.

Successful searches have been made for variations in the structures of other key molecules in the muscle fiber. For example, the regulatory protein, *troponin T,* has two slow and four fast isoforms (Schachat *et al.,* 1985). Troponin C and troponin I have fast and slow isoforms, as does

Table 12.7 Myosin Heavy Chain and Myosin Light Chain Content of Human Muscle Fibers

Fiber type	Myosin heavy chain(s)	Myosin light chains
Type I	MHC Iβ	LC1sa, LC1sb, LC2sa, LC2sb
Type IC (I/IIA)	MHC Iβ > MHC IIa	
Type IIC (IIA/I)	MHC IIa > MHC Iβ	
Type IIA	MHC IIa	LC1f, LC3f, LC2fa, LC2fb
Type IIAX	MHC IIa> MHC IIx	
Type IIXA	MHC IIx > MHC IIa	
Type IIX	MHC IIx	
Embryonic	MHC_{emb}	$LC1_{emb}$, LC1Vfetal
Neonatal	MHC_{neo}	

Adapted from Staron 2000; Staron and Johnson 1993.

tropomyosin (Schiaffino & Reggiani, 1996). Similarly there are at least two isoforms of Ca^{2+} ATPase in the *sarcoplasmic reticulum,* corresponding to fast (*SERCA* 1a) and slow (SERCA 2a) fibers respectively; and the enzyme *creatine kinase* has long been known to exist in fetal and adult forms and is now recognized to have mitochondrial and cytoplasmic isoforms (see chapter 14).

Fibers can also be distinguished from each other by their ultrastructure.

Ultrastructural Differences Among Fiber Types

Attempts have been made to differentiate mammalian muscle fibers on the basis of their ultrastructural appearances (Gauthier, 1986). The feature that can be correlated best with other properties of the fiber is the thickness of the *Z-line*. For example, in the guinea pig, the Z-line is more than twice as wide in the red fibers of the soleus as in the white fibers of the vastus muscle (Eisenberg, 1974; see also Padykula & Gauthier, 1967). In addition, the width of the *M-line* tends to follow that of the Z-line, being greater in the red fibers (Schiaffino *et al.,* 1970). The mitochondria also exhibit differences; in the red fibers they are large and spherical, with abundant *cristae,* and they tend to be grouped under the fiber *plasmalemma* as well as between *myofibrils.* In white fibers the mitochondria are thinner and longer, and fewer are found under the plasmalemma. Finally, the T-tubular system and the cisternae of the *SR* are more plentiful in type II fibers (Eisenberg & Kuda, 1976).

Classification of Motor Units

In the preceding sections, it emerged that there are many physiological and biochemical differences between muscle fibers, and some evidence was presented to suggest that these differences might be linked together. Indeed, many attempts have been made to correlate the various characteristics of muscle fibers and to use such associations as a basis for classifying motor units.

The histochemical and contractile properties of motor units are linked together.

Relationship of Muscle Fiber Types to Motor Unit Types

In one of the most comprehensive correlative studies, Burke *et al.* (1971; see also Burke *et al.,* 1973) stimulated spinal motoneurons through intracellular microelectrodes and recorded the twitch and *tetanic contractions* of the corresponding motor units. At the end of an experiment, they demonstrated the architecture of the motor unit that they had been investigating using the glycogen-depletion technique (see p. 181). By staining consecutive sections for various enzymes, the authors were able to establish other biochemical characteristics of the previously activated fibers. As part of this study, they were able to confirm that all the muscle fibers belonging to the same motor unit were histochemically similar in that they stained strongly for the same enzyme activities.

Burke and colleagues (1971) concluded that certain properties of the motor units are closely linked together and that three types of units can be recognized. One type (type *S*) has a slow twitch, develops relatively small tension because it contains relatively few fibers, and is resistant to fatigue (figure 12.8); the muscle fibers have high contents of mitochondrial enzymes, are poor in glycogen, stain weakly for myosin ATPase after alkaline preincubation, and have rich capillary networks. These fibers appear to be well equipped for aerobic metabolism and hence for prolonged activity.

A second type (type FF), in complete contrast, has a fast twitch, commonly develops large tensions indicating that it contains many fibers, and is susceptible to fatigue. These fibers have low contents of mitochondrial enzymes, are rich in glycogen, stain strongly for myosin ATPase after alkaline preincubation, and have poor capillary networks. They are suited for non-aerobic metabolism and are well suited for brief intense *effort.*

A third type (type FR) has intermediate properties. It has a fast twitch, develops moderate tensions, and is resistant to fatigue. The muscle fibers have high contents of glycogen and mitochondrial enzymes and have considerable myosin ATPase activity; they are well supplied with capillaries. These fibers would be expected to be well suited for prolonged exercise, but in addition would have capability to contribute high power output.

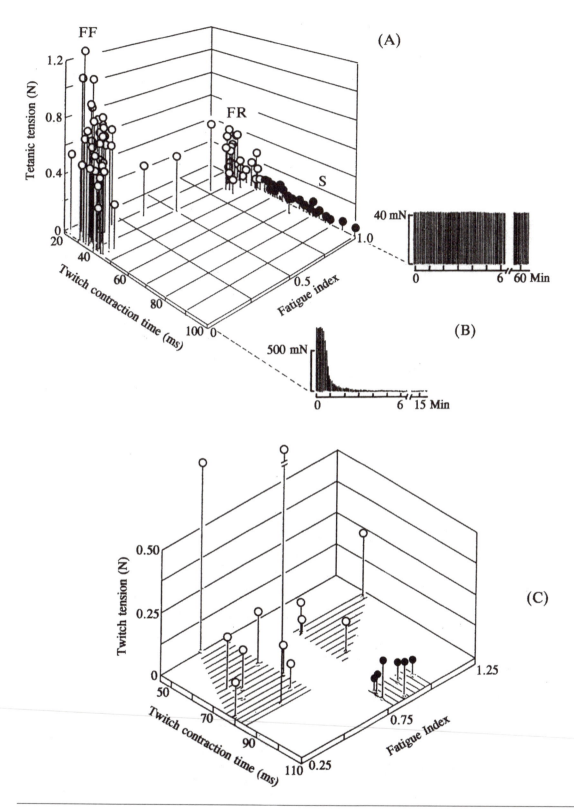

Figure 12.8 Physiological properties of different motor unit types. (A) A simulated three-dimensional display of the results for 81 cat medial gastrocnemius units. It can be seen that all but two of the units fall into one of three clusters (see text). (B) Examples of the responses of a type S unit (upper trace) and a type FF unit (lower trace) to intermittent tetanic stimulation. Note that force is sustained in the type S unit but not in the type FF unit. (C) The results of a similar type of study undertaken in the human gastrocnemius; although fewer units were examined than in A, there is a tendency for the values to fall into three groups.

A and B: Reprinted, by permission, from R.E. Burke et al., 1973, "Physiological types and histochemical profiles in motor unites of the cat gastrocnemius," *Journal of Physiology* 234: 735; C: Reprinted, by permission, from R.A.F. Garnett, 1979, "Motor unit organization of human medial gastrocnemius," *Journal of Physiology* 287: 37.

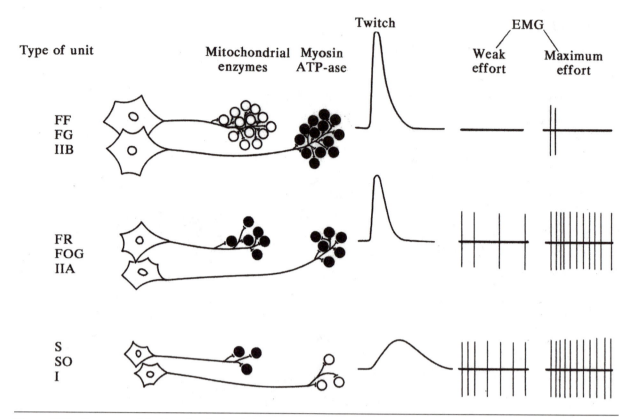

Figure 12.9 Summary of the histochemical and physiological properties of the three main types of motor unit, based largely on the animal studies of Burke *et al.* (1973).

Reprinted, by permission, from A.J. McComas, 1977, *Neuromuscular Function and Disorders* (London: Butterworth Publishing Co.), 66.

An attempt to summarize the contrasting histochemical and physiological characteristics of the three main motor unit types is presented in figure 12.9; the different terminologies employed are described in the following section.

Different classifications exist for the motor unit (and muscle fiber) types.

Schemes Used to Classify Muscle Fiber Types and Motor Unit Types

As already noted, Burke and colleagues (1971) termed the three types of motor unit S (slow contracting), FF (fast contracting, fast fatigue), and FR (fast contracting, fatigue resistant), respectively. A fourth type of unit, *FInt,* with properties in between those of FF and FR units, was subsequently recognized by Burke (1975); the existence of such intermediate units has been emphasized by McDonagh *et al.* (1980).

In a detailed biochemical investigation involving hindlimb muscles of the rabbit and guinea pig, Peter *et al.* (1972) have recognized three types of fibers. These authors have proposed a different classification, and the FF, FR, and S units become *fast-glycolytic (FG)*, *fast-oxidative-glycolytic (FOG)*, and slow-oxidative (SO) types, respectively. In table 12.8, the classification of Burke and colleagues has been reconciled with that of Peter *et al.* (1972) and with the widely used histochemical scheme of Brooke and Kaiser (1970), discussed earlier.

Motor units tend to overlap in their properties.

In reviewing table 12.8, it must be emphasized that, even though certain contractile properties tend to be associated in a particular motor unit, there are considerable overlaps between units. For example, although the motor units generating the largest tensions always have fast twitches, other fast-twitch units may develop tensions similar to those of slow-twitch units. This overlap is not surprising, since parameters such as twitch and tetanic tensions, and contraction and half-relaxation times, are distributed unimodally for a given muscle. One contractile parameter that does not show

Table 12.8 Fiber Type Classification

Property	Motor unit type		
	I (S) (SO)	IIA (FR) (FOG)	IIB (FF) (FG)
Twitch speed	Slow	Fast	Fast
Twitch force	Small	Intermediate	Large
Fatigability	Low	Low	High
Red color	Dark	Dark	Pale
Myoglobin	High	High	Low
Capillary supply	Rich	Rich	Poor
Mitochondria	Many	Many	Few
Z-line	Intermediate	Wide	Narrow
Glycogen	Low	High	High
Alkaline ATPase	Low	High	High
Acid ATPase	High	Low	Moderate
Oxidative enzymes	High	Medium-high	Low

overlap is *fatigability,* which enables fatigue-resistant units (S, FR) to be distinguished from fatigue-sensitive ones (FF). Related to fatigability is the presence or absence of the *"sag"* phenomenon in short unfused tetani, which differentiates between fast-twitch and slow-twitch units (Burke *et al.,* 1973). Although the sag phenomenon is a convenient way to distinguish type F from S motor units in cat muscles, it is not as useful in distinguishing motor units in muscles of other species, including the human (Bigland-Ritchie *et al.,* 1998).

Most properties of human motor units can also be correlated.

Relating Histochemical and Contractile Properties of the Motor Unit

In the case of human muscle, Buchthal and Schmalbruch (1970) were the first to attempt a systematic correlation between histochemical properties and contractile behavior. In their studies, small bundles of muscle fibers were stimulated, and the time courses of the twitches were recorded by a needle-mounted strain gauge inserted into the tendon. This technique did not enable the responses of single motor units to be analyzed, but the results were nevertheless important. Buchthal and Schmalbruch (1970) found that long contraction times predominate in muscles that have a large proportion of fibers rich in mitochondria (soleus, gastrocnemius), but that short contraction times are common in muscles consisting mainly of motor units poor in mitochondrial enzymes (lateral head of triceps brachii and platysma).

A more complete examination of human motor unit and muscle fiber types was undertaken by Garnett *et al.* (1979) in the medial gastrocnemius. The axons of single motor units were stimulated through needle electrodes in the innervation zone, and recordings were made of the twitch and tetanic responses. At the end of the experiment, the same motor unit was tetanized to exhaustion, and a needle biopsy was taken from the region of the muscle belly that had previously shown the greatest viable contractions. By staining alternate sections for glycogen and for myosin ATPase activity, the authors were able to combine the histochemical and physiological properties of the stimulated units. The results obtained, although fewer in number, were remarkably similar to those of Burke *et al.* (1973) for the medial gastrocnemius in the cat (compare A and C in figure 12.8). Another large human muscle, the tibialis anterior, also shows correlations between the contraction times and the tensions developed in single motor unit twitches (Andreassen & Arendt-Neilsen, 1987).

**In some
human
muscles,
correlations
are absent.**

Not all human muscles lend themselves to clear demarcations of motor unit properties. In small muscles, such as the first dorsal interosseous and abductor pollicis brevis muscles of the hand, and the extensor digitorum brevis muscle of the foot, there are no significant correlations between contraction time and twitch force (Sica & McComas, 1971; Milner-Brown *et al.,* 1973b; Thomas *et al.,* 1990b; Young & Mayer, 1981).

Further, in the adductor pollicis, there is an unusually high proportion of type I fibers, and yet the contraction time of the muscle is brief (Round *et al.,* 1984). It should be noted, however, that the contraction time is determined by the availability of cytosolic Ca^{2+} ions to the *myofilaments* in addition to the rate of ATP hydrolysis (see chapter 11).

**The
electrical
properties of
motoneurons
differ
according to
motor unit
type.**

Properties of Motoneurons of Different Motor Unit Types

In animal experiments, the use of intracellular stimulation and recording techniques in motoneurons provides the opportunity to establish whether or not there are features of the motoneuron itself that can be correlated with the type of muscle fiber that it innervates. The motoneurons of S motor units are those that are most readily excited by weak currents and have the highest input resistances (change in *membrane potential* ÷ applied current). Injections of HRP confirm that the S motoneurons are the smallest (Burke *et al.,* 1982; Ulfhake & Kellerth, 1982). In addition, the *action potentials* of the S motoneurons are followed by *after-hyperpolarizations (AHPs)*; because the motoneurons are refractory to excitation during the AHPs, the AHPs impose a limit on the discharge rate of a motoneuron and its motor unit (Eccles *et al.,* 1958a).

In humans, one accessible piece of information concerning the physiological properties of a motoneuron is the axonal conduction velocity. On the basis of animal studies, the motoneurons with the largest cell bodies would be predicted to be those with axons having the highest conduction velocities (Kernell, 1966; Burke, 1967) and therefore the greatest diameters (Hursch, 1939). Dengler *et al.* (1988) found that the conduction velocities of axons supplying the first dorsal interosseous muscle of the hand ranged from 40 to 63 m · s^{-1} and were highly correlated with the twitch forces of the respective motor units. Thus, it would appear that, in humans as in other mammals, the largest motoneurons supply the greatest numbers of muscle fibers (see, however, Thomas *et al.,* 1990b).

Also in humans, several investigators have attempted to derive the AHP durations of motoneurons by determining the relationships between firing frequencies and standard deviations of the interpulse intervals. For example, when this relationship is determined for various frequencies of firing of a motor unit during voluntary contraction, a frequency can be calculated above which firing becomes more regular. This "break-point" from linearity is thought to reflect the AHP of the innervating motoneuron, in light of the importance of the latter in determining firing frequency (Kudina & Alexeeva, 1992). Another more recent technique of estimating the AHP of the motoneuron involves analysis of the long interspike interval frequencies at low frequencies of firing—the so-called interval death-rate transform (Matthews, 1996). While these two techniques appear to be robust enough to distinguish motoneurons innervating muscles that differ in fiber/motor unit type, their use in distinguishing motor unit types in mixed muscles, or in determining pathological or physiological adaptations in motor unit properties, remains to be demonstrated.

Applied Physiology

Motor units may undergo striking alterations in disease or following nerve injury. For example, a motoneuron disorder or nerve damage results in very large motor units, while muscle diseases have the opposite effect. Further information on some of these changes is given in chapter 17, and in the next chapter we see how the abnormal motor units can be recognized in the *EMG* laboratory. Before closing this chapter, though, we will discuss how the number of motor units in a human muscle can be determined automatically within a few minutes (see "Estimation of Human Motor Units").

Estimation of Human Motor Units

There is a simple way to estimate the number of motor units in a living human muscle. All that is required is a stimulator, amplifier, and storage oscilloscope or computer with data acquisition capabilities. A recording electrode is applied to the skin overlying the motor innervation zone of the muscle, and a reference is attached elsewhere. The motor nerve is then stimulated with weak electrical shocks, starting with an intensity that is just subthreshold. As the stimulus intensity is gradually raised, the amplified compound action potential recorded from the muscle *(M-wave)* is seen to grow in discrete steps on the oscilloscope screen (figure 12.10). The assumption is made that each of these steps is due to the excitation of an additional motor unit. If the average amplitude or area of the putative motor unit potentials is then compared with the maximum M-wave of the muscle, an estimate of the number of motor units in the muscle is obtained.

Obviously, there are many assumptions in the method, not the least of which is that each increment in the graded response results from the discharge of an additional single motor unit. Nevertheless, the method does seem to work and has recently been made easier by entrusting the stimulation and response analysis procedures to a computer program (Galea *et al.,* 1991). A subroutine deals with the vexatious problem of "alternation," that is, fictitious motor unit responses caused by axons with similar thresholds discharging in different combinations as the stimulus is repeated. Another subroutine

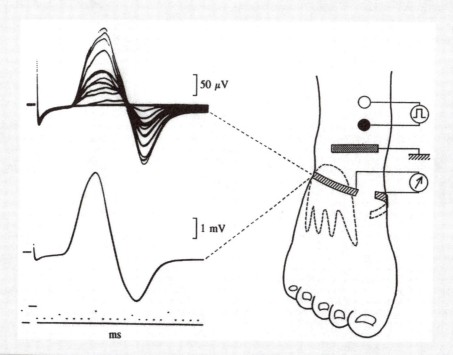

Figure 12.10 Original, "manual" method of motor unit estimation as applied to the human extensor digitorum brevis muscle. The placement of the stimulating and recording electrodes is shown on the right; and on the left are the superimposed, incremental responses to increasing stimulus intensities, together with the maximum M-wave. This subject was estimated to have just over 200 motor units in the muscle.

(continued)

arranges the evoked potentials in order of increasing area and subtracts each response from the next largest, so as to yield the potentials generated by single putative motor units. Figure 12.11 shows an example of the automated procedure as applied to a medial vastus muscle; the estimate of 286 units was well within the normal range for this muscle—clearly, motoneuron aging had not started in this healthy 68-year-old man!

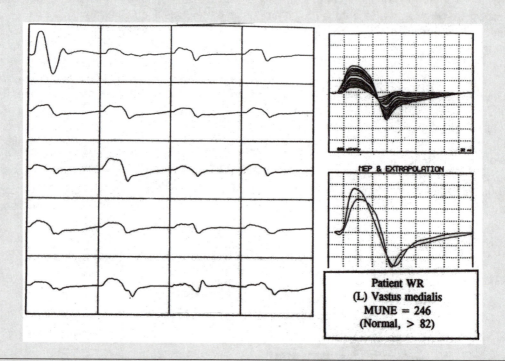

Figure 12.11 Automated motor unit estimation in the medial vastus muscle of a 68-year-old man. A computer software program increases the stimulus intensity and analyzes the progressively larger muscle responses (top, right). The potentials of individual motor units can be derived by successive subtractions of the incremental responses and are shown at the left. The two traces at bottom right compare the shapes of the sample of motor unit potentials and of the maximum M-wave. Further descriptions of this methodology can be found in Galea, de Bruin, Cavasin, and McComas (1991).

13

Motor Unit Recruitment

Having dealt with the various characteristics of *motor units,* it is opportune to consider how the units are used in voluntary and reflex contractions. In the course of this chapter, we will find out about the *impulse* firing rates of *motoneurons* and the orderly sequence in which the different types of motor units are typically recruited. There are also interesting aberrations from this orderly sequence that are worth considering. Issues related to neural connections and their implications for motor control and force-sharing among synergists are beyond the scope of this book.

Detection of Motor Unit Activation

The very fact that the motor units exhibit such striking differences among themselves in terms of size, speed of contraction, and biochemical capability suggests that their involvement is unlikely to be random but suited to the task demanded of them. Indeed, to some extent the task will shape the unit (see chapters 18 and 20). In order to comprehend the principles of motor unit *recruitment,* it is important to keep in mind that each motor unit has its own *force–frequency relationship,* the maximum force of which depends on the size of that motor unit (number of fibers and average fiber diameter), as well as any systematic differences in tension-generating capacity per unit cross section (which are small, but still worth noting, in the order type I < IIA < IIX). Furthermore, the ability of a motor unit to continue to contribute in exercise is related to the biochemical properties of the unit. Are we able to select the appropriate motor unit(s) for the task at hand, or are we committed to randomly activate motor units within a muscle? To address this question, it is necessary that we be able to detect the activation of individual motor units and identify the unique fiber type properties of that unit.

Electromyography (EMG) is a technique analogous to electrocardiography that can be used to monitor certain aspects of skeletal muscle activation. As with the electrocardiogram, knowledge of the EMG wave-forms does not yield any information concerning the mechanical properties of the muscle contraction. Factors such as load (shortening velocity), muscle length, fatigue, and *activity-dependent potentiation* can alter the mechanical response, independent of the measured EMG. There are two distinct types of EMG recording that have been used in an effort to understand muscle activation: surface and intramuscular. Each of these will be briefly described and the advantages and limitations outlined. (For a more extensive discourse on EMG, see Basmajian and DeLuca, 1985.)

The electromyogram provides a picture of muscle activation.

Surface Electromyography

Surface EMG indicates timing and magnitude of activation.

Surface EMG measurement is obtained by placing electrodes on the skin over the belly of the muscle. Most such recordings use a bipolar arrangement of electrodes, where two electrodes are placed on the prepared skin in line with the muscle fibers. Ideally both electrodes should be placed to one end of the fibers (both on the same side of the neuromuscular junction). The skin is usually shaved and abraded to improve electrical conductance. Recordings of surface EMG can be analyzed for timing,

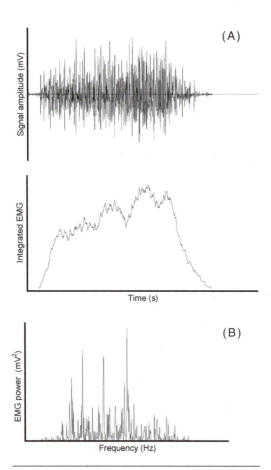

Figure 13.1   (A) A raw EMG tracing is shown during a maximal voluntary contraction. The timing of activation can be obtained from this signal. (B) Integration of the recorded raw EMG signal permits determination of the magnitude of the signal. Fourier analysis of the raw signal permits determination of the frequency content of the EMG signal. The median frequency decreases with sustained *effort,* and this has often been used as an indicator of fatigue.

magnitude, and frequency content, as illustrated in figure 13.1. The most elementary use of this signal is to identify when a given muscle is active. The magnitude of the EMG signal is typically evaluated by either integration or calculation of the root-mean-square value.

The frequency content of the surface EMG signal is determined by *fast Fourier transform (FFT)* and involves representing the EMG signal by a series of sine waves. Factors that influence the frequency content of the EMG signal include muscle fiber conduction velocity and a low-pass filter effect of the soft tissue between the active muscle fibers and the recording electrodes. It has been noted, at least in animals, that fast-twitch fibers have a higher conduction velocity than slow-twitch fibers. However, this changes during repetitive activation, so the predictive value of this feature of the EMG technique may be limited. In spite of this, wavelet analysis (FFT of sequential short segments [wavelets] of the EMG signal) has been proposed as a method to estimate the involvement of motor units of different fiber types (Wakeling *et al.,* 2001). To obtain more direct estimates of the frequency of firing of individual motor units, intramuscular recording is required.

Intramuscular Electromyography

The earliest study of motor unit function in humans appears to be that of Adrian and Bronk (1929), who were able to record the activity of single motor units by means of a fine wire positioned within, but insulated from, a hypodermic needle (see "Needle Electromyography—From Research Lab to Bedside"). Adrian and Bronk found that their subjects were able to increase the force of their muscle contractions by raising the firing frequency of the motor units (referred to as *rate-coding*) and by calling additional motor units into activity (recruitment). Since this important study, there have been many investigations of firing rate in which a variety of fine wire and needle electrodes have been used to detect the activities of single motor units (see, for example, Bigland & Lippold, 1954; Lindsley, 1935; Petajan &

Individual motor units can be detected with special electrodes.

Philip, 1969). Because of the large movements of the muscle fibers that take place during muscle contraction, some workers have preferred to introduce the fine recording wires through a hypodermic needle and then to withdraw the needle, allowing the wires to flex with the distortions of the muscle belly. From all these studies, it is evident that some motor units are consistently activated more easily than others and that those are employed in weak contractions. To increase the force of contraction, additional motor units are recruited, and the force exerted by each unit is increased with a higher rate of impulse firing (see discussion of the force–frequency relationship in chapter 11).

The Size Principle

In a steady contraction, small motor units are recruited before large ones.

It has long been recognized that force output during a voluntary contraction can be increased either by increasing the firing frequency of active motor units or by increasing the number of active motor units. Just how the motor system utilizes these two approaches is still a topic of research. One of the earliest studies to address this issue dealt with reflex mechanisms. Using anesthetized and spinalectomized cats, Henneman measured the amplitudes of impulses recorded from a *ventral root* filament by an extracellular electrode. As the stimuli increased, the amplitude of the reflex *(efferent)* impulses also increased. It was known that axon diameter determined the amplitude of these impulses, and it has since been confirmed that the *soma* size and axon diameter are closely related.

Needle Electromyography—From Research Lab to Bedside

Needle EMG had its origin in 1929 when Edgar Adrian and Detlev Bronk published a report in the *Journal of Physiology* on "The Discharge of Impulses in Motor Nerve Fibres." In order to obtain their results, Adrian and Bronk made a special recording electrode from a hypodermic needle into which they had inserted a fine insulated wire, which was then glued in place (figure 13.2). This electrode proved capable of detecting the discharges of individual motor units within a muscle, and hence provided information as to the frequency of impulse firing in motor nerve fibers. The recordings were displayed on a capillary-electrometer, a much less satisfactory instrument than the cathode ray tube, which began to be used in electrophysiology laboratories in the United States at about this time. Although Adrian is said to have doubted whether his simple electrode would have any clinical applications, it subsequently proved capable of detecting *motor unit action potentials* that were reduced in number but prolonged in duration in patients with chronic denervation—though the significance of this finding was initially misinterpreted (Buchthal & Clemmesen, 1941). The large amplitudes and complex wave-forms of the potentials in denervation were recognized afterward. An example of this kind of activity, in a patient with *spinal muscular atrophy,* is shown in the bottom trace of figure 13.2.

Rather later, Kugelberg (1949) described potentials that were unusually brief and "polycyclic" in patients with muscular dystrophy. Such potentials are now known to characterize all degenerative muscle disorders. In the middle trace of figure 13.2, such activity is shown from a patient with limb-girdle muscular dystrophy. It was not until the 1960s, however, that needle EMG became widely practiced as a diagnostic test for the investigation of neuromuscular disorders. Since then the field has expanded to such an extent that one or more international meetings are held each year; and in North America, several training courses are offered annually. The American Association of Electrodiagnostic Medicine, to which many American and Canadian electromyographers belong, has over 3,000 members.

What became of the EMG pioneers? Edgar Adrian continued his distinguished career at Cambridge University and was awarded the Nobel Prize in 1932 (with Charles Sherrington) for demonstrating how perception and movement depended on patterns of electrical impulses conducted in nerve fibers. He was elevated to the peerage in 1955. Detlev Bronk returned to the United States and pursued wide-ranging studies of nervous system function. Kugelberg and Buchthal were leaders in the EMG field for many years.

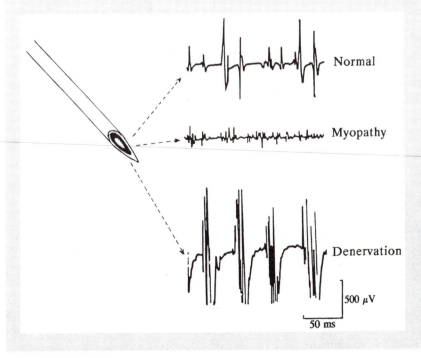

Figure 13.2 A concentric (coaxial) needle electrode. The central wire is usually 100 μm in diameter, and the outside diameter of the needle varies but is typically about 0.6 mm. At a given site in a normal muscle, the central wire can detect activity in some of the fibers belonging to 20 or so motor units. During a strong contraction, however, the impulse patterns are so dense and complex that the discharges of individual units cannot be easily distinguished. The three recordings displayed in the figure were obtained during weak effort (see text).

The observations on impulse amplitude were interpreted to indicate that small motoneurons were more easily activated than large ones (Henneman, 1957).

Estimation of Motor Unit Size

While investigating the responses to stretch of cat hindlimb muscles, Henneman *et al.* (1965) confirmed that the recruitment of motor units was orderly; that is, the smaller motor units had lower thresholds than the larger units. Numerous experiments have since been conducted on human muscles, during both voluntary and reflex contractions, to ascertain whether the same principle applies. One method of estimating motor unit size is to measure the amplitude of the *motor unit action potentials,* since the amplitude is proportional to the numbers of muscle fibers in each unit. Such studies have shown that the majority of units are, in fact, recruited voluntarily in order of increasing size (Monster & Chan, 1977; Tanji & Kato, 1973). This property of motor unit recruitment has come to be known as the *size principle.* There is general agreement among investigators that the size principle applies when either slow-ramp forces are exerted or constant low forces are compared with constant higher forces.

A more accurate indication of motor unit size is the *twitch* tension, and this may be determined by the technique of *spike-triggered averaging* during steady voluntary contractions (figure 13.3A). Figure 13.3B shows the results of multiple observations in the first dorsal interosseous muscle of the same subject, plotted on a log-log scale; it can be seen that there is a nearly linear relationship between the twitch tensions of the units and the voluntary forces at which they are recruited. Similar results, indicative of linear or curvilinear relationships, have been obtained by Monster and Chan (1977) in the extensor digitorum communis and by Yemm (1977) and Goldberg and Derfler (1977) in human jaw muscles. The impulse conduction velocities of the motor axons are also related to recruitment order, the most rapidly conducting axons belonging to the units with the highest thresholds (Freund *et al.,* 1975; Grimby, 1984). If the voluntary contractions are made more quickly, motor units appear to become active at lower forces. As Freund (1983) points out, this is illusory because the background force, increasing rapidly, will have reached the same level by the time that the newly recruited unit would have developed its maximal twitch tension. This delay is due to the time for nerve conduction, neuromuscular transmission, and *excitation-contraction coupling.*

Two additional features of voluntary motor unit recruitment are worth noting. Motor units are activated asynchronously, and the rates of firing are not uniform. Although the frequency of firing is given a number (e.g., 20 Hz), this should be considered a mean value, since the intervals between impulses show some variation.

> **Spike-triggered averaging can be used to estimate motor unit size.**

Advantages of Orderly Recruitment of Motor Units

The orderly recruitment of motor units seems logical when consideration is given to the biochemical properties of small versus larger motor units. Typically, small motor units are type I units, and unit size increases with progression through the fiber types: type I < type IIA < type IIX. Therefore, when low force is required, only type I motor units will be active. Only when force or power is high will recruitment demand involvement of the larger motor units.

A further advantage of the size principle is that the increments in force with new recruitment will be small at low efforts, and will increase in proportion to the absolute force expected. This permits a rather smooth increase in force. Considering that the range of motor unit size can be greater than 100-fold (see chapter 12), this can be an important consideration.

The Mechanism for the Orderly Recruitment of Motor Units

How is the order of recruitment established? Henneman *et al.* (1965) suggested a simple explanation. The smallest α-*motoneurons* in the ventral horn are likely to correspond to the smallest motor units, because the metabolic activity and therefore the soma size of a motoneuron are probably proportional to the number of muscle fibers and nerve endings that it is required to maintain through its *trophic* action (see chapter 18). Suppose that the motoneuron pool receives an excitatory input from the motor cortex and that roughly the same number of synapses is activated on all the moto-

> **The low thresholds of the type I motor units depend on the small sizes of the motoneuron cell bodies and on the densities of synapses.**

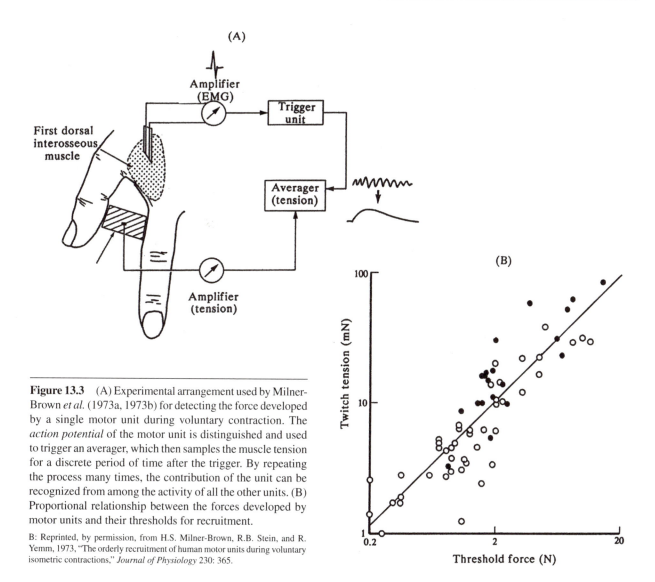

Figure 13.3   (A) Experimental arrangement used by Milner-Brown *et al.* (1973a, 1973b) for detecting the force developed by a single motor unit during voluntary contraction. The *action potential* of the motor unit is distinguished and used to trigger an averager, which then samples the muscle tension for a discrete period of time after the trigger. By repeating the process many times, the contribution of the unit can be recognized from among the activity of all the other units. (B) Proportional relationship between the forces developed by motor units and their thresholds for recruitment.

B: Reprinted, by permission, from H.S. Milner-Brown, R.B. Stein, and R. Yemm, 1973, "The orderly recruitment of human motor units during voluntary isometric contractions," *Journal of Physiology* 230: 365.

neurons. This means there will be a similar amount of current crossing the membranes of small and large motoneurons. The voltage change that can be expected for a given current is proportional to the input resistance of the motoneuron. Small cells have a high input resistance, so they will be expected to have a greater change in *membrane potential* for a given current. Small cells will reach threshold with less current than larger cells. Another way to look at this is to consider that the density of the synaptic current flowing between the excitatory synapses on a motoneuron and the *axon hillock* will differ among the motoneurons. The smallest cells will have the highest densities because the current is concentrated in a smaller membrane area. A large current density will, in turn, produce a correspondingly large *depolarization,* and so the smallest cells should have the lowest thresholds for excitation.

   The idea that an orderly and relatively predictable recruitment of motor units can be obtained with a common current applied to the complete pool of motoneurons has resulted in the conceptualization of several simulations, an example of which is the Heckman-Binder model of motor unit recruitment (Heckman & Binder, 1991). In this model, the threshold for recruitment (magnitude of current required) is determined by the size of the motor unit, and the frequency of firing of any recruited unit increases in proportion to the applied current in excess of the threshold. This model is illustrated in figure 13.4. The frequency of firing of motor units at three levels of current injection is shown. It is clear from this simulation that at near maximal effort, the frequency of activation of the slow (small) units will be higher than the frequency on the large (fast) units. However, the

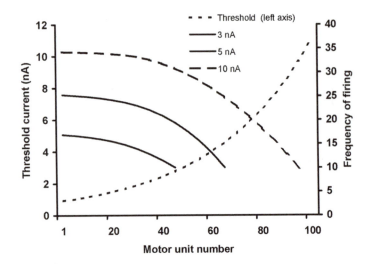

Figure 13.4 The Heckman-Binder model of motor unit recruitment is illustrated. The small dashes represent the threshold current needed to activate a given motor unit. The other three lines represent the firing frequency for any motor unit with constant current injection of 3, 5, or 10 nA. Note that the lower the threshold, the higher the frequency of firing at any level of current injection. With 3-nA injection, motor units up to about #45 are recruited. This represents all units with a threshold below this level of current. Note also that the frequency of firing when first recruited is about 10 Hz.

maximum frequency for any motoneuron is determined to some extent by the *refractory period,* which is a function of its after-hyperpolarization. After-hyperpolarization is more prolonged in type I units; therefore these units will have a lower maximal rate of firing.

| In steady or repetitive contractions, the type I (slow-twitch) motor units are recruited first. |

Motor Unit Recruitment During Exercise

In chapter 12, it was seen that the smallest motor units in a human or mammalian muscle are those with the slowest twitches and the greatest resistance to fatigue, which are called type *S* motor units (Burke *et al.,* 1971). These S units are composed primarily of fibers described, depending on the classification employed, as type I (Brooke & Kaiser, 1970), or type SO (Peter *et al.,* 1972). In view of the ordered recruitment of motor units during voluntary contractions, described earlier, the type S units are those that should be called into action first.

To examine this supposition, Gollnick *et al.* (1974a, 1974b) exercised volunteers on a bicycle or had them sustain isometric contractions of the quadriceps muscles. Specimens of muscle were taken periodically using a needle biopsy technique (Bergström, 1962). Sequential slices of muscle were stained histochemically for fiber type classification (myosin ATPase) and for glycogen content (periodic acid-Schiff technique). By correlating the absence of glycogen with the staining reactions of the fibers for myosin ATPase, it was possible to determine whether type I (slow-twitch) or type II (fast-twitch) fibers had been employed in the exercise. In the cycling exercise, it was found that slow-twitch fibers were the first to become depleted of glycogen at all workloads requiring less than the maximal oxygen consumption. At supramaximal levels, the fast-twitch fibers were also involved. A similar result was obtained for the isometric contractions of the quadriceps. These results have been confirmed by Vøllestad and colleagues (Vøllestad *et al.,* 1984; Vøllestad & Blom, 1985; Vøllestad *et al.,* 1992), who have extended these findings to show that the order of recruitment of muscle fiber types is type I, then IIA, then IIAX, and finally IIX. This order of recruitment was in place whether the comparison was of sustained contractions at different intensities, or at different times during exercise at the same submaximal force sustained until exhaustion.

The preferential involvement of type I units has also been demonstrated in animal experiments. For example, Gillespie *et al.* (1974) showed that these units were those used most heavily by the bush baby, *Galago senegalensis,* in steady running of moderate intensity. Similarly, in reflex contractions of cat plantaris muscles evoked by muscle stretch, Zajac and Faden (1985) found that recruitment invariably followed the order:

$$S \rightarrow FR \rightarrow FInt \rightarrow FF \text{ units}$$
$$(\text{or } I \rightarrow IIA \rightarrow IIint \rightarrow IIB)$$

The Catch-Like Property of Skeletal Muscle

There is an interesting refinement of the motoneuron discharge pattern that has been described by Denslow (1948) in human motor units. An initial short interpulse interval was observed before the activation frequency decreased to a steady level. This has subsequently been observed by many investigators, including Zajac and Young (1980a), who observed doublets in cats with stimulus-initiated locomotion. The functional advantage of this maneuver is that the first two twitches summate, bringing the motor unit tension rapidly toward the tetanic level, where it stays for the remainder of the discharge in spite of a subsequent large decrease in the firing frequency (Zajac & Young, 1980b). This behavior has been termed the *"catch-like property,"* since similar behavior was first reported in invertebrate muscle fibers. It is not clear what it is about the doublet that allows such *summation* to occur. Abbate *et al.* (2002) have reported that it is not due to a sustained elevation of free $[Ca^{2+}]$. Increased binding of Ca^{2+} to troponin remains a possibility.

Recruitment Versus Rate-Coding

Adrian and Bronk (1929) noted that increased force was produced by increasing the impulse firing rate and by recruiting extra motor units. But which of the two mechanisms is the more powerful? One approach is to examine the relationship between impulse firing frequency (or stimulus repetition rate) and force for individual motor units. Typically, when the force–frequency relationship is presented, it is the peak force for each contraction that is given; but during an incompletely fused tetanic contraction, it is the mean force that is more relevant. At the minimum voluntary firing frequency of 8 Hz, the contractions of a muscle such as tibialis anterior are only partly fused, so that the mean tension is approximately half the peak tension of the twitch. At the maximum steady firing frequency of 35 Hz, the contraction is fused and the tension developed is about 5 to 10 times larger than the peak twitch force (see figure 11.20 in chapter 11). Thus, the motor units in a muscle can increase their mean force 10-fold simply by adjusting their firing rates.

On the other hand, recruitment must also be a very effective way of increasing force, since some motor units can develop up to 100 times as much tension as others within the same muscle belly (see chapter 12).

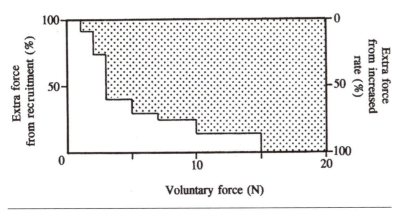

Figure 13.5 The relative importance of recruitment of additional motor units and of increase in firing frequency during increasingly strong contractions of the first dorsal interosseous muscle of one subject.

Reprinted, by permission, from H.S. Milner-Brown, R.B. Stein, and R. Yemm, 1973 "The orderly recruitment of human motor units during voluntary isometric contractions," *Journal of Physiology* 230: 365.

It was Adrian and Bronk's opinion that, at relatively low forces, recruitment was the more important mechanism but that increased firing rates were increasingly employed to bring contractile force closer to the maximum possible. These conclusions were confirmed by the spike-triggered averaging study of Milner-Brown *et al.* (1973a, 1973b); in the first dorsal interosseous muscle of the hand, these researchers observed that half of the motor units had already been recruited when only 10% of the maximal force had been developed (figure 13.5). They conceded, however, that the highest-threshold units might not have been detected by their averaging technique, and other investigators have argued that recruitment is an important factor at all force levels (e.g., Bigland & Lippold, 1954). It should be kept in mind that several small units could contribute only a small force, even if they were activated at a high frequency. It is not surprising that a substantial portion of the motor units would be active before reaching 50% of maximal force.

The explanation for the apparent confusion between experimental results is that the two strategies, recruitment and firing rate, are employed to different extents from one muscle to another.

This point was nicely demonstrated by Kukulka and Clamann (1981), using fine wires inserted into the muscle through a hypodermic needle. In the biceps brachii, recruitment was observed from 0% to 88% of maximum voluntary force, while in the adductor pollicis, no additional motor units were recruited at forces greater than 50% of maximal. Similar differences between the first dorsal interosseous muscle of the hand and the deltoid muscle (DeLuca *et al.*, 1982) suggest that small, distal muscles rely more on increases in firing rate for the development of large forces, whereas large, proximal muscles continue to recruit additional motor units.

An interesting refinement of motor unit firing rate has been found by Broman *et al.* (1985), who observed that the recruitment of an additional motor unit was associated with a transient drop in the firing rates of those units previously active. Such an effect would help to ensure that the increase in force associated with recruitment was smooth rather than abrupt.

Can the Size Principle Be Overridden?

One can test the size principle hypothesis experimentally by impaling motoneurons with micro-electrodes and measuring their input resistances during current pulses; the input resistance of a motoneuron will be inversely proportional to its size. It turns out that, although the largest axons arise from the largest motoneurons, as predicted (Burke, 1967), they do not necessarily supply the greatest numbers of muscle fibers, at least in the large muscles of the cat hindlimb (Stephens & Stuart, 1975; but see Bagust & Lewis, 1974); in the small distal muscles, a better correlation exists. On the basis of these observations, it would appear that motoneuron size may not be the sole factor determining excitation threshold during voluntary or reflex contractions. Rather, it would appear that there is some specialization of the synaptic input among the motoneurons, with those neurons belonging to the type I and small IIA motor units receiving the heaviest projections from the *muscle spindles* and from the corticospinal pathway (Burke, 1986). This arrangement favors activation of small motor units during reflex activation. Some forms of voluntary activation may not adhere to the size principle.

The order of motor unit recruitment changes only if the motor task is modified.

Basmajian (1963) reported that when human volunteers were given feedback from the oscilloscope screen and a loudspeaker, some learned to alter the order of recruitment among motor units in the thumb muscles. This demonstration appears to be an example of willful suppression of the size principle, since activation could be selected among at least three units. This challenging finding has been the source of much controversy and has received some support from experiments in which recruitment order was altered by sensory stimuli applied simultaneously to the skin (Garnett & Stephens, 1980; Grimby & Hannerz, 1968). It is also quite clear that recruitment order can vary if the same muscle is used for different purposes or in slightly different ways. For example, in the extensor digitorum communis, the recruitment order will depend on which of the four fingers is to be extended (Schmidt & Thomas, 1981). Similarly, in the biceps brachii, the threshold of a motor unit depends on whether the muscle is being used to flex the elbow, supinate the forearm, or externally rotate the humerus (Gielen & van der Gon, 1990). Passive factors, such as joint position, may also affect recruitment order, as was noted in the rectus femoris by Person and colleagues (1974, 1972).

There are additional examples of violation of the size principle.

During sustained submaximal muscle contraction, there is evidence that individual motor units will periodically be given a rest. This adaptation is thought to permit rotation of contribution between a number of motor units, thereby delaying the time when the task cannot be continued. This motor unit rotation strategy has been demonstrated during sustained isometric contraction of biceps brachii at 10% of maximum, but is not apparent at 40% (Fallentin *et al.*, 1993).

Lengthening Contractions Use a Special Control Strategy

In isolated muscle, *lengthening contractions* with supramaximal stimulation generate more force than isometric contractions. In contrast, when torque is measured during maximal voluntary lengthening contractions, the magnitude of torque is not typically greater than that of isometric torque. This disparity probably occurs as a result of inhibitory reflexes initiated by *Golgi tendon organs* (see chapter 4). Do these inhibitory reflexes affect motor units with the lowest threshold? If this is the case, then lengthening contractions would represent a situation in which there was selective recruitment of high-threshold motor units.

This question is not easily answered, for the same reasons that it becomes difficult to evaluate motor unit firing frequency during maximal effort contractions. The signal-to-noise ratio is not sufficient. However, assessment of motor unit activation during submaximal lengthening contractions has revealed a strategy that is different from that observed for submaximal *shortening contractions.* Thus, Nardone and colleagues found that selective recruitment of large motor units occurred during lengthening contractions. In one study (Nardone & Schieppati, 1988), they showed two patterns of EMG response in the triceps surae during lengthening contractions. In some subjects, the soleus muscle was active during shortening contractions (plantarflexion) but inactive during lengthening contractions (resisted dorsiflexion), while the gastrocnemius muscle increased activity during lengthening contractions. In other subjects, the soleus remained active during lengthening contractions, but with what appeared to be large motor unit action potentials. Gastrocnemius activity also increased during lengthening contractions in these subjects. The first pattern of response (soleus *derecruitment*) was evident in subjects with long *contraction times* for their soleus muscle, while the second pattern was evident in subjects with a relatively short soleus contraction time. These observations suggest that fast-twitch motor units are preferentially activated during lengthening contractions, but the fact that surface EMG was used for these experiments sheds some degree of uncertainty on this conclusion.

In their second study (Nardone *et al.,* 1989), these workers used intramuscular EMG recordings to permit identification of individual motor units. They identified units that were active only during shortening, or only during lengthening, or during both shortening and lengthening. The key observation here is that there were motor units active during lengthening but not during shortening. These units were identified as large units since they were not activated by reflex responses and had high axon conduction velocities. These data can be interpreted to indicate that fast-twitch motor units are selectively activated during lengthening contractions.

Despite the preceding observations, the point must be emphasized that, when the same motor task is undertaken in exactly the same way, the order in which motor units are recruited remains fixed (see also Henneman *et al.,* 1976).

In sudden movements, the fast-twitch units may be selectively activated.

Selective Fast-Twitch Unit Activation

What about rapidly executed movements? Under these circumstances, there is evidence from both human and animal experiments that type II units may sometimes have the lowest thresholds. For example, in the bush baby, Gillespie *et al.* (1974) found that jumping made greatest demands on the type IIB units. Also, when cats shake their hindpaws, the gastrocnemius muscles, containing both type I and type II units, are active while the soleus muscles, comprising only type I units, are silent (Smith *et al.,* 1980). The results of Grimby (1984), in the human short extensor muscle of the toes, are of particular interest, since he was able to follow the activities of single motor units during walking and running as well as in controlled toe extension. He found that, as expected, the lowest-threshold units have the lowest impulse conduction velocities. However, there are some high-threshold units that do not participate in walking or even in running, but only in rapid corrective movements—such as accelerations or sudden changes in direction. Even then, these units fire very few impulses in a high-frequency burst.

In conclusion, then, there is good evidence that the type I motor units are mostly employed during weak or moderately strong contractions of a sustained or repetitive nature. As the contractions become stronger, type IIA units are recruited in addition to the type I units and, finally, the type IIX units are also activated. In certain very rapid or sudden corrective movements, however, the type II units may be selectively activated.

Maximal Voluntary Contraction

There is agreement among investigators that the lowest frequency at which motor units can discharge steadily is from 8 to 12 Hz and that this is found during weak contractions (figure 13.6; Freund, 1983; Monster & Chan, 1977). Lower frequencies can be generated only in repetitive voluntary contractions of a muscle or in certain movement disorders (e.g., Parkinsonian tremor).

It has been much more difficult to determine the maximum firing frequencies, partly because of the considerable overlap of motor unit territories in any region of the muscle (see chapter 12). During a maximum contraction, as many as 20 motor units may be discharging in the vicinity of the recording electrode, and it is virtually impossible to recognize the firing pattern of an individual unit. Even if a tungsten wire microelectrode is employed, so as to improve the signal-to-noise ratio, the movement of the muscle belly during the contraction will invariably dislodge the tip from its proximity to the active fiber.

Maximal Firing Rates

For the reasons just given, it is not surprising that considerable disagreement has existed among investigators as to the maximum firing frequency. On the one hand, Norris and Gasteiger (1955) reported rates as high as 140 Hz; but others, such as Dasgupta and Simpson (1962), have insisted on much lower values. Another factor that may have contributed to the differences in experimental results is the speed of contraction. Tanji and Kato (1972) found that a motoneuron will discharge at a much higher rate if a given tension is to be reached rapidly rather than slowly. They observed many units that could fire initially at rates of 70 to 90 Hz. Similar high rates were also noted by Marsden *et al.* (1971), who made use of the anomalous innervation of some units within the adductor pollicis by median nerve fibers. On blocking of the major innervation (ulnar nerve) with local anesthesia, the activity of single (median nerve) units could be followed even during maximum contractions. Rates of 100 Hz or more were observed, but only at the start of a contraction. Firing rates decreased during sustained maximal contractions.

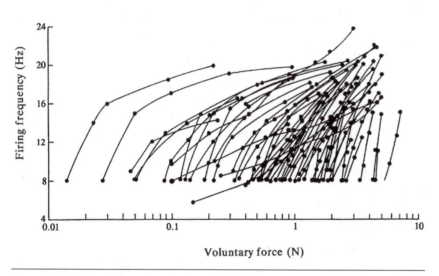

Figure 13.6 Firing frequencies of motor units in the human extensor digitorum communis muscle during steadily increasing contractions. It can be seen that most units commence discharging at approximately 8 Hz and that the frequency rises to 16 to 24 Hz as the contractions become stronger. Unlike the high-threshold units, the lowest-threshold units increase their firing rates over a relatively large range of forces.

Reprinted, by permission, from A.W. Monster and H. Chan, 1977, "Isometric force production by motor units of extensor digitorum communis muscle in man," *Journal of Neurophysiology* 40: 1434.

Steady impulse firing rates depend on the tension required of the muscle and on the contraction speed of the muscle.

Decreasing Firing Rates

Within a few seconds of activation, the maximal motor unit firing frequencies fall to values that depend both on the muscle and on the individual. Bellemare *et al.* (1983) observed that the mean firing rates for units in the adductor pollicis and biceps brachii were approximately 30 Hz, but for the soleus they were only 11 Hz (figure 13.7). The authors pointed out that the soleus has a much slower twitch than the other two muscles and that the reduced firing rate would still have been sufficient to achieve a fused contraction.

Leaving aside the very high frequencies that can be generated at the start of a maximal contraction, it would appear that there is a three- to fourfold range (i.e., 8-30 Hz) within which units can modulate their steady firing rates, depending on the force required (figure 13.6). As the effort continues and mechanical fatigue sets in, the maximum firing rate declines further (e.g., to 8-15 Hz, see figure 13.7); nevertheless, the rate still remains appropriate for ensuring a fused contraction, because the fatiguing muscle relaxes more slowly. It is thought that an inhibitory reflex from the fatiguing muscle is responsible for decreasing the motoneuron discharge rate (see chapter 15).

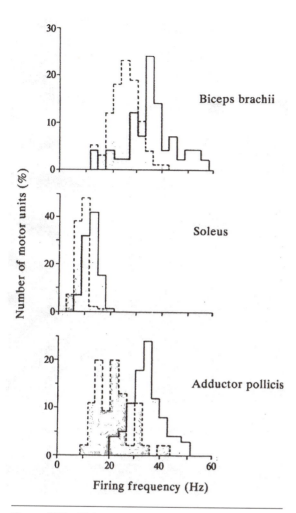

Figure 13.7 Firing rates of motor units in the biceps brachii, soleus, and adductor pollicis muscles of two subjects (solid and dashed lines, respectively) soon after the onset of maximum voluntary contractions; the recordings were made with tungsten microelectrodes. Note that the firing frequencies are consistently higher in one subject than in the other, and that the values are significantly lower for soleus than for the other muscles.

Reprinted, by permission, from F. Bellemare et al., 1983, "Motor-unit discharge rates in maximal voluntary contractions of three human muscles," *Journal of Neurophysiology* 50: 1386.

Does the Decreasing Frequency Represent Muscle Wisdom?

An interesting observation was made in 1979. It was reported that electrical stimulation of the motor nerve at a constant high frequency resulted in greater decrease in force (more fatigue) than occurred with voluntary effort (Bigland-Ritchie *et al.,* 1979). The mechanism for the force decrease was apparently intermittent neuromuscular junction transmission failure. Although the applied frequency was high, failure resulted in a lower frequency of *action potentials* on the muscle fibers. Decreasing the frequency of activation during this electrical stimulation actually permitted better maintenance of force, like that which was evident with voluntary effort. It was then realized that the decreasing frequency of firing of individual motor units during maximal effort probably contributed to maintenance of force. In recognition of the idea that it is rather "clever" of the motor system to anticipate this problem, this property of motor unit recruitment has been called *muscle wisdom.* Furthermore, it is recognized that prolongation of the twitch contraction during repetitive muscle use results in a shift in the force–frequency relationship such that fusion occurs at a lower frequency (see chapter 15), presumably taking advantage of the decreasing frequency of firing.

Recently, it has been suggested that muscle wisdom is not required to maintain force (Fuglevand & Keen, 2003). The original observation of greater force decline with constant electrical stimulation was done with 80-Hz stimulation. As we have seen, maximal sustained rate of firing of motor units does not typically reach this level. When constant 30-Hz stimulation was used, force was maintained better than with voluntary activation or with decreasing frequencies of electrical stimulation. This study suggests that the property of muscle that we call muscle wisdom actually contributes to decreased muscle performance typically called central fatigue (see chapter 15).

Testing Maximal Voluntary Activation

> In most muscles, all the motor units can be maximally activated.

As previously described, Grimby (1984) found that there were some motor units in the short extensor muscle of the toes that discharged during sudden corrective movements but not during sustained contractions. What of the other muscles? One experimental approach has been to compare the force developed during a maximal voluntary contraction with that generated by tetanic stimulation of the same muscle. In the small muscles of the hand, the results obtained to date suggest that the forces are similar, implying that all motor units can be recruited voluntarily and made to discharge at optimal rates for tension development (Bigland-Ritchie *et al.,* 1983; Merton, 1954).

There are, however, technical problems that make such experiments difficult and cloud the interpretation of the results. For one thing, the muscles are usually stimulated through their nerves. Not only is tetanic stimulation painful, but there are few accessible nerves that innervate a single muscle or even a group of muscles having only one action at the same joint. Also, some muscles have rather complex actions, and it is difficult to measure all the resultant forces. Rather than use

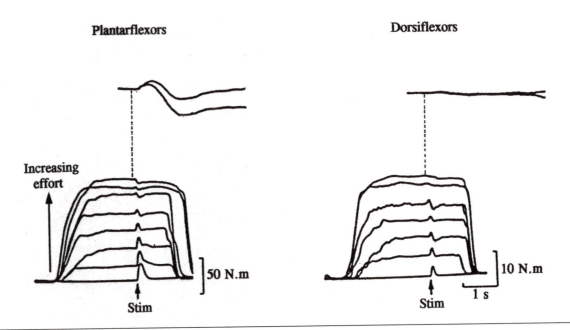

Figure 13.8 *Interpolated twitch* technique. The two sets of superimposed traces at the bottom of the figure show the responses of the dorsiflexor and plantarflexor muscles to single stimuli (Stim) delivered during relaxation and during the course of increasingly forceful contractions. In the strongest contractions (uppermost traces), no interpolated twitch can be detected in the dorsiflexors, but a small one appears to be present in the plantarflexors. The paired traces at the top of the figure were made at higher amplification and on a faster sweep; they confirm the persistence of a small twitch in the plantarflexors and the absence of one in the dorsiflexors.

Adapted from Bélanger and McComas (1981).

tetanic stimulation, an alternative approach has been to inject a single (or double) maximal stimulus to the motor nerve in the course of a maximal voluntary effort. If all the motor units have been recruited and are firing at optimal frequencies, no additional tension will be detected by an appropriately mounted strain gauge. If some motor units are silent or are firing at low frequencies, the interpolated stimulus will produce a twitch on top of the voluntary force record (see figure 13.8). This relatively painless technique, originally described by Merton (1954), has shown that during isometric contractions, motor units can be fully activated in some muscles but not in others. Activation is complete in the small muscles of the hand, the dorsiflexors of the ankle, the quadriceps, and the diaphragm (Bélanger & McComas, 1981; Bellemare & Bigland-Ritchie, 1984; Edwards *et al.*, 1975; Merton, 1954). In some subjects, however, there are motor units in the triceps surae that cannot be fully recruited in steady contractions (Bélanger & McComas, 1981); and as already noted, the same is true of the short extensors of the toes.

There is indirect evidence that activation is less than maximal during simple movements. When a person initiates a weight training program, typically there are rapid gains in maximal strength (1-repetition maximum) before any change in muscle size can be detected. These changes are thought to occur through neural adaptations that permit greater torque generation during critical stages of movement (Sale, 1992). This issue is discussed further in chapter 20, but is considered here for the implication that during movements such as the squat, bicep curl, and bench press, improved neural activation of the appropriate muscles can contribute to improved performance. This implies that prior to the training these individuals were incapable of fully activating the necessary muscle groups in the appropriate sequence. However, one cannot discount the possibility that changes in the intramuscular matrix that transmits muscle fiber forces to the distal tendon may occur rapidly (within days), thus accounting for some of this apparent "neural" component of training (Jones *et al.*, 1989).

Applied Physiology

Motor unit potentials, as recorded with an intramuscular needle electrode, can be greatly altered by nerve and muscle diseases. Indeed, these changes can be used to help diagnose patients complaining of weakness or increased fatigability. More information about this aspect of motor unit function was given in "Needle Electromyography—From Research Lab to Bedside."

People who suffer spinal cord injury lose partial or complete motor control of parts of their body distal to the injury. This change results in loss of mobility and impaired physical capabilities. These changes also predispose spinal cord-injured patients to chronic disease for which inactivity is a risk factor: osteoporosis, obesity, diabetes, and cardiovascular disease. Electrical stimulation of the paralyzed muscles (referred to as *functional electrical stimulation, FES*) provides a potential mechanism to restore some physical capabilities (Popovic, 2003), and in some cases limited mobility (Barbeau *et al.,* 2002).

There are several problems that still need to be overcome before FES becomes practical, but research has provided several recent advances that make this application of our knowledge of motor unit recruitment and muscle contractile properties feasible. Improved locomotion and muscle adaptive responses have been reported (Barbeau *et al.,* 2002), but jerky movements and accelerated fatigue of the stimulated muscles continue to be problems (Wise *et al.,* 2001). Variable-frequency trains of stimulation (Russ & Binder-Macleod, 1999) or even repeated doublets (Overgaard & Nielsen, 2001) may be the key to minimizing the impact of fatigue.

The physiology of muscle excitation and contraction has now been explored, both at the level of the muscle fiber and at that of the motor unit. One more feature of the working muscle remains to be examined. This feature is the biochemical reactions in the muscle fiber that provide the energy required for muscle contraction. These reactions are discussed in the next chapter.

14

Muscle Metabolism

This chapter is largely devoted to *adenosine triphosphate (ATP)*, the source of energy for many enzymatic reactions in the muscle fiber. The cellular processes that require ATP will be presented, along with a discussion of the determinants of how much ATP is needed. We will see how the energy associated with the molecular structure of glucose and lipid is transferred to ATP, and how some ATP synthesis may continue without the use of oxygen. The special role of the mitochondrion in cell metabolism will also be considered. The biochemistry presented in this chapter is rather cursory, to give the reader a basic understanding of the metabolic pathways that support the energy requirements of muscle contractions. For a more detailed presentation of these metabolic pathways, the reader is referred to textbooks that specialize in this topic, like Houston's *Biochemistry Primer for Exercise Science* (2001).

Energy Required for Muscle Contraction

There are many processes in living cells that require energy. Typically the immediate source of this energy is ATP. Many enzymes in cells throughout the body make use of chemical energy released by ATP hydrolysis in the following reaction:

$$ATP \rightarrow ADP + Pi + E$$

where ATP is adenosine triphosphate, *ADP* is *adenosine diphosphate,* Pi is *inorganic phosphate,* and E is energy that is made available to perform work. The amount of energy that is made available by ATP hydrolysis varies according to the energy charge: $[ATP] / [ADP] \cdot [Pi]$. Skeletal muscle can use ATP at a rate that would deplete available supplies in seconds if ATP were not replenished. Metabolic processes are available in these fibers to replenish ATP at a rate that can match the incredible rate of degradation, thereby preventing or minimizing the otherwise inevitable decrease in ATP concentration. A severe decrease in ATP concentration will compromise the energy charge and is incompatible with viability of the cell, so this concentration is strongly defended by the metabolic pathways.

Biological work is the term used to describe processes within living cells that require energy. This includes chemical reactions of synthesis, transfer of ions or molecules across membranes against *electrochemical gradients,* and mechanical work (displacement against a force). When a muscle is activated, there are three principal reactions that require ATP. These reactions are catalyzed by ATPase enzymes:

Three processes require ATP.

- Na^+-K^+ ATPase
- Actomyosin ATPase
- Ca^{2+} ATPase

The Ca^{2+} ATPase is also referred to as *sarco-endoplasmic reticulum calcium ATPase (SERCA).* These ion pumps account for much of the energy that is required at rest as well.

Some energy is needed for activation processes.

Each time a muscle fiber is activated, an *action potential* is transmitted across the *sarcolemma* and into the *T-tubules*. The action potential is a passive event and, as such, it is propagated along the sarcolemma without the use of ATP. This action potential elicits release of Ca^{2+} from the *terminal cisternae,* which is also passive. These events of *activation* disturb the concentration gradients for Na^+ and K^+ across the sarcolemma and for Ca^{2+} between the *sarcoplasmic reticulum* and the *sarcoplasm.* Adenosine triphosphate is required to restore these concentration gradients by active transport, using the ion pumps.

Cross-bridges transform chemical energy of ATP to force generation and work.

Chapter 11 deals with the contractile responses of muscle, and knowledge of the contractile processes will facilitate an understanding of the energetics of muscle contraction. When the myosin head binds with *actin,* forming a *cross-bridge, myosin ATPase* is activated and the cross-bridge advances to a force-generating state. This transition requires energy that is derived from hydrolyzing ATP. If the load imposed on the muscle is less than the force exerted by the cross-bridges, then shortening of the muscle will occur. The energy released by hydrolysis of ATP is transformed to heat and work.

Replacing Adenosine Triphosphate

As mentioned previously, the rate at which ATP can be used by the ATPase reactions is incredibly fast. The rate of ATP use can increase more than 100-fold above the resting level for short periods of time. The quantification of energy use in muscle will be addressed later in this chapter. Now, let's consider the mechanisms available to replenish ATP in each muscle fiber. These mechanisms effectively minimize the decrease in ATP concentration that would otherwise occur when a muscle is activated.

Immediate Energy System

Phosphocreatine maintains ATP concentration.

Even for relatively brief periods of activity, the muscle must be able to renew its supply of ATP. In this respect, the muscle fiber differs from other cells in having a substantial energy reserve in the form of *phosphocreatine* (PCr). As ATP is hydrolyzed, it is instantly replenished from PCr by the action of the enzyme *creatine kinase* (CK):

$$ADP + PCr \Leftrightarrow ATP + Cr$$
$$CK$$

So effective is PCr in maintaining ATP levels that, when a muscle fiber is exercised to fatigue, substantial amounts of ATP remain, even though the supply of PCr may be virtually exhausted. In a resting muscle, this store of PCr is four to eight times larger than that of ATP. The role of PCr does not end with replenishment of ATP. Although PCr has often been considered to be a storage supply of energy available to immediately replenish ATP, it is now recognized that *creatine* functions in the muscle to activate oxidative metabolism (Walsh *et al.,* 2001) and shuttle energy between the mitochondria and the *myofibrils* (Bessman & Geiger, 1981). The creatine shuttle will be further discussed later in this chapter, but first additional mechanisms of ATP repletion will be considered.

Adenosine triphosphate can be synthesized from ADP in another way. The enzyme myoadenylate kinase or *myokinase* (MK) can convert two molecules of ADP to one of ATP and one of adenosine monophosphate (AMP):

$$2ADP \Leftrightarrow ATP + AMP$$
$$MK$$

AMP can then be deaminated:

$$AMP \Leftrightarrow IMP + NH_3$$

where IMP is inosine monophosphate and NH_3 is ammonia. This reaction helps preserve the energy charge of ATP by keeping the concentration of ADP low. This pathway for the production of ATP probably plays a relatively minor role, though in exercise of moderate or high intensity, the concentration of ammonia in the arterial plasma rises severalfold (Banister & Cameron, 1990). The myoadenylate reaction is not the only source of ammonia during sustained exercise. When carbohydrate supplies become low, protein catabolism contributes to *gluconeogenesis,* and amino acids can be deaminated for direct entry into the Krebs cycle (see later). Deamination of amino acids results in the release of ammonia.

Glycolysis provides a rapid supply of ATP.

The Short-Term Energy System

Glycolysis is the sequence of reactions beginning with the substrates glucose or glycogen and resulting in the formation of pyruvic acid or *lactic acid.* This sequence is illustrated in figure 14.1. The key features of glycolysis include

- rapid, local formation of ATP;
- regulation by Ca^{2+}, energy charge, hormones, and end-products of the sequence of reactions; and
- stimulation of oxidative metabolism by end-products.

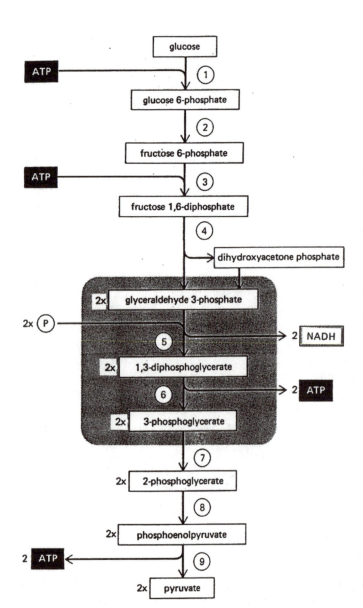

The enzymes of glycolysis are not simply floating freely in the *cytoplasm.* They are bound, and are located in specific regions where ATP is required. Enzymes of glycolysis have been identified in association with glycogen granules, the sarcolemma, and sarcoplasmic reticulum. It has been suggested that these enzymes are organized in such a manner that glucose and the intermediates (see figure 14.1) are passed from one enzyme to the next to prevent a buildup of these molecules and to increase the probability of reaction with the appropriate enzyme. This channeling effectively increases the concentration of the glycolytic intermediates in the location of the enzyme (Hochachka, 2003). This theory has been used to explain why flux through the glycolytic pathway can increase without a corresponding increase in concentration of reactants.

The advantage of this channeling can be envisioned if the relationship between concentration gradient and distance is considered in determining the possible rate of an enzymatic reaction. If two enzymes in a sequence are located some distance apart, then a gradient of concentration is required to permit delivery of

Figure 14.1 Steps in the breakdown of glucose to pyruvic acid. At step 4, the six-carbon sugar molecule is split, and steps 5 and 6 are those responsible for the net synthesis of ATP and NADH molecules.

Reprinted, by permission, from B. Alberts et al., 1983, Small molecules, energy, and biosynthesis. In *Molecular Biology of the Cell* edited by B. Alberts et al. (New York: Garland Publishing, Inc.), 57.

the substrate at a given rate. The closer the two enzymes are, the smaller this concentration gradient needs to be. If the enzymes are adjacent, and the chemical reactants are handed from one enzyme to the next, then virtually no concentration gradient is needed.

The mechanism by which this channeling or cytoplasmic circulation occurs has not been elucidated (or confirmed). It has been suggested that *tubulin* or actin may be involved. This seems unlikely since *molecular motors* would be required to transport the intermediates from enzyme to enzyme. Unfortunately, the metabolic cost of this process would result in a net loss of ATP. However, the enzymes may be bound collectively to some internal structure, such as a membrane or cytoskeletal element.

One of the most important features of the glycolytic pathway is that it can supply ATP very quickly, and very close to the site where ATP is required. A consequence of rapid glycolysis is an increase in pyruvate concentration and reduced nicotinamide adenine dinucleotide (NADH) concentration. The enzyme lactate dehydrogenase plays an important role in converting pyruvic acid to lactic acid. In conjunction with this reaction (see later), NADH is reconverted to NAD^+.

The Long-Term Energy System

The oxidation of glucose and fat provides a long-term supply of ATP.

At rest, virtually all of the energy used in the muscle cells is provided by oxidative metabolism, with a mix of carbohydrate and fat for substrate. When muscle contractions are initiated, products of the immediate and short-term energy systems (creatine, pyruvic acid, and NADH) will stimulate oxidative metabolism, the long-term metabolic system.

Oxidative metabolism includes four processes, which occur in the mitochondria: *β-oxidation, citric acid cycle* (also called *Krebs cycle* or *tricarboxylic acid cycle*), *electron transfer chain,* and *oxidative phosphorylation.*

Mitochondrial Structure

Final processing of energy substrates occurs in the mitochondria.

The mitochondria are specialized organelles that generate ATP using chemical energy from NADH, pyruvic acid, lactic acid, fats, and amino acids. The mitochondrial volume relative to total muscle cell volume can be less than 2% in the case of fast-twitch glycolytic fibers to over 40% in specialized flight muscles of bumblebees, hummingbirds, and other endurance athletes of the animal kingdom.

In electron micrographs of muscle prepared by longitudinal section, the mitochondria appear to be discrete spheroid or ellipsoidal organelles. However, several authors have identified connections between mitochondria, particularly when tissue is prepared with transverse sections (Kirkwood *et al.,* 1986). These observations have led to the conclusion that what appear to be discrete mitochondria in longitudinal sections are actually interconnected, forming a mitochondrial reticulum or network.

The detailed structure of mitochondria is relevant to the understanding of how mitochondria process the available substrates and rephosphorylate ADP to form ATP. The mitochondrial reticulum has a double membrane. The inner membrane, which separates the intermembrane space from

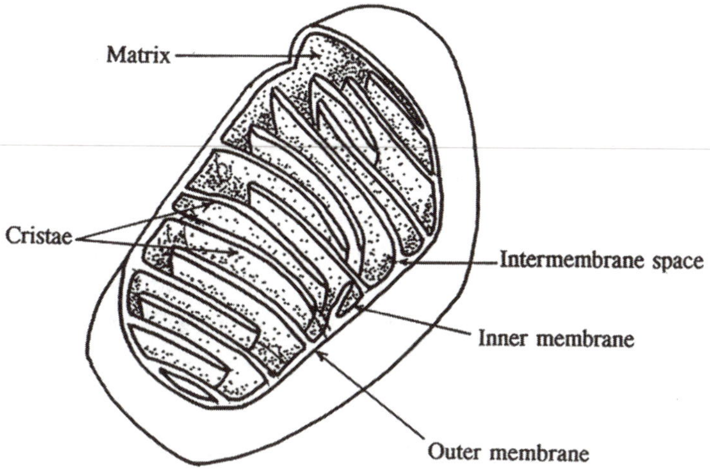

Figure 14.2 Structure of a muscle mitochondrion.

the mitochondrial matrix, is convoluted, having many folds. The projection of the intermembrane space into the middle of the mitochondrion forms a *crista* (see figure 14.2).

The Citric Acid Cycle

The citric acid cycle feeds the electron transfer chain.

The citric acid cycle is illustrated in figure 14.3. This sequence of reactions is referred to as a cycle because it begins and ends with oxaloacetate. *Acetyl-CoA,* which is formed by β-oxidation or decarboxylation of pyruvic acid, is added to the four-carbon oxaloacetate to form six-carbon citrate. Subsequent decarboxylations result in the release of CO_2 and the reformation of oxaloacetate.

In addition to acetyl-CoA and oxaloacetate, other important molecules are required for the citric acid cycle to proceed, including *flavin adenine dinucleotide (FAD)*, nicotinamide adenine dinucleotide (NAD^+), *guanosine diphosphate (GDP),* and inorganic phosphate (Pi). Of these, NAD^+ appears to be the most important. At rest, mitochondrial NAD is primarily available in the reduced form (NADH) (Jobsis & Stainsby, 1968); NADH is required for the electron transfer process to occur (see later). When metabolic energy demand is increased, as it would be during exercise, NADH is oxidized to NAD^+, which activates the citric acid cycle. Water (H_2O) is also required here, but it can be assumed to be in unlimited supply.

The Electron Transfer Chain

Energy is trapped in the respiratory chain.

The energy from the oxidation of acetyl-CoA is harnessed in two ways. One of these is through the creation of a high-energy phosphate bond, in *guanosine triphosphate (GTP)*. In turn, GTP donates its phosphate to ADP, forming ATP. However, most of the energy captured in the citric acid cycle comes from the transfer of electrons. The electrons are first received by the hydrogen receptor molecules, *nicotinamide adenine dinucleotide (NAD)* and flavin adenine dinucleotide (FAD), and then passed along a special series of molecules (cytochromes of the electron transport chain or *respiratory chain*) in the mitochondria. These cytochromes are bound to the inner mitochondrial membrane.

As the electrons are passed along the respiratory chain, their energy levels become progressively lower; the energy released is used to actively transport protons across the inner mitochondrial membrane against a concentration gradient (the chemi-osmotic process: Mitchell, 1961). Since the protons become more concentrated in the intermembrane space, and since the matrix of the mitochondrion will have a surplus of negative charges, the protons will tend to diffuse back across the membrane down their electrochemical gradient. This reentry of protons to the matrix is regulated at the F-complex of the inner mitochondrial membrane and is coupled to ATP synthase, a membrane-bound enzyme that converts ADP and

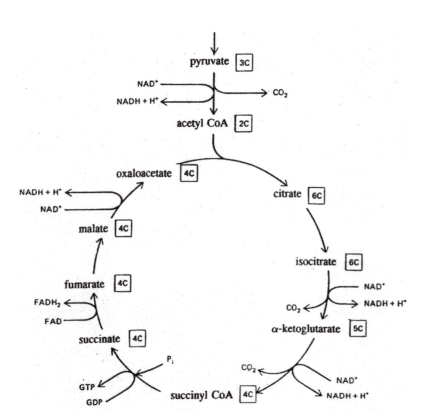

Figure 14.3   The citric acid cycle. The number of carbon (C) atoms in each intermediate molecule is shown. The C atoms are removed as CO_2, while the H atoms are taken up by NAD^+ and FAD, two carrier molecules in the electron transfer (respiratory) chain.

inorganic phosphate to ATP. At the end of the respiratory chain, the electrons are accepted by oxygen. For each molecule of acetyl-CoA oxidized in the citric acid cycle, 12 molecules of ATP are formed. Since 2 molecules of acetyl-CoA are produced from each glucose molecule, the citric acid cycle creates 24 ATP molecules. The glycolytic pathway (glucose to pyruvate) can add another 8 molecules of ATP if it is assumed that cytoplasmic NADH can transfer *reducing equivalents* to the mitochondria, and 6 more can be made available from NADH formed in association with pyruvate dehydrogenase.

Substrates for Energy Metabolism

The short-term and long-term metabolic pathways require fat, carbohydrate, and protein to permit the provision of ATP in the muscle. The provision of these substrates will now be considered.

Glycogenolysis

> **In the muscle fiber, glucose is obtained from glycogen.**

Although muscle fibers can oxidize glucose entering from the bloodstream via the interstitial fluid, glucose or, more correctly, glucose-1-phosphate is also provided by the hydrolysis of glycogen during contractile activity. As the storage form of glucose, glycogen can be seen in electron micrographs of muscle fibers as granules, 10 to 40 nm in diameter, in the *cytosol*. Bound to the surfaces of the granules are the enzymes necessary to synthesize glycogen from glucose (glycogen synthetase) and also those required to degrade it back to glucose (glycogen phosphorylase). Each glycogen molecule is formed from the linking together of 10,000 to 30,000 glucose units in a branching pattern, giving the molecule a molecular weight of around 2,600 kD (see figure 14.4A). The various enzymatic reactions involved in the synthesis and degradation of glycogen are shown in figure 14.4B; note that glucose must be phosphorylated before it can be used to manufacture glycogen and also that some of the glycogen is hydrolyzed in the lysosomes.

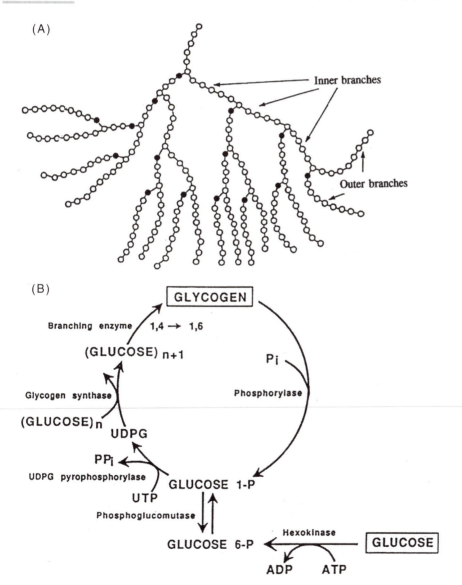

Figure 14.4 (A) Branching structure of glycogen; each circle represents a glucose residue. (B) Enzymatic reactions involved in the synthesis and degradation of glycogen.

Adapted, by permission, from T. DeBarsy and H.G. Hers, 1990, Normal metabolism and disorders of carbohydrate metabolism. In *Baillier's Clinical Endocrinology and Metabolism*, edited by J.B. Harris and D.M. Turnbull (Toronto, Canada: Bailliere Tindall), 500.

Fat oxidation proceeds with the conversion of fatty acids to acetyl-CoA.

Fatty Acids for β-Oxidation

As already noted, fat (lipid) metabolism is used to supply the energy requirements of resting muscle; it also provides much of the energy in exercise, especially when the exercise is of low intensity and long duration. In prolonged exercise, such as distance running, it is possible to consume virtually all of the glycogen, and continued effort will depend largely on lipid metabolism (Åstrand, 1967). Most of the fat is available to the muscle fiber as fatty acids, which are transported across the *plasmalemma* from the bloodstream; some fat, however, is in the form of fine lipid droplets in the cytosol. Within the droplets, each fat molecule is composed of three fatty acid molecules linked to *glycerol* and hence is termed a triglyceride. The lengths of the fatty acid chains (tails) vary considerably; palmitic acid (see figure 14.5) and stearic acid are considered to have "long" chains with 16 and 18 carbon atoms, respectively.

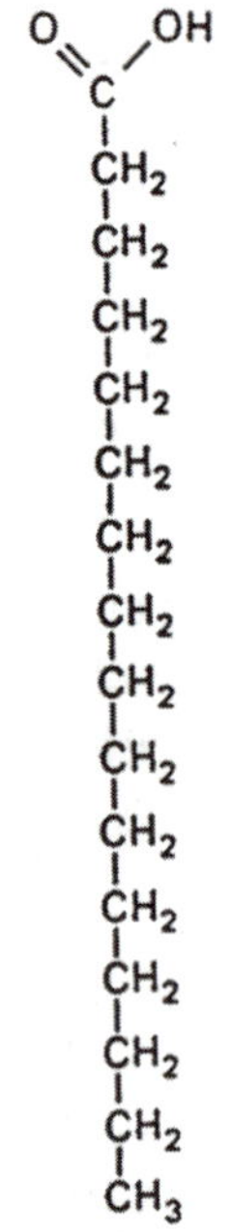

The breakdown of a triglyceride molecule begins with its being split into glycerol and the three component fatty acids. The fatty acids are then taken into the matrix of the mitochondrion for metabolic processing. While medium- and short-chain fatty acids can pass directly through the inner mitochondrial membrane, the long-chain fatty acids require a special transport system. The long-chain fatty acids are first esterified at the outer mitochondrial membrane with CoA; the latter is in turn converted to acylcarnitine, by the enzyme *carnitine palmityltransferase* at the outer surface of the inner mitochondrial membrane. Acylcarnitine is then passed through the inner membrane by the enzyme acylcarnitine translocase in exchange for free *carnitine* from the matrix (see figure 14.6).

The long-chain acylcarnitine, like the short- and medium-chain fatty acids that have passed directly into the matrix space, is next converted to fatty acyl-CoA (figure 14.6). The latter enters a sequence of reactions that cuts off two

Figure 14.5 Palmitic acid. This example of a fatty acid has 16 C atoms.

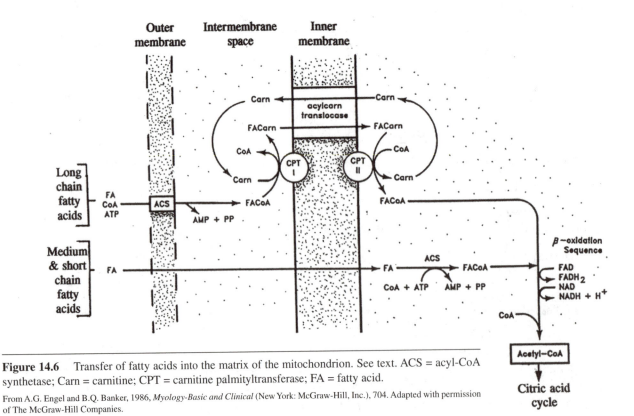

Figure 14.6   Transfer of fatty acids into the matrix of the mitochondrion. See text. ACS = acyl-CoA synthetase; Carn = carnitine; CPT = carnitine palmityltransferase; FA = fatty acid.

From A.G. Engel and B.Q. Banker, 1986, *Myology-Basic and Clinical* (New York: McGraw-Hill, Inc.), 704. Adapted with permission of The McGraw-Hill Companies.

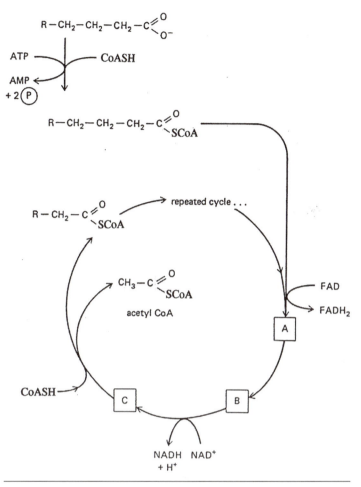

carbon atoms from the fatty acid tail and produces one molecule of acetyl-CoA; the cycle is repeated with two carbons being removed during each turn. It can be seen from figure 14.7 that hydrogen atoms, each with an additional electron, are given to the carrier molecules, NAD and FAD. These electrons are then passed down the respiratory chain and enable ATP to be produced. Meanwhile, the acetyl-CoA enters the citric acid cycle (figure 14.3) and is oxidized to CO_2 and H_2O, with the creation of GTP, NADH, and $FADH_2$.

Integration of the Metabolic Systems

We have seen that there are essentially three metabolic systems that function in a cohesive manner to preserve skeletal muscle ATP concentration. These have been referred to as the immediate, short-term, and long-term metabolic systems. However, they should not be considered discrete systems that operate in isolation from each other. Rather, the terms used to describe these systems should be considered in relation to the speeds with which the systems can be turned on.

Thus the immediate systems, which include PCr as well as the myokinase reaction, can replenish ATP as quickly as ADP is produced. Glycolysis is a bit slower to be activated, but can achieve a relatively high rate of ATP regeneration. Oxidative metabolism increases slowly to a level where it is replenishing ATP at a substantial rate. However, the rate of oxygen use in a muscle fiber will begin to rise immediately upon initiation of exercise, and the maximal rate of ATP regeneration is slower than that associated with glycolysis.

Figure 14.8 illustrates a scheme of integration of the three metabolic systems. Whether ATP is hydrolyzed by the myosin ATPase or one of the main ion pumps, the enzyme CK is located in proximity to these ATPases, and PCr can rapidly replenish ATP. The result of these reactions is increased concentration of creatine and inorganic phosphate. These initial metabolic reactions serve to accelerate the other metabolic systems.

Figure 14.7 The fatty acid oxidation cycle. Each cycle removes two C atoms from the fatty acid in the form of acetyl-CoA and donates H atoms to the NAD and FAD carrier molecules of the respiratory chain.

Reprinted, by permission, from B. Alberts et al., 1983, Small molecules, energy, and biosynthesis. In *Molecular Biology of the Cell* edited by B. Alberts et al. (New York: Garland Publishing, Inc.), 489.

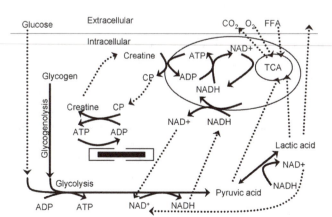

Figure 14.8 Integrated scheme of the various processes by which ATP is replenished in the muscle cell. The immediate energy system results in replacement of ATP with hydrolysis of phosphocreatine (CP). Glycolysis represents the short-term energy system, and oxidative processes in the mitochondria are the long-term energy system. The three systems are integrated in that products of one system are needed for another.

The Creatine Shuttle

The enzyme that facilitates the transfer of a high-energy phosphate bond from PCr to ADP, forming ATP and free creatine, is creatine kinase (CK). It is interesting to note that there are several isoforms of this enzyme, and skeletal muscle contains at least two of them. One was introduced in chapter 1. This cytoplasmic CK is bound at the *M-region* (CK$_c$). Other CKs are probably bound in locations close to the other sites of ATPase activity, the sarcoplasmic reticulum, the I-band, and sarcolemma. These enzymes are critically located where they can rapidly replenish ATP as it is hydrolyzed.

There is also a mitochondrial CK (CK$_{mi}$) that appears to satisfy a very different role. It is located on the outer surface of the inner mitochondrial membrane, in association with adenylate transferase that is embedded in the same membrane. This CK$_{mi}$ rephosphorylates cytoplasmic creatine with mitochondrial ATP, which is made available by the adenylate transferase. This strategic location of CK permits creatine to provide a valuable shuttle service. When the cytoplasmic ATPases hydrolyze ATP to ADP, CK$_c$ can rapidly rephosphorylate ADP to ATP, forming creatine in the process. Creatine will diffuse according to its concentration gradient away from the CK$_c$ and toward the mitochondria (see figure 14.9). At the mitochondrial membrane, CK$_{mi}$ will rephosphorylate creatine, resulting in the formation of mitochondrial ADP and cytoplasmic PCr.

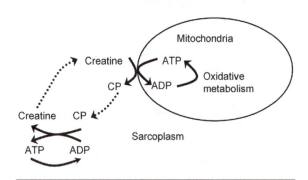

Figure 14.9 The creatine shuttle is illustrated. Hydrolysis of ATP to ADP by any of the cellular ATPases results in activation of creatine kinase, which is located close to the major ATPases. Phosphocreatine is hydrolyzed, donating its phosphate group to ADP, forming ATP and creatine. Creatine diffuses to the mitochondria, where another isoform of creatine kinase rephosphorylates this to PCr, using mitochondrial ATP. The resulting increase in mitochondrial ADP is a key factor in accelerating mitochondrial oxidative phosphorylation.

This creatine shuttle, which was first proposed by Bessman and Fonyo (1966), can help maintain cytoplasmic ATP concentration while transferring high-energy phosphate from the mitochondria to the cytoplasmic ATPases. If there were not a mitochondrial CK available in muscle, then ADP formed in the cytoplasm would have to diffuse to the mitochondria to be rephosphorylated. This would require a buildup of ADP in the cytoplasm to establish a concentration gradient to encourage this diffusion. Increased ADP concentration would occur at the expense of ATP and would contribute to a decreased energy charge. Yet ADP is maintained in the micromolar concentration range, and ATP concentration does not substantially decrease. In contrast, creatine concentration can rise to 30 mM. Two features permit creatine to provide a more rapid shuttle service than ADP could. The rate of diffusion of a molecule in the cytoplasm is proportional to the concentration gradient and inversely proportional to its molecular weight. Because creatine is a smaller molecule than ADP, it will diffuse through the cytoplasm more rapidly than ADP.

Glycolysis Drives Oxidative Metabolism

The glycolytic enzymes are also located in close proximity to the ATPases that are using energy. When muscle contraction is initiated, glycolysis is activated, and the rate of ATP yield can increase over the first 20 to 40 s. In addition to ATP, glycolysis will result in the formation of NADH, pyruvic acid, and lactic acid. These products play a role in activating oxidative metabolism.

It can be seen in figure 14.8 that creatine from the immediate energy system, as well as NADH, pyruvic acid, and lactic acid from the short-term energy system, can be transferred to the mitochondria. Here, these products, particularly creatine, accelerate oxidative metabolism. At the mitochondrial membrane, creatine reacts with mitochondrial ATP and CK to form PCr that can then diffuse into the cytoplasm, as well as ADP that will stimulate oxidative phosphorylation. At the mitochondria, creatine decreases the ratio of ATP to ADP. Phosphocreatine in the cytoplasm has the opposite effect (Greenhaff, 2001).

Cytoplasmic reducing equivalents can be transferred to the mitochondria.

Shuttling Reducing Equivalents

We have seen that NADH is produced in the cytoplasm by glycolysis. Additional ATP can be synthesized via the electron transfer chain and oxidative phosphorylation if this NADH can be transported into the mitochondria. As indicated in figure 14.10, this transfer is not direct. There are three pathways by which it can occur: glycerol phosphate shuttle, malate-aspartate shuttle, and lactate shuttle. The glycerol phosphate shuttle transfers electrons to FAD, forming $FADH_2$, which enters the electron transfer chain at a later point than NADH and generates only two ATP. This appears to be the dominant method of transfer of reducing equivalents in skeletal muscle (see figure 14.10). The malate-aspartate shuttle can transfer electrons from cytoplasmic NADH to mitochondrial NAD^+, thus forming NADH in the mitochondrial matrix. This process can yield three ATP molecules per NADH via the electron transfer chain and oxidative phosphorylation.

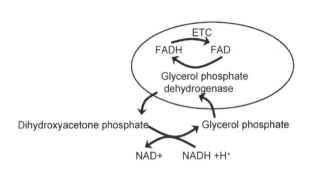

Figure 14.10 The glycerol phosphate shuttle is illustrated. This is one of three mechanisms by which reducing equivalents from glycolysis in the cytoplasm can be transferred to the mitochondria. ETC is the electron transfer chain.

Lactate May Also Serve to Transfer Reducing Equivalents

The lactate shuttle is a relative newcomer with respect to our awareness of the possible transfer of reducing equivalents from the cytoplasm to the mitochondria. The history of our knowledge of the formation and disposal of lactate is long and lively (Needham, 1971; Svedahl & MacIntosh, 2003). Although lactic acid has long been considered an undesirable end-product of glycolysis, this reputation may not be warranted (Brooks, 2000).

It can be seen in figure 14.8 that lactic acid is formed in glycolysis, and that cytoplasmic NADH is oxidized to NAD^+ in the process. The enzyme that catalyzes this reaction is lactate dehydrogenase. This reaction provides two very important changes in the cytoplasm that permit glycolysis to continue: decreased pyruvic acid concentration and increased NAD^+ concentration. Removal of pyruvic acid prevents product inhibition from slowing glycolysis, and NAD^+ is required at two steps in this reaction sequence (see figure 14.1).

Although there is still some controversy (Sahlin *et al.,* 2002; Rasmussen *et al.,* 2002; Van Hall, 2000; Brooks *et al.,* 1999), lactate dehydrogenase has been identified in mitochondria. The presence of this enzyme in the mitochondria suggests the possibility that lactic acid can move across the mitochondrial membrane and be converted back to pyruvic acid. In this process, mitochondrial NADH would be formed:

$$\text{Lactic acid} \xrightarrow[\text{LDH}]{NAD^+ \quad NADH} \text{Pyruvic acid}$$

The pyruvic acid formed in this reaction can undergo decarboxylation and enter the citric acid cycle (see figure 14.3), and the NADH can enter the electron transfer chain.

Regulation of the Metabolic Systems

There are two forms of regulation that are used to accelerate the metabolic systems to replenish ATP: feedback and *feedforward*. A feedback system relies on a change component (substrate, product) to activate the control system. This is like a home heating system that has a thermostat to detect the house temperature. When the temperature falls below the set point, then the furnace is triggered to turn on. If ATP is regulated similarly, then a fall in [ATP] could be the signal to turn on the metabolic system to replenish the supply. Feedforward control uses a parallel activation system such that when ATPase is turned on, the metabolic system is simultaneously activated by the same signal. Metabolic pathways experience both of these forms of regulation.

Enzymatic Control

Enzymatic activity is regulated by feedback, feedforward, and allosteric mechanisms.

Although the activity of an enzyme or the rate at which the reaction proceeds depends on the concentrations of the reactants and products, it can also be modulated by *allosteric regulators*. An allosteric regulator is a specific molecule that binds to the enzyme and either activates or inhibits it. Although many enzymes will catalyze a reaction in either direction, the net change in concentration of reactants and products will be toward an equilibrium ratio. For example, if the equilibrium ratio of reactant to product is 1:4, then the reaction will proceed whenever the concentration ratio of these molecules is greater than 1:4 (e.g., 2:4 or 1:3). The rate of the reaction will therefore increase if reactant concentration increases or product concentration decreases. The reverse reaction will occur if the ratio changes in the opposite direction.

In the case of the cytoplasmic CK reaction, the equilibrium ratio for ADP:ATP is very low. A small increase in ADP concentration or a decrease in ATP concentration will accelerate the reaction. In contrast, the equilibrium ratio for mitochondrial CK is extremely large and irrelevant. Adenylate transferase provides ATP and removes ADP, keeping the ratio of these two essentially infinite. An increase in creatine concentration will favor formation of PCr at the expense of (mitochondrial) ATP.

Phosphofructokinase (PFK) is a key enzyme of glycolysis and is under allosteric regulation. Activity of the enzyme is inhibited by ATP and by H^+. Activity of PFK is increased by Ca^{2+}, Pi, and ADP. In the case of Pi or ADP, this can be considered a form of feedback regulation. An increase in concentration of these molecules is indicative of a need for more ATP. Activation of PFK will result in higher enzymatic activity, which translates to faster synthesis of ATP. Calcium is an example of a feedforward control system. Activation of the muscle fiber results in increased $[Ca^{2+}]$. This activates actomyosin ATPase and SERCA, and consequently increases the rate of ATP hydrolysis. At the same time the increased $[Ca^{2+}]$ activates PFK, resulting in more rapid formation of ATP.

It is worth pointing out that much of what we know about enzyme regulation relies on results of experiments conducted *in vitro*. The behavior of these enzymes may be different *in vivo* where temperature is higher and other conditions may vary from the typical *in vitro* circumstances. For example, it has been difficult to demonstrate feedback inhibition of glycolysis by H^+ *in vivo* (Krause & Wegener, 1996).

Regulation of mitochondrial metabolism involves control of three interdependent processes: citric acid cycle, electron transfer chain, and oxidative phosphorylation. To understand the regulation of these pathways, it is useful to consider the reactants and products of these reactions. The citric acid cycle requires acetyl-CoA, oxaloacetic acid, and NAD^+. The electron transfer chain needs NADH and oxygen. Oxidative phosphorylation requires ADP and Pi. At rest in skeletal muscle, the redox status of the mitochondria favors NADH (Jobsis & Stainsby, 1968). The rate of flux through the citric acid cycle will accelerate when acetyl-CoA and NAD^+ are made available. The limiting reactant for oxidative phosphorylation, and perhaps for flux through the three mitochondrial metabolic pathways collectively, appears to be ADP. When creatine concentration increases in the cytoplasm, activating CK_{mi}, there will be an increase in mitochondrial [ADP]. Thus, creatine drives oxidative metabolism.

Glycogen hydrolysis is stimulated by epinephrine.

For glycogen to be broken down to glucose, it must first be phosphorylated by glycogen phosphorylase; however, this enzyme must itself be phosphorylated by another enzyme, phosphorylase kinase, which is itself phosphorylated by *protein kinase*. The latter, in turn, is switched on by *cyclic AMP (cAMP)*, which is released from the inner face of the muscle fiber plasmalemma, following the combination of epinephrine (adrenaline) with its receptor; *G protein* acts as an intermediary. The complete sequence of events, culminating in the degradation of glycogen, is set out in figure 14.11. Cyclic AMP is an example of a *second messenger,* that is, a molecular signal that works in the interior of the cell, having been generated by an event at the cell membrane. In the case of cAMP, this event is the combination of epinephrine with its receptor. This binding of epinephrine with its receptor also triggers the opening of an ion channel that allows Ca^{2+},

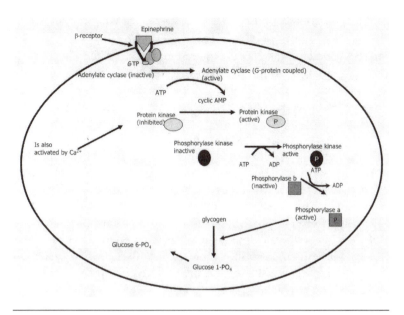

Figure 14.11 Steps in the activation (by *phosphorylation*) of the enzyme glycogen phosphorylase, after epinephrine reaches the outer face of the muscle fiber plasmalemma.

another second messenger, to invade the muscle fiber directly. Other known second messengers are arachidonic acid and *inositol triphosphate.*

Why Is Lactic Acid Formed in Muscle?

It is a popular belief that production of lactic acid in a muscle fiber is evidence of inadequate supply of O_2. This belief probably stems from the designation of glycolysis as an "anaerobic" metabolic pathway. This term was originally invoked when biochemists discovered that *in vitro* muscle, or muscle extract, in the absence of O_2 could still undergo glycolysis with lactic acid as the final product (Hill & Lupton, 1923).

While it is true that an inadequate supply of O_2 will result in accelerated glycolysis and accumulation of lactate,

Lactic acid formation is not necessarily a symptom of limited O_2 supply.

lactate output should not always be construed as evidence of inadequate O_2 supply. Regulation of the metabolic systems dictates that increasing concentration of pyruvic acid will necessarily result in lactic acid formation. Accumulation of pyruvic acid will occur if production exceeds removal by mitochondrial metabolism. Therefore, either accelerated glycolysis or limited uptake of pyruvic acid by the mitochondria can result in a net formation of lactic acid. Considering that glycolysis can be accelerated by sympathetic nervous system activation (which stimulates glycogenolysis), it seems likely that there are circumstances in which lactic acid formation occurs in spite of adequate O_2 availability. This topic is explored more fully in the *Handbook of Physiology* (Gladden, 1996) and in a review on anaerobic threshold by Svedahl and MacIntosh (2003).

We have seen that the muscle fiber is capable of replenishing ATP as fast as it can be hydrolyzed, and that three integrated metabolic systems serve this function. Now, it is time to consider what factors contribute to the magnitude of energy required by the muscle.

How Much Energy Is Needed?

The energetics of muscle contraction has been a topic of intensive research for over a century. Several approaches have been taken, including measurement of heat release during a muscle contraction and measurement of oxygen uptake.

It was mentioned earlier in this chapter that three processes in an active muscle require energy: Na^+-K^+ ATPase, Ca^{2+} ATPase, and actomyosin ATPase. At an elementary level, the energy required by the ion pumps is determined by the number of activations, and the energy required by the actomyosin ATPase is determined by how many cross-bridges are activated as well as the frequency and duration of cross-bridge cycling. On the basis of these considerations, Stainsby and Lambert (1979) have proposed an equation to describe the energy requirement of a muscle contraction. A simplified version of the equation is as follows:

$$Ec = BA + (f1 \cdot A \cdot f2 \cdot L \cdot f3 \cdot v^{-1} \cdot tc)$$

where Ec is the energy cost per contraction; L is muscle length, relative to optimal length; v is velocity of shortening and tc is the cross-bridge cycle rate; B is a constant relating level of activation (A) to energy cost; and f1, f2, and f3 are unspecified functions relating A, L, and 1/v to the rate of use of energy by cross-bridges. How each of these factors contributes to the energy cost of a muscle contraction is described next.

The first term in the equation (BA) relates to the energy required for ion pumping. This energy has also been referred to as *"activation energy,"* as it is the energy related to the muscle activation processes and is independent of cross-bridge activity. The energy used by the ion pumps to restore the concentration gradients for Na^+, K^+, and Ca^{2+} depends on how many action potentials have been elicited and how many *motor units* are involved.

The most direct method by which activation energy has been measured is heat measurement for contractions at different lengths. This approach is illustrated in figure 14.12. When heat output is measured at progressively longer lengths, the total heat decreases in proportion to the decrease in developed force. If this relationship is extrapolated to a length where developed force is zero (no overlap of thick and thin filaments), then all of the estimated heat output will be associated with ion pumping. Using this approach, it has been estimated that 20% to 30% of the total energy for a muscle contraction at optimal length is required for activation processes. Figure 14.12 also shows that the *heat of activation* decreases in fatigued muscles. This fact has implications for the mechanism of fatigue presented in the next chapter.

As already indicated, the amount of energy used by the cross-bridges is dependent on the level of activation, muscle length, velocity of contraction, and rate of cross-bridge cycling. Within a given fiber, the level of activation is a function of the frequency of *impulses* on the sarcolemma. This will determine the free $[Ca^{2+}]$ and hence the occupancy of *troponin* C by Ca^{2+}. This, in turn, determines what proportion of the myosin-binding sites on the actin filaments are available for cross-bridge formation. Muscle length determines the overlap of the filaments and therefore the proportion of myosin heads that oppose actin filaments. The impact of velocity of shortening on the energetics of muscle is rather complex. At slow rates of shortening, the energy cost is increased, presumably because displacement of the filaments brings more myosin heads into a position favorable for cross-bridge formation. At higher velocities there is a decreased probability of cross-bridge formation. This property explains why fewer cross-bridges are generating force during a dynamic contraction.

The final property of cross-bridges that contributes to the energy cost of a muscle contraction is the rate of cycling. As long as Ca^{2+} is bound to a given troponin C, the myosin heads in that region can interact with the corresponding actin molecules. Each myosin head in this region will undergo cycles of attaching, generating force, and detaching. With each cycle, it is generally assumed that one ATP is hydrolyzed. If a cross-bridge remains attached for a long time, then the amount of energy needed will be relatively small. However, if the cross-bridge undergoes several cycles in a short time, then more energy is needed. Type I myosin is slow to cycle, while type II myosin cycles more quickly. For this reason, the energy cost to maintain a given isometric force is less if type I myosin is involved. This is one of the factors that makes slow-twitch fibers advantageous as postural muscles. The measure of the ability to maintain force with a low energy requirement is referred to as "economy of force maintenance." The muscles best suited to this function are the catch muscles of mollusks. As long as the catch muscles are activated, the cross-bridges remain in a semipermanent bound state. With no cycling of cross-bridges, the energy requirement is kept very low.

The determinants of energy cost include level of activation, length, velocity, and fiber type.

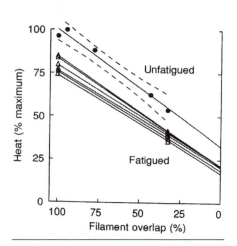

Figure 14.12   Heat production is measured to permit estimation of the energy use in a muscle cell or bundle of fibers. Through stretching of the fiber and measuring the heat at different lengths, the heat associated with transport of ions by the Na^+-K^+ ATPase and Ca^{2+} ATPase can be estimated (intercept at 0 filament overlap). Note that fatigue results in less energy for these processes of activation.

Adapted, by permission, from C.J. Barclay, N.A. Curtin, and R.C. Woledge, 1993 "Changes in crossbridge and non-crossbridge energetics during moderate fatigue of frog muscle fibres," *Journal of Physiology* 468(1): 543-555.

Applied Physiology

There are several metabolic diseases that result in altered muscle function. Here we consider a genetic abnormality that results in storage of glycogen without the capability to get it out of storage, as well as lipid metabolism disorders that result in fatigue and cramps. We will also present a special procedure that athletes can use to increase the availability of glycogen for endurance events.

In McArdle's
syndrome,
lack of
myophos-
phorylase
causes
exercise
intolerance.

McArdle's Syndrome

In 1947 Dr. Bryan McArdle investigated a 30-year-old man who had been admitted to Guy's Hospital in London with a lifetime history of pain, weakness, and stiffness of his muscles during exertion (McArdle, 1951). Even mild exercise, such as walking 100 m, was sufficient to induce the symptoms, which were always relieved by rest. When the muscles were stiff and swollen, it was impossible to record any *EMG* activity with a needle electrode—hence the stiffness was a physiological *contracture*. McArdle carried out a number of biochemical tests on this patient and showed that the lactate concentration fell in the venous blood during ischemic exercise, whereas in a normal subject a substantial rise occurred. On the basis of this observation, McArdle postulated that this patient was unable to break glycogen down to pyruvate and thence to lactate. Since glycogen breakdown depends successively on a debranching enzyme, muscle glycogen phosphorylase *(myophosphorylase)*, and phosphoglucomutase (figure 14.4B), a hereditary deficiency of any of these enzymes could have been responsible for the symptoms described.

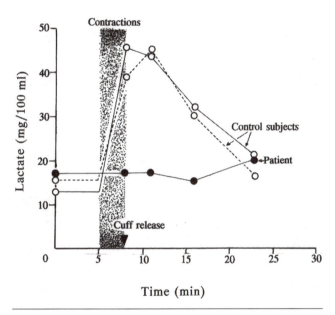

Figure 14.13   Lactate concentrations in venous blood taken from the forearm after repetitive handgrips, performed under ischemic conditions by control subjects and patients with McArdle's syndrome. The absence of phosphorylase prevents the breakdown of glycogen and hence formation of lactic acid in the muscles.

Adapted, by permission, from C. Pearson et al., 1961, "A metabolic myopathy due to absence of muscle phosphorylase," *American Journal of Medicine* 30: 506.

The mystery was solved 10 years later when a group of investigators in Los Angeles had the opportunity to study a 19-year-old man who gave a similar history and also showed no rise in blood lactate during ischemic muscle contractions (figure 14.13). Pearson *et al.* (1961) further showed that the ability to exercise was greatly improved if glucose was given intravenously—another pointer to the presence of a glycogen breakdown disorder. Finally, by histochemical staining of a muscle biopsy, the investigators were able to demonstrate that the enzyme that was deficient was myophosphorylase. If patients with *McArdle's syndrome* are made to keep on exercising, despite their pain, something very interesting happens—the pain disappears, as does the fatigue. This "second wind" phenomenon, which is present to a lesser extent in normal subjects, is due to the muscle fibers switching from glycogen to fatty acids as a source of fuel and to the arrival of glucose from the liver.

Further light on the pathophysiology of McArdle's syndrome has been obtained by Richard Edwards and his collaborators at the University of Liverpool (Cooper *et al.*, 1989). They have shown that not only does force decline more rapidly than normal in patients with this disorder, but so does the muscle compound action potential (figure 14.14), the latter indicating a loss of muscle fiber excitability. It would be tempting to ascribe these various changes to insufficient ATP not only for the myosin–actin interactions but also specifically for Na^+-K^+ pumping. This explanation may be too simple, however, for in McArdle's patients, as in normal subjects, there is only a modest decline in muscle [ATP] during exercise. However, this observation gives credence to the idea that compartmentation of adenine nucleotide is important and that local glycolytic enzymes normally provide adequate ATP for the Na^+-K^+ ATPase. This observation is also consistent with the notion that the cell is able to sense impending collapse of energy supply and down-regulate activation to prevent a serious decrease in ATP concentration. This notion is further explored in chapter 15.

After McArdle's syndrome had been described and named after its discoverer, a number of other enzyme deficiencies were recognized in which the formation or utilization of glycogen was

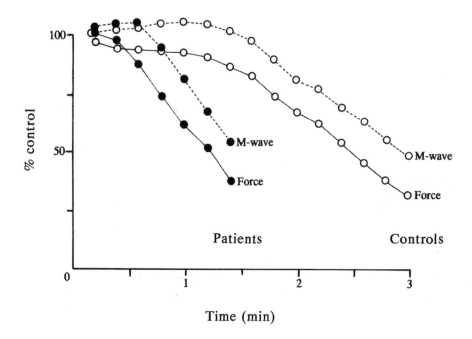

Figure 14.14 Changes in mean force and muscle excitation *(M-wave)* in seven patients with McArdle's syndrome and nine control subjects during intermittent repetitive stimulation at 20 Hz under ischemic conditions. Note the much greater declines in force and excitation in the patients. This occurs in spite of the lack of formation of lactic acid (see figure 14.13).

Adapted from Cooper *et al.* (1989, p. 5).

impaired; all of these conditions are extremely rare and appear to be inherited through autosomal recessive genes (see figure 2 of DeBarsy & Hers, 1990).

Lipid Disorders

Lipid disorders can also cause fatigability and muscle cramps.

Twenty years after McArdle's pioneering study, reports began to appear of occasional patients in whom muscle cramps, weakness, and fatigability were associated with excessive amounts of triglyceride in the affected muscle fibers. Engel and Angelini (1973) showed that the inability of one such patient to oxidize long-chain fatty acids could be corrected by adding carnitine to her diet; indeed, her disorder was one of hereditary carnitine deficiency. In the same year, DiMauro and DiMauro (1973) reported a patient in whom there was a deficiency of the lipid-handling enzyme carnitine palmityltransferase (see figure 14.7). In such patients, prolonged exercise causes aching and excessive fatigability of muscles; in severe episodes, the muscle fibers become necrotic and release myoglobin into the bloodstream and thence into the urine.

Like carnitine deficiency, the absence of carnitine palmityltransferase is an inherited mitochondrial disorder. However, mitochondria can be abnormal in other respects; for example, in rare instances, patients are found in whom the energy from glucose and lipid oxidation, instead of being used for the synthesis of ATP, is released entirely as heat, causing perspiration, heat intolerance, and increased thirst and hunger. This situation would arise with an uncoupling of oxidative phosphorylation from the electron transfer chain. In other mitochondrial disorders, there may be abnormal function in one of the molecules in the respiratory chain or in one of the enzymes in the citric acid cycle. Some of these conditions are associated with mitochondria that appear abnormal in electron micrographs; for example, they may be greater in size or number, or distorted in shape and internal structure.

Glycogen Supercompensation

Building up glycogen stores in muscles improves athletic performance.

In view of the importance of glycogen as a source of energy, a muscle that has its glycogen store increased would be expected to fatigue more slowly than before. This proposition was first examined by Christensen and Hansen (1939), who showed that a carbohydrate-rich diet was associated with better performance in prolonged exercise. Some 30 years later, Bergström *et al.* (1967) repeated this work, with similar results. In addition, they analyzed samples of muscle taken with a biopsy needle and showed that muscle glycogen had been increased by the carbohydrate-enriched diet. The principle of glycogen loading has now been adopted by many athletes, especially those in endurance events. The initial strategy was to exercise the muscles to exhaustion 1 week before the competition so as to deplete the muscle of glycogen. Over the next 3 days, the athlete prevents glycogen synthesis by eating a carbohydrate-free diet; he or she then consumes a carbohydrate-rich diet to stimulate glycogen synthesis. Such a procedure can increase muscle glycogen from the normal 1% of muscle weight to as much as 3% to 4%. It has been found that similar enhancement of glycogen storage can be achieved with an appropriate taper period (decreased volume of exercise) while the athlete consumes a diet that is rich in carbohydrates. This procedure avoids the need for glycogen depletion and a high-fat diet.

But does this strategy work? Doubts of its effectiveness have come from a recent study by Bangsbo *et al.* (1992), which showed that glycogen breakdown was not enhanced in quadriceps muscles with double their normal amounts of glycogen. This study involved single-leg exercise rather than whole body exercise. Perhaps this difference detracts from the potential benefit associated with *glycogen supercompensation.* It can be estimated that performance of a marathon with an economy of 1.0 $kcal \cdot km^{-1} \cdot kg^{-1}$ will require 2,940 kcal of energy for a 70-kg person. This exceeds the total caloric equivalent of carbohydrate stores without glycogen supercompensation. Inadequate carbohydrate stores will relegate the athlete to completing the event at a lower intensity because fat will have to be used to supplement the supply of energy during the event.

This chapter has completed our study of the electrical and chemical processes that underlie the maintenance of resting muscle and its transformation into a device for generating force or movement. Our study has ranged from the survey of individual enzymatic reactions to the performance of muscle fibers, motor units, and entire muscles. In the third part of the book, we will see how the muscle fibers and their *motoneurons* can adjust to a variety of stresses.

III

The Adaptable Neuromuscular System

One of the remarkable features of skeletal muscle is its adaptability. This adaptability is seen in the short term, in those situations in which muscles must continue to develop force for as long as possible. The adaptations are diverse, for they include changes in blood flow, ion pumping, and motoneuron impulse firing. Were it not for these adaptations, muscle fatigue, considered in chapter 15, would set in much sooner than it usually does. In the longer term, adaptations take place when muscle fibers lose their nerve supply, as discussed in chapter 16. Even in these adverse circumstances, many of the adaptations appear advantageous; and some of them, such as the presumptive release of chemical attractants by the muscle fibers, have evolved to encourage restoration of the nerve supply. We will discover in chapter 17 that this restoration can be achieved by regeneration of the injured nerve fibers or by sprouting of the remaining healthy nerve fibers.

The study of denervation leads naturally to consideration, in chapter 18, of the trophic effects of muscle and nerve. Thus, the nerve not only excites the muscle fiber, but also sustains it; were it not so, the muscle fiber would not shrink and degenerate after the nerve supply is lost. And the sustaining influence works in the reverse direction also: the motoneuron is dependent on trophic signals, in the form of chemicals, secreted by the muscle fibers.

Next, in chapter 19, is the topic of disuse. Although the effects on the muscle fibers are not as profound as those resulting from denervation, they can be severe nonetheless, affecting not only the sizes and functional properties of the muscle fibers, but the impulse firing patterns of the motoneurons too.

The subject of chapter 20 is training, as intended for the development of muscle strength, power, and endurance. The huge gains in muscle mass following strength training in humans are remarkable, and have become the inspiration for bodybuilding competitions and fitness magazines. In contrast, muscles trained for endurance usually appear normal, and it is only during continuous activity that the improved function is evident. We are now beginning to understand the way in which different adaptations are achieved and to appreciate the importance of muscle fiber activation as a stimulus.

Chapter 21 deals with the adaptations of muscle fibers in response to injury and emphasizes their effectiveness in being able to repair themselves rapidly. The injury is not necessarily from some external agent, for it will be shown that untrained muscles can be damaged by unaccustomed contractions, especially those taking place as the muscle is forced to lengthen.

Finally, and perhaps fittingly, we conclude the survey of skeletal muscle with a consideration of aging. Even though an older adult may not be able to compete as effectively as a younger one, force generation is usually maintained until the sixth decade. Moreover, the elderly muscles retain some degree of plasticity and can respond to training programs and even compensate for partial loss of innervation.

15

Fatigue

In this chapter, we begin a survey of situations in which muscles and *motoneurons* exhibit adaptations. The choice of fatigue is perhaps surprising, but, as will be seen, the adaptations can involve processes as disparate as changes in Na^+-K^+ pumping and motoneuron firing rate. Even the loss of force that characterizes fatigue can be viewed as an adaptation; for without it, serious muscle damage would undoubtedly occur. Fatigue will be shown to have both central and peripheral components; although more is known about the factors involved in peripheral fatigue, there is still uncertainty over their relative contributions. The chapter closes by considering the way in which muscle recovers from fatigue.

Defining Fatigue

In everyday language, the word "fatigue" is used to describe any reduction in (perceived or real) physical or mental performance. For the physiologist, however, the word has a more restricted meaning; muscle fatigue is defined as the "failure to maintain the required or expected force" (Edwards, 1981). In the case of isometric contractions, "force" is a suitable outcome measure. However, when one considers dynamic contractions, "contractile response" can replace "force." This would include shortening, power, or velocity of shortening as outcome measures. The rate of fatigue depends on the muscles employed, the relative intensity of the exercise, and whether or not the contractions are continuous or intermittent.

Figure 15.1 shows, as an example of fatigue, the decline in force of ankle plantarflexors or dorsiflexors during a sustained contraction. In general, up to half the force can be lost in the first minute of a maximal *effort* contraction. The simple definition of fatigue just given is adequate to distinguish fatigue from *weakness,* in which there is an inability to develop an initial force appropriate to the circumstances. If necessary, the definition of muscle fatigue could be broadened to include an inability to sustain rapidly executed movements as, for example, when the fingers are tapped as quickly as possible. This kind of fatigue has not been adequately studied, and it is possible that the most important component is a reduction in the "intensity" of the commands developed in the central nervous system.

Fatigue is also interpreted as the phenomenon whereby the effort required to maintain a constant submaximal task gradually increases. This definition would allow for the gradual appearance of "fatigue" during a task such as a marathon run or Ironman competition, in which muscle contractile force is maintained constant by gradually increasing effort. Increased motor unit recruitment or rate of firing, or both, would overcome the decreased contractile response for a given activation that occurs due to fatigue.

Most of our knowledge of fatigue has come from the study of prolonged voluntary or stimulated contractions, made at a constant muscle length (i.e., isometric), and it is this body of work that will be reviewed. An adequate understanding of muscle fatigue requires determination of the site of fatigue, the contractile consequences, and the cellular factors involved. In relation to the site, fatigue may affect either central or peripheral elements in the motor system.

Fatigue is "the failure to maintain the required or expected force."

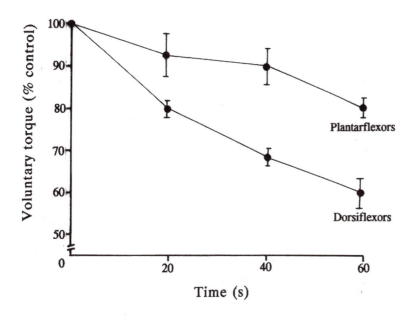

Figure 15.1 Rates of fatigue during continuous maximal isometric contractions of ankle dorsiflexor and plantarflexor muscles in healthy adults; values are means ± SEMs. Note that the slow-twitch plantarflexors fatigue less rapidly than the faster-twitch dorsiflexors.

Adapted from Bélanger and McComas (1983, p. 629).

The failure to maintain the expected or required contractile response could be due to the development of inadequacies in any of the parts of the motor command pathway: premotor cortex, motor cortex, descending pathways, developing inhibitory pathways, motoneuron activation, neuromuscular junction, muscle fiber membrane, T-tubules, voltage sensors, ryanodine receptors, sarcoplasmic reticulum Ca^{2+} availability, Ca^{2+} binding, and finally, *actin*–myosin interaction. This sequence can be divided into central (associated with the central nervous system) and peripheral (the motor units). Each of these will be addressed next.

Central Fatigue

The elements in the central nervous system that may affect fatigue during voluntary exercise include the emotions and other psychological factors responsible for the sense of effort, as well as the various descending motor pathways and the interneurons and motoneurons in the brainstem and spinal cord. For many years, the descending motor pathways have been known to comprise the corticospinal, rubrospinal, tectospinal, vestibulospinal, and reticulospinal tracts. Similarly, the topographic representation of movements in the motor cortex is well understood. In contrast, almost nothing is known about the identities of the neurons involved in the desire to move or in the generation and assessment of volitional effort.

Central fatigue can be estimated by stimulus interpolation.

Identifying Central Fatigue

Recordings of brain activity using scalp electrodes have shown that a surprisingly large area of cerebral cortex is involved in the initiation of movement. It has long been known that the basal ganglia and cerebellum also participate in the preparatory events. Once the movement starts, however, it is probable that the brain activity becomes more spatially focused. Although the central mechanisms are complex (and beyond the scope of this book), it is relatively easy to determine whether central fatigue is present during a maximal voluntary contraction. The simplest method is to deliver a maximal electrical stimulus to the motor nerve or the contracting muscle and to look for a *twitch* superimposed on the torque recording during a maximal voluntary contraction (MVC) (Bélanger & McComas, 1981; Merton, 1954). If a twitch is observed, then either not all motoneurons have been recruited or else some are not firing *impulses* at the optimal frequency for force generation. The sensitivity of the technique can be improved by applying two or more stimuli close together, rather than a single one, so as to make the elicited force response more evident.

An alternative strategy for detecting central fatigue is to compare the voluntary force with that developed by *tetanic stimulation* of the same muscle. Unfortunately, tetanic stimulation of human peripheral nerves is invariably painful; further, the same nerve may innervate antagonist as well as agonist muscles, and some synergistic muscles may be supplied by other nerves. These anatomical variations make this approach less desirable.

Evidence of Central Fatigue

Central fatigue becomes increasingly prominent as the exercise continues.

The first observation, on applying the twitch interpolation technique, is that even at rest, some subjects appear unable to activate all their *motor units* fully. Bélanger and McComas (1981) found this to be true of the ankle plantarflexor muscles in half of their subjects, although the ankle dorsiflexors, in contrast, could always be fully activated. An inability to activate some motor units fully during a sustained isometric contraction is perhaps not surprising, since some motoneurons may participate only in rapid, brief (phasic) types of movement, as demonstrated by means of single-unit recordings in the extensor digitorum brevis muscle (Grimby *et al.*, 1981).

Regardless of whether or not all motor units are optimally recruited at the start of a contraction, central fatigue becomes an increasingly important factor the longer that the contraction is maintained (McKenzie *et al.*, 1992; Thomas *et al.*, 1989). The simplest demonstration of the presence of central fatigue is that force can be momentarily enhanced by giving encouragement or a sudden loud command to the subject. An interesting example of central fatigue was reported by Asmussen (1979), who had a subject perform repeated contractions on a finger ergograph with his eyes closed. As fatigue developed, the excursion of the finger became smaller (see figure 15.2). When the excursion was considerably diminished, the subject was instructed to open his eyes. This simple change in circumstance resulted in an astounding recovery of contractile performance, presumably due to distraction and increased neural activity associated with vision. Decreased attention to the physical task or perhaps to the unpleasant sensations associated with the task resulted in better performance.

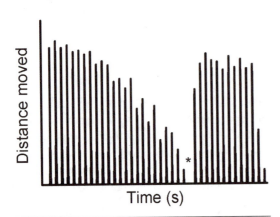

Figure 15.2 Motion achieved on a finger ergograph, with eyes closed initially. Asterisk indicates time at which the subject was permitted to open his eyes. The remarkable "recovery" from the apparent fatigue cannot be explained by any change that would be expected in the muscles. It has been suggested that the sensory distraction provided by visual input allowed the subject to overcome a central inhibition of performance and to continue contracting the muscles for several more contractions when it had appeared that exhaustion was imminent.

Adapted from Asmussen (1979).

Taylor *et al.* (1996) demonstrated the presence of "central fatigue" using the technique of *transcranial magnetic stimulation* (TMS). In their experiment, subjects were asked to generate a maximum isometric contraction of the elbow flexors, and during the contraction the experimenters stimu-lated the motor cortex. Surprisingly, they found that, although TMS stimulation did not produce any added force to the elbow flexors during a "fresh" MVC, it did add force to the contractile response during the fatiguing contraction. Since TMS stimulates the cortical cells trans-synaptically, the interpretation was that a site "upstream" from the motor cortex was declining in its excitation of the cortex during voluntary MVCs.

Distraction can overcome central fatigue.

Peripheral Fatigue

Although there is clear evidence that central fatigue can contribute to the failure in persisting with a physical task (Gandevia, 2001), there is also considerable evidence that peripheral fatigue is substantial (Westerblad *et al.*, 2000; Bigland-Ritchie *et al.*, 1986a). The central component can be bypassed by studying the contractile response to electrical stimulation of the peripheral nerves or direct stimulation of the muscle. This approach has provided information regarding the changes in the contractile response as well as the possible mechanisms for these changes.

The events at peripheral sites in the motor system that could contribute to fatigue include impulse conduction in the motor axons and their terminals, neuromuscular transmission, conduction of impulses in the muscle fibers, *excitation-contraction coupling (E-C coupling)*, and the contractile process itself. In well-motivated people, the major contributions to fatigue come from these peripheral mechanisms. Several sites can be involved, singly or together, and these will be considered in the sequence in which they are normally involved during muscle contractions. As will be seen, the susceptibility of a site is affected by the nature of the experiment.

The Contractile Manifestation of Peripheral Fatigue

When electrical stimulation is used to study muscle, the response depends on the nature of the stimulus provided and the prevailing conditions for the muscle (load, length, temperature). The relevant parameters of stimulation are the frequency and train duration (see chapter 11). Considering these conditions, the definition of fatigue can be modified to "less than the anticipated contractile response for a given stimulus." This definition can include consideration of dynamic contractions where velocity or magnitude of shortening is changing rather than force.

The contractile response of a muscle can be assessed by obtaining the *force–length, force–frequency,* and *force–velocity relationships* (see chapter 11). Each of these relationships is affected by fatigue, and the changes that are seen will provide some indication of possible mechanisms of fatigue.

The twitch can be larger in fatigue.

Twitch Potentiation During Fatigue

Isometric twitch contractions have been evaluated during and following fatiguing exercise. It will be recalled that the twitch is the elementary contractile response of the motor system, and with supramaximal motor nerve stimulation, all motor units should be activated. This condition is not always met in studies with human muscle, because supramaximal stimulation can be painful. However, this is a common approach in animal studies.

When a twitch contraction is obtained periodically during intermittent tetanic contractions, the amplitude of the twitch will at first increase, even if the tetanic force is already declining (Vandenboom & Houston, 1996; Pasquet *et al.,* 2000). This initial increase is the result of *activity-dependent potentiation* (see chapter 11), and the subsequent decrease is fatigue, not simply dissipation of the enhancement associated with activity-dependent potentiation (MacIntosh *et al.,* 1993; MacIntosh & Rassier, 2002). Interestingly, when the fatiguing contractions are finished, the twitch response will appear to recover very quickly (within seconds), then will decrease in amplitude over the next few minutes. It is unclear what permits the early rapid recovery, but it seems reasonable to assume that the subsequent decrease represents dissipation of potentiation. The amplitude of the twitch will level off at a value that is substantially reduced in comparison to the amplitude before the exercise (MacIntosh *et al.,* 1994).

Relaxation Is Slowed in Fatigue

One of the common observations in fatigue studies is that relaxation is slowed. This is evident when evaluating twitch or tetanic contractions (increased *half-relaxation time*). Slowing of relaxation can be detected subjectively, especially if rapid oscillatory movements are performed. Quick relaxation is essential for this kind of movement, which becomes impossible when relaxation is slowed.

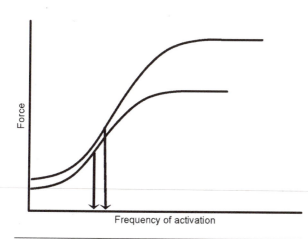

Figure 15.3 Force–frequency relationships for a control muscle (upper curve) and a fatigued muscle are shown. The lower curve shows a combination of reduced contractile capacity (reduced maximal force) and a shift to the left due to slowing of the contractile response. This slowing is predominantly due to slowing of relaxation and results in a lower fusion frequency. The vertical arrows show the location of the frequency at half-maximal force.

Although slowing of relaxation is a sign of fatigue, it can also be considered a mechanism that diminishes the impact of fatigue on the contractile response. Slowing of relaxation causes a leftward shift in the force–frequency relation due to greater *summation* at a given frequency of activation (see figure 15.3). Unfortunately, this is coupled with a reduced force-generating capacity; and although the frequency for half-maximal force generation is lower, the force-generating capability at any frequency is compromised. It can be seen in this figure that the decrease in force is less in the midrange of stimulation frequencies. When this prolongation of relaxation is coupled with activity-dependent potentiation, low-frequency contractions will be even less affected.

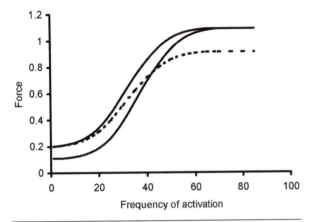

Figure 15.4 Three force–frequency curves are shown. The solid line in the top left position is the control relationship. The dashed line represents high-frequency fatigue. Force is depressed at high frequencies, but not at low frequencies. Recovery from this type of fatigue occurs quickly. The lower solid line represents low-frequency fatigue. Force is depressed for twitch contractions and at low frequencies of activation, but maximal force is not affected.

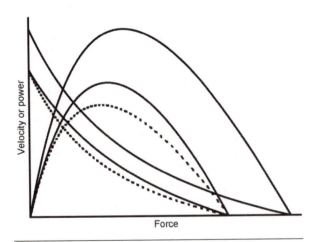

Figure 15.5 Force–velocity (concave) and power–velocity (convex) relationships are shown. The control condition is represented by the upper line of each pair. The lower solid line represents the changes due to fatigue. In this case, maximal isometric force and maximal velocity have decreased, and the curvature for the force–velocity relationship is decreased. The fact that fatigue results in less curvature permits a higher peak power than would otherwise be the case. The dashed line presents the fatigued force–velocity relationship that would be expected if the curvature was the same as the control relationship. Power is less than is the case in the fatigued muscle.

The Force–Frequency and Force–Velocity–Power Relationships in Fatigue

Fatigue has been categorized as either low-frequency or high-frequency fatigue according to whether contractile responses are affected more at low or high frequencies of testing stimulation. This categorization is not simply a designation of the frequency of stimulation that has caused the fatigue. These two categories of fatigue are shown in figure 15.4.

Low-frequency fatigue is more common than high-frequency fatigue, and this form of fatigue may be more physiological than high-frequency fatigue. To detect low-frequency fatigue, some recovery should be permitted to allow dissipation of activity-dependent potentiation. Low-frequency fatigue persists for hours, and has been detected 24 hr after a strenuous bout of exercise (Edwards *et al.*, 1977).

Most studies that evaluate fatigue in skeletal muscle deal strictly with isometric contractions, and only a few have evaluated the impact of fatigue on the force–velocity properties. A major problem is that the force–velocity properties are difficult to evaluate immediately after a bout of exercise, since multiple contractions are required, and the rapid recovery alters the state of the muscle during testing.

The force–velocity relationship can be characterized by measurement of three parameters: maximal isometric force (P_o), maximal velocity of shortening (V_{max}), and the degree of curvature of the relationship. Changes in these parameters and their influence on power output are shown in figure 15.5.

The Length Dependence of Force May Be Altered in Fatigue

In most cases in which fatigue is studied, isometric contractions are obtained prior to, during, and after the repetitive contractions that cause fatigue. Typically, the force–length relationship is obtained initially and the muscle is set at optimal length (the length that gives the greatest active force). Usually the length dependence of force is not given further consideration. However, in the few studies that have addressed this issue, disparate results have been found. Fitch and McComas (1985) found greater fatigue after stimulation at 20

Hz conducted at optimal length than with the same stimulation conducted at a short length. The muscle compound *action potential (M-wave)* amplitude was changed very little in either case. It should be recognized that the energy cost per contraction would be greater at optimal length, due to the involvement of a greater number of *cross-bridges*. However, the ion pumping should have been comparable between the two conditions. This observation indicates that the fatigue that was observed was related to the metabolic or cross-bridge cycling differences between the two length conditions. There should not have been differences between the two cases with respect to disruption of ion concentration gradients. This may be an oversimplified resolution for these observations. It has been noted that the magnitude of fatigue depends more on what length the change is measured at than on the length at which the fatigue occurred (Aljure & Borrero, 1968). These issues still need to be resolved.

Membrane Properties in Peripheral Fatigue

Impulse conduction can fail at the branch points or endings of motor axons.

Sometimes, when a motor nerve is stimulated repetitively, the M-waves undergo step-like fluctuations in amplitude, suggesting that action potentials intermittently fail to propagate beyond major bifurcations in the axons of the largest motor units. This observation is also consistent with failure to activate a given axon in the nerve.

Krnjevic and Miledi (1958) showed evidence of probable failure at a branch point in the rat diaphragm motor nerve. By stimulating single motor axons in the phrenic nerve and recording *end-plate potentials* with microelectrodes, they were able to compare the responses in two fibers belonging to the same motor unit. On some occasions they found that a normal-sized end-plate potential in one fiber was not associated with any detectable response in the other fiber. This type of defect could best be explained by a failure of the impulse to invade the axon terminal. A similar defect is known to occur in mice with hereditary motor end-plate disease (Duchen & Stefani, 1971).

Krnjevic and Miledi's (1958) experiments showed that the conduction failure was very sensitive to anoxia, but it is difficult to escape the conclusion that this susceptibility may have been induced by the *in vitro* conditions of the experiment. For example, in human single-fiber *EMG* (see chapter 11) in which impulse propagation can be studied simultaneously in two or more muscle fibers belonging to the same motor unit, it is unusual to see one of the fibers "dropping out." Again, when single cat motoneurons are stimulated *in vivo,* the summated muscle fiber action potentials may show no significant decrement even after several thousand stimuli (Burke *et al.,* 1973). Presumably the intact blood supply is able to maintain the preterminal axon in satisfactory condition.

Muscle Action Potential Changes

It is known that the giant axon of the squid can continue to propagate action potentials for long periods even when the axoplasm has been extruded, and it would be surprising if the membrane of an intact muscle fiber was any less efficient. Yet there is strong evidence that under certain circumstances, the muscle fiber membrane may become inexcitable because of the associated repetitive activation.

In frog muscle fibers *in vitro,* action potentials are diminished by contractions. Muscle action potentials may also diminish in mammalian muscle *in vivo.*

Lüttgau (1965) isolated single fibers of the frog and stimulated them *in vitro*. Lüttgau found that, at a stimulation rate of 100 Hz, some action potentials began to drop out within 2 s. Rather unexpectedly, Lüttgau found that if stimulation was given at lower rates until the fiber became exhausted and could twitch no longer, the action potential mechanism recovered. Membrane excitability was also maintained if the contractile responses were abolished by metabolic inhibitors such as sodium cyanide and iodoacetate, or by immersion of the fiber in a hypertonic bathing solution. Lüttgau concluded that, in fatigue, the action potential mechanism could be affected by the metabolic changes associated with contraction. It should be pointed out, however, that a change in the action potential does not necessarily mean that the contractile response will change. Others have reported minor changes in the action potential in fatigue, and changes in the action potential, independent of the contractile response (Grabowski *et al.,* 1972; Hanson, 1974).

Interesting though Lüttgau's findings are, one must exercise caution before translating these results into the human *in vivo* situation. Not only is there a species difference, but the relatively large volume of fluid bathing the muscle fibers in Lüttgau's experiments would eliminate any significant changes in the composition of the extracellular milieu. On the other hand, the findings of Burke *et al.* (1973) that mammalian motor units are capable of firing many thousands of impulses when stimulated individually may give a misleadingly favorable impression of the action potential-generating mechanism during fatigue. During strong effort, most or all of the motor units will be recruited rather than a single unit, and the rise in intramuscular pressure will be sufficient to occlude the circulation, rendering the muscle ischemic (Barcroft & Millen, 1939). Also, there will be a large impulse-mediated efflux of K^+ from the muscle fibers, causing the concentration of K^+ to rise in the interstitial fluid and affect the excitability of the muscle fiber *plasmalemma* (see "The Potassium Challenge"). There may even be opening of K^+ channels that would increase the rate of loss of K^+ from the cells (Renaud, 2002). This possibility relies on the assumption that the *equilibrium potential* for K^+ remains more negative than the *membrane potential.*

Leaving theory aside, there is surprising disagreement over the results of experiments in which the M-wave is evoked by interpolated stimuli during the course of MVCs. For example, Stephens and Taylor (1972), employing a double-pulse technique, found that there was an early decline in the M-wave of the first dorsal interosseus muscle of the hand during isometric contractions. However, this result was at variance with the earlier one of Merton (1954) in the adductor pollicis muscle (see the next section). Subsequently, Bigland-Ritchie *et al.* (1982) reexamined this problem and concluded that the M-wave is well preserved at a time when fatigue has set in and there is already a decline in voluntary EMG activity.

Delayed Action Potential Decline

The general consensus at present is that, although the muscle fiber action potential may ultimately begin to fail during voluntary contraction, the attenuation is relatively modest and occurs only after there has been a significant reduction in the ability to develop force. This point was well made in the classic study of Merton (1954) on the adductor pollicis muscle. After repeated MVCs of this muscle (usually his own), performed with the arm ischemic, Merton found that the twitch could be completely abolished at a time when the M-wave had barely changed. Similar findings are obtained if, instead of using voluntary contractions, muscles are fatigued by motor nerve stimulation at modest rates (10-30 Hz), as shown in figure 15.6. It is found that approximately

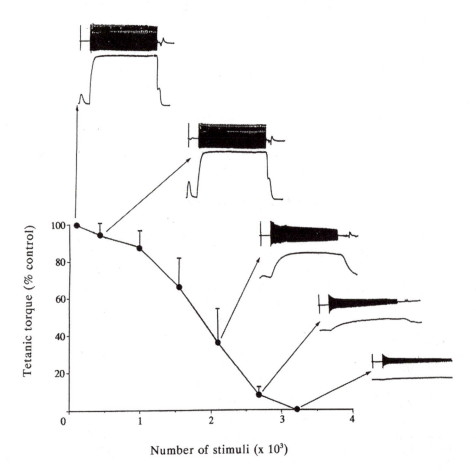

Number of stimuli (x 10³)

Figure 15.6 Mean fatigue (±SEM) of ankle dorsiflexor muscles in healthy subjects during intermittent 30-Hz stimulation with ischemia. Each pair of sample traces shows M-waves above and torque below; a single stimulus, for eliciting a twitch, is delivered before each tetanus. Note the slowing of torque development and of relaxation as fatigue sets in. In the final pair of traces, M-waves are still present, though much reduced, although no torque can be detected.

Adapted from Garland *et al.* (1988b, pp. 90-91).

1,000 impulses can be conducted without decrement in the M-wave, even in the presence of ischemia. By this time the twitch is greatly reduced, although tetanic tension is better preserved. In the experiment illustrated in figure 15.6, the depression of the twitch, together with marked slowing of the relaxation following the *tetanus,* is clearly seen after 2,000 stimuli. If higher rates of stimulation are employed (e.g., 80-100 Hz), the decline in the M-wave is much more rapid and does indeed become the limiting factor in force development, since reducing the stimulation frequency leads to an improvement in force (Edwards *et al.,* 1977). Furthermore, it has been shown that *"muscle wisdom,"* the decreasing frequency of activation during maximal voluntary effort (see p. 205), is not as effective at maintaining force as constant-frequency stimulation at 30 Hz (Fuglevand & Keen, 2003).

The conclusion from the various studies is that when muscle is maximally activated, either voluntarily or following physiological rates of stimulation, the reduction in force either is of central origin or begins when there is failure of a cellular mechanism subsequent to the muscle fiber action potential; that is, the peripheral failure must involve either E-C coupling or the contractile machinery. Both possibilities will now be considered.

Excitation-Contraction Coupling Failure

Before considering changes in E-C coupling, it is necessary to review briefly the normal physiology of the coupling process. The first step in the E-C coupling process is the propagation of an action potential into the interior of the muscle fiber from the surface membrane (see chapter 11), with *T-tubules* providing this pathway. The *dihydropyridine receptors (DHPR)* in the wall of the T-tubules act as *voltage sensors* and, in response to the action potential, signal the *ryanodine receptors (RYRs)* linking the T-tubule to the *sarcoplasmic reticulum* (SR). The activated RYRs permit the rapid escape of Ca^{2+} from the lumen of the SR into the *cytosol,* where some of the Ca^{2+} combines with *troponin* on the *actin* filaments. The Ca^{2+} is then pumped into the SR, by a membrane-bound Ca^{2+} ATPase, in anticipation of the next excitation of the fiber.

Fatigued muscle can still develop force in contractures.

Clear Evidence for Failure of Excitation-Contraction Coupling

Perhaps the clearest demonstration of E-C coupling failure in fatigued muscle has come from experiments in which part of the coupling mechanism has been bypassed. In the first of these, Eberstein and Sandow (1963) treated isolated frog muscle fibers, which had been stimulated to exhaustion, with either 0.1 M KCl solution or *caffeine.* Under both circumstances, sustained *contractures* of the muscle fibers resulted (figure 15.7). The presumed effect of the K^+ would be to produce a maintained *depolarization* of the T-tubules, whereas caffeine is known to directly

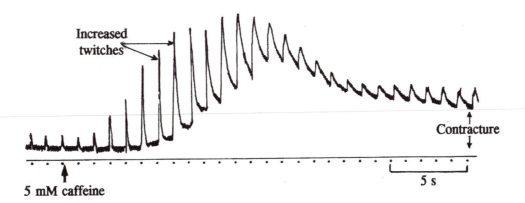

Figure 15.7 Effect of adding caffeine to the fluid bathing a highly fatigued frog muscle fiber. Not only do the twitches become much larger, but a steady tension (contracture) develops. This experiment illustrates that the muscle is capable of contractile response if sufficient stimulus is given.

Adapted, by permission, from A. Eberstein and A. Sandow, 1963, Fatigue mechanisms in muscle fibres. In *The effects of use and disease on neuromuscular function,* edited by E. Gutmann and P. Hnik (Prague, Czech Republic: Institute of Physiology, Czech Academy of Sciences), 524.

promote the release of Ca^{2+} from the SR. This experiment shows that in the fatigued muscle fibers, the contractile machinery was still capable of generating force but lacked the signal to do so with direct electrical stimulation. Eberstein and Sandow's important observations have been confirmed in amphibian muscle (Grabowski *et al.,* 1972) and more recently in mouse muscle fibers (Lännergren & Westerblad, 1991).

T-Tubules

Impulse propagation may fail in the T-tubules.

If the major factor in fatigue is indeed failure of the E-C coupling mechanism, which component is likely to be at fault? There are good reasons for supposing that the inwardly propagated action potential would be vulnerable because of pooling of K^+ in the T-tubules during impulse activity. Hodgkin and Horowicz (1959) produced rapid changes in the chemical composition of the fluid bathing single frog muscle fibers and found that the effects of Cl^- were more rapid than those of K^+. They reasoned that a sizable fraction of the K^+ channels were less accessible and were more likely to be in the T-tubular membrane; diffusion of K^+ from the T-tubules would be delayed because of their small lumens. Calculations of K^+ efflux during repetitive activity show that large changes in K^+ concentration would be expected in the extracellular fluid (see "The Potassium Challenge"). Whereas the surface membrane can prevent K^+-induced depolarization by enhancing the activity of the electrogenic sodium pump, the density of pump sites is relatively low in the T-tubular membrane (Fambrough *et al.,* 1987). Thus, the combination of a high K^+ concentration in the T-tubules, due to slow diffusion, and only modest electrogenic Na^+-K^+ pumping would be expected to result in sustained depolarization of the tubular membrane and block of the local action potential mechanism. Failure of inward spread of the action potential is suggested by the experiments of Edman and Lou (1992). These authors rapidly froze single fibers that had been fatigued by tetanic stimulation and examined the appearance of the *myofibrils* with the electron microscope. They found that the myofibrils in the periphery of a fiber had been activated (myofibrils straight), whereas the inner myofibrils had not (myofibrils wavy).

If the inward spread of activation along the T-tubules is compromised by altered ion distribution across the membrane, then allowing the muscle fiber to rest for a few seconds should permit recovery. The time constant for diffusion from the T-tubules is about 1 s. There are circumstances where such a rest does not permit full recovery of the contractile response. Therefore, other factors that persist beyond this time must also contribute to fatigue.

A logical question would be whether the T-tubules show any structural changes in fatigued muscle fibers. In single frog muscle fibers, González-Serratos *et al.* (1978) observed *vacuoles* in T-tubules; and by means of electron probes, these were shown to have higher Na^+ concentrations than the muscle fiber *cytoplasm.* Transient vacuolation, disappearing within seconds, has also been seen in tetanized rat muscle fibers, but the vacuoles are associated more with the SR than with the T-tubules (Landon, 1982). Longer-lasting T-tubule vacuoles have been observed in mouse and frog single fibers. Such disruption of the T-tubule membrane could contribute to impaired E-C coupling, but it is not known if this occurs *in vivo.* This structural aberration may be a feature of inappropriate osmotic conditions in the T-tubules specific to the *in vitro* preparation.

Ca^{2+} Release From the SR

Ca^{2+} release from the SR is reduced in fatigue.

Impairment of any of several steps in E-C coupling could compromise the release of Ca^{2+} from the *terminal cisternae:* compromised conduction of the action potential down the T-tubules, inhibition of the voltage sensor, impaired connection between the DHP and the RYRs, inhibition of the RYRs, and decreased (free) Ca^{2+} in the terminal cisternae. It is now possible to evaluate Ca^{2+} levels within the muscle fibers. The principle of this method is to inject a Ca^{2+}-sensitive fluorescent compound, such as Indo-1 or Fura-2, into a single muscle fiber and to measure the response to photo-excitation with a photomultiplier. This approach is similar to that of Blinks *et al.* (1978), who used aequorin, a bioluminescent protein; the fact that these workers found little correlation between tension and intracellular Ca^{2+} concentration may have been due to the fact that the Ca^{2+} concentration was sometimes higher than that needed to saturate the troponin-binding sites. In more

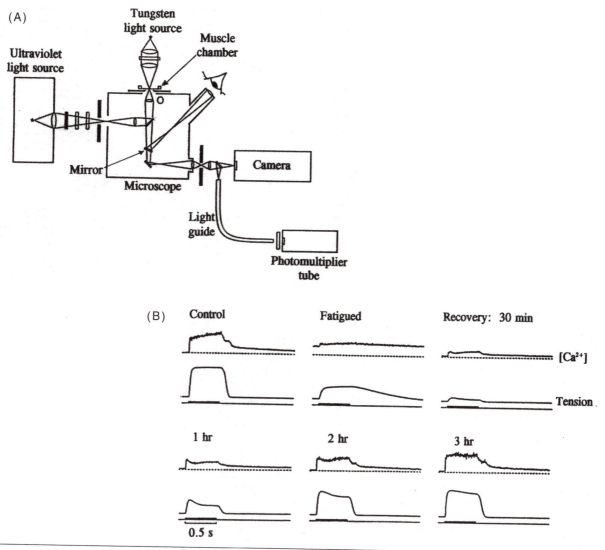

Figure 15.8 Measurement of Ca^{2+} transients in a single amphibian muscle fiber injected with Fura-2 and subjected to tetanic stimulation. (A) The fiber in the muscle chamber is transilluminated by the ultraviolet light via the dichroic mirror; the mirror reflects light of one wavelength and transmits light of another. In the muscle fiber, Fura-2 fluoresces in the presence of Ca^{2+}, and the intensity of the fluorescence is measured with the photomultiplier tube. The tungsten lamp is used to view the fiber by conventional microscopy. (B) Comparison of tension and intracellular $[Ca^{2+}]$ in the control period, after fatiguing stimulation, and at various times in the recovery period. Note that although the baseline (resting) $[Ca^{2+}]$ is raised at the end of fatigue, only a small amount of Ca^{2+} is released by the tetanic stimulation. Both $[Ca^{2+}]$ and tension are depressed for the first 2 hr of the recovery process.

Adapted, by permission, from J.A. Lee, H. Westerblad, and D.G. Allen, 1991, "Changes in tetanic and resting [Ca 2+]i during fatigue and recovery of single muscle fibres from Xenopus laevis," *Journal of Physiology* 433: 307-326.

recent studies, employing aequorin or Fura-2 in single toad muscle fibers but with lesser amounts of fatigue, a close relationship between Ca^{2+} concentration and tetanic force has been found (Allen *et al.*, 1989; Lee *et al.*, 1991). Figure 15.8 shows one of the sets of recordings made from a single fiber, injected with Fura-2, during fatigue and recovery; it illustrates the parallel changes in Ca^{2+} release and muscle tension.

Lee *et al.* (1991) also observed a second factor that would contribute to E-C uncoupling; the force at a given free Ca^{2+} concentration was lower in fatigued than in nonfatigued muscles. This is known as decreased *Ca^{2+} sensitivity*. Changes in Ca^{2+} sensitivity, with modification of the intracellular environment to mimic the conditions of fatigue, have been studied extensively with skinned fiber experiments (Cooke & Pate, 1985; Cooke *et al.*, 1988; Pate *et al.*, 1992). This issue will be considered further when the biochemical changes associated with fatigue are presented.

Biochemical Changes in Muscle Fibers

In addition to generating less force, the contractile machinery of the fatigued muscle fiber develops tetanic tension more slowly than normal and prolongs the relaxation phase following twitch and tetanic contractions. We have already seen how this change causes a shift in the force–frequency relationship (figure 15.3). At the same time, there are marked changes taking place in the chemical composition of the muscle fiber cytoplasm (figure 15.9). These include

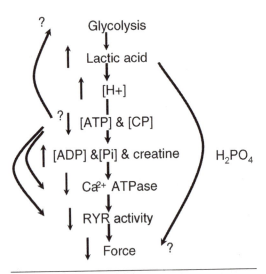

- an accumulation of H^+ and lactate from the breakdown of glucose and muscle glycogen (see chapter 14);
- a rise in *inorganic phosphate* (Pi) from the splitting of *ATP* by *myosin* and membrane ATPases, and an increase in diprotonated *phosphate* because phosphate contributes to the buffering of H^+;
- a fall in *phosphocreatine* and a corresponding rise in *creatine* and Pi;
- a rise in Ca^{2+} concentration, due to impaired pumping of Ca^{2+} back into the SR; and
- a gain of water, some of it in the form of vacuoles (see earlier).

Figure 15.9 Interrelationship of main biochemical changes during development of fatigue. Increased metabolic activity related to repeated or sustained muscle contractions results in increased lactate, H^+, *ADP*, Pi, and creatine concentrations. Adenosine triphosphate and phosphocreatine (CP) concentrations decline. The amount of Ca^{2+} released through the RYRs is decreased in fatigue, but the mechanism of this is not known. The inhibitory effects of H^+ and Pi on sensitivity to Ca^{2+} are apparently evident only at low temperatures (hence the "?"). Adenosine triphosphate concentration does not decrease substantially when measured globally, but localized changes may be substantial.

Techniques to Measure Chemical Change

In animal muscles, it has long been possible to study the chemical events in fatigued muscles by analysis of whole muscles following stimulated or natural contractions. As long ago as 1807, Berzelius detected increased amounts of *lactic acid* in fatigued muscle (described in Needham, 1971). In the case of Ca^{2+} measurements, a recent refinement has been to measure the appearance of free Ca^{2+} in the cytosol of single muscle fibers, using one of the fluorescent dyes (Lee *et al.*, 1991; see p. 235).

For human muscles various techniques have been employed for sampling muscle metabolic status, including the withdrawal of venous blood during ischemic contractions and, in some laboratories, the taking of small specimens of muscle (50-100 mg) with a biopsy needle at different times during the fatiguing procedure. The former technique was useful in demonstrating the efflux of H^+, K^+, and lactate during repeated muscle contractions and remains an important diagnostic approach for patients suspected of having *myophosphorylase* deficiency (*McArdle's syndrome;* see "Applied Physiology" section of chapter 14). One of the main contributions of the needle biopsy technique was to show, in fatigued muscle, that there is a great decrease in the amount of phosphocreatine, the substrate necessary for phosphorylating ADP to ATP (see chapter 14). Although there was some controversy initially, the results of biopsy sampling indicate that, even in pronounced fatigue, there is only a modest reduction in the level of ATP. With the advent of *magnetic resonance spectroscopy (MRS)*, it has become possible to investigate changes in the phosphorus compounds of the muscle fibers during fatigue, as well as alterations in H^+ concentration (see the next section). Some of the chemical changes and their possible effects on the contractile machinery will now be considered in more detail.

Various techniques have been used to measure the chemical changes.

The concentration of H^+ rises in fatigue and can depress force generation.

Acidosis and Fatigue

Glycolysis results in the formation of lactic acid. The fall in cytoplasmic pH can now be detected in contracting muscle fibers *in situ* by means of MRS (see figure 15.12). The lowest pHs observed in human muscle fatigue are around 6.5. If rabbit muscle fibers are skinned and placed in a fluid

The Potassium Challenge

With each impulse, there is a gain of Na^+ by the muscle fiber and a loss of K^+. The K^+ efflux into the interstitial fluid has been measured with isotopes for single frog muscle fibers (Hodgkin & Horowicz, 1959) and by chemical analysis of the fluid bathing the stimulated rat diaphragm (Creese et al., 1958). Both studies yielded a value of approximately 10 pmol $K^+ \cdot cm^{-2}$ membrane $\cdot$ impulse^{-1}. Suppose that a strong isometric contraction is performed; the intramuscular pressure will rise above the arterial systolic level and will prevent the capillary circulation from removing K^+ from the interstitial spaces of the muscle (Barcroft & Millen, 1939). Also, ignore for the moment any effects of the Na^+-K^+ pump. Then if the mean muscle fiber

diameter is taken as 50 μm, the interfiber distance as 1 μm, and the excitation frequency as 25 impulses $\cdot$ s^{-1}, it can be calculated that the interstitial K^+ concentration would be large enough to reduce the resting potential of the muscle fiber by approximately 15 mV and to block impulse conduction. Using ion-sensitive microelectrodes, it is now possible to measure the K^+ concentration in the interstitial spaces during or following muscle contractions; although values of 8 to 10 mM are usually found, concentrations as high as 15 mM have been observed in human forearm muscles (Vyskocil et al., 1983; figure 15.10).

Why is it that, despite the potential for such high K^+ concentrations, a maximally stimulated muscle, even under ischemic conditions, can conduct a thousand or so impulses in each fiber? Similarly, how is the M-wave (muscle compound action potential) so well maintained during isometric voluntary contractions?

The answer lies in the Na^+-K^+ pump. This important transporter molecule quickly enhances its activity when a muscle is first activated and, because of its electrogenic nature, is able to maintain the resting potential at, or even above (more negative than), the control level (figure 15.11). Among its effects, the hyperpolarizing action of the pump is responsible for the enlargement of the M-wave that is seen in the first 30 s of a voluntary or stimulated contraction (pseudofacilitation; see figure 15.11). Under the influence of the pump, the resting potential may become 30 mV or so more negative than E_K (the potassium equilibrium potential). In this circumstance, the strong negative charge will permit K^+ to enter the cell "down" its electrochemical gradient. This passive flux will become greater if, in the contracting muscle, the calcium-gated K^+ channels and also the ATP-sensitive K^+ channels open (see chapter 7; also Rudy, 1988).

Suppose now that the contraction is submaximal. What would be the effect of the K^+ efflux from the active fibers on the noncontracting ones? It might be thought that the latter would become depolarized in accordance with the GHK equation (see chapter 9). In fact, studies of mammalian muscle show that Na^+-K^+ pump activity is increased in the noncontracting fibers also and can cause them to hyperpolarize (Kuiack & McComas, 1992). The advantages

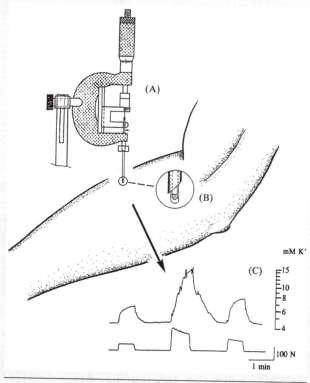

Figure 15.10 Measurement of interstitial $[K^+]$ in intact human muscle. A sharp metal cannula is inserted through a small skin incision into the brachioradialis muscle. Within the cannula is an ion-sensitive glass electrode, the enlarged tip of which is shown in (B); the tip of the electrode can be gently driven out of the cannula into the muscle tissue by a micrometer screw, seen in (A). During voluntary contractions, the interstitial $[K^+]$ rises, the highest value corresponding to the strongest contraction [middle section in (C)].

Adapted, by permission, from F. Vyskocil et al., 1983, "The measurement of K+ concentration changes in human muscles during volitional contractions," Pflügers Archive 399: 236.

(continued)

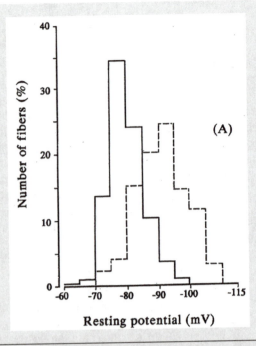

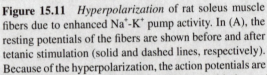

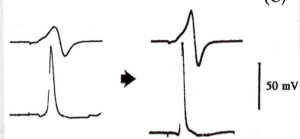

Figure 15.11 *Hyperpolarization* of rat soleus muscle fibers due to enhanced Na^+-K^+ pump activity. In (A), the resting potentials of the fibers are shown before and after tetanic stimulation (solid and dashed lines, respectively). Because of the hyperpolarization, the action potentials are enlarged in the stimulated fibers, compared with the resting ones [dashed and solid lines, respectively, in (B)]. An example of action potential enlargement is shown in (C); the upper trace in each pair of records shows the muscle M-wave, and the distance separating the upper and lower traces equals the resting potential.

Reprinted, by permission, from A. Hicks and A.J. McComas, 1989, "Increased sodium pump activity following repetitive stimulation of rat soleus muscles," *Journal of Physiology* 414: 339.

of such a muscle strategy are obvious, since the previously quiescent fibers will remain available for *recruitment* as more force is required. Also, the quiescent fibers, by increasing their Na^+-K^+ pumping, will moderate the rise in interstitial K^+ concentration.

What is the stimulus responsible for facilitating pump activity? Probably the most powerful one for the contracting fibers is the impulse-mediated rise in Na^+ concentration in the fibers, because there are many studies, in a variety of cells, that show the importance of intracellular Na^+ in regulating pump activity (see Thomas, 1972). However, it also appears that epinephrine and norepinephrine promote pump activity and are largely responsible for boosting the pump activity in the noncontracting fibers (Kuiack & McComas, 1992). It is intriguing that noradrenergic nerve fibers have been observed not only in the walls of the muscle arterioles but also terminating on some of the muscle fibers (Barker & Saito, 1981).

There is a different type of physiological challenge in situations in which a large fraction of the body muscle mass is contracting, as in vigorous cycling or treadmill running. Because the muscle contractions are intermittent, K^+ can enter the capillary circulation from the interstitial spaces. The K^+ efflux may be so large that the arterial concentration may double the resting value (Medbø & Sejersted, 1990); there is then the danger that cardiac excitability will be affected, causing death from ventricular *fibrillation* or atrial paralysis. Once again, it would seem that the Na^+-K^+ pump comes to the rescue by increasing its activity in muscle fibers throughout the body and assisting in K^+ *homeostasis* (Clausen, 1990). Evidence for such a response has come from the work of Sjøgaard (1986), who showed that, if only one leg is exercised, the K^+ level in the arterial blood of the other leg may be higher than the venous level, suggesting that the muscle fibers in the resting leg have been extracting K^+ from the circulation. In keeping with the results of Kuiack and McComas (1992), it is probable that catecholamines are implicated in stimulating the pump, especially since epinephrine and norepinephrine are released from the suprarenal gland by elevations in plasma K^+ concentration (Popham *et al.*, 1990).

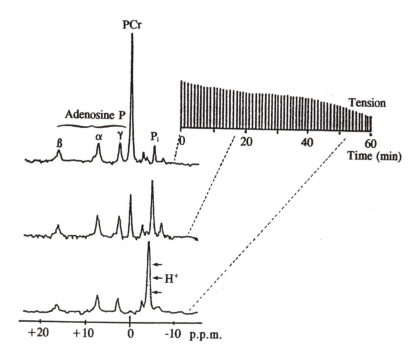

Figure 15.12 The first published report of MRS applied to the investigation of fatigue, by Dawson *et al.* (1978). Frog gastrocnemius muscles were made anaerobic and were stimulated for 1 s every minute. The trace spectra show the changes in the various phosphorus compounds in the muscle as fatigue sets in. There is a complete loss of phosphocreatine (PCr), but substantial amounts of ATP remain (α, β, and γ in figure). Inorganic phosphate (Pi) increases, due to hydrolysis of PCr and ATP, and the corresponding peak is shifted to the right under the influence of the increasing [H^+].

Adapted, by permission, from M.J. Dawson et al., 1978, "Muscular fatigue investigated by phosphorus nuclear magnetic resonance," *Nature* 274: 862.

medium with a similarly low pH, there is a 30% reduction in the force generated by fast-twitch fibers and a 10% loss in slow-twitch fibers (Bolitho-Donaldson & Hermansen, 1978). The force-reducing effects of H^+ are thought to be brought about in at least two ways:

- The force per cross-bridge is decreased (Edman & Lou, 1991).
- The kinetics of actin–myosin interaction are altered so that fewer cross-bridges are engaged at a given submaximal [Ca^{2+}] (Cooke *et al.,* 1988; i.e., Ca^{2+} sensitivity is reduced).

On the other hand, since Ca^{2+} buffering may be reduced by the presence of more H^+ and since Ca^{2+} ATPase is inhibited, the magnitude of the Ca^{2+} transient in response to a single activation can be enhanced.

The inherited disorder, myophosphorylase deficiency (McArdle's syndrome; see chapter 14), is relevant to the effects of H^+. In this disorder, the breakdown of muscle glycogen is prevented, and hence there is minimal production of lactic acid and no drop in muscle pH. Nevertheless, the muscles of affected patients fatigue much more rapidly than normal. Does this observation mean that pH changes are of little significance in the fatigue of healthy muscles? Alternatively, are other metabolic changes in the myophosphorylase-deficient patient so powerful that they supersede any effects of H^+ concentration?

Several reports in recent years have argued against the importance of acidosis as a mechanism of fatigue. It should be pointed out that the research that has demonstrated an effect of H^+ on Ca^{2+} sensitivity was conducted *in vitro* on skinned fibers at room temperature. This may be relevant for poikilothermic amphibians, but temperature is an important issue with respect to mammalian muscle behavior. When skinned fiber experiments were done with mammalian muscle at 35° C, it was found that acidosis does not decrease Ca^{2+} sensitivity (Pate *et al.,* 1995). In fact, recent work with intact mammalian muscle has suggested that lactic acidosis is good for the muscle (Nielsen *et al.,* 2001).

Inorganic phosphate and diprotonated phosphate increase in fatigue.

Inorganic Phosphate and Fatigue

There are two types of change in phosphate concentration in muscle fatigue. First there is a rise in Pi through the splitting of ATP by myosin ATPase; one molecule of ATP is hydrolyzed during each cycle of a myosin cross-bridge (see chapter 11). Additional Pi will, of course, come from the consumption of ATP by ionic pumps in the surface membrane (Na^+-K^+ ATPase) and in the SR (Ca^{2+}

ATPase). Although the *creatine kinase* reaction permits re-*phosphorylation* of ADP, there is still a net release of Pi, corresponding to the net decrease in [PCr]. If skinned muscle fibers are exposed to different concentrations of Pi in the bathing fluid, force is reduced by one-third at a value of 10 mM Pi (Cooke & Pate, 1985). Although Pi levels as high as 40 mM can be seen in muscle fatigue, most of the increase occurs relatively early in contraction (within 30 s), by which time there is only a modest loss of force, during either voluntary or stimulated contraction (cf. Cady *et al.*, 1989). In fact, much of the decrease in [PCr], and the corresponding increase in [Pi], occurs during the time when activity-dependent potentiation results in increased force development (MacIntosh *et al.*, 1993).

The second type of change in phosphate involves the buffering reaction:

$$HPO_4^{2-} + H^+ \Leftrightarrow H_2PO_4^-$$

At the normal muscle pH of 7.0, the amount of monoprotonated phosphate (HPO_4^{2-}) is approximately twice that of the diprotonated form ($H_2PO_4^-$). As the pH falls, however, the proportions reverse so that, at pH 6.5, there is twice as much of the diprotonated form. An increased diprotonated phosphate to 20 mM was found to reduce force by half (Nosek *et al.*, 1987). Since this concentration is similar to that observed in fatigue, it is quite possible that the production of diprotonated phosphate is an important factor in the loss of force. Further, there is a satisfactory temporal correspondence between the change in protonated phosphate and the loss of force in human adductor pollicis muscles examined by MRS (Miller *et al.*, 1988). Once again, however, temperature may be an issue (Coupland *et al.*, 2001).

The Immediate Energy Supply

Phosphocreatine declines markedly in fatigue, but ATP does not.

One of the earliest results in the biochemical investigation of muscle fatigue was the demonstration of a loss of PCr. This finding, initially made on whole animal muscles and human muscle biopsy specimens, has now been confirmed by MRS (Dawson *et al.*, 1978; Miller *et al.*, 1987; see also figure 15.12). Since PCr provides the phosphate for the phosphorylation of ADP to ATP, it might be anticipated that the concentration of ATP would become similarly reduced. A loss of ATP would have important consequences for the cross-bridge cycling mechanism, because it would prevent the detachment of the actin filaments from the myosin cross-bridges and would cause a contracture to develop. However, although there is a slowing of muscle relaxation during fatigue, the affected muscles are still able to relax completely.

Another consequence of ATP deficiency would be the loss of energy for ion pumping, both for Na^+-K^+ exchange at the surface membrane and for the return of Ca^{2+} to the SR. In fatigued muscle, there is certainly evidence of defective Ca^{2+} pumping, since the cytosolic concentration of Ca^{2+} remains elevated (see figure 15.8B). Indeed, defective pumping could be the major cause of the prolongation of the relaxation that is so characteristic of fatigued muscle.

The rate of ATP hydrolysis is reduced in fatigue.

Given these theoretical considerations, the results of ATP measurements are, at first sight, disappointingly normal. Thus, estimates made on muscle biopsies or by MRS show little, if any, change in ATP concentration. However, Dawson *et al.* (1978) have stressed that it is not so much the total amount of ATP that is critical, but rather the availability of free energy through the hydrolysis of ATP. Since the reaction products of ATP hydrolysis (ADP and Pi) increase in fatiguing muscle, there will be a reduction in the free energy of hydrolysis: [ATP] / [ADP] · [P_i].

The background (resting) concentration of Ca^{2+} rises in fatigue.

Dawson *et al.* (1978) calculated the relative availability of free energy during a fatiguing contraction and found that, although it did not correlate well with the reduction in force, there was good correspondence with the slowing of muscle relaxation. Thus, the latter might result from defective Ca^{2+} pumping brought on by reduced energy of hydrolysis of ATP.

Resting Calcium Concentration

The fluorometric measurements of intracellular Ca^{2+} have already been considered in relation to the failure of E-C coupling. However, Lee *et al.* (1991), using Fura-2, made two other significant observations in fatigue. The first was that the slowing of relaxation, which is a characteristic property of fatigued muscle, is associated with, and presumably largely due to, a similar slow fall in $[Ca^{2+}]$ at the end of each activation (figure 15.8B). The slowness of the decline in $[Ca^{2+}]$ could be due to reduced Ca^{2+} pumping by the SR or to saturation of *parvalbumin* with Ca^{2+}. Considering that slowing of relaxation is a characteristic feature of muscle of larger mammals (dog, human) that lack parvalbumin, it seems reasonable to assume that Ca^{2+} pumping is the culprit here.

The other finding was that the background concentration of Ca^{2+} in the cytoplasm rises steadily during fatigue (figure 15.8B). The main factor in producing the rise is not K^+-induced depolarization, but either increased leakiness of the SR or the reduced uptake of Ca^{2+} by the membrane pumps in the SR. Lee *et al.* (1991) have pointed out that an increased concentration of cytosolic Ca^{2+} could have important consequences for enzyme reactions in the fibers, in addition to those involved in cross-bridge attachment and release. For example, Ca^{2+} operating as a *second messenger* increases muscle metabolism (Solandt, 1936), stimulates *protease* activity (Turner *et al.*, 1988), and uncouples oxidative reactions in the mitochondria from ATP production (Chapman & Tunstall, 1987). Furthermore, Ca^{2+} deposits in the mitochondria are one of the first signs of fiber degeneration (Wrogemann & Pena, 1976). The activation of proteases like *calpain* could have important implications for deactivation of RYRs or DHP receptors.

Water content of muscle increases in fatigue.

It has been known for a long time that exercising muscles take up water from the plasma (Fenn, 1936). The plasma loss can be detected by rises in the concentrations of red blood cells and plasma proteins (Sejersted *et al.*, 1986). Sjøgaard *et al.* (1985) have shown, by chemical analysis of human muscle biopsies, that during submaximal contractions, most of the increase in water content occurs in the interstitial spaces of the muscle. As the effort intensifies, however, water also moves into the fibers. The uptake of water is responsible for the swelling of the muscle during strenuous exercise and, in the short term, is also the cause of the intracellular vacuoles that can be transiently observed under the microscope (see p. 234).

Does the muscle fiber *down-regulate* contraction to preserve cellular integrity?

Attempts by various investigators to demonstrate that fatigue is caused by a failure to supply energy for contraction have failed. It is as though the muscle anticipates impending compromise of the energy supply and down-regulates activation to avoid the situation. This idea is consistent with the frequent observation that Ca^{2+} transients are diminished in fatigue. Reduced Ca^{2+} release would decrease the rate of energy requirement in each active muscle fiber. The question is, how does the fiber detect the challenge to energy supply and how does it go about reducing Ca^{2+} release?

Impaired E-C coupling likely is the key factor in peripheral fatigue.

As we have observed, muscle fatigue is an extremely complex state, and it is therefore useful to summarize the main features before proceeding to consider recovery from fatigue. Leaving aside central causes of fatigue, there are a number of peripheral factors that have been incriminated. In some cases, the evidence is purely circumstantial and comes from the finding of a change in concentration of a particular metabolite. In other instances, the likelihood of a causal relationship has been strengthened by showing that muscle force is diminished if muscle fibers, especially skinned ones, are exposed to concentrations of metabolites similar to those found in fatigue. Nevertheless, it is possible to produce fatigue, through the use of repeated submaximal contractions, in such a way that there are only minor changes in metabolite concentration (Vøllestad *et al.*, 1988). This observation and the close correspondence between force and the intracellular Ca^{2+} concentration in fatigue suggest that impaired E-C coupling is likely to be the most important of the peripheral factors. This conclusion is reinforced by Eberstein and Sandow's (1963) finding that force can be restored in a fatigued muscle by the application of caffeine. Table 15.1 summarizes the physiological and biochemical changes known to take place in muscle fibers during fatigue.

Recovery From Fatigue

For human subjects and animals, the recovery from fatigue is as important as the development of fatigue. Further, it is possible to learn something more of the interrelationship of factors in producing fatigue by studying their associations as recovery takes place.

Ischemia prolongs muscle fatigue.

The recovery from fatigue can be investigated in various ways, but there is an attractive simplicity in studying the aftermath of repetitive stimulation of human muscle performed under ischemic conditions. The first point to make is that even though the repetitive stimulation has stopped, the

Table 15.1   Summary of Changes in Muscle Fibers During Fatigue

Mechanical	Electrical	Biochemical	
Decline in force	Early hyperpolarization	Increased $[Na^+]$	Increased P_i
Slowed force development	Late depolarization	Reduced $[K^+]$	Increased ADP
Slowed relaxation	Slowed impulse conduction		
	Reduced EMG activity (central fatigue), usually during maximal effort	Increased $[H^+]$	Reduced Ca^{2+} fluxes
	Increased EMG activity (during submaximal contraction)	Increased lactate	Reduced Ca^{2+} sensitivity
		Increased H_2PO_4	Increased resting Ca^{2+}

manifestations of fatigue remain for as long as the circulation is occluded. That is, there continues to be a diminution in force, a slowing of twitch and tetanic relaxation, a fall in the size of the M-wave (muscle compound action potential), and a reduction in voluntary EMG activity; at the same time, the pain associated with fatigue persists and is felt in the muscle (see p. 244).

Muscle excitability recovers rapidly after restoration of the circulation.

As soon as the arterial cuff is released, the pain is relieved and is succeeded by a pleasant feeling of warmth in the previously ischemic limb; this sensation is accompanied by a large blood flow well above that in the resting state (reactive hyperemia), due to dilation of blood vessels in the skin and muscle. The M-wave enlarges rapidly, usually returning to its control amplitude within seconds of cuff release. This increase has been attributed to the "washing out" of K^+ from the interstitial spaces of the muscle, producing a rise in E_K, the potassium equilibrium potential, and hence a recovery of the resting and action potential amplitudes (see "The Potassium Challenge"). In some muscles, and especially in the human brachial biceps, the M-wave continues to enlarge under the influence of electrogenic Na^+-K^+ pumping (figure 15.13), returning to control values after about 15 min. After fatigue produced by tetanic stimulation, there may then be a late depression of the M-wave, due either to reduced Na^+-K^+ pumping or to a failure of Na^+ channel inactivation (Galea et al., 1993). The significance of the late depression is not clear; similar changes of the M-wave have also been observed in single motor units of the rat (Lännergren et al., 1989). Voluntary EMG activity also increases quickly after cuff release, due to recovery of the individual muscle fiber action potentials and to removal of the reflex inhibition of motoneurons (see "An Inhibitory Reflex?").

Twitch tension and contractions at low frequency recover more slowly.

The changes in muscle force lag behind those in electrical excitability, suggesting the presence of defective E-C coupling or cross-bridge contribution to force during the early part of recovery from fatigue. However, maximal voluntary force and tetanic response typically recover over the course of minutes after a bout of vigorous exercise. In contrast, twitch force shows an early rapid increase that may become larger than was present prior to the exercise. It subsequently decreases during the next few minutes, and recovers only very slowly after this. The enhanced twitch has been attributed to persistence of activity-dependent potentiation (Garner et al., 1989). Dissipation of this mechanism reveals the depressed twitch. The contractile response to low-frequency stimulation may require more than 24 hr for recovery to the level observed prior to the exercise (Edwards et al., 1977).

In single frog muscle fibers, depression of the twitch can also be demonstrated in the recovery period. Fluorometric measurements with Fura-2 show that the reduced force is associated with a smaller delivery of Ca^{2+} from the SR (figure 15.8B; see also Lee et al., 1991). Further, normal force can be restored by application of caffeine to the fluid bathing the fibers (Allen et al., 1989), and this substance is known to release Ca^{2+} from the SR. It therefore appears that low-frequency fatigue during the recovery period resembles the fatigue due to prolonged muscle activity in that both are due to impaired E-C coupling.

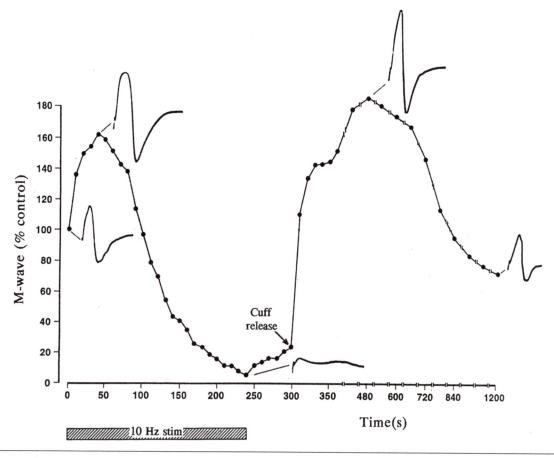

Figure 15.13 Changes in M-wave amplitude in the biceps brachii muscle of a healthy subject during and following 10-Hz stimulation under ischemic conditions. Immediately after the arterial cuff is deflated, at 300 s, the M-wave rapidly increases as K^+ is flushed out of the interstitial spaces in the muscle. The later increase in M-wave amplitude, at 480 s, is due to enhanced Na^+-K^+ pumping.

McComas, A.J., Galea, V., Einhorn, R.W., Hicks, A.L., Kuiack, S.: 'The role of the Na+, K+ -pump in delaying muscle fatigue, p. 39 in Sargeant, A.J., Kernel, D.: 'Neuromuscular Fatigue', North Holland / Royal Netherland Academy of Arts and Sciences, Amsterdam, 1993.

An Inhibitory Reflex?

As a steady voluntary contraction is continued, there is a reduction in the impulse (EMG) activity generated by the muscle. To investigate this observation further, recordings have been made from single motor units using fine wires or tungsten microelectrodes or, in particularly favorable circumstances, surface electrodes.

Such recordings reveal that the impulse firing frequencies fall from initial values that may be as high as 100 impulses · s⁻¹ to values as low as 10 to 15 impulses · s⁻¹ (see chapter 13). As pointed out by various workers (Bigland-Ritchie *et al.*, 1983;

Marsden *et al.*, 1969), this decline in frequency does not, by itself, reduce the forces generated by the motor units. The reason is that the relaxation of the contractile mechanism becomes considerably prolonged during fatigue, such that the twitch half-relaxation time may be from two to three times the starting value. This prolongation enables a steady force to be maintained with fewer excitations of the muscle fiber than in the rested state. In this respect, the reduction in firing frequency improves the efficiency of the fatiguing motor units; the phenomenon has been aptly described as an example

(continued)

of muscle "wisdom." In addition to the fall in firing frequency, it is probable that some motoneurons stop discharging altogether as fatigue sets in. As indicated earlier in this chapter, this decrease in firing frequency probably contributes to the fatigue that is observed.

How is the reduction in EMG activity brought about? Brenda Bigland-Ritchie has suggested that there is an inhibitory reflex such that afferents from the fatiguing muscle depress the excitabilities of the motoneurons supplying the same muscle, thereby rendering them less responsive to motor commands from the brain (Bigland-Ritchie et al., 1986b).

There are a number of observations that are certainly compatible with this proposal. For example, Bigland-Ritchie et al. (1986b) found that if muscles are fatigued by voluntary contractions under ischemic conditions, the reduction in motor unit firing rate will persist for as long as the ischemic cuff is maintained, suggesting that metabolites are accumulating in the muscle belly and exciting inhibitory afferent fibers. Further, Garland et al. (1988a) showed that voluntary EMG activity was depressed if fatigue had been achieved by electrical stimulation of the muscle rather than by voluntary contraction. In these circumstances, the loss of EMG activity could not have been produced by exhaustion or refractoriness of the descending motor pathways but must have required some other mechanism, possibly a reflex. Other observations have come from human fatigue experiments in which the firing rates of single motoneuron axons were determined after impulse conduction in muscle afferents had been blocked by local anesthetic (Macefield et al., 1993). The finding that the decline in firing rate was less than in the intact state was further evidence that, during fatigue, there is normally an inhibitory influence mediated by muscle afferents.

Which afferent fibers might participate in such a reflex? One possibility would be mechanoreceptors in the muscle belly with increased sensitivities to stretch during fatigue. These receptors include not only the *muscle spindles* and *Golgi tendon organs,* but also the free nerve endings supplied by the slowly conducting (type III and type IV) nerve fibers (see chapter 4). Although some mechanoreceptors do indeed discharge more frequently during fatigue, others do not, and overall there is a decline in impulse activity (Hayward et al., 1991). More plausible candidates for the afferent limb of the reflex would be those metaboreceptors responsive to muscle hypoxia or to increased concentrations of H^+, K^+, bradykinins, and prostaglandin (Mense, 1986; Rotto & Kaufman, 1988).

Applied Physiology

Muscle pain and tenderness are associated with severe or unaccustomed exercise.

In this section, we will first consider some of the sensations associated with fatigue, not only during the performance of the fatiguing activity but in the next few days as well. See "An Inhibitory Reflex?" (p. 243) for a discussion of an inhibitory reflex that may come into play during fatiguing contractions and may serve to regulate motoneuron firing frequency.

During the development of muscle fatigue, especially under ischemic conditions, pain becomes an increasingly prominent distraction and, indeed, may contribute to central fatigue. Part of the pain can be described as "dull" or "hard" and is felt diffusely in the muscle belly; there is another component, with a burning quality, that is localized to the myotendinous region and is exacerbated with each relaxation. During repeated contractions without ischemia, pain builds up in those muscles that move joints and are allowed to shorten. In some situations, however, muscles may contract while being lengthened; examples are the gluteal and quadriceps muscles that support the body weight during walking down stairs, the calf muscle during descending a ladder backward, and the elbow flexors when a heavy weight is being lowered. In muscles that have been used strenuously in such ways, there is commonly pain and a feeling of stiffness when similar activities are performed on the next day (*delayed-onset muscle soreness* or *DOMS*). The muscles are also tender to palpation and may feel swollen and rather firmer than usual due to tissue edema. These symptoms may last several days and, apart from the influx of water in the muscle, have been attributed to stretching of the *connective tissue* around the muscle belly and in the myotendinous junctions (Jones & Round, 1990). In addition, there may have been damage to the muscle fibers (see chapter 21).

Our study of muscle fatigue is complete. Despite the large number of experimental observations, there are still some uncertainties; and, in particular, it is not clear to what extents the various biochemical perturbations contribute to the loss of force. The answers may come from the combination of results obtained by increasingly refined *in vitro* techniques and further examination of human subjects by MRS. Now we must leave fatigue and consider other adaptations of muscle.

16

Loss of Muscle Innervation

Injury to a limb may result in a motor nerve's being crushed or divided. Alternatively, nerves may degenerate or become inflamed as part of a disease process. Regardless of the cause, nerve dysfunction has serious consequences not only for the *motoneurons* and their *axons* but also for muscle fibers.

Changes in Motor Axons and Neuromuscular Junctions

We will start by considering the changes that take place in an axon following section or crush; it will be shown that the main trunk of a motor axon differs rather sharply from the terminal arborization on the muscle fiber in both the onset and the rapidity of response. Partly for this reason and partly on account of their contrasting functions, the trunk and termination of the axon will be analyzed separately in relation to their degenerative processes. Chapter 17 deals with the remarkable regenerative capacity of the motor axons that, in successful instances, enables them to make new synaptic connections with the muscle fibers.

In the distal stump, the myelin sheath retracts and then breaks up.

Changes in the Distal Nerve Stump Following Injury: Wallerian Degeneration

Following nerve section or severe crush, the distal nerve stump remains capable of propagating *action potentials* for many hours. In rats, usually 24 hr or so elapses before electrical activity is impaired and about 80 hr before all the axons become inexcitable. In humans and in baboons, *impulse* conduction in some fibers may persist for as long as 200 hr (Gilliatt & Hjorth, 1972; see also table 16.1). When the axon commences to break up, the process is termed *Wallerian degeneration,* in recognition of Waller (1850), who was the first to describe the changes in the axon that were visible under the light microscope. The observations of Waller have been confirmed on many occasions, and the nature of the degenerative changes has been studied with the electron microscope (a good account is given in Williams & Hall, 1971a, 1971b).

Retraction of *myelin* at the *nodes of Ranvier* is one of the earliest changes. This process spreads along the distal stump from the site of the injury and by 1 hr may have involved 20 mm or so of axon. Within the next few hours, the nodes show other evidence of increased metabolic activity in the accumulation of mitochondria and *lysosomes*. At 24 hr the degenerative changes within the axon cylinder are well advanced, with disruption of the *microtubules,* endoplasmic reticulum, and *neurofilaments*. Another early change, already visible within a few minutes of the nerve lesion, is extreme dilatation of the *Schmidt-Lanterman incisures.*

Table 16.1   Time to Conduction Failure Distal to Nerve Section

Species	Nerve	Time to failure (hr)	Authors
		Nerve trunk recording	
Rabbit	Peroneal	71-78	Gutmann & Holubár (1950)
Rat	Peroneal	79-81	Gutmann & Holubár (1950)
Guinea pig	Peroneal	72-82	Gutmann & Holubár (1950)
Cat	Sciatic	72-101	Rosenblueth & Dempsey (1939)
Dog	Phrenic	96	Erlanger & Schloepfle (1946)
Baboon	Peroneal	120-216	Gilliatt & Hjorth (1972)
		Muscle recording	
Rabbit	Peroneal	30-32	Gutmann & Holubár (1952)
Rat	Sciatic	24-36	Miledi & Slater (1970)
Guinea pig	Sciatic	40-45	Kaeser & Lambert (1962)
Cat	Sciatic	69-79	Lissák et al. (1939)
Human	Median, ulnara*	85-128	Landau (1953)
Baboon	Peroneal	96-144	Gilliatt & Hjorth (1972)
Human	Facial	120-192	Gilliatt & Taylor (1959)

*From observation of muscle twitch—no electrical recording.

Adapted from Gilliatt and Hjorth (1972).

The myelin sheath, having already withdrawn slightly, expanding the nodes of Ranvier, now begins to break up, with the lamellae peeling off from the nodes. Both at the nodes and at the dilated Schmidt-Lanterman incisures, the axons undergo progressive constriction; eventually a series of large ellipsoids are formed, each being bounded by the degenerating myelin sheath and cut off from the remainder of the axon. In time the ellipsoids become partitioned internally by unfolding of the myelin sheath, while at the ends of the ellipsoids, small globules of degenerating myelin become pinched off.

The Schwann cells multiply and help to digest the myelin sheath.

The *Schwann cells* become highly active during the degenerative process, first spreading over the denuded nodes of Ranvier and then undergoing mitotic division to form a maximum number of cells at about 15 to 25 days. The Schwann cells hasten the disintegration of the myelin sheath and engulf some of the lipid droplets. It is probable that a portion of these cells are transformed into *macrophages;* however, autoradiographic studies have shown that the majority of macrophages are carried to the degenerating nerve in the bloodstream (Olsson & Sjöstrand, 1969). These hematogenous cells are first seen at about 3 days following the lesion, and they are especially active in removing the degenerating myelin. The macrophages, and the myelin debris that they ingest, also make the Schwann cells multiply (Baichwal *et al.,* 1988). With the passage of time, the lipid-laden macrophages appear to migrate from the endoneurial spaces toward the periphery of the nerve trunk.

Schwann cells are also involved in providing *trophic* factors that are important for the eventual *regeneration* process. For example, the expression of *nerve growth factor (NGF)* and its receptor is *up-regulated* in nonneural cells in the distal segment of *axotomized* motoneurons. Other factors, such as *brain-derived neurotrophic factor (BDNF)* and neurotrophin-4 (NT-4), are also regulated in nonneural, presumably Schwann, cells, during the post-axotomy period. The induction of this expression may be the result of factors secreted by invading macrophages (Funakoshi *et al.,* 1993).

If the axon has been crushed rather than cut, the surrounding *basement membrane* will remain intact and act as a scaffolding for the columns of dividing Schwann cells *(bands of Bungner).*

If the macrophage invasion is delayed, as it is in a mutant strain of mouse described by Lunn *et al.* (1989), the degeneration of the fibers is slowed considerably, so that action potentials can be conducted for up to 14 days, compared with the 3-day period for transected axons in normal mice.

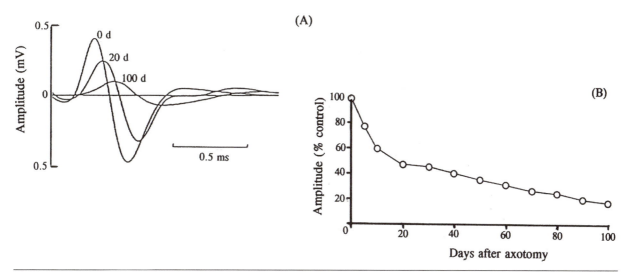

Figure 16.1   (A) Recordings of compound action potentials in a cat common peroneal nerve at 0, 20, and 100 days after distal section. (B) Mean sizes of compound action potentials at different times after distal nerve section.

Adapted, by permission, from T. Gordon et al., 1991, "Axotomy-induced changes in rabbit hindlimb nerves and the effects of chronic electrical stimulation," *Journal of Neuroscience* 11: 2157-2169.

Changes in the Proximal Nerve Stump Following Injury

In the proximal nerve stump, the axis cylinders shrink.

While the distal nerve stump is degenerating, dramatic changes start to take place in the proximal stump. Gordon *et al.* (1991) investigated these changes by implanting stimulating and recording electrodes around the sciatic and common peroneal nerves of cats and then cutting the nerves distally. They found that the amplitude of the compound action potential in the proximal stump begins to decrease within a few days of nerve section, and by 80 days is less than 20% of the initial value (figure 16.1).

Similar reductions were observed in the ulnar nerve stumps of patients who had suffered amputation of the hand months or years previously (McComas *et al.*, 1978). In the study by Gordon *et al.* (1991), there was also considerable *atrophy* of the axons (from a mean fiber diameter of 6.6 μm to one of 4.4 μm at 227 days). Figure 16.2 illustrates the same phenomenon in a cat sciatic nerve after distal section. Neither the atrophy nor the reduced action potential amplitude could be prevented by chronic intermittent stimulation of the proximal nerve stump. Since the atrophy is greatest in the largest axons, it has the effect of converting the normal bimodal distribution of fiber diameters into a unimodal one. Not only are the nerve fibers smaller, but many appear to have collapsed after distal nerve section, assuming a flattened rather than a circular profile on cross section; in some the *axis cylinder* is detached from the myelin sheath (Gillespie & Stein, 1983). In view of the proportionality between impulse conduction velocity and fiber diameter (see chapter 9), it would be expected that the impulse conduction velocities would be slower following nerve section, and a number of studies have shown that this is indeed the case (for example, see Milner & Stein, 1981).

Degeneration of Motor Nerve Terminals

The onset of degeneration in the motor nerve terminals depends on stump length and species.

It is an interesting observation that, after an axon has been divided, the first severe degenerative changes are seen at the neuromuscular junction rather than in the separated stump of axon. The onset of the changes in the axon terminal depends on two factors: the length of the distal stump of axon and the species of animal affected. Miledi and Slater (1970) studied the effect of stump length in the rat by cutting the phrenic nerve either as close as possible to the diaphragm or else in the neck. They found that when the nerve is divided distally, about 8 hr elapse before there is any evidence of neuromuscular degeneration in the diaphragm. When the greater length of nerve is left, the onset of neuromuscular failure is delayed by approximately 1 hr for each 1.5 cm of axon remaining. This intriguing relationship between the length of the nerve stump and the onset of *denervation* phenomena

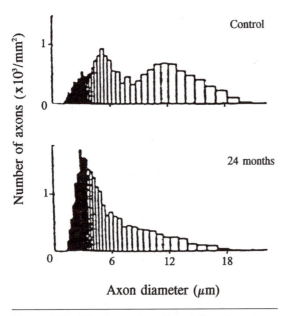

Figure 16.2 Diameters of myelinated fibers in an intact cat sciatic nerve and in a nerve examined 24 months after distal section.

Adapted, by permission, from P.J. Dyck et al., 1984, Permanent axotomy: A model of chronic neuronal degeneration preceded by axonal atrophy, myelin remodelling and degeneration. In *Peripheral neuropathy*, edited by P.J. Dyck, P.K Thomas, E.H. Lambert, and R. Bunge (Philadelphia, PA: W.B. Saunders), 678.

is of significance for an understanding of the nature of the neurotrophic controlling system (see chapter 18).

As already stated, the onset of *end-plate* failure also depends on the species studied, being much later in humans and baboons than in smaller mammals such as rats (see table 16.1). Once the degenerative changes start, however, they proceed rapidly, and only 3 to 5 hr are required for complete disruption of the end-plate. During this time, the synaptic vesicles form clumps; the mitochondria swell, and their *cristae* break up into small vesicular pieces; membrane whorls and glycogen bags appear in the axoplasm, and lysosomes become evident. At this stage the whole end-plate may become fragmented; meanwhile, the Schwann cell sends a cytoplasmic process into the *primary synaptic cleft,* and this completely envelops the degenerating end-plate. It is probable that the Schwann cell assists in the final dissolution of the end-plate. Subsequently the Schwann cell withdraws, and after a month or so, only the *sole-plate* is left of the original neuromuscular junction.

The electrophysiological behavior of the end-plate parallels the morphological changes just described. For the first 8 hr or more, depending on the nerve stump length, there is the usual spontaneous discharge of *MEPPs,* and electrical stimulation of the axon results in normal neuromuscular transmission. From these observations it would appear that the synaptic vesicles contain normal numbers of *ACh*

molecules and that there is no impediment to the mobilization and release of transmitter from the vesicles. After this latent period is over, neuromuscular function fails abruptly; the MEPPs cease, and no postsynaptic response can be detected following nerve stimulation.

Intact neuro-muscular junctions may cease trans-mission (silent synapses).

Following injury or disease, muscle fibers may become inexcitable without other features of denervation, such as spontaneous *fibrillation* and positive sharp wave potentials (see p. 253). In such instances, when excitability returns, it does so without any electrophysiological signs typically associated with denervation/reinnervation or remyelination: that is, temporal dispersion of evoked responses. This observation is consistent with changes at the neuromuscular junction resulting in the arrest and return of transmission, rather than axon regeneration or remyelination contributing to return of function. Thus, because there is no axon restructuring and accompanying increased distribution of conduction velocities, the evoked responses retain their normal synchronous pattern.

This peculiar behavior has occasionally been seen following crush or stretch injuries to peripheral nerves or nerve roots (McComas *et al.,* 1974a; McComas, 1977) and, more frequently, in patients recovering from overactivity of the thyroid gland (thyrotoxicosis; McComas *et al.,* 1974b). To explain these puzzling findings, it has been necessary to postulate the existence of silent *synapses,* that is, of neuromuscular junctions that are largely intact but in which there is insufficient *depolarization* of the muscle fiber membranes to generate action potentials. Under more natural circumstances, the phenomenon is seen when muscle fibers have been multiply innervated during normal embryonic development (see chapter 6, "Polyneuronal Innervation"), while in animal experiments, silent synapses have been produced by irradiation of motoneurons in the spinal cords of rats (Fewings *et al.,* 1977).

Motoneuron cell bodies change following axotomy.

Responses of Motoneuron Cell Bodies to Axotomy

Within days following axotomy, the *soma* swells, the *nucleus* translocates to a peripheral position, and the Nissl substance, which is the orderly array of endoplasmic reticulum with its proteosynthetic machinery, disperses. These changes are known as *chromatolysis.* In addition, *RNA* and protein synthesis increase (Watson, 1974). Electrophysiological properties change such that motoneurons become less "differentiated," or more homogeneous in several properties as opposed to the clear fast–slow distinction in healthy motoneurons (Kuno *et al.,* 1974a; Gustafsson, 1979). Interestingly, these electrophysiological changes in motoneurons also occur when neuromuscular transmission is blocked with *botulinum toxin,* suggesting that successful communication with muscle fibers is a prerequisite to the maintenance of these properties (Pinter *et al.,* 1991). Axotomized motoneurons also gradually experience a decrease in the number of synaptic connections onto the soma and *dendrites,* as well as a loss of dendritic branching (Brannstrom & Kellerth, 1998).

Changes in Muscle Fibers

If the nerve to a muscle is severed, the muscle will gradually waste over a period of weeks. This process is termed denervation atrophy; it demonstrates that the muscle fibers are dependent upon the motoneuron for the maintenance of their normal structure. This sustaining action depends on the contractile activity, which the motoneuron induces through impulse bombardment of the muscle fiber. This is not the whole story, however, for the motoneuron also sends chemical messages to the muscle fiber via the microtubules and neurofilaments that run along the axon (see chapter 8). These influences are referred to as trophic; it will be seen in chapter 18 that the trophic influence is mutual, because the motoneuron is itself affected by the muscle fibers to which it is connected. The gross changes that take place in a muscle following denervation have been studied on many occasions (see, for example, Lu *et al.,* 1997; Sunderland & Ray, 1950; Tower, 1939; Viguie *et al.,* 1997; Adams *et al.,* 1962). The changes, which probably involve every part of the muscle fiber, have been summarized in figure 16.3.

• **Denervation causes marked atrophy of all muscle fibers.** The most obvious change in the muscle following denervation is its reduction in size. This atrophy can be detected at about the third day and is rapid in the ensuing 2 months. At the end of this period, only 20% to 40% of the original muscle mass remains (Sunderland & Ray, 1950); much of this will be *connective tissue,* since the latter accounts for 10% to 25% of the initial muscle weight. Although further atrophy may occur, it is now a much slower process. Most, if not all, of the loss in muscle mass associated with denervation can be attributed to the lack of contractile *activation* alone, since prolonged paralysis without nerve damage produces atrophy that is equal in magnitude to that with denervation (Buffelli *et al.,* 1997). According to Stonnington and Engel (1973), the denervation atrophy affects red and white

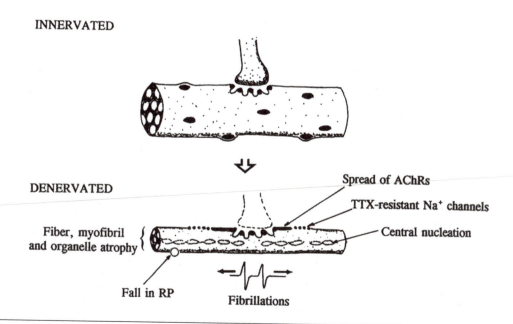

Figure 16.3 Some of the structural and functional changes in muscle fibers that follow denervation.

fibers equally. It is of interest that in the hemidiaphragm, denervation atrophy is preceded by a 1- to 2-week phase of *hypertrophy,* in which the cross-sectional areas of the fibers are almost doubled and new *myofibrils* are formed (Miledi & Slater, 1969). This curious response is due to excessive stretch of the denervated half of the diaphragm by the contractions of the innervated remainder. This interesting response to passive stretch is discussed further in chapter 20.

• **Myonuclei take up central positions in the fibers.** As early as the second day after denervation, the *myonuclei* start to become rounded instead of narrow and elongated; at the same time, the *nucleoli* become enlarged and more prominent. Many of the nuclei then move into the centers of the fibers where they line up to form chains. There is dispute as to whether the number of nuclei actually increases; much of the apparent preponderance of the nuclei is due to their preservation within atrophying fibers, but a true increase has also been reported (Bowden & Gutmann, 1944). After prolonged periods (up to 18 months, in rats), the number of myonuclei per muscle fiber decreases dramatically, as does the number of *satellite cells* (after an initial rise) (Viguie *et al.,* 1997). In late atrophy (e.g., 6 months), all that remains of some fibers are chains or clumps of nuclei surrounded by thin cylinders of *cytoplasm.*

• **Necrosis (death) of fibers may also occur.** In addition to the changes of "simple" atrophy, previously described, a variable proportion of the muscle fibers undergo further degenerative changes after several months. These changes are probably irreversible and may result in the death of the fiber; not uncommonly they are restricted to part of a fiber. The affected fibers swell; the nuclei also enlarge and then begin to fragment. *Vacuoles* appear in the cytoplasm of the fiber, and elsewhere the cytoplasm stains darkly because of increased *basophilia.* The cross-striations become less distinct, and the *plasmalemma* thickens before disintegrating. Mononuclear cells appear at the site of *necrosis* and subject the accumulating fiber debris to *phagocytosis.* Although complete destruction of the muscle fiber may eventually occur, Schmalbruch *et al.* (1991) have suggested that, in long-term denervation, there are repeated cycles of regeneration and necrosis, with the necrotic phase occupying only 2 days or so. As the muscle fibers atrophy or degenerate, the connective tissue within the muscle becomes increasingly prominent, with large adipocytes (fat cells) occupying spaces between the surviving fibers and fiber bundles.

• **Ultrastructural studies show that all components of the muscle fibers become smaller.** Several excellent studies have been published of the fine structure of the denervated muscle fiber as examined with the electron microscope (Miledi & Slater, 1969; Romanes, 1941; Schmalbruch *et al.,* 1991; Stonnington & Engel, 1973). The account by Stonnington and Engel is notable for the measurements of cross-sectional areas of the various fiber organelles (see figure 16.4). In this last study, the soleus and superficial medial gastrocnemius muscles of rats were studied between 1 and 84 days after section of the sciatic nerve. Such studies show that during the period of atrophy, the change in fiber area is matched by a fall in the mean area of the myofibrils. The atrophy begins at the periphery of a myofibril; but after the first month, degeneration also becomes visible in the interior. During the first week the mitochondria in both red and white fibers enlarge in the longitudinal axis of the fiber. Subsequently, the organelles shrink and form clusters; some mitochondria undergo frank degeneration and inclusion in *autophagic* vacuoles. Similarly, the *sarcoplasmic reticulum* at first enlarges and then diminishes, though to a lesser extent than the fiber itself. Other changes observed with the electron microscope include abnormalities of the *Z-disk,* irregularities and small papillary projections of the plasmalemma, and focal dilatations of the *transverse tubules* and sarcoplasmic tubules. Increased numbers of ribosomes can be found between the myofibrils and under the surface membrane.

• **Enzyme activities decrease in the denervated fibers.** There have been several studies of the effect of denervation on the biochemical and histochemical properties of muscle, and those of Hogan and colleagues (Romanul & Hogan, 1965; Hogan *et al.,* 1965) in the rat are particularly thorough. There is a rapid decrease in the high enzyme activities typical of a particular type of fiber; as a result fiber types can no longer be distinguished. In addition, the red muscle fibers (as in soleus) become paler, partly due to a reduction in myoglobin concentration. Gundersen *et al.* (1988) have demonstrated, in rat soleus and extensor digitorum longus muscles, large reductions

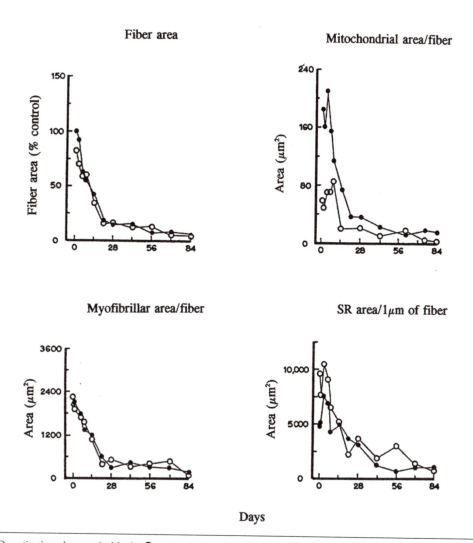

Fiber area

Mitochondrial area/fiber

Myofibrillar area/fiber

SR area/1μm of fiber

Days

Figure 16.4 Quantitative changes in black (●, soleus) and white (○, superficial medial gastrocnemius) muscle fibers of the rat following denervation (see text).

Adapted, by permission, from H.H. Stonnington, and A.G. Engel, 1973, "Normal and denervated muscle," *Neurology* 23: 716.

in the activities of *creatine kinase,* several glycolytic enzymes (glycogen phosphorylase, lactic dehydrogenase, phosphofructokinase, pyruvate kinase, and glyceraldehyde-3-phosphate dehydrogenase), and two enzymes of the *citric acid cycle* (citrate synthase and malate dehydrogenase). The only exceptions to these decreases are in the soleus muscles, in which slight increases in Ca^{2+} ATPase and *parvalbumin* occur. With respect to the *myosin heavy chains (MHCs)* that contain the ATPase enzyme and that therefore govern contractile speed, it appears that the expression of MHC types I and IIb decreases, with corresponding increases in the types IIa and IIx (Huey & Bodine, 1998; Jakubiec-Puka *et al.,* 1999). Interestingly, some of the changes seen in MHC expression with denervation can be attenuated by immobilizing the muscle in a lengthened position, thus demonstrating that passive stretch is a significant component in control of the expression of MHCs (Loughna & Morgan, 1999).

• **Denervated muscle fibers also produce trophic factors that influence the reinnervation process.** Denervated muscle fibers produce a number of chemical factors that may play a collective role in the regeneration/*reinnervation* process. These substances include glial cell-derived neurotrophic factor (GDNF; Lie & Weis, 1998); neurocresin, a neurite-outgrowth factor (Nishimune *et al.,* 1997); *insulin-like growth factors (IGFs;* Caroni, 1993); and BDNF (Koliatsos *et al.,*

1993). Neurotrophin-4 (NT-4) is usually expressed in activated fibers, and therefore its absence during denervation may also provide a negative signal that promotes reinnervation (Funakoshi *et al.*, 1995).

- **Twitch contractions become slower in denervated muscles, and the twitch–tetanus ratio increases.** Even though the nerve supply to a muscle may have been interrupted, it is still possible to stimulate the muscle directly. As would be expected, the tension developed during a single *twitch* or during a *tetanus* is reduced, for the muscle will be undergoing denervation atrophy and will have lost some of its myofibrils (see p. 250). In addition, the twitch becomes significantly slower, and this is true for a slow-twitch muscle such as soleus as well as for fast-twitch muscles like the flexor hallucis longus (FHL). Lewis (1972) has shown that the slowing occurs relatively abruptly during the third week of denervation; even in chronically denervated animals, however, the twitches of the fast muscles remain considerably faster than those of slow muscles studied in control animals. Lewis also found that in both types of muscle the *twitch–tetanus ratio* increased and that in the FHL, but not in the soleus, the maximum rate of rise of tetanic tension was somewhat diminished. Although some of the slowing of the twitch might have been due to a reduction in impulse conduction velocity on the *sarcolemma*, the most satisfactory way of reconciling the various findings was to attribute them to an increased amplitude and prolongation of activation following a single stimulus (see chapter 11). This, in turn, could have been due to such factors as the increased duration of the muscle fiber action potential and the greater proportion of sarcoplasmic reticulum (especially *terminal cisternae*) relative to the myofibril mass in the atrophied fibers.

- **The electrical properties of the muscle plasmalemma change after denervation.** Following denervation, important changes take place at the surface of the fiber in addition to those in the fiber interior. The irregularity of the plasmalemma and the development of small papillary projections have already been described. Their relationship to the altered electrophysiological properties of the membrane described next is uncertain. These additional changes occur:

1. First, there is a fall in resting *membrane potential* at 2 hr (Albuquerque *et al.*, 1971). This depolarization begins at the end-plate and spreads outward, eventually amounting to about 20 mV; it is caused by inhibition of the Na^+-K^+ pump (Bray *et al.*, 1976). A decrease in K^+ *permeability* (see change listed next) also contributes to this membrane depolarization.

2. The permeability of the muscle fiber membrane alters, becoming smaller for both K^+ and Cl^- (Klaus *et al.*, 1960; Thesleff, 1963).

3. The muscle action potential, which can be initiated by direct stimulation of the muscle fiber, has a lower rate of rise and a longer duration. Unlike the situation in a normal fiber, the action potential can still be elicited in the presence of *tetrodotoxin* (*TTX*; see figure 16.5); this resistance to TTX is first evident at about 36 hr (Harris & Thesleff, 1972). The impulse has a reduced propagation velocity, and the *refractory period* that follows is prolonged.

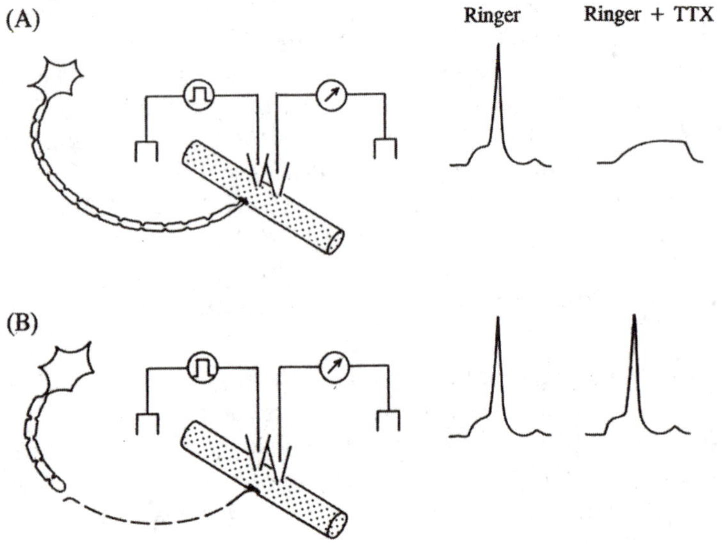

(A)

Ringer   Ringer + TTX

(B)

Figure 16.5 The development of tetrodotoxin (TTX) resistance following denervation. (A) A still-innervated muscle fiber has been impaled by stimulating and recording microelectrodes. In a standard Ringer solution, the fiber responds to direct stimulation with an action potential, but after TTX has been added, only an electrotonic local potential is seen. (B) Following denervation, an action potential can be elicited by direct stimulation in the presence of TTX.

Reprinted, by permission, from A.J. McComas, 1977, *Neuromuscular Function and Disorders* (London: Butterworth Publishing Co.), 74.

4. The sensitivity of the muscle fiber to ACh, though remaining highest in the end-plate region, after about 24 hr can be detected in the remainder of the fiber membrane (Axelsson & Thesleff, 1959). This development results from the formation of new *acetylcholine receptors* in the membrane areas outside the end-plate.

5. After an interval of about 1 week, the muscle fiber membrane becomes spontaneously excitable; the resulting impulses are termed fibrillation potentials (see figure 16.6D). These impulses are fully developed action potentials that recur with a frequency of 0.5 to 3 Hz. Belmar and Eyzaguirre (1966) analyzed this activity using an extracellular electrode at different points along the fiber and, from the configuration of the potentials, argued that it must have arisen at the site of the original end-plate. Further evidence supporting this conclusion was that they could alter the frequency of the fibrillations by passing a current through the membrane in this region but not elsewhere. Intracellular recordings have shown that this region of the membrane is electrically unstable, developing spontaneous oscillations that may build up and become large enough to initiate action potentials (Purves & Sakmann, 1974; Thesleff & Ward, 1975; see also figure 16.6C). A second type of generator activity consists of irregularly occurring sudden depolarizations (fibrillatory origin potentials or FOPs). These latter potentials are sometimes too small to trigger action potentials (figure 16.6, A & B) and may possibly correspond to the positive sharp waves (figure 16.6E) that can be recorded with a coaxial *EMG* electrode from denervated muscle. This second type of activity is also encountered most commonly at the old end-plate zone, but, like the oscillatory activity, can sometimes occur elsewhere. Since fibrillation potentials persist after curarization, the underlying membrane depolarizations cannot have resulted from the action of ACh. The spontaneous depolarizations do, however, depend on transient increases in the Na^+ permeability of the membrane, for they are abolished by removing Na^+ from the bathing solution.

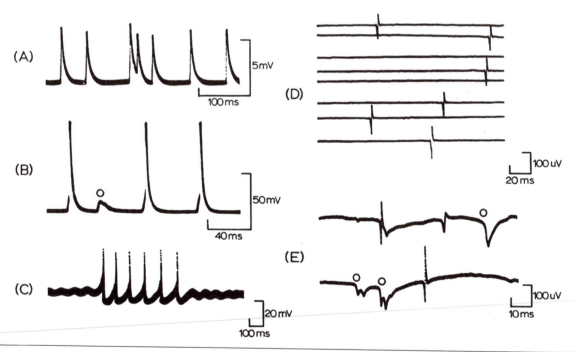

Figure 16.6   Spontaneous electrical activity in denervated muscle fibers. (A, B, C) Intracellular recordings from denervated rat diaphragm muscle maintained in organ culture. (A) Subthreshold FOPs (see text). (B) The FOPS are large enough to initiate action potentials except on one occasion (O); note that the amplification is lower than in A. (C) Spontaneous oscillation of muscle fiber membrane potential, resulting in a burst of action potentials. Adapted from Purves and Sakmann (1974). (D, E) Recordings made with coaxial needle electrodes from two patients with severe denervation. Fibrillation potentials are evident in both records, but in E there are also simple and complex positive sharp waves (O).

Reprinted, by permission, from D. Purves and B. Sakmann, 1974, "Membrane properties underlying spontaneous activity of denervated muscle fibres," *Journal of Physiology* 239: 133.

6. The concentration of *cholinesterase* in the end-plate falls by a variable amount that, in the rat sternomastoid muscle, is as much as 50% to 70% of the original value; this change takes place in the first few days following denervation (Guth *et al.,* 1964).

7. Fast-twitch muscles become sensitive to *caffeine* and respond to this drug with a *contracture* (Gutmann & Sandow, 1965).

8. The denervated muscle fiber stimulates neighboring intact motor axons to *sprout.* Unlike the situation in an innervated muscle fiber, the denervated muscle fiber membrane becomes receptive to the arrival of an axonal sprout and will permit a new neuromuscular junction to form. Usually the site of the original end-plate is chosen, but if this is prevented by mechanical factors, other regions of the fiber engage in *synaptogenesis.* This phenomenon may be due to the production by denervated fibers of substances that promote neural outgrowth from nearby axons (see earlier).

Applied Physiology

Of the hundreds of causes of peripheral nerve damage and disease, one of the most fascinating is an autoimmune disorder termed *idiopathic polyradiculoneuritis* or, in honor of those who described it first, the *Guillain-Barré syndrome (GBS).*

The Guillain-Barré syndrome is an autoimmune disorder that may have a precipitating event.

This condition, which can occur at any age, may arise either spontaneously or, in approximately half of the cases, following a viral infection some 2 to 4 weeks earlier. In some cases the precipitating event may be a surgical operation or an immunizing procedure; many cases occurred after the use of an influenza vaccine prepared from pigs (Schonberger *et al.,* 1979). In addition to the suggestive history, evidence that the disorder is immunologically based comes from the finding of circulating antibodies to peripheral nerve (Melnick, 1963). Further, sera from patients will cause unfolding and degeneration of the *myelin sheaths* of axons maintained in tissue culture (Cook *et al.,* 1969; Dubois-Dalcq *et al.,* 1971) or of sciatic nerve fibers following intraneural injection in rats (Feasby *et al.,* 1982).

The circulating lymphocytes can also be shown to be myelinotoxic (Arnäson *et al.,* 1969) and will transform on exposure to peripheral nerve *antigen* (Knowles *et al.,* 1969). Finally, there is the finding of immunoglobulin deposits and of lymphocytes in the nerve trunks of affected patients. The likeliest hypothesis for the pathogenesis of the disorder is that during the preceding injection, viruses enter the Schwann cells and replicate. The emergent viruses may then contain Schwann cell antigen, presumably myelin basic protein or ganglioside, on their outer surfaces; alternatively, the virus and myelin sheath may possess an identical antigenic complex. At any rate, the exposed antigen promotes the production of antibody and lymphocytes by the patient; these then react with the myelin sheath and cause its degeneration. The immunological disorder is discussed further in relation to *experimental allergic neuritis (EAN),* which closely resembles GBS in humans (Willison & Kennedy, 1993; see "A GBS-Like Illness Can Be Produced in Animals," p. 255 and also Rostami, 1993).

Guillain-Barré syndrome patients develop ascending paralysis and numbness.

Patients who are developing GBS neuropathy typically become aware of numbness and *weakness* of the legs. These symptoms start in the feet and then ascend to involve the remainder of the legs; a day or so later, similar symptoms commence in the hands and spread along the arms. In some cases there may be involvement of the cranial nerves and the spinal cord. Occasionally the progression of the disease is so rapid and severe that patients require urgent tracheostomy and artificial ventilation to compensate for the paralysis of the respiratory muscles. Recovery usually starts in 2 to 4 weeks and may eventually be complete, but in some patients it is delayed and some numbness and weakness persist.

One useful diagnostic test is the examination of the cerebrospinal fluid, for in about half of the cases there are significant rises in protein without any accompanying increases in white cells. Nerve conduction studies are easy to do and will demonstrate involvement of peripheral nerves; the reduced

A GBS-Like Illness Can Be Produced in Animals

Further insights into the Guillain-Barré type of nerve disorder have come from the production of an animal model. In 1955, Waksman and Adams reported that they had been able to induce an inflammatory disorder of peripheral nerves and nerve roots by an immunological method, and they noted its resemblance to the GBS. Their technique was to make a suspension of rabbit sciatic nerve and to inject small samples, together with Freund's adjuvant (containing heat-killed tubercle bacilli), into the footpads of other rabbits (figure 16.7). They observed that after 12 days nearly all the injected animals developed the clinical signs of a polyneuropathy. The animals "tended to lie in a splayed position, with all extremities extended and the head resting on the floor. . . . In hopping they were unsteady and erratic. . . . Upon landing they would often stagger or lurch to one side. . . . The musculature of extremities and trunk was weak and slack" (pp. 215-216). Most of the affected animals recovered spontaneously after several weeks of illness.

In the Animal Model, One of the Myelin Basic Proteins Is the Antigen

These important results have been confirmed in many laboratories, and it is now possible to understand in more detail the immunological mechanisms involved in the pathogenesis. In the first place, it appears that the antigenic component of peripheral nerve is one or both of the two basic proteins contained in myelin. The chemical compositions of these proteins have now been determined. The P1 protein has a molecular weight of 18 kD and contains 168 amino acids; it is identical to one of the two basic proteins found in myelin from the central nervous system (Brostoff & Eylar, 1972). The antigenic property of the P1 protein is confined to a small part of the molecule containing only nine amino acids (Westall et al., 1971). When injected by itself, the P1 protein, or the critical nine-amino acid residue, will induce a demyelinating disorder in both the central and peripheral nervous systems. However, if intact peripheral nervous system (PNS) myelin is administered instead, only the peripheral nerves and roots are affected, and the CNS is spared. Wísniewskí et al. (1974) have explained this curious result by suggesting that the antigenic site on

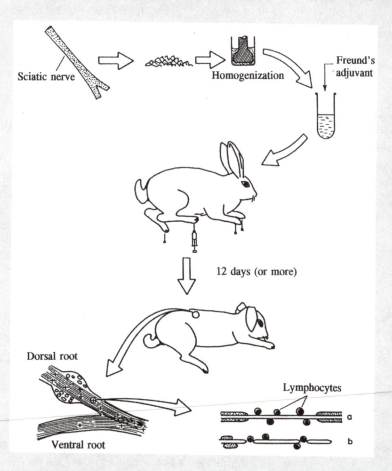

Figure 16.7 Experimental allergic neuritis (see text). At bottom left, the relative involvement of the dorsal, ventral, and mixed roots is indicated (+). At bottom right is shown (a) a fiber with segmental *demyelination* and (b) a more severely affected fiber undergoing Wallerian degeneration.

Reprinted from A.J. McComas, 1977, *Neuromuscular Function and Disorders* (London: Butterworth Publishing Co.), 236.

(continued)

the P1 protein is either hidden within the PNS myelin substructure or is conformationally altered so as to be immunologically inert. They further propose that it is the second basic protein of PNS myelin, the P2 protein, that is responsible for EAN. This protein has a molecular weight of 11 to 12 kD and contains 101 amino acids; when purified and injected with Freund's adjuvant into an animal, it can be shown to produce an inflammatory polyneuropathy (Brostoff *et al.*, 1972).

Both Lymphocytes and Antibodies Participate in Nerve Destruction

There is good evidence that although antibodies to myelin protein can be demonstrated in EAN, the degenerative changes are mostly cell mediated. For example, lymphocytes removed from an affected animal will produce demyelination when injected into another animal (Aström & Waksman, 1962), and they are also active when applied to myelinated axons in tissue culture (Lampert, 1969). According to Arnäson *et al.* (1969), the lymphocytes cross into the nerve parenchyma through the small veins and, upon exposure to myelin antigen, are transformed into larger cells. These then undergo repeated mitoses before commencing the demyelinating process.

With the electron microscope, the mononuclear cells can be seen to have penetrated the basement membranes of the Schwann cells with processes that eventually surround the myelin sheaths. The Schwann cells are now isolated from their axons but are not themselves attacked by the mononuclear cells. Instead, the mononuclear cells appear to pry apart the tightly bound spiral of the myelin sheath and cause it to undergo a bubbly dissolution (Lampert, 1969). Eventually, the myelin becomes completely stripped from the axis cylinder and is then removed by phagocytes (figure 16.7, bottom).

In most animals, recovery takes place and is achieved by remyelination of the affected axons. In a small proportion of animals, the demyelinating process assumes a chronic course. In such animals the nerve fibers become abnormally thickened so as to form onion bulbs; the thickening is caused by excessive numbers of Schwann cells and their processes (Pollard *et al.*, 1975).

impulse conduction velocities indicate that demyelination has taken place. In figure 16.8 there is evidence not only of slowed impulse conduction, but also of complete block in a proportion of the median nerve fibers, presumably because of a demyelinating plaque in the forearm.

As we have seen, paralysis is only one of the consequences when a skeletal muscle is deprived of its innervation. Profound as these denervation changes are, they can be reversed, partially or completely, if a new nerve supply can be obtained by the muscle. The next chapter describes the events that take place in the muscle and nerve fibers during reinnervation and the possible stimuli that bring these changes about.

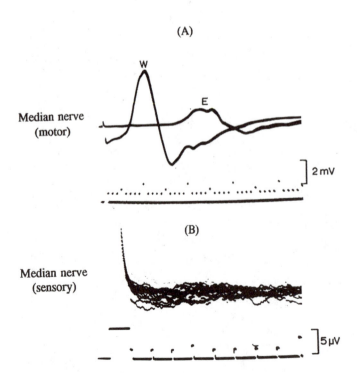

Figure 16.8 Electrophysiological findings in a patient with Guillain-Barré syndrome. (A) Recordings made from the thenar muscles with surface electrodes following stimulation of the median nerve at the wrist (response *W*) and at the elbow (response *E*). (B) Absence of detectable sensory nerve action potential when the thumb was stimulated maximally and the recordings were made from the medial nerve at the wrist with surface electrodes (several recordings following stimulation are shown superimposed). For significance of findings, see text.

Reprinted, by permission, from A.J. McComas, 1977, *Neuromuscular Function and Disorders* (London: Butterworth Publishing Co.), 235.

17

Recovery of Muscle Innervation

Muscles can recover from *denervation* in different ways. Either the motor nerve will regenerate and grow back to the muscle or, in the case of partial denervation, surviving motor *axons* will *sprout* and establish synaptic connections with the denervated muscle fibers, a process termed *collateral reinnervation.* Both of these processes will now be considered.

Nerve Regeneration

If an axon is divided and viewed under the microscope, axoplasm can be seen to bulge out of the end of the central stump; this extruded material probably reflects the continuing *axoplasmic transport* from the cell body (see chapter 8). After a few minutes the extrusion ceases, possibly because some of the *myelin lamellae* have slipped over each other and sealed the exposed end of the cut axon. The end of the fiber now gradually swells; initially this is the result of degenerative changes, but the later distension is due to the formation of a *growth cone.* This important structure was encountered in chapter 6, in which the outgrowth of axons to their targets in the embryo was discussed. Studies of nerve *regeneration* have revealed more of the properties of the growth cone.

The Growth Cone

The growth cone is formed at the cut end of the axon by local mechanisms.

The growth cone still forms if the divided axon is separated from the cell body by a second cut. Hence there must be local mechanisms at the tip of the divided axon that control the formation of the cone (Brown & Lunn, 1988). The changes that subsequently take place in the growth cone were described with memorable clarity by Cajal (1928), who also summarized much of the earlier work in this field. In most of his studies, Cajal stained the regenerating fibers with silver. His drawings not only are elegant but also display his findings in remarkable detail, especially when it is remembered that all the observations were made with an ordinary light microscope. Within 24 hr of nerve section, the growing axons are seen to have penetrated into the exudate produced by the wound; sometimes the dilated endings remain single, but in other instances they divide into two or more processes. In some nerve fibers additional sprouts form above the cut ends at *nodes of Ranvier.*

GAP-43 and intracellular Ca²⁺ help to direct the growth cone.

Studies with the electron microscope have shown that the growth cones are rich in mitochondria, vesicles, dense bodies, and lamellar figures. The growth cones also contain a novel protein, *GAP-43* (or B-50), which is synthesized by the neuronal cell body and then conveyed to the tip of the growing axon. Expression and axonal transport of GAP-43 are dramatically induced in axons during normal developmental growth and during regeneration (Pekiner *et al.,* 1996). GAP-43 has a molecular weight of about 24 kD, and antibodies labeled with gold show that it is associated with the membrane. Exactly what GAP-43 does in the *plasmalemma* is not clear—PC12 cells deficient in GAP-43 are still capable of extending neurites. GAP-43 has *calmodulin*-binding properties, and is also phosphorylated by *protein kinase* C. One suggestion is that it transforms the interactions of receptors with external guidance factors, within the *cytoplasm* of the growth cone (see later). This signal would, in turn, control the laying down of new cytoskeleton in the cone (Gordon-Weeks, 1989).

Expression of molecules essential for *anterograde* and *retrograde* axonal transport, *kinesin* and *dynein*, is up-regulated to ensure an increase in traffic up and down the axon (Su *et al.*, 1997). Plasma membrane needed for the extension of the growth cone appears to be transported to the cone in the form of vesicles, which fuse with the cone membrane (Ide, 1996).

Initially the growth cones are free of surrounding *Schwann cell* cytoplasm; but if they encounter the peripheral nerve stump, they will grow preferentially down the columns formed by the Schwann cells of the stump *(bands of Bungner)*. The growing axon tips are guided down the stump by the same guidance factors that play similar roles in embryogenesis. These molecules include adhesive substances such as *laminin, fibronectin, neural cell adhesion molecule (N-CAM)*, and N- and E-*cadherins*, as well as *trophic* factors such as *nerve growth factor (NGF)*, basic *fibroblast growth factor* (bFGF), and *brain-derived neurotrophic factor (BDNF)*, all of which are produced by Schwann cells distal to the axon injury site (Ide, 1996). Both proliferating Schwann cells and the *macrophages* that invade the distal site secrete factors important in the axon regeneration process (Terenghi, 1999).

Schwann cells provide pathways for regenerated axons.

Axon Elongation and Contact With Muscle

The rate of axon growth slows distally.

Axon growth is at first slow but subsequently quickens. Gutmann *et al.* (1942) calculated an average velocity of 4.3 mm · day^{-1} during a period between 13 and 25 days following nerve crush. Jacobsen and Guth (1965) measured the extent of regeneration by stimulating the nerve at the site of crush and recording the compound *action potential* at different positions in a distal direction. They found that the rate of growth increased until it had reached 3.0 mm · day^{-1} by the 18th day. In humans, Sunderland (1947) was able to make observations on the recovery of nerve function in injured military personnel; he noted that the speed of nerve growth depends not only on the time after injury but also on the site of the lesion, being highest after a proximal wound. For example, in the thigh and upper arm, the average rate is about 3 mm · day^{-1}, whereas in the hand and foot, values of only 0.5 mm · day^{-1} are found. Another factor influencing nerve regeneration is previous injury, for recovery is then less complete; in such cases the Schwann cells divide to form large groups associated with multiple axons of varying diameters (Thomas, 1970).

Axon regeneration is altered by several conditions and interventions.

Nerve regeneration rate can be altered by a variety of interventions that target the metabolic processes involved. For example, the regeneration rate can be increased using short-term electrical stimulation (Brushart *et al.*, 2002; Al Majed *et al.*, 2000), and treatment with exogenous *neurotrophins* such as BDNF (Novikova *et al.*, 1997), *glial cell-derived neurotrophic factor* (Chen *et al.*, 2001), *ciliary neurotrophic factor* (CNTF; Oyesiku & Wigston, 1996), neurotrophin-3 (Sterne *et al.*, 1997), neurotrophin-4 (Yin *et al.*, 2001), and NGF (Fine *et al.*, 2002). Regeneration is hampered in diabetes (Pekiner *et al.*, 1996) and with advanced age (Black & Lasek, 1979). All of these treatments function via changes in the metabolic machinery involved in the regeneration process, and thus involve the cell body as well as the elongating axon itself. However, the particular mechanisms in each case remain to be elucidated.

Growth cones are directed to old end-plates.

Once the regenerating axons reach the denervated muscle, the growth cones appear to seek out the sites of the old *end-plates* in order to establish new neuromuscular junctions. The situation does not arise simply because the axons have been guided back by the peripheral nerve stump; even if the axons are forced to take other pathways, they will still run along the denervated muscle fiber until the old end-plate is found (Bennett *et al.*, 1973a). As described later, it is the *basement membrane* investing the former end-plate that expresses some of the molecular signals needed to attract the growth cone. In addition, terminal Schwann cells that sit atop the neuromuscular junction send out extensions to guide regenerating nerves to the end-plate when normal transmission at that end-plate is interrupted because of denervation or pharmacological paralysis (Son *et al.*, 1996).

Regenerated axons are thinner and conduct impulses more slowly.

Once an axon has established successful synaptic connections with the previously denervated muscle, the diameter of the *axis cylinder* increases (Aitken *et al.*, 1947). Schröder (1972) found that the *myelin sheath* also became thicker, though to a lesser extent than the axis cylinder, in studies on the regeneration of sciatic nerves in dogs. At 12 months after suture, the mean diameter of the axis cylinder was 79% of normal, whereas that of the myelin sheath was only 57%. Since the speed of *impulse* propagation is directly proportional to the diameter of the axis cylinder and is also dependent on the insulating properties of the myelin sheath (see chapter 9), low velocities would be expected in regenerated nerves. Hodes *et al.* (1948) found values as low as 50% to 60% of normal in human nerves 3 to 4 years after repair by suture. Rather higher, but still reduced, values were noted in a study by Ballantyne and Campbell (1973), which included observations on both sensory and motor fibers. A further change in the appearance of the new axons is that the nodes of Ranvier are situated closer together, though in an irregular manner; this alteration is caused by the proliferation of Schwann cells that takes place during nerve regeneration.

Collateral Reinnervation

There is another way in which the nerve supply to a muscle can be restored, but this mechanism is applicable only in instances in which the denervation has been incomplete. In this *reinnervation,* the surviving healthy axons form sprouts that grow across to denervated muscle fibers in their vicinity and form new neuromuscular junctions.

Axon Sprouting

Sprouts can develop at several sites in spared axons.

The sprouts commonly grow out from the nodes of Ranvier of the healthy axon (nodal sprouts), but they can also arise distally, either from the neuromuscular junction itself (terminal sprouts) or from the region of motor axon immediately in front of it (pre- or subterminal sprouts). These three branching systems are all examples of collateral reinnervation (figure 17.1). The existence of this type of innervation was first postulated by Exner (1885) in order to explain the absence of degeneration in the rabbit cricothyroid muscle following its partial denervation. Much later, interest

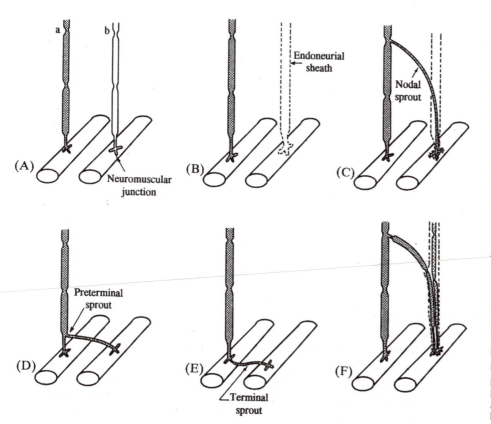

Figure 17.1 Different types of collateral reinnervation. (A) Axon *b* is about to undergo *Wallerian degeneration* while axon *a* remains healthy. (B) Degeneration of *b* has taken place, leaving the endoneurial sheath intact. (C) The denervated muscle fiber has been innervated by a nodal sprout from *a*, which has entered and followed the endoneurial sheath. (D) The axon branch has arisen from *a* prior to the neuromuscular junction; (E) the sprout has formed from the terminal axon expansion. (F) Axon *b* has regenerated so that the previously denervated muscle fiber is now supplied by two *motoneurons;* the respective axons share the same endoneurial sheath.

Reprinted, by permission, from A.J. McComas, 1977, *Neuromuscular Function and Disorders* (London: Butterworth Publishing Co.), 224.

in collateral reinnervation was revived by the experiments of Weiss and Edds (1945) in the rat, in which partial denervation of leg muscles had been effected by dividing *ventral roots* entering the lumbosacral plexus. Using a similar preparation, Hoffman (1951) subsequently showed that a nodal sprout might sometimes enter the endoneurial tube vacated by a degenerating axon and follow it down to the original end-plate zone.

Terminal sprouting has already been mentioned as one of the mechanisms by which an axon can undertake collateral reinnervation of a denervated muscle fiber in its vicinity. The process has a further significance in that it enables a motoneuron to either renew or else enlarge existing synaptic connections with its own colony of muscle fibers. As a result of terminal sprouting, the new fine axon twigs ramify on the surface of the muscle fiber and enlarge the territory of the neuromuscular junction; at the same time, the terminal axonal expansions are often swollen. Many examples of the bizarre and distorted innervations that can result, especially in patients with neuromuscular disorders, were described by Cöers and Woolf (1959) using the methylene blue staining reaction.

Particularly clear examples of terminal sprouting have been observed by Duchen and his colleagues (Duchen, 1970a, 1970b; Duchen *et al.*, 1972) in muscles poisoned with *botulinum* or *tetanus toxins*. The first evidence of sprouting can be seen as early as 6 days after the injection of botulinum toxin into soleus muscles of mice and well before any degenerative changes are visible in the motor nerve terminals. Figure 17.2 and the accompanying legend are reproduced from Duchen's (1970a) paper. Figure 17.2A shows a normal nerve ending with a single preterminal axon (arrow) innervating the end-plate. Terminal sprouting (figure 17.2B) develops after neuromuscular transmission has been blocked by the toxin. The nerve sprouts grow in all directions and branch, while the original end-plate may disappear (figure 17.2C). New nerve terminals are then formed randomly on some of the nerve sprouts (figure 17.2D). It can be seen that the nerve fibers that were at first terminal sprouts in figure 17.2B have now become branched preterminal axons in figure 17.2D.

Having now dealt with the main features of the neural response to muscle denervation, it is appropriate to consider some of the underlying phenomena in more detail. So far no attention has been given to two of the most fascinating problems in neurobiology. First, what is the stimulus that causes an axon to sprout, and second, how do the sprouts find their way to the denervated muscle fibers? As will be seen, rather more is known about the second question than the first.

Terminal sprouting can enlarge existing neuromuscular junctions.

Terminal sprouting has been studied after poisoning with botulinum toxin.

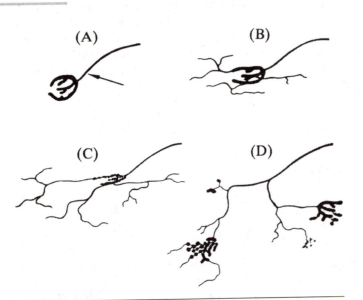

Figure 17.2   Remodeling of muscle fiber innervation following terminal sprouting, as seen in mouse soleus muscle fibers after the injection of botulinum toxin.

Reprinted, by permission, from L.W. Duchen, 1970, "Changes in motor innervation and cholinesterase localization induced by botulinum toxin in skeletal muscle of the mouse: differences between fast and slow muscles," *Journal of Neurology, Neurosurgery, and Psychiatry* 33: 51.

• **Muscle fibers may normally suppress motoneuron sprouting.** First consider the nature of the stimulus that initiates the sprouting at the central end of a divided nerve fiber:

-A signal from the denervated muscle cannot be involved, since sprouting will still occur even if the nerve is cut at a distance from the affected muscles. For example, if the sciatic nerve is cut in midthigh, sprouting will still take place, although the central nerve stump is surrounded by the still-innervated hamstring muscles and is well away from the denervated muscle fibers below the knee. An alternative suggestion is that the nerve is stimulated to sprout by some factor(s) released by the degenerating peripheral stump. Yet this is also unlikely, for sprouting continues at the cut central end of the nerve after the peripheral stump has been removed. Furthermore,

Cajal (1928) and others before him had found it impossible to induce sprouting in intact axons by placing a length of degenerating nerve in their vicinity.

-A more attractive hypothesis suggests that the sprouting from the central end of a cut axon takes place because the motoneuron is no longer receiving instruction "not to sprout" from the muscle (see figure 17.3). A carrier for such instructions would be the retrograde axoplasmic transport, discussed in chapter 8. As would be expected with this mechanism of communication, the surge in production of *RNA* in the *nucleolus* of the motoneuron occurs sooner if the axon has been divided close to the cell body. This observation could be interpreted to mean that the synthesis of protein, necessary for axonal sprouting, is normally repressed by a factor traveling in the axon toward the motoneuron *soma*.

• **Muscle inactivity may stimulate collateral reinnervation.** The sequence of events responsible for collateral reinnervation is equally puzzling. It will be recalled that, in this process, sprouts arise from intact axons, either at the nodes of Ranvier or at the motor nerve terminals. If the hypothesis outlined in figure 17.3 is correct, the motoneuron will still be receiving instructions not to sprout from the muscle fibers of its own intact *motor unit*. These instructions must now be overridden by local signals from the denervated muscle fibers. One such signal is associated with inactivity of the fibers. One can show this role of inactivity by paralyzing innervated muscle fibers with *tetrodotoxin (TTX)* applied to the nerve fibers or by one of the other methods described in chapter 19. Regardless of the method employed, axonal sprouts will form in the terminal arborization and will seek

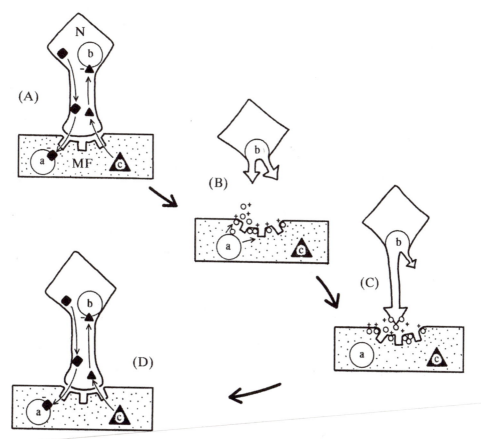

Figure 17.3   Postulated mechanism for axon regeneration and muscle fiber reinnervation following nerve section. (A) In the intact nerve-muscle preparation, the sprouting apparatus "b" of the motoneuron "N" is inhibited by centripetal delivery of molecular signals from "c" in the muscle fiber "MF." Similarly, "a" in the muscle fiber is prevented from secreting axon-attracting molecules by centrifugal inhibition from the motoneuron. (B) The axon has been sectioned; "b" is no longer repressed and starts to produce axon sprouts, and "a," also unrestrained, secretes attractants that (C) will guide an axon sprout to the site of the original end-plate. (D) Once the motor axon has reestablished a successful neuromuscular connection, the mutual inhibiting mechanism is reimposed.

Reprinted, by permission, from A.J. McComas, 1977, *Neuromuscular Function and Disorders* (London: Butterworth Publishing Co.), 226.

new sites for synaptogenesis on the surfaces of the inactive fibers. The terminal sprouting can be prevented if the muscle fibers are stimulated directly (Van Essen & Jansen, 1974).

- **Factors may be released from degenerating axons.** A second stimulus for sprouting is likely to be nerve fiber degeneration. Indeed, terminal sprouting can occur in muscles in which the motor nerve fibers were left intact but the muscle sensory fibers were made to degenerate (Brown *et al.*, 1978). It is not known whether the products of nerve degeneration act directly on adjacent neuromuscular junctions or whether other steps are involved (e.g., growth factors might be released by the macrophages digesting the nerve stump).

- **Factors may be released from denervated muscle fibers.** Yet another possibility is that the denervated muscle fibers release an agent that induces terminal or nodal sprouting. Early, but unsuccessful, attempts to identify such a substance were made by Van Harreveld (1947) and Hoffman (1950). Subsequently, Tweedle and Kabara (1977) reported positive results, probably due to the lipid component of their muscle extracts. Henderson *et al.* (1983) denervated chicken muscles and found that they started to produce a *neurite-promoting factor* within 1 day when tested on motoneurons in tissue culture. This factor attained a maximal concentration at 3 days and then declined as the muscle fibers began to degenerate. Other experimental results, consistent with the release of a sprouting factor from denervated muscle, have been obtained by Pockett and Slack (1982) and Slack and Pockett (1982). More recently, proteins that are up-regulated in denervated muscles, such as *insulin-like growth factor (IGF)*, glial cell-derived neurotrophic factor (GDNF), and neurocrescin (a neurite outgrowth factor) have been put forward as possible candidates for such an agent secreted by muscle, based on their nerve regeneration-promoting activities (Glazner & Ishii, 1995; Nishimune *et al.*, 1997; Lie & Weis, 1998).

Role of the Basement Membrane

The basement membrane and terminal Schwann cells attract the growth cone and direct the morphological changes at the new synapses.

The next stage in the reinnervation process is better understood, for there is very good evidence that the basement membrane surrounding the muscle fiber directs the neural growth cone to form a new terminal arborization. It also instructs the muscle fiber to develop synaptic folds. These key roles of the basement membrane were first demonstrated by McMahan and colleagues (Sanes *et al.*, 1978), who, in one type of experiment, denervated the frog cutaneous pectoris muscle and also removed a segment of the muscle; the surrounding regions were prevented from regenerating into the gap by a dose of irradiation. Despite the absence of underlying muscle, the motor nerve fibers regenerated to the sites of the old end-plates and developed terminal arborizations containing *ACh* and synaptic vesicles (figure 17.4). Conversely, if the nerve fibers were stopped from growing back but the muscle fibers were allowed to regenerate, the membranes of the latter underwent the complex infolding characteristic of the *secondary synaptic clefts* and formed *AChRs* and *acetylcholinesterase* (figure 17.4; McMahan & Slater, 1984). Biochemical analysis of the basement membrane shows that an essential component is the protein *agrin*, which can be produced by both the nerve and the muscle fibers and inserted into the basement membrane. Other important components of the basement membrane are synaptic (S) laminin, N-CAM, and *terminal anchorage protein (TAP$_1$)*. In addition, the nerve terminals produce *calcitonin gene-related peptide (CGRP)*, which stimulates the synthesis of AChRs in the muscle fiber and promotes their incorporation into the muscle fiber plasmalemma.

As mentioned previously, terminal Schwann cells that sit atop the neuromuscular junction also play a role in the reinnervation process by sending out extensions to guide regenerating nerves to the end-plate (Son *et al.*, 1996).

Multi-Innervated Muscle Fibers

Reinnervated muscle fibers may have more than one neuromuscular junction.

In normal mammalian *extrafusal* muscle, each fiber is usually supplied by only one motoneuron. Consider, however, a partially denervated muscle in which the twin phenomena of axon regeneration and collateral reinnervation proceed together. What happens if a regenerating axon returns to a muscle fiber that, through collateral reinnervation, has already formed a *synapse* with a foreign motoneuron? This situation is likely to arise in those disorders in which the peripheral nerve fiber

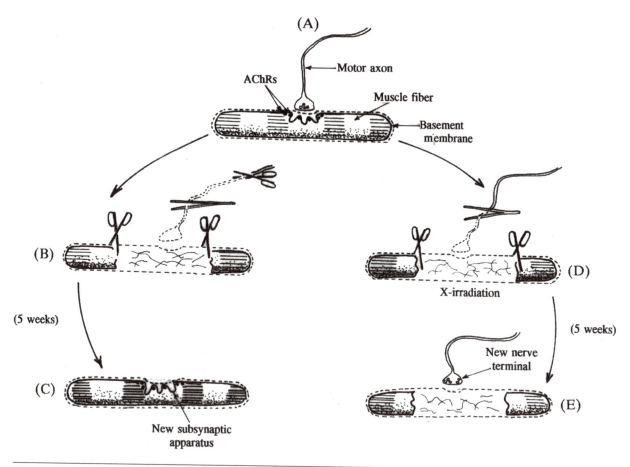

Figure 17.4 Experiments demonstrating the key role of the basement membrane in directing regeneration of the neuromuscular junction. (B) The muscle fibers are cut on either side of their innervation zones and then allowed to regenerate. Although reinnervation is prevented by cutting and crushing of the motor axons, (C) normal subsynaptic apparatus develops in the muscle fibers. (D) Regeneration of the muscle fiber is blocked by irradiation, but (E) a new motor nerve terminal is still able to form. A normal fiber and axon are shown in (A).

Based on experiments of Sanes *et al.* (1978) and McMahan and Slater (1984).

stump is left in continuity even though the axis cylinder has degenerated. It will occur, for example, after a nerve crush or following motoneuron recovery from a "dying-back" type of neuropathy. In humans the observations are few, but in animals several investigations have been directed to this problem. The critical experimental maneuvers and observations may be summarized in the following way.

First of all, provided the original nerve supply to a muscle is left intact, that muscle is unable to accept innervation from a foreign nerve (Elsberg, 1917). Once the original nerve supply is interrupted, either by cutting, crushing, or the application of botulinum toxin, the muscle will readily form new neuromuscular junctions with a foreign nerve presented to it (Fex *et al.,* 1966; see figure 17.5). If the original nerve is now allowed to grow back to the muscle, it is able to reestablish synaptic connections with a large proportion of fibers, even though these fibers already have an ectopic innervation from the foreign nerve. In the study by Frank *et al.* (1974), the proportion of doubly innervated fibers in the muscle was as high as 57% after a nerve crush.

The next question is whether the double innervation persists, for in embryonic and neonatal muscle fibers of mammals, any accessory innervation is ultimately rejected (see chapter 6). There is also evidence from cross-innervation experiments on the extraocular muscles of fish that foreign innervation ceases over a period of 2 days after the original nerve supply has been reestablished (Marotte & Mark, 1970). However, the results of single muscle fiber studies on this type of preparation are at variance with this interpretation (Scott, 1977). As far as mammalian muscles are concerned,

Double innervation may persist for long periods.

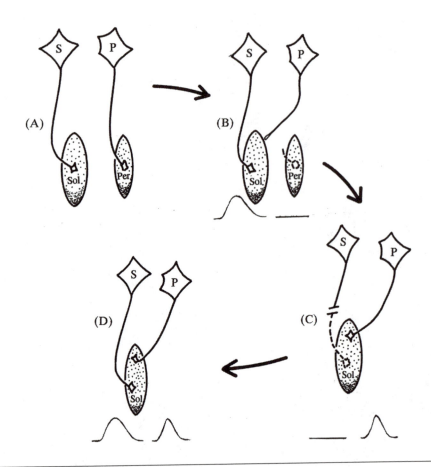

Figure 17.5 Hyperinnervation experiment. (A) The normal soleus *(Sol)* and peroneal muscles *(Per)* with their corresponding motoneurons *(S* and *P,* respectively). (B) The peroneal nerve has been cut and the central stump applied to the soleus; no innervation takes place (note absent *twitch* response below). If the soleus nerve is now divided or crushed, as in (C), the peroneal nerve rapidly forms neuromuscular junctions on the soleus muscle. (D) If the soleus nerve is now allowed to grow back, many muscle fibers will acquire a double innervation. Based on studies of Fex *et al.* (1966) and others.

Reprinted, by permission, from A.J. McComas, 1977, *Neuromuscular Function and Disorders* (London: Butterworth Publishing Co.), 228.

most of the evidence available is firmly in favor of the persistence of foreign synapses without any suppression by restoration of the original nerve supply (Fex *et al.,* 1966; Tonge, 1974). The longest period of study to date appears to be that of Frank and his colleagues (1974, 1975), who found that foreign innervation of rat soleus muscles by the superficial fibular nerve persisted for 9 months or longer. The induction of these doubly innervated fibers in adult mammalian muscles raises interesting questions. For example, do the two axons belong to motoneurons of the same type (see chapter 12)? If not, does the muscle fiber contain two regions of contrasting biochemical and contractile properties? If, on the contrary, the muscle fiber has identical properties throughout its length, which of the two motoneurons is the dominant one?

Even though the evidence for the maintenance of doubly innervated fibers is convincing, it is possible that in some circumstances suppression of synapses occurs in reinnervated mammalian muscle. The results of Brown and Butler (1974) on reinnervated *intrafusal muscle fibers* may prove to be a case in point. These authors found that, in *muscle spindles* of the cat, the pattern of sensory fiber discharges following motor fiber stimulation indicated that the initial reinnervation came exclusively from α-motor axons. Subsequently, the *spindle* responses changed to a form that suggested that the slower-growing γ-*axons* had now reached the spindle and restored the correct innervation of the intrafusal fibers.

In some conditions, super-numerary innervation may be suppressed.

Changes in the Muscle Fibers Following Reinnervation

As far as is known, all the effects of denervation on muscle fibers are reversible, provided the new nerve is introduced within a reasonable period of time. Exactly what constitutes a "reasonable" period is not clear, for the requisite experiments do not appear to have been performed.

Delay makes successful reinnervation less likely.

One factor that is time dependent is the *necrosis* that takes place in a small portion of the denervated muscle fibers. In some of the other fibers, the *myonuclei* may survive with only vestigial collars of cytoplasm, and it is doubtful if these fibers can recover their function. In the rat, for example, delaying reinnervation of hindlimb muscles for 6 months results in a profound decrease in recovery of muscle mass and motor function in spite of very successful axon regeneration (Kobayashi *et al.,* 1997). Another factor is the effect of prolonged denervation on the integrity of the intramuscular nerve sheaths. After 6 months of denervation in rat hindlimb muscles, these intramuscular nerve sheaths deteriorate, thus limiting the reinnervation of muscle fibers by regenerating axons (Fu & Gordon, 1995).

Another reason why reinnervation is more likely to be successful when attempted early rather than late is that fewer motoneurons will have undergone irreversible retrograde degeneration (see chapter 18). Also, with increasing time the peripheral nerve stump retracts, the endoneurial spaces decrease in size, and the Schwann cells become less active (Gutmann, 1971). With these qualifications in mind, there is no doubt that an excellent return of muscle function can result if reinnervation is able to take place within a few months.

Reinnervation after nerve crush may be complete in 1 month.

McArdle and Albuquerque (1973) produced temporary denervation in the hindleg of the rat by crushing selected nerves as close as possible to the muscles. In such animals there is a return of *MEPPs* 9 days later, initially with a low discharge frequency. At about this time, the resting *membrane potential* starts to rise and then reaches normal values at 30 days (see figure 17.6). The ionic *permeability* of the membrane and the sensitivity of the membrane to ACh assume normal levels by about 30 days.

The new neuromuscular junctions are initially nonfunctional.

In the early stages of reinnervation, the new neuromuscular junctions pass through a phase in which they are unable to excite the muscle fibers, even though adequate supplies of transmitter are available in the axon terminals. This point was first demonstrated by Miledi (1960a), who found that MEPPs are present at a time when no evoked response can be detected in the muscle fiber following nerve stimulation. Originally shown in the frog, the presence of nontransmitting synapses during reinnervation has since been established in the rat (Dennis & Miledi, 1974) and in the fowl (Bennett *et al.*, 1973b), though not in the mouse (Tonge, 1974). The transmission failure may be caused by a block of impulse conduction

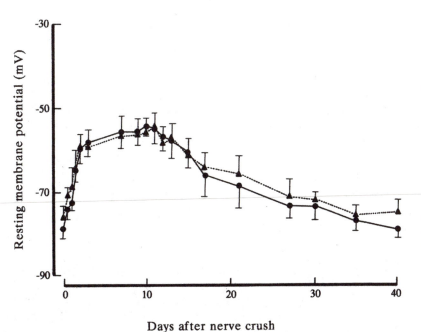

Days after nerve crush

Figure 17.6 Fall of resting membrane potential following temporary denervation produced by nerve crush. After reinnervation has taken place, the resting potential is eventually restored to its original level for both the extensor digitorum longus (˜) and soleus (p) muscles.

Adapted from *The Journal of General Physiology,* 1973, 61, 11, by copyright permission of The Rockefeller University.

in the distal motor axon, since in some instances a second electric shock to the nerve, given 4 ms after the first, is able to elicit a muscle fiber response (Dennis & Miledi, 1974). In the frog, the nontransmitting stage appears to last only a day or two, and in the rat, its duration is shorter still. Of significance for an understanding of trophic mechanisms (chapter 18) is the observation that even before the synapse has become operational, some of the denervation phenomena are already beginning to disappear from the muscle fiber. For example, the ACh sensitivity of the membrane starts to resume its normal pattern, and a change can be detected in the *contracture* responses of frog slow muscle fibers to K^+ and ACh (Elul *et al.*, 1968). These findings can be explained only by postulating that some neural factor other than impulse-induced activity in the muscle fiber is exerting a trophic action on the muscle.

Denervated muscle fibers will accept innervation from autonomic axons.

The reversal of denervation characteristics described in the previous section depends only upon the reappearance of a functioning neuromuscular junction and is not affected by the origin of the motor axon. Suppose that, instead of a motor axon, a *cholinergic* autonomic axon is presented to a denervated skeletal muscle fiber. Would the muscle fiber be able to accept the new axon because, like the original motor nerve, the autonomic axom is cholinergic—or would its autonomic origin in some way preclude it from establishing a functional synapse? In an ingenious experiment, Landmesser (1971, 1972) was able to introduce the (autonomic) preganglionic fibers of the frog vagus nerve to a denervated sartorius muscle transplant. She found that the vagal fibers can make functional synaptic connections and usually do so at the site of the original end-plates. The vagal neurons are able to maintain the normal structure and electrical properties of the sartorius fibers but cannot induce the formation of *cholinesterase* at the neuromuscular junction. Because of the discrepancy between the cholinesterase and other properties of the reinnervated muscle, more than one nerve trophic factor must be involved (see chapter 18).

Motoneurons can convert muscle fibers from one type to another.

For a more detailed analysis of the effects of cross-innervation, it is necessary to return to studies of mammalian skeletal muscles. What happens if muscle fibers capture axons from motoneurons that normally supply a different type of fiber (e.g., type I as opposed to type II)? Initially it was assumed that a reinnervated muscle fiber would regain all its old characteristics (color, enzyme profile, twitch speed, etc.). In an elegant experiment, which is described in chapter 18, Buller *et al.* (1960b) showed that this supposition was incorrect, and that motoneurons were capable of transforming the contractile properties of the muscle fibers. Other pioneering experiments, performed a few years later, demonstrated that the histochemical staining properties and chemical compositions of the muscle fibers are also modified by cross-innervation (Drahota & Gutmann, 1963; Dubowitz, 1967; Romanul & Van der Meulen, 1966). In order to bring about these changes, the motoneuron evidently regulates the expression of genes in the myonuclei that code for those proteins (e.g., *myosin heavy chains*) characteristic of the different fiber types. The nature of this regulation is discussed in chapter 18.

Interestingly, there is some evidence that muscle fibers can change the properties of the innervating motoneurons, emphasizing that communication between nerves and muscles is in both directions. Foehring and colleagues (1987), for example, noted that when cat soleus was reinnervated by the lateral gastrocnemius nerve, more motoneurons showed tendencies to become "slow-like" than when the same nerve reinnervated its own muscle. In other words, the soleus muscle had an influence on its innervating motoneurons; however, the effect was not seen when the soleus nerve reinnervated the lateral gastrocnemius. These results have raised the suggestion that muscle fibers have a significant influence on the properties of their innervating motoneurons through a retrograde influence of some kind, and that this influence is stronger for slow muscle fibers than for fast muscle fibers (Mendell *et al.*, 1994).

Motor Unit Properties Following Reinnervation

It will be recalled that in normal muscle the fibers of the different motor units overlap (see chapter 12). Further, because several types of muscle fiber can be distinguished histochemically, the normal appearance of a stained cross section of muscle is a mosaic (figure 17.7A). Once reinnervation

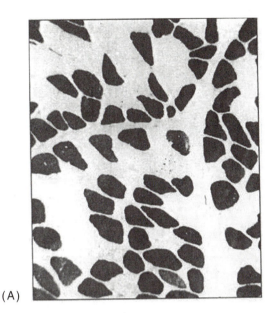 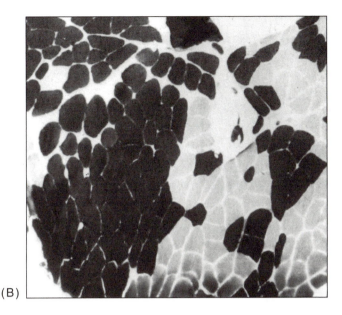

Figure 17.7 (A) Normal mosaic pattern of muscle fiber types in a human quadriceps muscle, stained for *myosin ATPase* activity at pH 10.0. The type II fibers are dark, but the intervening type I fibers are unstained. (B) Fiber type grouping in a patient with *spinal muscular atrophy;* the staining method was the same as in A.

Courtesy of Dr. John Maguire.

Sizes of new units are proportional to the numbers of denervated fibers available.

The enlarged motor units develop more tension.

begins in a partially denervated muscle, many neighboring fibers may come to be innervated by the same axon. Since the motoneuron controls the biochemical composition of the muscle fibers, all the fibers in the same unit will exhibit similar histochemical staining. Thus, instead of the normal mosaic appearance of the muscle, large groups of fibers will be found with similar staining characteristics. This fiber type grouping, when seen in a muscle biopsy, is one of the most useful signs of previous denervation followed by collateral reinnervation (figure 17.7B).

A leg muscle such as the rat tibialis anterior muscle can be partially denervated by sectioning of one of the ventral roots. Following reinnervation by fibers of the remaining root, it is found that, on average, the new units are four times larger (Kugelberg *et al.,* 1970). Since the enlarged motor unit boundaries correspond to those of the muscle fiber *fascicles,* it seems probable that *connective tissue* septae within the muscle belly impede the growth of the axonal sprouts. One can find more evidence of the power of collateral reinnervation in patients with motoneuron disease and other chronic denervating disorders by taking advantage of the fact that the mean sizes of the motor unit potentials are proportional to the numbers of component muscle fibers. In such patients a hyperbolic relationship between the sizes and numbers of surviving motor units is observed; that is, the sizes of the new units are always proportional to the numbers of denervated fibers available for adoption (see figure 17.8; McComas *et al.,* 1971b).

When the maximal twitch tensions of partially denervated muscles are measured, it is found that the values remain within the normal range until less than 20% of the usual population of units is left (see figure 17.9). In the most severely denervated muscles, the surviving units generate, on average, seven times the normal tension (Gordon *et al.,* 1993; McComas *et al.,* 1971b). Among these enlarged motor units may be found others that are unusually small and may be those that, having previously sprouted, are now beginning to fail (Dengler *et al.,* 1989).

Abnormally large motor unit tensions can also be demonstrated in animal muscles following reinnervation. Values can be 16 to 19 times normal (Luff *et al.,* 1988). In such muscles, as in those of patients with chronic denervation, there may also be some very small units (Bagust & Lewis, 1974). Whether a partially denervated muscle is as efficient in continuous work, as opposed to single twitches, is less clear, since Milner-Brown *et al.* (1974a) found no evidence of increased motor unit

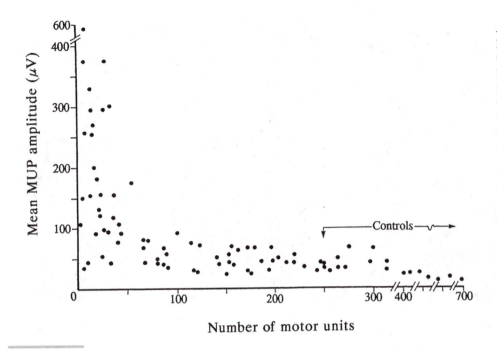

Figure 17.8 Sizes of motor units in partially denervated hypothenar muscles of ALS patients. The sizes have been expressed in terms of the mean potential amplitudes of the surviving motor units, since the greater the number of fibers in a motor unit, the larger will be the potential of the motor unit. Note the hyperbolic relationship between the numbers and sizes of units.

From "The extent and time course of motoneuron involvement in amyotrophic lateral sclerosis," by M. Dantes and A. J. McComas, 1991, *Muscle and nerve*, 14, p. 418. Copyright 1991 by John Wiley & Sons, Inc. Reprinted with permission of John Wiley & Sons, Inc.

After recovery from nerve division, the normal recruitment pattern is initially lost but may be restored.

forces when examining patients with peripheral nerve injuries. One factor that may have contributed to Milner-Brown *et al.*'s negative findings is that in some partially denervated muscles there may be impaired neuromuscular function. This abnormality may result from the inability of a parent motoneuron to maintain its increased population of neuromuscular junctions satisfactorily. Also, impulse conduction in the slender new preterminal axons also may be unusually hazardous.

In reinnervated or partially denervated muscles, the order of motor unit *recruitment* during voluntary contractions is of special interest. In a normal muscle, progressively larger motor units are called into activity with increasing *effort* (*size principle,* see chapter 13), but this pattern may be lost in muscles reinnervated after total nerve section (Milner-Brown *et al.,* 1974b). In the disease *amyotrophic lateral sclerosis (ALS)*, however, in which the muscles undergo progressive denervation and collateral reinnervation takes place, the orderly recruitment pattern is maintained. The explanation is that in nerve regeneration following axon division, the occurrence and extent of reinnervation will depend largely on such simple mechanical factors as the accessibility of the Schwann cells in the peripheral stump to the newly formed axon sprouts. For example, in patients in whom the ulnar nerve has been accidentally divided at the wrist and then rejoined, nerve fibers that had previously supplied the first dorsal interosseous muscle may now go to the adductor pollicis or to the lumbricals

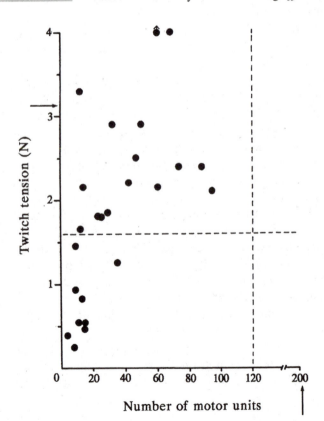

Figure 17.9 Maximum isometric twitch tensions of 25 extensor hallucis brevis muscles with varying degrees of denervation. The horizontal and vertical dashed lines indicate the lower limits of the respective normal ranges, and the arrows identify the control mean values. Note that muscles with severe denervation may still develop normal force.

Reprinted, by permission, from A.J. McComas et al., 1971, "Functional compensation in partially denervated muscles," *Journal of Neurology, Neurosurgery, and Psychiatry* 34: 456.

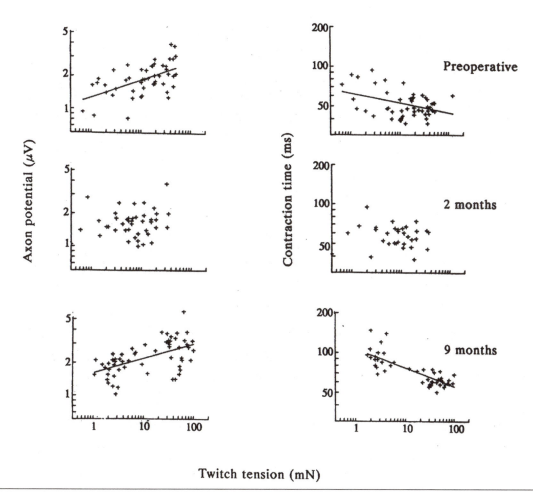

Twitch tension (mN)

Figure 17.10   Recovery of normal motor unit properties in cat triceps surae muscles after nerve section and suture. The amplitude of the axon potential (left) is an indirect measure of axon diameter and hence the size of the motoneuron cell body. Note that the normal correlations among axon potential, *contraction time,* and twitch tension, which are seen in the preoperative *(control)* muscles, are missing at 2 months of recovery but have reappeared at 9 months.

Adapted, by permission, from T. Gordon and R.B. Stein, 1982, "Reorganization of motor-unit properties in reinnervated muscles of the cat," *Journal Neurophysiology* 48: 1181.

and other interossei (Thomas *et al.,* 1987b). However, the situation is not simple because, in the case of median nerve regeneration following section at the wrist, the normal relationship between motor unit size and recruitment threshold is eventually restored—despite many axons having been misdirected from other muscles.

In animal muscles, too, there is evidence that, under favorable circumstances, restoration of motor unit properties may take place after random reinnervation. For example, 2 months after nerve section and resuture, cat triceps surae motor units have mixed populations of muscle fibers. By 9 months, however, the different types of motor unit can be easily recognized (Gordon & Stein, 1982). Thus, the largest motoneurons (as gauged by their extracellular potentials) develop the largest forces and have the fastest twitches, as would be expected of type *FF* motor units (figure 17.10). Such studies even suggest that muscle fibers may be redistributed among the motor units—like the intrafusal muscle fibers after reinnervation (Brown & Butler, 1974).

Applied Physiology

In this section we will briefly consider the relevance of collateral reinnervation to neuromuscular disorders and its implications for the diagnosis and prognosis of patients.

• **Collateral reinnervation is prevalent in neuromuscular disorders.** The importance of collateral reinnervation lies in the fact that it is not only a very powerful compensatory mechanism but also an extremely common phenomenon after nerve injuries and in neuromuscular diseases. Evidence of its presence is found in almost every case of partial denervation, irrespective of etiology; thus, it is to be seen in patients with ALS (see chapter 22), spinal muscular atrophy (SMA) (see chapter 6), and old poliomyelitis, as well as in cases of trauma or disease of peripheral nerves. It must not be imagined that axonal regeneration and collateral reinnervation are mutually exclusive recovery processes. On the contrary, they occur simultaneously in any reversible neuromuscular disorder in which muscle denervation has been incomplete. Thus, while damaged axons are growing back toward the muscle, intact healthy axons are already sprouting and undertaking collateral reinnervation. This situation is one that will occur, for example, after nerve damage or in the *Guillain-Barré syndrome* (see chapter 16). In a chronic disorder, such as ALS or SMA, collateral reinnervation may be so effective that it may mask the presence of the disease until it is far advanced. Not only is there no wasting of the muscles, but strength may be preserved until fewer than 20% of the axons remain (see p. 267).

• **Reinnervated motor units are especially susceptible to the effects of aging.** An interesting question is whether reinnervated motor units are as viable as normal ones. The answer appears to be no, at least in those instances in which the motor units are much larger than normal. This conclusion was reached by Pachter and Eberstein (1992), who examined rat plantaris muscles 12 months after severe partial denervation. At this time many signs of recent denervation are present; these include degenerating end-plates, muscle fibers without terminal arborization, angulated muscle fibers, and grouped muscle fiber *atrophy*. In those fibers that still retain their innervation, the numbers of terminal nerve branches are reduced at the end-plates. These important findings could well explain why, for example, human patients may develop *weakness* and wasting of skeletal muscles 15 to 20 years after recovery from poliomyelitis (Cashman *et al.,* 1987).

We have now completed our study of the capabilities for nerve regeneration in the neuromuscular system. The next topic to be covered is *trophism* in this system. This is a topic that has been touched on in the present chapter, but is given in-depth attention in the next chapter.

18

Trophism

In this chapter we will try to understand how a *motoneuron* and the colony of muscle fibers that it innervates exert a sustaining action on each other. It will be seen that, in the case of the muscle fibers, the effects of the motoneuron are mediated in part by *impulse*-induced stretching of fibers. The remainder of the effects, and those achieved by the muscle on the motoneurons, are brought about by chemical messengers.

Motoneuron Effects on Muscle

When the nerve supply to a muscle is interrupted by section, crush, or disease, a number of important changes take place in the muscle fibers (see chapter 16). The fibers become thinner and develop smaller tetanic tensions, while their surface membranes become more widely sensitive to *ACh* and start discharging *action potentials* spontaneously *(fibrillations)*. These *denervation* phenomena can be viewed in a different way, however; for they indicate that, under normal circumstances, the intact motoneurons prevent such changes from taking place. That is, the motoneurons, in addition to exciting muscle fibers and causing them to contract, have a sustaining, or myotrophic, action that regulates many of the structural, biochemical, and physiological features of the muscle fibers.

Motoneurons exert a sustaining influence on muscle fibers.

Further evidence of the existence of *trophism* comes from cross-innervation experiments of the kind pioneered by Buller *et al.* (1960b). These authors had intended to examine the possible role of the muscle in determining the shape of the motonueron action potential. Their experiments involved selecting two muscles in the cat hindlimb with contrasting contractile properties, cutting their respective nerves, and then cross-connecting the nerves (figure 18.1). Thus, the flexor digitorum longus (FDL) was connected to the proximal stump of the nerve that had previously supplied the soleus (SOL) muscle, while the latter muscle was connected to the spinal cord through the nerve to FDL. After several months had elapsed, to allow the nerve fibers to regenerate, the authors stimulated the reconstructed nerves and found, to their surprise, that the FDL, which normally has a fast *twitch,* now had a slow twitch. Conversely the SOL, which is the slowest-contracting muscle in the hindlimb, now had a much faster twitch (figure 18.1). Clearly, the contractile properties of the muscle fibers had been transformed by the new nerve supply. This classic cross-innervation experiment has been repeated many times, most often in the cat and rat, and the same results have always been obtained. Subsequent histochemical analyses of cross-innervated muscles, first reported by Dubowitz (1967) and by Romanul and Van Der Meulen (1967), also indicated that nerves determine the phenotypic characteristics of postnatal muscle fibers. The extent of conversion of muscles (fast to slow or slow to fast) by cross-innervation is not complete, however, and shows considerable variation among species (Edgerton *et al.*, 1996).

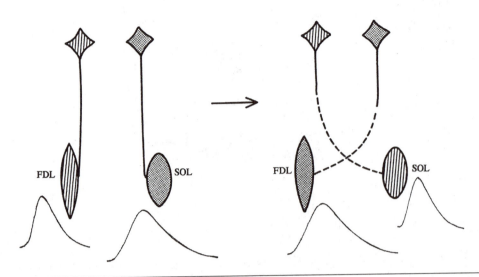

Figure 18.1   Cross-innervation experiment of Buller *et al.* (1960b) in which the twitch durations of the cat soleus *(SOL)* and flexor digitorum longus *(FDL)* muscles were largely reversed. See text.

A sustaining influence is largely brought about by impulse patterns.

Role of Impulse Patterns

How is the *trophic* action of the motoneurons exerted? An obvious clue comes from the results of disuse. If a muscle is no longer used fully, either as the intended result of an experiment or, in human subjects, because of an injury to a limb, the muscles may undergo striking *atrophy* and become weaker (see chapter 19). On the basis of such observations, it is safe to conclude that, in some way, the nerve impulse patterns must normally be responsible for maintaining muscle bulk and strength. In a second type of experiment, Lømo and Rosenthal (1972) showed that impulse activity could prevent many of the features of denervation from taking place. These authors cut the nerve to the soleus and extensor digitorum longus (EDL) muscles of the rat and then stimulated the muscle fibers directly for 5 days. They found that they were able to reduce the muscle atrophy and loss of contractile force that would otherwise have occurred. In later experiments, Hennig and Lømo (1987) showed that direct electrical stimulation of denervated muscle improved the tetanic tension in the soleus 37-fold and that of the fast-twitch EDL 8-fold. In addition, direct electrical stimulation prevented the spread of ACh sensitivity along the muscle fibers—another consequence of denervation (Lømo & Rosenthal, 1972).

A third type of observation, also pointing to the importance of impulse activity, comes from experiments that are the logical extension of the cross-innervation studies already described, and involve changing the nerve impulse patterns received by the muscle fibers. The design of these experiments is influenced by the fact that when an animal is standing quietly, soleus motoneurons tend to discharge steadily at low frequencies, thereby keeping the soleus muscle under sufficient tension to prevent the ankle from flexing under the weight of the animal. The motoneuron discharge to a fast-twitch muscle is quite different and consists of high-frequency bursts of impulses as the animal moves forward (Salmons & Vrbová, 1969).

In the first such investigation, Salmons and Vrbová fastened stimulating electrodes around the nerve to the tibialis anterior (TA) and EDL muscles in the rabbit and cat and attached the leads to a stimulator implanted beneath the skin of the abdomen (figure 18.2). The stimulator was programmed to deliver an intermittent stream of shocks at 10 Hz. At the end of 6 weeks, Salmons and Vrbová found that the twitches of the TA and EDL had become considerably slower. Confirmation of the importance of the impulse pattern was obtained by Al-Amood *et al.* (1973), who attached a stimulator to the spinous process of a cat vertebra and used it to excite *ventral root* fibers supplying the FDL muscle. The shocks were delivered at a frequency of 10 Hz for 8 weeks; and, once again, the stimulated muscle was found to have become slower.

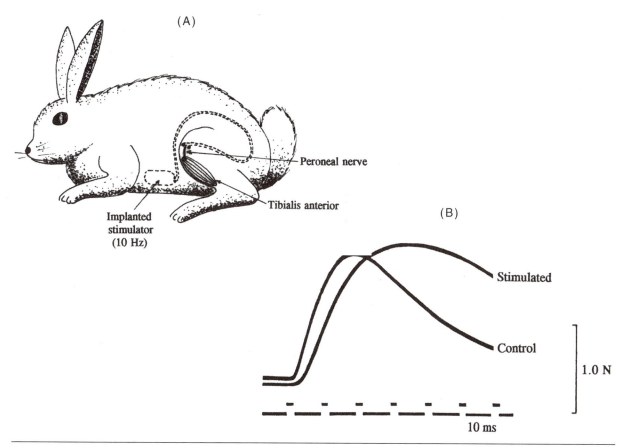

Figure 18.2 (A) Experimental arrangement for chronic stimulation of rabbit tibialis anterior muscle. (B) Examples of twitches from a muscle that had been stimulated at 10 Hz for 6 weeks, and from the control, unstimulated muscle in the opposite leg; note the much slower twitch in the stimulated muscle.

Adapted, by permission, from S. Salmons and G. Vrbová, 1969, "The influence of activity on some contractile characteristics of mammalian fast and slow muscles," *Journal of Physiology* 201: 542.

Impulse patterns determine the speed of contraction and fatigability.

Effects on Contractile Speed and Fatigability

In the chronic stimulation experiments considered in the previous section, was it the frequency of the impulses or simply the extra number of impulses that was responsible for transforming a muscle from fast-twitch to slow-twitch? This issue has been thoroughly explored by Lømo and colleagues using different frequencies and periods of stimulation applied to denervated rat hindlimb muscles. The experiments of Gorza *et al.* (1988) included the use of bursts of high-frequency stimuli (60 shocks at 100 Hz every 60 s) and of short or long epochs of low-frequency stimuli (10 Hz). The authors concluded that

- low-frequency (10-15 Hz) stimulation induces slow-twitch characteristics and
- high-frequency (100 Hz) stimulation induces fast-twitch characteristics.

In a separate study, Westgaard & Lømo (1988) demonstrated that

- the total number of impulses determines the *fatigability* of the fibers.

Thus, the greater the number of stimuli delivered during the day, regardless of frequency, the more resistant the fibers become to fatigue (figure 18.3).

It is important to note, however, that the ability of stimulus frequency to convert muscle fibers from one type to another is not always as clear-cut as in the case of the rat soleus; both the species and the previous phenotypic expression of the muscle will influence the responsiveness of the fibers to changes in excitation pattern (for discussion, see Gorza *et al.,* 1988).

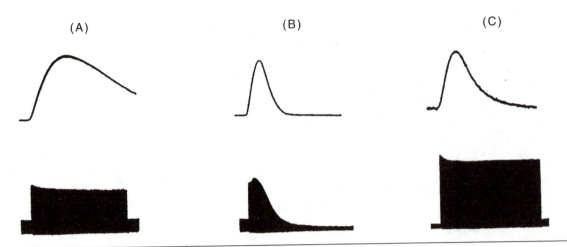

Figure 18.3   Contractile properties of three surgically denervated soleus muscles following 2 months of direct stimulation. The muscles received either many pulses per day at 15 Hz (muscle A), few pulses per day at 100 Hz (muscle B), or many pulses per day at 100 Hz (muscle C). The top row of recordings shows the respective twitches, and the lower row the responses to 2 min of intermittent *tetanic stimulation* at 77 Hz. The sweeps are slower and the force amplification is less in the lower row of records than in the upper row. Note that each muscle differs from the other two; for example, although B and C both have fast twitches, only B fatigues rapidly.

Adapted, by permission, from R.H. Westgaard and T. Lomo, 1988, "Control of contractile properties within adaptive ranges by patterns of impulse activity in the rat," *Journal of Neuroscience* 8: 4417.

Effects on Muscle Fiber Histochemical and Immunocytochemical Properties

Impulse patterns also determine histo-chemical and immuno-cytochemical features of the muscle fibers.

The speed of contraction and the susceptibility to fatigue are not the only properties of the muscle fiber that are controlled by impulse patterns. For example, Pette *et al.* (1975) have shown that the histochemical staining properties are also affected, in that chronic stimulation of rabbit TA muscle converts the fibers from type II to type I. Similarly, high-frequency stimulation will change type I fibers to type IIC (Gundersen *et al.*, 1988; note that IIC is actually hybrid I, IIA) The alterations in staining and in the speed of contraction are due to the substitution of a different *isoform* of the *myosin heavy chain (MHC)*. Thus, in fibers acquiring the fast isoform, the cycling of the myosin *cross-bridges* with the *actin* filaments becomes quicker, because these heavy chains can split *ATP* faster (see chapter 11). In the experiments of Lømo and colleagues (Gorza *et al.*, 1988), immuno-cytochemical techniques demonstrated that the fast MHC isoform appears in slow-twitch fibers by the seventh day of appropriate stimulation and increases thereafter, even though the fibers still retain small amounts of slow isoforms. Similar results have been obtained by other workers studying transformations between slow- and fast-twitch muscles (see chapter 19). Not only the isoforms of the MHCs, but also those of the light chains and *troponins,* and the isoforms of the ATPase of the *sarcoplasmic reticulum (SERCA*1 of type II fibers and SERCA2 of type I fibers), are altered by patterned stimulation. In addition, the fibers that acquire slow-twitch characteristics develop greater mitochondrial volumes and increased capillary densities. All of these changes make the slow-twitch fibers better equipped to carry out prolonged work under aerobic conditions.

Sutherland and colleagues (1998) have demonstrated quite elegantly the dose-response nature of the *adaptation* of muscles to chronic electrical stimulation. They stimulated rabbit muscles at frequencies of 2.5, 5, and 10 Hz for 24 hr per day for 10 months. Their results showed that it was possible to cause significant increases in mitochondrial enzymes with very little change in myosin heavy or light chains. They also showed that, at 5 Hz, some muscles showed dramatic changes in contractile protein isoforms while others did not—thus, there is a different threshold of *activation* that is required for adaptation of mitochondrial energy systems and contractile proteins.

Role of Muscle Stretch

The trophic effects of impulse patterns are mediated by stretch of the muscle fiber.

The appearance of new isoforms in a muscle fiber following patterned stimulation indicates that protein synthesis may have been modified at the level of gene *transcription.* Until recently it was difficult to imagine what sort of coupling might be involved between the electrical and genetic events. It now seems probable that the link is stretching of the muscle fiber membrane (or of the extracellular matrix) and that, through the release of soluble factors, several *second messenger* systems are activated. In addition, Ca^{2+}, released into the *cytosol* from the sarcoplasmic reticulum, may also act as a second messenger. The second messengers then activate immediate-early genes in the *myonuclei;* and the latter, in turn, permit transcription of the genes coding for the new isoforms of the various contractile proteins to take place. This subject is discussed more fully in chapter 20.

Some trophic effects are not dependent on impulse patterns.

Trophic Effects by Factors Other Than Impulse Patterns

It would be surprising if all the trophic effects of nerve on muscle could be accounted for solely on the basis of impulse patterns, since some *motor units* have such high thresholds that they would rarely be excited during everyday life. This situation applies particularly to muscles of the human upper arm, for many individuals may pass an entire day without having to lift the arm above shoulder level, let alone make a maximal contraction. Even Lømo, the strongest protagonist of the trophic influence of impulse activity, has drawn attention to the small amounts of activity exhibited by some motor units; and it is possible that there may have been other units, with still higher thresholds, that would not have been detected in his experiments (Hennig & Lømo, 1985).

There is further reason why impulse patterns are unlikely to be the sole explanation for all the trophic effects of motoneurons on muscle fibers. In experiments on animals, for example, the consequences of disuse on various muscle properties are usually less than those of surgical denervation (see chapters 16 and 19). In addition to these general observations, five other specific findings point to the presence of trophic influences not mediated by impulses:

• Block of *axoplasmic transport.* Using an implanted cuff technique, it is possible to block axoplasmic transport with *colchicine* without interrupting impulse propagation (Albuquerque *et al.,* 1971; Cangiano, 1973; Hofmann & Thesleff, 1972). The deprived muscle fibers will then acquire features of denervation—a fall in resting *membrane potential,* a spread of ACh sensitivity, and development of tetrodotoxin *(TTX)-resistant sodium channels.*

• Effect of nerve stump length. The onset of denervation features can be delayed if a nerve is cut proximally so as to leave an appreciable stump attached to the muscle (figure 18.4). This interesting phenomenon was first shown for fibrillation activity (Luco & Eyzaguirre, 1955) and subsequently for the resting membrane potential, spread of *AChRs,* and development of TTX-resistant Na^+ channels (Albuquerque *et al.,* 1971; Harris & Thesleff, 1972). The importance of this observation is that it cannot be explained by an absence of impulse activity, because the latter will occur as soon as the nerve is cut and will be independent of the site of section. An explanation can be provided in terms of axoplasmic transport, however, by suggesting that, following nerve section, the muscle fiber is able to receive the trophic material remaining in the distal nerve stump. A longer stump will contain a greater amount of material, and the onset of denervation phenomena will be correspondingly delayed.

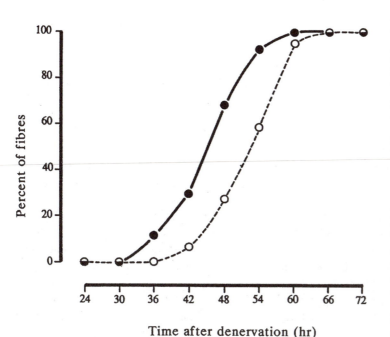

Time after denervation (hr)

Figure 18.4 The development of resistance to TTX in mammalian muscle fibers that have been denervated by nerve section either close to the muscle (●) or else well away from the muscle (○). The ordinate shows the percentage of TTX-resistant fibers.

Adapted, by permission, from J.B. Harris and S.Thesleff, 1972, "Nerve stump length and membrane changes in denervated skeletal muscle," *Nature New Biology* 236: 60.

• Effects of cross-innervation in the absence of activity. Roy and colleagues (1996) cross-innervated the cat soleus (slow muscle) with the nerve normally innervating the fast flexor hallucis longus muscle, in cats in which the motor nerve was silent as a result of the surgical procedure of *"spinal cord isolation."* In the latter model, the spinal cord above and below the motoneurons innervating the hindlimbs was severed and all dorsal roots were cut, thus rendering the spinal cord in these segments electrically "silent." Eight months later, the inactivity of the soleus resulted in the appearance of the contractile and biochemical properties of a fast-twitch muscle. Interestingly, this "speeding up" was more pronounced in the soleus innervated by the fast nerve. Thus, the motor nerve, even when silent, has a trophic effect on muscle properties.

• Partial denervation of single fibers. The muscle fibers of the frog sartorius muscle receive a double innervation from the obturator nerve. If impulse patterns were the sole trophic mechanism, division of one nerve branch should not affect the activity of the muscle appreciably, because the fibers would still be excited through the remaining nerve branch. In such fibers, however, there is a definite increase in ACh sensitivity around the denervated *end-plates* (Miledi, 1960b), suggesting that some factor other than muscle fiber activity is normally controlling the distribution of AChRs.

• Fibrillations and positive sharp waves. When the nerve supplying a muscle is severed, there is an interval of several days, and then the denervated muscle fibers begin to discharge spontaneously. As recorded with a needle electrode, this spontaneous activity appears as either fibrillations or positive sharp waves (see chapter 16). However, if a human muscle is subjected to disuse rather than denervation, the spontaneous activity does not develop (an example of this absence is given in "A Clinical Example of Neurapraxia" in chapter 19).

The non-impulse trophic effects are brought about by myotrophic factors.

What is the nature of the trophic mechanism that is independent of impulse patterns? The general consensus is that it consists of different types of trophic molecules (trophins), which are synthesized in the motoneuron *soma* and transported along the *axon* to its terminal branches. In some instances, one can establish the transport velocity by observing the effects of cutting the nerve at different levels. For example, the fast fraction of axoplasmic transport appears to contain those factors responsible for maintaining the motor nerve terminals and for preventing fibrillations and the synthesis of TTX-resistant Na^+ channels in the muscle fibers. Until recently, one of the arguments against the existence of myotrophic molecules was that none had been isolated and characterized. To be accepted as a myotrophic substance, Lømo and Gundersen (1988) insist that the following criteria be satisfied, namely, that the putative agent must

- be released from motor nerve endings;
- not be supplied by blood or nonneuronal cells; and,
- through its absence, result in an abnormality of skeletal muscle fibers that persists despite electrical stimulation.

While the first two criteria are reasonable, the third requires qualification, for the electrical stimulation should be sufficient only to match the naturally occurring excitation of the innervated fibers. If some motor units hardly ever fire, the results of bombarding their denervated fibers with stimuli may be misleading.

Application of the second criterion would exclude *sciatin.* Isolated from sciatic nerve, sciatin was found to promote maturation and survival of muscle cells *in vitro* and to induce the synthesis of AChRs and *acetylcholinesterase (AChE;* Markelonis & Tae Hwan, 1979). Subsequent studies showed that sciatin was identical to *transferrin,* a plasma protein that regulates iron transport into cells. Certain other substances, however, appear as stronger candidates for myotrophic mediators. Two of these are *agrin* and *ARIA,* both peptides that have been found to regulate the aggregation of AChRs and AChE at the neuromuscular junction (Fischbach *et al.,* 1989; Godfrey *et al.,* 1984). Another candidate is *calcitonin gene-related peptide (CGRP),* which also acts at the neuromuscular junction and controls the numbers of AChRs (Fontaine *et al.,* 1986). Thesleff *et al.* (1990) have speculated that ACh may have a facilitatory role by assisting in the uptake of neuropeptides by the muscle fiber. With more sensitive collection and purification procedures, it is possible, if not probable, that other trophic proteins and peptides will be identified; possibly mRNA is itself a trophic molecule.

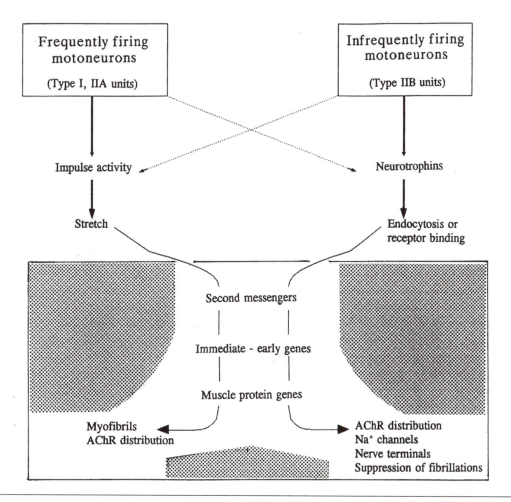

Figure 18.5 Nerve-mediated trophic mechanisms for different muscle fiber types. Type I and IIA fibers are heavily dependent on impulse activity, and the resulting stretch is especially important for the synthesis of myofibrillar proteins. In contrast, type IIB fibers probably rely more on myotrophins, and these are more effective in suppressing fibrillations, maintaining the motor nerve terminal, and controlling Na⁺ channel synthesis. Both stretch and myotrophins regulate AChR distribution.

How do the trophic agents enter the muscle fiber? Libelius and Tågerud (1984) have reported that, following denervation, muscle fibers show increased endocytotic activity in the vicinity of the end-plate, and it is possible that this is the means by which access to the fiber is obtained. Alternatively, other trophic agents might combine with receptors on the surface of the muscle fiber membrane. Regardless of the means of access, it is probable that the *myotrophins* use second messenger systems to activate immediate-early genes in the myonuclei, thereby controlling the expression of muscle protein genes (see figure 18.5).

In assessing the relative importance of myotrophic molecules and impulse activity, it is important to bear in mind the possibility that the two may be related, at least at some junctions; thus Musick and Hubbard (1972) found that the release of substances from the nerve endings was enhanced by electrical stimulation of the nerve.

Muscle Effects on Motoneurons

Can muscle influence nerve? In view of the extensive use of feedback systems by the body, the expected answer would be yes. Indeed, it would be unnecessarily wasteful if, following a peripheral nerve injury, a motoneuron continued to manufacture trophic material for a colony of muscle fibers to which it was no longer connected. Nor could a motoneuron gear its metabolic activity to a higher level, in order to support an enlarged muscle fiber colony, in the absence of a message from the periphery informing the cell that *collateral reinnervation* was taking place (see chapter 17). The very existence of *microtubules* within the axon suggests that there might be a *retrograde,* as well as an *anterograde,* transport of information between the motoneuron and the muscle fiber. Thus, if only anterograde signals were required, the cell could achieve this simply by bulk movement of axoplasm, though this would be a rather slow process. The need for two-way traffic virtually rules out the practical use of such bulk movement of fluid (see chapter 8).

The Cell Body Response to Axotomy

Axonal damage alters the appearance of the motoneuron cell body.

The possible existence of two-way traffic in trophic effects between the spinal cord and the periphery has interested a number of investigators, among whom one of the earliest was Nissl (1892). Nissl avulsed facial nerves in rabbits and studied the changes visible in neurons of the facial *nucleus* under the light microscope (figure 18.6). He noticed especially striking alterations in the granular material situated in the region of the *axon hillock*. This material, since called Nissl substance, is now known to consist of *RNA*. Following *axotomy,* the granules become smaller and are dispersed to the periphery of the motoneuron soma *(chromatolysis)*. At the same time, the neuron becomes swollen and rounded in outline. The nucleus is usually pushed to the edge of the soma opposite the axon hillock; within the nucleus the *nucleolus* enlarges. These changes are complete at 7 to 10 days and are sufficiently prominent for the affected neurons to be readily distinguished from the other cells.

Subsequent workers have confirmed Nissl's observations and have described additional features. For example, the glial cells in the vicinity of the affected motoneurons swell (Watson, 1972), and *lysosomes* may appear in the neuronal *cytoplasm*. In addition, there is retraction of the *dendrites* and loss of the synaptic connections that they normally make with other neurons; these changes probably underlie the longer latencies and increased temporal dispersion of reflex discharges.

After a while, however, the motoneuron dendrites extend and even develop *growth cones,* as if they were searching for *afferent* nerve fibers with which to make synaptic connections (Rose

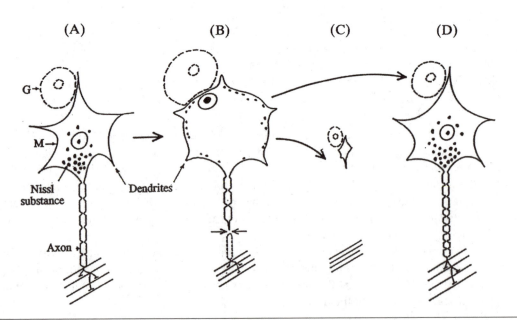

Figure 18.6   Changes in motoneuron *(M)* following interruption of its axon. (A) A normal motoneuron innervating several muscle fibers. (B) Representation of the situation about 7 days after axotomy (at arrows); note the swollen motoneuron soma with its displaced nucleus and enlarged nucleolus. The Nissl substance is dispersed, and the *dendrites* have retracted. By this time the distal axon stump is degenerating, and the muscle fibers have started to atrophy. The glial cell (G) close to the motoneuron is also enlarged. (D) If *reinnervation* of muscle is successful, normal neuronal architecture is restored, though the axon now has shorter internodal segments. (C) Failure to innervate is associated with progressive atrophy of the motoneuron and, in some instances, leads to its eventual disappearance.

Reprinted, by permisison, from A.J. McComas, 1977, *Neuromuscular Function and Disorders* (London: Butterworth Publishing Co.), 90.

& Tourond, 1993). Microelectrode recordings from axotomized motoneurons have revealed little change in resting potential, although the overshoot of the action potential becomes larger and there is a considerable reduction in impulse conduction velocity along the axon stump (Eccles *et al.,* 1958b; Kuno *et al.,* 1974a). Normally the motoneurons supplying the fast- and slow-twitch muscles can be differentiated on the basis of the *hyperpolarization* that follows the action potential; after axotomy this difference becomes less marked, but it is restored as soon as reinnervation takes place (Kuno *et al.,* 1974b; Titmus & Faber, 1990).

> **RNA increases in the motoneuron cell body after axotomy.**

In a series of elegant experiments on the hypoglossal nerve and tongue muscles of the rat, Watson analyzed the nature and course of some of the biochemical events taking place within the motoneuron following axotomy. He showed that the first detectable change within the motoneuron is an increase in ribosomal RNA within the nucleolus (Watson, 1968b). It is probable that this RNA is required for the synthesis of proteins as part of the forthcoming repair process in the axon. The latency of this change in RNA depends on the site of nerve injury, being least following a lesion close to the hypoglossal nucleus. In another type of experiment, Watson (1969) injected *botulinum toxin,* a substance that is known to interfere with the release of ACh (and probably other trophic factors) at the end-plate. Since a rise in nucleolar RNA could be detected in these animals also, the phenomenon cannot have been due to axonal injury but rather must have resulted from an interruption or change in the signals from the muscle fiber. The absence of activity of the muscle, due to block of neuromuscular transmission, could alter the signals transmitted from the muscle fibers.

Gutmann (1971) has pointed out that the vigorous metabolic response of the motoneuron to disconnection from its target organ hardly deserves the term "retrograde degeneration" and that Nissl's own description, "primare Zellreizung" (primary cell excitation), is more appropriate. Subsequent studies (Hoffman & Lasek, 1980) have shown that the increased messenger RNA codes for proteins, such as actin and *tubulin,* that are associated with slow axonal transport, as well as for *GAP-43* (B 50) and other novel proteins found in the growth cones of the regenerating axons (for review, see Skene, 1989). In contrast, levels of messenger RNA for *neurofilament* proteins and for transmitter-synthesizing enzymes decrease after nerve section (Grafstein & McQuarrie, 1978).

In an extension of his earlier work, Watson (1970) was able to show that if a motoneuron is allowed to reinnervate muscle fibers through axonal sprouting, a second rise in its metabolic activity takes place. During this phase, the dendrites return to their original lengths and acquire fresh synaptic connections. Successful reinnervation enables the motoneuron to resume a healthy appearance in other respects—the cell size becomes normal, the nucleus regains its central position, and the Nissl substance re-forms at the axon hillock; at the same time the regenerating axon enlarges its diameter (Sanders & Young, 1946). In contrast, if reinnervation of muscle is prevented, many motoneurons will disintegrate or else gradually atrophy, particularly in young animals.

Role of Axoplasmic Transport

> **Signals can be sent to the motoneuron cell body from the muscle and motor axon by axoplasmic transport.**

Experimental data have accumulated to show that several different types of substance can be conveyed from muscle to nerve soma by axoplasmic transport (see chapter 8). For example, Watson (1968a) demonstrated that the injection of [^{3}H]-lysine into muscle was followed after a short interval by its appearance and proximally directed movement in motor nerves. Glatt and Honegger (1973) coupled the fluorescent dye, *Evans blue,* with albumin, and within 12 hr of injection of this marker into the rat triceps muscle, motoneurons in the cervical cord could be seen to fluoresce; the length of the neural pathway was 30 mm.

In the study of Kristensson and Olsson (1971), *horseradish peroxidase (HRP)* was injected into the gastrocnemius muscles of mice. It is known that this electron-dense material is taken up by *pinocytosis* at the neuromuscular junction and can then be seen with the electron microscope to be contained in coated vesicles within the axon terminal (Zacks & Saito, 1969). Kristensson and Olsson found that HRP, after being transported in the axons to the soma, was distributed in small cytoplasmic granules around the nucleus within 24 hr of injection.

In conclusion, the demonstration that an axon can take up exogenous proteins and transport them to the neuron has some important implications. It provides a possible mechanism for the trophic influence of muscle on nerve, in the absence of which chromatolysis occurs. The process also sheds light on the way in which certain toxins and viruses may spread from the periphery to the central nervous system.

The Effects of Chronic Stimulation, Muscle Inactivity, and Muscle Stretch

Czeh and colleagues (1978) showed that inactivation of soleus by pharmacological blockade of the passage of nerve impulses in the axon resulted in a speeding up of the muscle contraction and a shortening of the *after-hyperpolarization* of the innervating motoneurons. When they stimulated the muscle during the period of paralysis, they were able to prevent not only the speeding up of the muscle, but also the electrical change in the innervating motoneuron. Stimulating proximal to the blockade, which is ineffective in activating the muscle fibers, did not result in these changes. This is evidence for an effect of muscle activity "state" on the properties of innervating motoneurons. These experiments also demonstrated that the length at which muscle is immobilized has an effect on motoneuron properties (Gallego *et al.*, 1979). A more recent experiment by Munson and colleagues (1997) showed a similar effect, but using the model of chronic low-frequency stimulation of the cat medial gastrocnemius. In this study, when the medial gastrocnemius was almost entirely converted to slow fibers as a result of chronic stimulation, the innervating motoneurons became more like those innervating slow fibers.

Neurotrophins, Neurotrophin Receptors, and Polypeptides

With the recognition that muscle was in some way necessary for the upkeep of motoneurons, several attempts have been made to find the factors (neurotrophins) involved. One successful approach followed the purification and molecular characterization of a trophic factor found in pig brain. This factor, *brain-derived neurotrophic factor (BDNF)*, was known to "save" sensory neurons in embryonic rats from naturally occurring cell death and was shown by Leibrock *et al.* (1989) to be a polypeptide with 119 amino acids (Thoenen, 1991). Other groups then looked for neurotrophins with structures similar to BDNF and found two, termed NT-3 and NT-4 (also known as NT-5 and NT-4/5). All three neurotrophins strongly resemble a factor that had been discovered much earlier, was known to be necessary for the growth and maintenance of sympathetic and sensory neurons, and is produced in large amounts by the salivary glands; this is *nerve growth factor (NGF;* Levi-Montalcini, 1987).

Unlike NGF, however, BDNF and neurotrophins 3 and 4/5 are produced by muscle fibers and can rescue motoneurons from naturally occurring cell death in the embryo and from the cell death that follows axotomy (Henderson *et al.*, 1993; Yan *et al.*, 1992). By analogy with NGF (Meakin & Shooter, 1992), the muscle neurotrophins are thought to combine with receptors on the nerve terminals and then to cluster together. Next, the clusters are taken into the axoplasm within membrane-bound vesicles via *endocytosis*. While in the nerve terminals, the neurotrophin-receptor complexes may exert local effects, such as the potentiation of *synaptic transmission* in developing neuromuscular junctions of the embryo (Lohof *et al.*, 1993). Otherwise, the neurotrophin-receptor complexes are transported along microtubules to the motoneuron cell bodies to act on the genetic machinery in the nucleus.

The receptors for the neurotrophins on the motor nerve terminals have been identified recently and consist of high- and low-affinity types (Meakin & Shooter, 1992). The high-affinity receptors appear to be the ones necessary for the biological actions of the neurotrophins and are *tryosine kinases* (Trks), produced by Trk proto-oncogenes. Brain-derived neurotrophic factor and NT-4/5 are bound by TrkB, while NT-3 combines with TrkC; the binding results in *phosphorylation* of the tyrosine kinase (figure 18.7).

It is reasonable to suggest that muscle-produced BDNF, NT-3, and NT-4/5, and their motoneuron-based receptors, may constitute an important multidimensional signaling component that keeps the innervating motoneurons informed not only of the state of innervation/denervation of the muscle, but also of its activity state (Skup *et al.*, 2002; Gomez-Pinilla *et al.*, 2001, 2002).

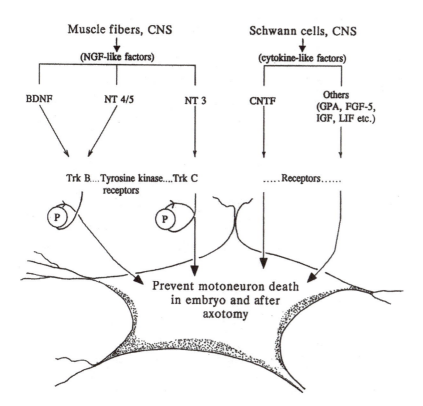

Figure 18.7 Neurotrophins that act on motoneurons. The neurotrophins fall into two classes, those resembling NGF and those resembling *cytokines*. BDNF = brain-derived neurotrophic factor; CNTF = ciliary neurotrophic factor; FGF-5 = fibroblast growth factor-5; GPA = growth-promoting activity; IGF = insulin-like growth factor; LIF = leukemia-inhibiting factor; NT = neurotrophin; P = *phosphate*. Not shown is CDF, choline acetyltransferase development factor.

Although the situation has not been studied, it is safe to assume that the same kinds of neurotrophin receptor occur on the dendrites of the motoneuron, where they receive trophic inputs from other nerve cells. Loss of these signals may have a disastrous effect on the motoneuron (see "Motoneurons Become Dysfunctional After Strokes").

Other polypeptides can sustain motoneurons.

Muscle fibers also produce other polypeptides that, although unrelated to neurotrophins in their molecular structure, can nevertheless promote the survival of motoneurons, at least *in vitro* (figure 18.7). These polypeptides are *fibroblast growth factor-5*, *leukemia-inhibiting factor, insulin-like growth factor,* and *growth-promoting activity* (Mudge, 1993). Yet another polypeptide, *ciliary neurotrophic factor (CNTF)*, is known to have a powerful sustaining effect on motoneurons but is produced by *Schwann cells* rather than muscle fibers. Ciliary neurotrophic factor has been observed to rescue facial motoneurons in rats following axotomy (Sendtner *et al.*, 1990) and to increase the survival of motoneurons during embryonic cell death (Oppenheim *et al.*, 1991); it also prolongs the life expectancies of mice with hereditary motoneuron degeneration (Sendtner *et al.*, 1992). Another polypeptide possessing trophic activity, with a molecular weight of 22 kD, is *choline acetyltransferase development factor (CDF);* CDF increases the level of *choline acetyltransferase* activity in cultured spinal cord neurons and can protect embryonic motoneurons from naturally occurring cell death (McManaman *et al.*, 1990). Figure 18.8 shows the greater survival of motoneurons in the spinal cord of the chick embryo following treatment with CDF.

In view of the functional and trophic interactions between *glia* and neurons, it is not surprising that a trophic peptide should have been identified in glia. This factor, *glial cell-derived neurotrophic factor (GDNF)*, is as effective as BDNF in preventing motoneurons from undergoing cell death in the embryo or after axotomy (Henderson *et al.*, 1994; Yan *et al.*, 1995).

Applied Physiology

The trophic effects of motoneurons and muscle fibers on each other, considered in this chapter, are dramatic examples of a widespread phenomenon in the nervous system. It seems very likely that all neurons in the brain and spinal cord exert trophic influences on each other, and that similar

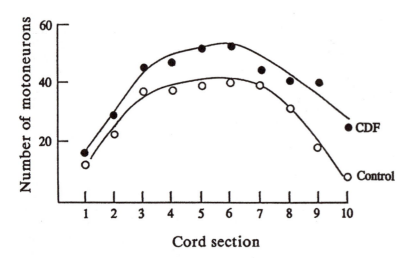

Figure 18.8 Motoneuron survival in chick embryos treated with CDF or saline *(control)*. The lumbar region of the cord was divided into 10 sections, and the number of motoneurons in the lateral motor column was counted in each; section 1 is most rostral.

Adapted, by permission, from J.L. McManaman et al., 1990, "Rescue of motoneurons from cell death by a purified skeletal muscle polypeptide: Effects of the Ch AT development factor, CDF," *Neuron* 4: 893.

influences exist between neurons and glial cells. It is because of the loss of trophic effects that a lesion in one tract of nerve fibers may cause secondary neuronal degeneration "downstream." For example, damage to the optic nerve produces degeneration not only of the optic nerve fibers, as expected, but also of the neurons in the lateral geniculate nucleus, with which the optic nerve fibers have synaptic connections (Cook *et al.,* 1951). In "Motoneurons Become Dysfunctional After Strokes" we review an experiment of nature that appears to demonstrate the trophic effects of neurons in the cerebral cortex on motoneurons in the spinal cord.

It is not known why some motoneurons should be more resistant to loss of descending input from the motor cortex than others. What is clear, however, is that there are regions of the brain that can compensate for the disruption of connections from the motor cortex. Thus, if motor unit estimates are performed on patients with severe spinal cord injuries, in whom all descending motor pathways have been divided, the reduction in functioning motor units is very much greater (Brandstater & Dinsdale, 1976). These findings are of more than theoretical interest, because they suggest that one therapeutic approach to the problem of stroke would be to prevent the secondary loss of functioning motoneurons.

Now that we have studied denervation and neurotrophism, it is logical to examine the effects of disuse on muscle. This is the subject of chapter 19.

Motoneurons Become Dysfunctional After Strokes

Are motoneurons trophically dependent not only on muscle, but also on other neurons from which they receive synaptic connections? One way of exploring this possibility would be to examine patients who have had cerebral hemorrhages or thromboses resulting in *weakness* of the muscles on the opposite side of the body (hemiplegia). Vascular lesions of this kind are known to destroy axons descending from the motor cortex to motoneurons of the brainstem and spinal cord. Patients with hemiplegia have not only weakness but also atrophy of muscles on the affected side. Although it is reasonable to attribute the wasting to disuse,

examination of such muscles with a needle recording electrode will often reveal spontaneous muscle fiber discharges, or fibrillations (see chapter 16; Goldkamp, 1967). Since this finding normally indicates the presence of muscle fiber denervation, it raises the possibility that some motoneurons have undergone dysfunctional or degenerative changes, secondary to loss of trophic input from the motor cortex. The definitive test would be to determine the number of functioning motoneurons supplying one of the muscles on the side of the hemiplegia and to relate it to the value for the corresponding muscle on the normal side. Such estimates of motoneuron numbers can be made by comparing the potentials or twitch tensions of single motor units with the response of the entire muscle (see "Estimation of Human Motor Units" in chapter 12). When this type of analysis is performed on hemiplegic patients, it is found that for the first 2 months after the stroke, there is no significant difference between the numbers of motor units (and functional motoneurons) on the two sides of the body (McComas et al., 1973b). Soon after that time, however, the number of motor units on the hemiplegic side drops off by half, and remains reduced (figure 18.9).

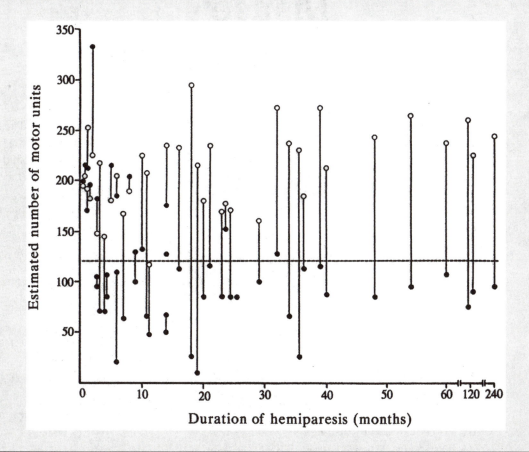

Figure 18.9 Numbers of functioning motor units in the extensor digitorum brevis muscles of hemiparetic patients. Filled and open circles indicate values for hemiparetic and normal sides, respectively, and the horizontal line indicates lower limit of normal range.

Reprinted, by permission, from A.J. McComas et al., 1973, "Functional changes in motoneurons of hemiparetic patients," *Journal of Neurology, Neurosurgery, and Psychiatry* 36: 185.

19

Disuse

No tissue in the body is more responsive to changes in usage than skeletal muscle. The ability of exercise programs to increase strength or endurance is described in chapter 20; the present chapter deals with the effects of disuse. Inevitably, studies in animals have provided more detailed information than those in humans and have allowed the pathophysiological mechanisms underlying some of the disuse effects to be analyzed more completely.

In view of the voluminous literature and the diversity of findings, some generalizations should be made at the outset:

- The effects of disuse are more marked in animals than in humans. The effects of disuse are, generally speaking, the opposite of those induced by *training,* although in many instances mechanisms specific to disuse are brought into play to alter muscle properties.
- The consequences of disuse depend, to a considerable extent, on the experimental model employed; the absence of load bearing is the most important factor.
- In animals, slow-twitch (type I) fibers are more susceptible to disuse than fast-twitch (type II) fibers. This has not been borne out in human studies.
- Not only do disused fibers become smaller, but they may exhibit other morphological changes.
- The motor innervation as well as the voluntary *activation* of the disused muscle or limb may also be affected by disuse.
- The effects of disuse are reversible.

Studies in Human Subjects

It is not difficult to find examples of the effects of disuse on human muscle because most people, at one time or another, will have had a relevant experience. The research literature contains reports of the effects of muscle disuse due to pathological conditions and, more recently, using experimental models, of the mechanisms associated with the muscular *adaptations* to altered use in a healthy system.

Models of Decreased Neuromuscular Activity

Bed rest and space travel affect muscles similarly.

Anyone who has been obliged to rest in bed because of illness or injury will have noticed a feeling of *weakness* on standing or walking; further, some leg muscles, the quadriceps in particular, show obvious wasting after as little as 3 to 4 days. A recent and more exotic source of data has been the observations made on astronauts returning from periods of space travel. It should be noted that the amount of *EMG* activity and joint movement generated by astronauts in performing their mission tasks inside the space shuttle may be considerable, and variable. In addition, the astronauts' preflight training levels are usually not known, or reported, and this factor can also be partially responsible for the considerable variability in results from spaceflight studies. However, clearly the lack of load bearing, due to the low gravitational forces in space, is the provocative factor in producing the loss of muscle mass and weakness that constitute the major consistent finding from these studies. Astronauts returning to earth also experience increased fatigue, impaired neuromuscular coordination, and a tendency toward *delayed-onset muscle soreness (DOMS)*. The muscle weakness is more pronounced than would be predicted from the *atrophy,* and is particularly marked for explosive contractions (Di Prampero & Narici, 2003). The rate of atrophy of knee extensors, knee flexors, and ankle extensors has been estimated at about 0.8% of muscle volume per day during a 16-day flight, using the technique of *magnetic resonance imaging (MRI)* to measure volumes (Akima *et al.,* 2000). Short periods in space (5 to 11 days) have been reported to result in some small but measurable fiber type changes in vastus lateralis from type I to type II, with atrophy in the order IIX > IIA > I (Edgerton *et al.,* 1995). Thin filament density and length also decrease in some fibers, rendering these fibers more prone to sarcomere damage during the reloading period, due to an increased force per thin filament (Riley *et al.,* 2000).

An attempt to simulate the effects of reduced gravity has been made by subjecting healthy individuals to bed rest with the head tilted downward by 6° (Dudley *et al.,* 1989; Hikida *et al.,* 1989). This procedure allows for larger numbers of subjects and longer periods of time of disuse to be studied, as well as allowing for more experimental control. It is obviously a different model from spaceflight, since subjects in bed rest studies are normally asked to keep contractions to a minimum, while this is not the case for astronauts. Nonetheless, most of the findings from bed rest studies support those that have been reported from spaceflights, once again emphasizing the importance of load bearing on the properties of muscles. During bed rest, whole body and muscle protein synthesis decrease by 14% and 50%, respectively, after 14 days (Ferrando *et al.,* 1996). Prolonged periods of bed rest have demonstrated that besides the atrophic patterns seen in spaceflight, there is a loss in the ability to maximally activate muscles, as measured by the difference between the force during a maximal voluntary contraction and electrical stimulation of the nerve (Koryak, 1995), and a prolongation of the evoked *twitch* response. There is also evidence for a decrease in the force-generating capacity per unit cross section *("specific tension")* of individual fibers (Larsson *et al.,* 1996). Fiber type changes that have been noted in spaceflight studies do not appear to occur in bed rest (Berg *et al.,* 1997).

A variation on the bed rest theme has been the use of "dry" water immersion, in which subjects are positioned horizontally in a special bath in which a highly elastic fabric film separates the subject from the water (Koryak, 1999). In this model, the subject is not supported by a bed, and thus the conditions are more similar to the zero-gravity state in space. The results from this model are similar to those of bed rest and spaceflight.

Systematic observations on the effects of disuse in humans have also been made on subjects whose limbs were immobilized as part of the treatment for injuries or as an experimental strategy. Variations on the immobilization model have included normal subjects using arm and foot slings to minimize limb use (Parcell *et al.,* 2000; Berg *et al.,* 1991). Of course, disuse studies, whether involving patients with bone fractures or healthy limbs of volunteers, have the added component that will influence the outcome; fixing of the muscle at a constant length during the period of relative disuse will permit changes in sarcomere number (see later).

Effects Observed in Human Studies of Decreased Neuromuscular Activity

The primary effect of decreased neuromuscular activity in humans is loss in muscle mass, which can be seen at the single muscle fiber as well as the whole-muscle level. This effect, which is more pronounced in muscles used extensively, such as those involved in posture, may be accompanied by qualitative changes in histochemical composition and in contractile and fatigue properties. A phenomenon that accompanies decreased neuromuscular usage, which is difficult to assess in animal studies referred to later in this chapter, is a decrease in the capability to maximally activate the muscle, thus implying nervous system alterations. These effects are now discussed in more detail.

Decrease in Muscle Mass

Muscle cross-sectional area can be estimated, rather crudely, by measuring the girth or volume of a limb and correcting for the thickness of the subcutaneous fat. More accurate determinations can be made by visualizing the muscles with *computerized axial tomography (CAT)*, MRI, or ultrasound. *Ultrasonography* is the preferred method because unlike the situation with CAT, X rays are not used, and it is much less expensive than MRI. The results of these various techniques show that limb immobilization produces rapid muscle wasting, first detectable by 3 days (Lindboe & Platou, 1984) and then gradually slowing. This rapid change in muscle girth also occurs in other disuse models, although it is less rapid and extensive in spaceflight, most likely due to the muscle activation and light load bearing associated with the tasks performed while aloft.

It should be noted that the amount of atrophy will depend on the usage of the muscles prior to immobilization and will therefore always be greater in antigravity muscles than in their antagonists; hence the quadriceps will show more wasting than the hamstrings. Second, specific to the immobilization model, because of *sarcomere* resorption (see p. 294), the length of the muscle at the time of fixation is critical. For example, wasting will be accentuated in the quadriceps because the leg is usually immobilized with the knee extended, a position that holds the muscle at a short length and therefore encourages loss of sarcomeres. A final point is that estimates of muscle belly shrinkage may underestimate the fiber atrophy if there is an increase in the amount of *connective tissue* (Jokl & Konstadt, 1983; Tomanek & Lund, 1974).

Some interesting comparisons can be made when one considers what happens to paralyzed limb muscles following a complete spinal cord injury—the ultimate disuse condition. In this case, atrophy of quadriceps and hamstrings is extreme, and very similar in extent (Castro *et al.*, 1999).

Atrophy at the Muscle Fiber Level

Using specimens of muscle, most often obtained by needle biopsy, attempts have been made to determine whether there is greater atrophy of slow-twitch (type I) or fast-twitch (type II) fibers following limb disuse. The results so obtained are conflicting, and probably depend on the disuse model, the choice of muscle, its degree of previous usage, and the amount of contractile activity (loaded or unloaded) that is present. As previously noted, atrophy will be greatest in antigravity muscles fixed in shortened positions; if only a small amount of isometric contraction is allowed, this will better maintain the fast-twitch fibers, which are normally used intermittently, than the more constantly employed slow-twitch fibers. In many studies, however, there is significant wasting of both fiber types. As an example of work in this field, Sargeant *et al.* (1977) found substantial atrophy in both type I and type II fibers in the quadriceps muscles of patients whose legs had been immobilized for fractures; the cross-sectional areas of the type I and type II fibers were reduced by 46% and 37%, respectively. Like other estimates of this kind, these values will be rather high because the fibers in the contralateral legs, which are used for comparison, will exhibit some degree of *hypertrophy* due to the extra weight bearing imposed on that limb. The investigation by MacDougall *et al.* (1980) was free from this limitation in that it was carried out on an arm muscle, the triceps brachii; also, unlike most other studies, it involved healthy subjects rather than patient volunteers. These authors found that in the long head of the triceps, 5 to 6 weeks of immobilization reduced the cross-sectional areas of type I and type II fibers by 25% and 30%, respectively. Since these studies were published, findings have generally supported the lack of systematic fiber type sensitivity to immobilization-induced atrophy in particular, and to disuse atrophy in general, in the human. Similarly, there is no apparent preferential fiber type atrophy following spaceflight and bed rest.

The one known example of preferential type I fiber involvement in humans involves soleus muscles in patients with ankles immobilized for 6 weeks after rupture of the Achilles tendon (Häggmark & Eriksson, 1979). Further, if the muscles are examined within 2 weeks of such an injury, some type I fibers can be seen to have developed central cores (Dr. J. Maguire, personal communication, 1994), in keeping with similar observations on the immobilized soleus muscles of small mammals (Karpati *et al.*, 1972).

Effects on Muscle Fiber Type Composition

There are now several reports of a decrease in the percentage of human type I fibers after immobilization. For example, this change has been noted in vastus lateralis muscles 4 to 6 weeks after knee immobilization for acute ligamentous injuries (Halkjaer-Kristensen & Ingemann-Hansen, 1985; Ingemann-Hansen & Halkjaer-Kristensen, 1983; Young *et al.*, 1982). Similarly, the proportion of type IIB (IIX) fibers was increased in soleus muscles of legs immobilized for similar periods of time (Häggmark & Eriksson, 1979). These results are supported by more recent studies in which normal healthy subjects agreed to have their left knee joints immobilized in casts for 3 weeks (Hortobagyi *et al.*, 2000). An increased expression of type IIx *myosin heavy chains,* and a decrease of type I, were found in the quadriceps muscles of these subjects. Although there were no differences in fiber type proportions, the large fiber type sampling error inherent in the needle biopsy technique must be taken into account (Elder *et al.*, 1982). In bed rest and in spaceflight, however, there is evidence of conversion of fiber types from slow to fast (Riley *et al.*, 1998; Widrick *et al.*, 1999). Moreover, in the paralyzed leg muscles of spinal cord injury patients, once again the epitome of long-term disuse, type I myosin heavy chain content and percentage of type I muscle fibers decrease gradually from the time of injury (Martin *et al.*, 1992; Jankala *et al.*, 1997).

Effects on Muscle Force Capability and Fatigue Characteristics

> **Muscle weakness may exceed atrophy because of poor motor unit recruitment.**

There have been few studies of stimulated force after a period of limb immobilization; but in general, the decrease in tension (see figure 19.1) is commensurate with the degree of muscle wasting or rather greater than that expected. White and Davies (1984) attribute the latter finding to a reduction in the specific tension of the atrophied fibers (i.e., force generated for a given cross-sectional area). In their study of the adductor pollicis muscle, Duchateau and Hainaut (1987) noted that although tetanic force was reduced by 33% after 6 weeks of thumb immobilization, the twitch force remained normal.

Of greater interest is the finding of a more severe reduction in voluntary strength than in stimulated force. An example of this discrepancy is the study of arm fixation in healthy subjects by Sale *et al.* (1982b), in which the median-innervated thenar muscles exhibited a normal twitch but a striking loss of voluntary force; significantly, these muscles did not appear wasted, despite having been in a cast for 5 weeks. Very similar results were found by Duchateau and Hainaut (1987), also in the thumb muscles. The explanation for this rather surprising finding is that after a period of disuse, the subject is unable to recruit the *motor unit* population fully—an example of the nervous system "forgetting" a motor task. Thus, Fuglsang-Frederiksen and Scheel (1978) observed that the density of the EMG interference pattern was reduced in comparison with that in the normal leg when maximal voluntary contraction of the quadriceps muscle was examined with a coaxial

Twitch responses - ankle plantarflexors

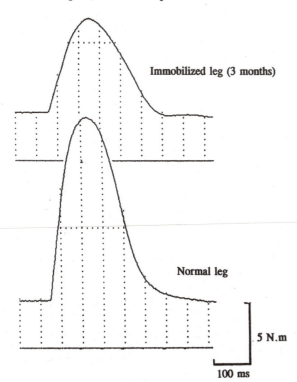

Immobilized leg (3 months)

Normal leg

5 N.m

100 ms

Figure 19.1 Reduced twitch force after immobilization of human plantarflexor muscles. The subject had sustained comminuted fractures of the right tibia and fibula; following surgery the leg was placed in a below-knee cast for 3 months. At the end of this time, the twitch force in the treated leg (top trace) was only half that of the opposite limb (lower trace); the similar contraction and *half-relaxation times* suggest that the considerable muscle wasting was due to atrophy of both fast-twitch and slow-twitch fibers.

Reprinted, by permission, from A.J. McComas, 1994, "Neuromuscular adaptations in man that accompany changes in activity," *Medicine and Science in Sports and Exercise* 26: 1499-1509.

recording electrode after a period of disuse. Similarly, Duchateau and Hainaut (1987) demonstrated quite convincingly that subjects were unable to generate high motor unit firing rates for first dorsal interosseus and adductor pollicis muscles following 6 to 8 weeks of immobilization of the corresponding joints. Other evidence of impaired motoneuron activation is the ability to elicit *interpolated twitches* (see chapter 13) in subjects who, for one reason or another, have reduced physical activity; clinical examples of this phenomenon are patients left moderately disabled after polio (Allen *et al.,* 1994). The functional nature of the problem is illustrated by the fact that motor unit *recruitment* improves after repeated attempts at maximal contraction. An additional result of immobilization, with implications for voluntary recruitment, is motor end-plate dysfunction; thus increased neuromuscular *"jitter"* is apparent in soleus muscle immobilized for 4 weeks (Grana *et al.,* 1996).

In summary, not only muscles are involved in disuse-associated deficits. The decrease in the ability to maximally activate muscles voluntarily, for a short period following disuse, is a well-accepted phenomenon that has been demonstrated in all disuse models.

Disused human muscle is more fatigable.

In their investigation of muscle changes after thumb immobilization, Duchateau and Hainaut (1987) observed that tetanic force declined more rapidly in the treated hand than in the contralateral control hand (see figure 19.2). This very clear difference between the two sides is in contrast to the mixed results obtained in animal experiments (see p. 293).

It is worth noting that muscles that are paralyzed as a result of spinal cord injury fatigue more rapidly with electrical stimulation than control muscles (Stein *et al.,* 1992; Gerrits *et al.,* 1999). This tendency, which gets worse with time after injury (Gaviria & Ohanna, 1999), is consistent with a decrease in the mitochondrial enzyme succinic dehydrogenase (SDH) and an increased type II fiber population.

The findings in disused human muscles are summarized in table 19.1.

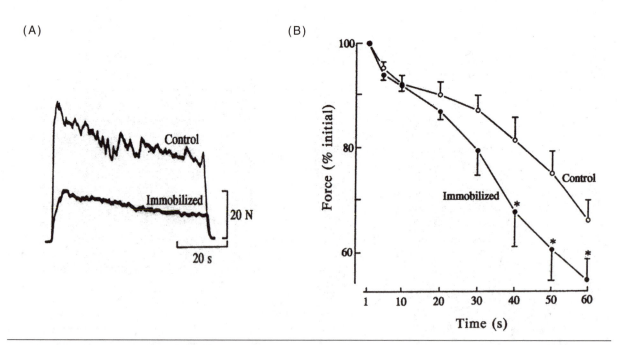

Figure 19.2 Disuse effects in the human adductor pollicis muscle following 6 weeks of immobilization (for forearm fractures). (A) Reduced voluntary torque during 1 min of maximum *effort* compared with the untreated hand *(control)*. (B) Increased *fatigability* in two subjects relative to the control hand. The forces were elicited by intermittent 30-Hz tetani. Values are means ± SEs, and those marked with asterisks differed significantly from controls.

Adapted, by permission, from J. Duchateau and K. Hainaut, 1987, "Electrical and mechanical changes in immobilized human muscles," *Journal of Applied Physiology* 62: 2171.

Table 19.1   Summary of Disuse Findings in Human Muscles

Property	Involvement
Muscle atrophy	Greatest for leg antigravity muscles, least for hand muscles
Fiber type atrophy	Both type I and type II fibers affected
Twitch force	May remain normal
Tetanic force	Reduced
Voluntary force	Very reduced
Fatigability	Increased

Studies in Animals

Studies using animals have allowed us to search systematically for the mechanisms that result in the primary effects of decreased neuromuscular activity observed in humans and discussed in the preceding sections. Some advantages of animal studies in the examination of neuromuscular disuse include relative homogeneity among subjects, shorter time course of physiological and biochemical changes, and experimental control of specific factors thought to play a role in the physiological response to disuse (such as loss of muscle activation vs. weight bearing, altered muscle length, the role of the motor nerve, and the effects of countermeasures). We will now consider the results emanating from animal studies in more detail.

Animal Models of Decreased Neuromuscular Activity

The experimental methods devised to examine the consequences of disuse in mammalian muscles are remarkable for variety and ingenuity. Some of the techniques have required diligent and skillful nursing of the animals on a daily basis, often for several months. The methods, summarized in figure 19.3, fall into three categories.

1. Removal of load bearing
   - Animals in space
   - Hindlimb suspension
2. Decreased muscle excitation (and contraction)
   - General anesthesia
   - Spinal cord isolation
   - Local anesthesia
   - Neurapraxia
   - Botulinum toxin application
   - Curarization
   - Bungarotoxin application
   - Tenotomy
3. Limb immobilization
   - Joint pinning
   - Casting

Many of the foregoing strategies were intended to clarify the nature of *trophism,* that is, whether contractile activity or the delivery of *trophic* factors by the motoneuron is responsible for maintaining the biochemical and physiological properties of the muscle fibers (see chapter 18). Consequently, observations were usually made of twitch and *tetanic contractions* and of muscle fiber histochemistry rather than of muscle morphology, though a notable exception was the pioneering study of Tower (1937) using the *spinal cord isolation* preparation. A second comment is that the mechanisms involved in the disuse model were often mixed; for example, in hindlimb suspension, there is not only an absence of weight bearing but also a reduction in EMG activity and muscle contraction. A final point is that virtually all the observations have been made on hindlimb muscles of the cat and rat, and there must be some hesitation in applying the findings to other species and to

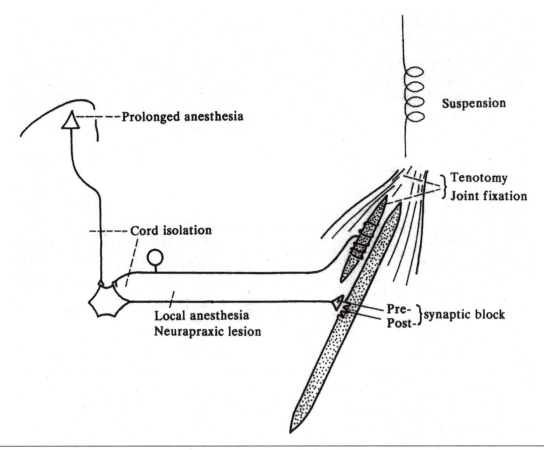

Figure 19.3   Summary of experimental strategies employed to study effects of disuse on animal muscles (see text).

Reprinted, by permission, from A.J. McComas, 1977, *Neuromuscular Function and Disorders* (London: Butterworth Publishing Co.), 82.

non-weight-bearing muscles, such as those of the human arm. The following commentary is given with these general considerations in mind.

Removal of load bearing affects slow-twitch (type I) fibers most.

Removal of Weight Bearing

In both the Russian COSMOS flights and those of the NASA program, rats have been subjected to space travel. These studies have shown that the slow-twitch muscle fibers, especially those in the soleus, are those most susceptible to the effects of weightlessness; enzyme assays suggest that there is a tendency for type I fibers to become IIA fibers (Martin *et al.,* 1988).

Coincident with the recognition of the muscular consequences of space travel has been the development of an animal model of weightlessness, in which the rear of the animal is suspended, compelling it to propel itself around the cage by the forelimbs (Musacchia *et al.,* 1980; figure 19.4A). It is evident from chronic EMG recordings with implanted electrodes that the loss of weight bearing is associated with considerable reduction in *impulse* activity in the hindlimb muscles (Blewett & Elder, 1993). Some of this residual activity produces periodic extension of the hindlimbs, but often no movement can be observed. The results of the various studies employing this model have been unanimous in identifying the type I (slow-twitch) fibers as having the greatest atrophy and loss of tetanic force. If young animals are treated in this way, the incidence of type I fibers is reduced in the soleus, and the muscle has a twitch that is briefer than normal (figure 19.4B; Elder & McComas, 1987).

Motoneuron quiescence produces muscle atrophy and degenerative changes.

Blockade of Motoneuron Activation of Muscle

One can achieve a decrease in EMG activity in a noninvasive manner by keeping an animal under prolonged general anesthesia. Davis and Montgomery (1977) found that, after 3 weeks, there was little change in the twitch speeds of cat hindlimb muscles. A demanding surgical technique is spinal

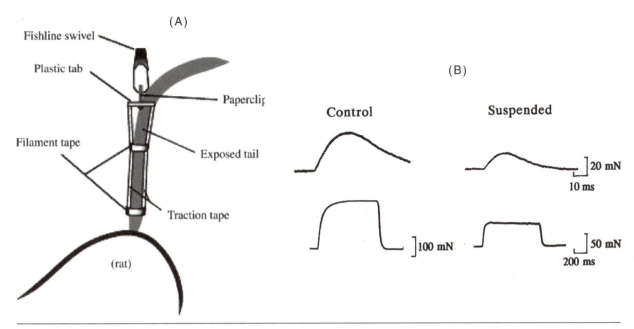

Figure 19.4 (A) Hindlimb suspension model of disuse. Tension tape is placed along two sides of the tail and connected to an overhead suspension system that allows mobility in such a way that the hindpaws are kept clear of the ground. The animal is still able to move around the cage using its forepaws and to feed, drink, and sleep; the body weight is rather lower than that of untreated animals, however. (B) Twitch and tetanic forces developed by soleus muscles in a suspended hamster and in a control of the same age. Note the smaller forces and faster twitches in the suspended animal; the fast-twitch plantaris muscle (not shown) was unaffected by suspension.

A: Adapted, by permission, from E.R. Morey-Holten and R.K. Globus, 2002, "Hindlimb unloading rodent model: Technical aspects," *Journal of Applied Physiology* 92: 1367-1377; B: Adapted, by permission, from K. Corley, N. Kowalchuk, and A.J. McComas, 1984, "Contrasting effects of suspension on hind limbs in the hamster," *Experimental Neurology* 85: 35.

cord isolation, conceived by Sarah Tower (1937). To deprive motoneurons of their normal synaptic bombardment, she divided the spinal cord above and below the segments of interest and also sectioned the dorsal root fibers. Tower observed marked decreases in the diameters of the paralyzed muscle fibers. The fibers failed to stain well, and there were reductions in the numbers and sizes of the *myonuclei;* the motor *end-plate* regions of the fibers were well preserved. Associated with the muscle fiber atrophy was considerable proliferation of connective tissue.

Use of the spinal cord isolation preparation was subsequently made by Johns and Thesleff (1961), who found only a small sensitivity to *ACh* outside the end-plate region of the muscle fibers. Atrophy of muscle fibers was noted by Klinkerfuss and Haugh (1970) and by Karpati and Engel (1968); the latter authors also observed features in the muscles that would normally be described as "myopathic." These features included *necrosis* of muscle fibers, proliferation of endomysial connective tissue, and the presence of muscles with obvious distortions of their myofibrillar architecture (ringed and snake-coil fibers).

In recent years, Eldridge has studied muscle disuse using cats with spinal cord isolation, maintained for as long as 3 years. In such animals, the neuromuscular junctions express extrajunctional *AChRs* and sprouting of the motor nerve terminals (Eldridge, 1984). In addition, the twitches of slow muscles become faster, due to the substitution of fast for slow isoforms of *myosin* light chains, *tropomyosin,* and *troponin* (Steinbach *et al.,* 1980). More recently, Roy and his colleagues have used this preparation to detail changes in the content of contractile and metabolic proteins and in the contractile characteristics of various hindlimb muscles in the cat and rat (Grossman *et al.,* 1998; Roy *et al.,* 2000; Zhong *et al.,* 2002).

When kittens, rather than adult animals, are subjected to spinal cord isolation, there is a failure of the normal differentiation of slow-twitch muscles; instead the muscles become fast contracting (Buller *et al.,* 1960a), containing abnormally high proportions of type II fibers (Karpati & Engel, 1968).

Robert and Oester (1970) introduced a technique for implanting a plastic cuff containing local anesthetic around a muscle nerve. The local anesthetic is gradually released from the cuff and is able to block impulse conduction along the motoneurons for a week or more. In rats treated in this way, there is extrajunctional ACh sensitivity in the muscle fibers (Lømo & Rosenthal, 1972); further, the paralyzed fibers will readily accept a fresh innervation (Jansen *et al.,* 1973). A possible drawback of such experiments is that the local anesthetic may interfere with *axoplasmic transport* (Bisby, 1975; Byers *et al.,* 1973), but this complication can be avoided if *tetrodotoxin* (*TTX;* see chapter 9) is given instead by intraneural injection, or using a cuff around the nerve that is fed via an osmotic minipump. In this treatment, atrophy occurs that is as severe as that seen with *denervation,* with drastic decreases in the content of most energetic enzymes and changes in contractile function reminiscent of immobilization in the human. Tetrodoxin is also found to increase the distribution of AChRs (Pestronk *et al.,* 1976) and to induce terminal sprouting at neuromuscular junctions (Brown & Ironton, 1977).

Nerve Compression

Either through accident or by experimental design, nerve fibers may be compressed to an extent that leaves their *axis cylinders* in continuity while temporarily depriving their membranes of excitability and hence of the ability to conduct impulses (neurapraxia). Weir Mitchell, a surgeon in the American Civil War, described transient palsies in some of his patients after bullet wounds: "This condition of local shock is very curious. A man is shot in the thigh, the ball passes near the sciatic nerve, and instantly the limb is paralyzed; within a few minutes, or at the close of a day or a week, the volitional control in part returns" (Mitchell *et al.,* 1864, 18).

Denny-Brown and Brenner (1944, p. 22) produced neurapraxic lesions in cats by compressing the sciatic nerve with a bag containing mercury at known pressures. In muscles that had been paralyzed for as long as 3 weeks, they observed that there was an absence of atrophy and of *fibrillation* activity. Therefore, they reasoned that "anatomic continuity of nerve, not receipt of impulses, prevents atrophy and fibrillation of muscle."

For a clinical illustration of prolonged impulse block, see "A Clinical Example of Neurapraxia" (p. 296). Extensive use of nerve compression as an experimental tool was subsequently made by Gilliatt and colleagues. They were able to demonstrate that impulse block results from the sliding of a *node of Ranvier* under the membrane of the adjacent segment of nerve (intussusception; Ochoa *et al.,* 1971). In a later study, also in the baboon, Gilliatt and colleagues showed that, following a neurapraxic lesion, the extrajunctional ACh sensitivity is considerably less than that after surgical denervation (Gilliatt *et al.,* 1978). The neurapraxic model has also been employed by Kowalchuk and McComas (1987) to investigate the effects of disuse on rat hindlimb muscles. After only 1 week of impulse block, there is marked wasting of slow- and fast-twitch muscles, with significant prolongation of the twitch in the fast-twitch muscle; in this instance, the changes are as severe as those reported in rat muscles after surgical denervation.

Neuromuscular Transmission Blockade

Botulinum toxin is produced by soil bacteria and is responsible for an especially dangerous type of food poisoning; it interferes with the emptying of ACh from synaptic vesicles in the motor nerve terminal (see chapter 10). Both fast-twitch and slow-twitch muscles undergo considerable atrophy after treatment with botulinum toxin. The fast muscles, in particular, show a number of degenerative features (Thesleff *et al.,* 1990). In addition, the resting *membrane potentials* of the fibers fall, and the twitch durations increase. At the surfaces of the fibers, botulinum toxin induces extrajunctional ACh sensitivity beyond the end-plate region (Thesleff, 1960); the fibers also synthesize Na^+ channels that are abnormally resistant to TTX. Outside the neuromuscular junction, *acetylcholinesterase* disappears from the fiber surface, but the synaptic moiety, in contrast, is largely retained. The situation can be summarized by stating that the effects induced by botulinum toxin are qualitatively similar to those following surgical denervation of the muscle fiber, but they are less pronounced.

Botulinum toxin was also used experimentally to demonstrate the importance of nerve–muscle communication in maintaining motoneuron properties. Pinter and colleagues (1991) demonstrated in the cat that, after botulinum toxin injection into cat hindlimb muscle, only those motoneurons in which neuromuscular transmission was completely eliminated showed *axotomy*-like characteristics. In those motoneurons in which even a small proportion of the muscle fibers were receiving neurotransmitter, axotomy-like properties did not appear.

Another toxin employed in the study of disuse is β-*bungarotoxin*, which is one of the toxins in the venom of the banded krait, *Bungarus multicinctus*, a poisonous snake indigenous to parts of southern Asia. Like botulinum toxin, β-bungarotoxin prevents ACh release and causes similar changes to take place in the muscle fiber membrane (Hofmann & Thesleff, 1972). It is possible that presynaptic blocking agents such as botulinum toxin and β-bungarotoxin, in addition to depriving the muscle fiber of ACh and impulse activity, also prevent the release of *neurotrophins*. This potential complication is avoided if postsynaptic blocking agents, such as d-*tubocurarine, succinylcholine,* and α-*bungarotoxin,* are used instead. Berg and Hall (1975) have used these substances to paralyze rats, maintaining the animals for 3 days with artificial respiration. At the end of this time, the muscle fibers show such denervation features as a fall in resting membrane potential, spread of ACh sensitivity, and resistance to TTX.

Tenotomy and Joint Immobilization

Striking changes follow tenotomy and joint fixation.

The purpose of *tenotomy* is to allow the muscle to shorten passively and thereby diminish any excitatory input to the motoneurons from the *muscle spindles;* the motoneurons should therefore become quiescent. Following tenotomy in the rabbit, there is pronounced atrophy, particularly of type I fibers, and this is associated with a speeding up of the twitch (Vrbová, 1963). However, the experimental situation is not ideal because the motoneurons may be subjected to increased, rather than reduced, sensory bombardment (Hník, 1972). Again, through depriving the muscle fibers of passive stretch, atrophic changes may be anticipated anyway (Gutmann *et al.,* 1971b).

Experimental immobilization achieved by bone pinning has been applied to animals by Fischbach and Robbins (1969). A very considerable reduction in background impulse activity is produced in muscles, and marked speeding of the twitch can be seen. At the single motor unit level, it has been possible to confirm the susceptibility of *S* units to immobilization, but it appears that the *FR* units are even more affected as judged by reductions in their twitch and tetanic tensions. Although smaller changes in force generation are observed in *FF* units, both these and S units exhibit a speeding up of their twitches (Mayer *et al.,* 1981; St.-Pierre & Gardiner, 1985).

Perplexing results from disuse experiments have been reported by Robinson *et al.* (1991), who immobilized the hindlimbs of cats for 3 weeks by unilateral casting. While tetanic force decreased in slow-twitch motor units of the tibialis posterior, as might have been anticipated from previous studies (e.g., St.-Pierre & Gardiner, 1985), the twitch and tetanic tensions actually increased in fast-twitch units. However, this apparent anomaly could have arisen, in part, from *collateral reinnervation* of functionally denervated fibers (see p. 267).

Physiological Responses

Disuse has produced variable effects on muscle fatigability.

In keeping with the discordant findings in single motor units (see the preceding sections), animal models of disuse have produced variable effects on muscle *fatigability.* Increased susceptibility to fatigue has been reported by Fell *et al.* (1985) after hindlimb suspension, although decreased fatigability was observed by others (Haida *et al.,* 1989; Robinson *et al.,* 1991). No significant changes were noted by Mayer *et al.* (1981) or by Witzmann *et al.* (1983). The picture is not any clearer at the present time. However, in those studies in which increased fatigability was seen, the findings are consistent with similar observations in humans (Duchateau & Hainaut, 1987) and with reports of decreased oxidative capacity in animal muscles after disuse (Booth, 1977; Rifenberick *et al.,* 1973).

Disuse-Induced Changes at the Neuromuscular Junction

After as few as 5 days of immobilization, motor *axon* terminals begin to *sprout* and to become distorted in their longitudinal axes (Fahim & Robbins, 1986). With the electron microscope, other

Muscle fibers get shorter due to decreased sarcomere numbers.

signs of degeneration can be found in the neuromuscular junctions of type I and type II fibers; these include exposure of the synaptic folds and disruption of nerve terminals (Pachter & Eberstein, 1984). The presence of several small axons over the same primary cleft suggests that some nerve terminals may also be sprouting or regenerating.

Tabary *et al.* (1972) showed that the fibers in the cat soleus muscle could respond to immobilization of the hindlimb in a plaster cast by altering the number of sarcomeres. If the muscles were fixed in a shortened position, sarcomeres were removed, while fixation at a long length caused sarcomeres to be added. Williams and Goldspink (1978) subsequently demonstrated that this remarkable plasticity was taking place at the ends of the fibers. The functional advantages of the adaptation are evident, if the artificiality of immobilization is ignored; since the modified fibers can once again operate at normal sarcomere lengths, the *actin* and myosin filaments will be able to overlap optimally and to generate maximal tensions.

It is important, in the rehabilitation of orthopedic patients, to appreciate that much of the stiffness of a limb newly removed from a cast is due to shortening of the muscle fibers, as well as to changes in the joint capsules and ligaments. As the shortened muscle is stretched, damage to the *myofilaments* will inevitably occur; however, the end result of stretching is myofibrillar protein synthesis and the eventual restoration of the missing sarcomeres.

Disuse Effects Brought About by Changes in Gene Expression

Since new proteins may be produced in the muscle fibers as a result of disuse, it is evident that gene expression must be modified at the transcriptional or translational levels. Microarray analysis has been applied to disused muscles in an attempt to determine changes in gene expression profiles. Generally speaking, disuse appears to cause an *up-regulation* of expression of genes for glycolytic enzymes, glucose transporters, *proteases, translation* factors, faster isoforms of myosin, and *ion channels,* while down-regulating the expression of fatty acid transporters, vesicle trafficking factors, neuroreceptors, slow *myosin isoforms,* and transmitters (Jankala *et al.,* 1997; Wittwer *et al.,* 2002). Clearly these are not all of the changes in gene expression that occur, and this information will yield a more complete picture as research continues. However, it does emphasize the fact that alterations in gene expression are not all in the same direction. Activation of the *ubiquitin-proteosome system* is also an important component during disuse atrophy, as it is in atrophy of other etiologies (Solomon *et al.,* 1998).

This dramatic change in gene expression may be facilitated to some extent by a decrease in the *cytoplasm* volume-to-*nucleus* ratio that occurs with disuse atrophy (Kasper & Xun, 1996a, 1996b).

The length at which the muscle is immobilized can inform us as to the role of decreased neuromuscular activation versus passive tension on these alterations in gene expression. In the case of the myofibrillar proteins, immobilization in a shortened position causes the fast (IIB) myosin heavy chain gene to be expressed in the slow soleus muscle of the rat (Loughna *et al.,* 1990). Since the same gene is repressed if the soleus is fixed in a lengthened position, it is clear that it is the absence of passive stretch that enables this gene to be expressed in this type of situation. In the case of cytochrome c, on the other hand, the decrease in expression is independent of the length at which the muscle is immobilized (Booth *et al.,* 1996).

Other Biochemical Changes in Disuse

Most of the biochemical studies of disuse have been conducted on animal rather than human muscles (for review, see St.-Pierre & Gardiner, 1987). As would be anticipated from the presence of muscle atrophy, there is a loss of myofibrillar proteins from the fibers; but, in addition, the concentration of these proteins is reduced in the residual muscle tissue. This last finding may explain why, in some studies, the twitch and tetanic forces developed by disused muscles are diminished in relation to the cross-sectional areas of the muscles (i.e., reduced "specific tension"). In contrast to the myofibrillar proteins, there are increases in connective tissue proteins (Herbison *et al.,* 1978; Jokl & Konstadt, 1983). The concentrations of glycolytic and oxidative enzymes may also be lessened by disuse, but the length at which the muscle was immobilized

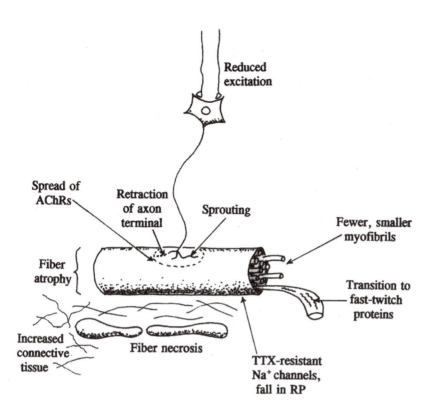

Figure 19.5 Summary of changes in muscle and nerve fibers that may follow disuse; RP = resting potential. See text.

is critical; thus, decreases in concentration are more likely to be found if the muscle is fixed in a shortened position (but see Booth *et al.*, 1996).

Atrophy As the Main Consequence of Disuse

It is time to summarize the effects of disuse on animal muscles, bearing in mind that there will be some variation depending on the methodological model employed as well as on the choices of muscle and species (figure 19.5):

• As far as the muscle fibers are concerned, the most striking consequence of disuse is atrophy, especially in fibers of the slow-twitch (type I) variety. In addition, a minority of fibers undergo necrosis, and there is an increase in the endomysial and perimysial connective tissue. The muscles develop smaller twitch and tetanic tensions, even smaller than those expected on the basis of fiber atrophy. There is also a tendency for slow-twitch fibers to be transformed into fast-twitch fibers, with attendant changes in the isoforms of the myofibrillar proteins.

• At the surfaces of the disused fibers, there is a spread of AChRs beyond the neuromuscular junction, and the resting membrane potential is diminished. The motor nerve terminals are abnormal in showing signs of degeneration in some places but also evidence of sprouting in others.

• Finally, there is a loss of motor drive after a period of disuse, such that the motor units cannot be recruited fully.

Applied Physiology

Secondary effects can be caused by disuse in patients with disabling illnesses.

In this section we will make some brief comments about the possible contributions of disuse to patient disability. Following this, "A Clinical Example of Neurapraxia" presents an example of muscle disuse in a young patient with prolonged impulse block after an unusual mishap.

Patients often complain of weakness and increased fatigability, and it is essential to recognize that much of the problem may be due to disuse rather than to the primary illness. Disuse will produce weakness through atrophy of muscle fibers and loss of myofibrillar protein. Of equal

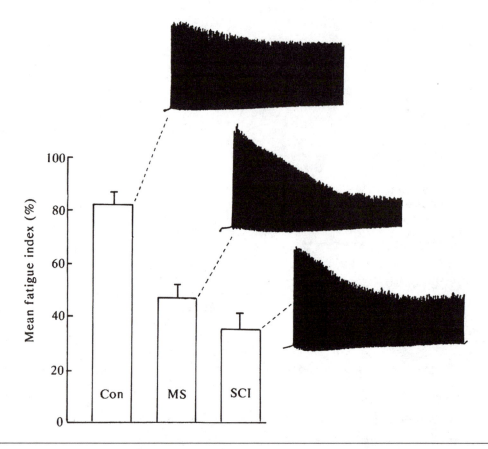

Figure 19.6   Increased fatigability of disused muscles in patients with injury or disease of the spinal cord. The histograms show the respective mean fatigue indices (+ SEMs) for patients with multiple sclerosis (MS) or spinal cord injury (SCI) and for control subjects (Con). Each histogram is linked to a typical recording of the forces developed by intermittent stimulation at 30 Hz for 3 min. The fatigue index is the amplitude of the last response as a percentage of that of the first response.

From "Muscle fatigue in some neurological disorders," by R. A. J. Lenman, F. M. Tulley, G. Vrbová, M. R. Dimitrijevic, and J. A. Towle, 1989, *Muscle and Nerve*, 12, p. 940. Copyright 1989 by John Wiley & Sons, Inc. Adapted with permission of John Wiley & Sons, Inc.

importance, however, is the inability of the motor centers in the brain to recruit motoneurons fully due to disuse of the descending motor pathways (see p. 287). Finally, disuse will increase the fatigability of those motor units that are still functional, as can be seen in figure 19.6.

The opposite of disuse is, of course, increased use, and this is the subject of the next chapter, which is devoted to the effects of training. We will see that, as with disuse, there have been some ingenious experiments in animals and some interesting, if incomplete, observations in humans.

A Clinical Example of Neurapraxia

This example of neurapraxia occurred in an 11-year-old schoolgirl who, 2 months before the EMG examination, had caught her left arm between the rollers of a clothes wringer; the arm had been drawn in to the midforearm level. Immediately after the accident, she had noticed numbness of the whole hand and muscle weakness; the latter affected all the intrinsic muscles of the left hand together with the extensors of the fingers and wrist. Both sensation and muscle strength had been improving up to the time of the EMG study.

Figure 19.7 shows the EMG responses evoked in the thenar and hypothenar muscles following supramaximal stimulation of the median and ulnar nerves. It can be seen that the muscle potentials were much larger when the nerves were excited at the wrist rather than at the elbow. The compact forms and normal configurations of the diminished potentials suggest that the discrepancy had not arisen from dispersion of the muscle responses due to slowed nerve impulse conduction. Instead the findings must have resulted from neurapraxic lesions in the forearm, which had caused local inexcitability of median and ulnar nerve fibers. In relation to an understanding of neurotrophic phenomena

(see chapter 18), it is significant that, in spite of the interruption of the conduction of impulses, the paralyzed muscle fibers showed no other evidence of denervation. Thus, the muscles did not appear wasted, nor could fibrillation potentials and sharp-wave activity be detected during exploration of the muscle with a needle electrode (figure 19.7, bottom). It also was unlikely that any collateral reinnervation had occurred, for the amplitudes of the potentials evoked from the still-functioning motor units were not enlarged.

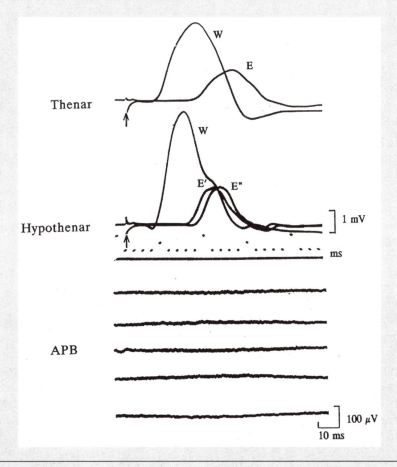

Figure 19.7 Responses evoked from the hand muscles of a patient with neurapraxia, using a stimulating and recording arrangement similar to that shown in figure 9.10. The thenar and hypothenar responses are much larger when the median and ulnar nerves are stimulated at the wrist *(W)* than at the elbow *(E)* or just below and above the elbow *(E', E")*. The differences in size indicate the presence of nonconducting nerve lesions in the forearm. The lower part of the figure shows the absence of fibrillation activity when a needle electrode is inserted into the abductor pollicis brevis (APB) muscle.

Reprinted, by permission, from A.J. McComas, 1977, *Neuromuscular Function and Disorders* (London: Butterworth Publishing Co.), 247.

20

Muscle Training

Skeletal muscle has a remarkable ability to adapt to the circumstances of its use. This *adaptation* can be with respect to size, enzyme activity, or *isoform* expression as well as changes in organelle and extracellular structures. This property of muscle to change its structure and function is referred to as plasticity. In chapter 19, we saw that the neuromuscular system deteriorates when it is not used. Now we will consider the consequences of increased use.

Physical *training* or the regular practice of a movement or task will lead to the enhancement of performance of that task. When we exercise, or more specifically when we perform repeated or sustained muscle contractions, there are various acute changes (in temperature and in the concentrations of ions and small molecules) in the muscle cell, in the extracellular space, and in the vascular fluids. For example, there is a decrease in intracellular $[K^+]$ and an increase in extracellular $[K^+]$ during a series of brief contractions. These changes represent a disturbance of *homeostasis*. With training, there is a reduction in the perceived *effort* required to perform the practiced task, less disturbance of homeostasis, improved tolerance for any such disturbance, and more rapid restoration of homeostasis following the exercise. These adaptations by the muscle facilitate performance of the type of exercise that is being practiced.

Thus, if force or power is required, the muscles become stronger, faster, or both. If, on the contrary, the muscle activity is of long duration, then the muscles become less readily fatigued. Although the muscle changes are most obvious, the *motoneurons* are also affected in terms of their patterns of *recruitment* and *impulse* discharges. In this chapter, we will survey the various outcomes of training and consider the mechanisms for the adaptations that occur in response to repeated practice of a physical exercise. Specific changes to enhance strength or power and endurance will be considered. As in so many other situations, the human and animal studies have different advantages and disadvantages and so tend to complement each other. Evidence from both types of studies will be considered.

Muscle Strength and Power

Muscle strength is usually expressed as the ability to exert force, which according to the force–velocity properties must be tested with isometric or slow *shortening contractions,* since under these conditions force is the highest. The maximum force exerted by a muscle is dependent on the cross-sectional area of the muscle, taking into account the angle of *pennation,* and on the pattern of excitation. Strength training results in larger muscles and enhanced excitation.

The properties of the *force–velocity relationship* will dictate the power that can be generated by a muscle (or synergistic muscle group). Maximal velocity of shortening is determined by fiber length and orientation (pennation) as well as the *myosin isoform* composition and the pattern of excitation. Improvements in power production by increased velocity of shortening against any load or resistance can be obtained by several mechanisms: addition of *sarcomeres* in series, reduced pennation, and transformation to faster myosin isoform. In addition, Van Cutsem *et al.* (1998) and Hakkinen *et al.* (1985) have observed increased *EMG* activity at the beginning of a ballistic contraction after training,

A higher firing frequency can contribute to improved performance.

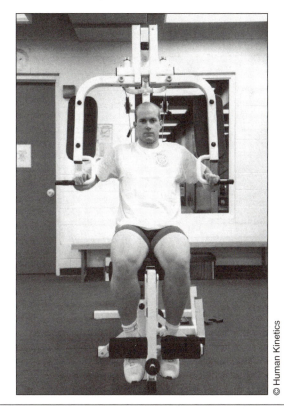

© Human Kinetics

© Human Kinetics

Figure 20.1   Methods for improving muscle bulk and strength.

indicating that a higher firing frequency can contribute to improved performance. It may be that a brief premotor silence, that is, an absence of EMG activity immediately before the contraction by shortening the *refractory period,* facilitates subsequent increase in firing (Walter, 1988).

Training for Improved Strength

Thomas DeLorme (1945), a captain in the U.S. Army during the Second World War, had many patients in his care with severe wasting and *weakness* of the quadriceps muscles caused by disuse following knee injuries. DeLorme reasoned that it was important to involve as much of the *motor unit* population as possible in the exercise program for their rehabilitation. Therefore, rather than subject his patients to many repetitions of low-intensity contractions, as had been the normal practice, he made them lift a heavy load instead. The load was one that could only just be raised 10 times in a sequence of contractions. This has come to be known as the 10-repetition maximum. As the muscles became stronger, the load was increased accordingly. This provided a simple progression to the intensity of the exercise, in proportion to the improvements made.

Rapidity of Strength Gains

Small numbers of intensive contractions are the most effective means of improving muscle bulk and strength.

The progressive *resistance training* program has since become accepted as the most effective means of improving muscle bulk and strength and, as such, has been adopted widely by bodybuilders and various other types of athletes. Whether exercising with weights or on one of the strength training machines designed for specific movements (see figure 20.1), one can obtain improvements in strength. So effective is the progressive resistance training program that no more than 10 repetitions a day, using loads from 60% to 90% of maximum, will increase strength by 0.5% to 1.0% per day over a period of several weeks (Jones *et al.,* 1989). An example is given in figure 20.2, taken from a study in which healthy subjects trained by lifting near-maximal loads on a leg extension machine. It can be seen that over a period of 12 weeks the mean maximum strength almost tripled (Rutherford & Jones, 1986).

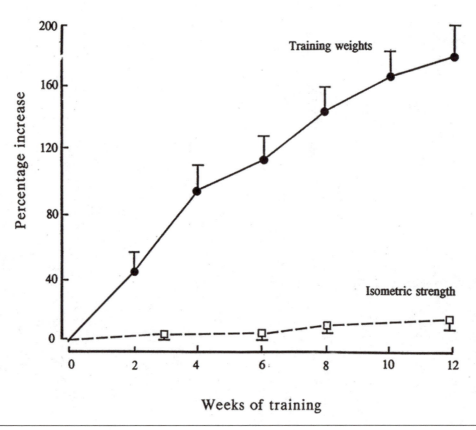

Figure 20.2   Marked improvement in the ability of the knee extensors to lift weights following the onset of a training program (upper curve). The lower curve shows the relatively small increase in isometric strength, measured with the knee at a right angle, in the same subjects.

Adapted, by permission, from O.M. Rutherford and D.A. Jones, 1986, "The role of learning and coordination in strength training," *European Journal of Applied Physiology* 55: 102.

The majority of improvements are specific to the movement practiced.

Neural adaptation can explain specificity of strength gains.

Specificity of Strength Gains

Two puzzling but related findings from progressive resistance training are that during the early stages, the increase in strength precedes the change in muscle girth and is largely restricted to the training task (Enoka, 1988; Sale, 1988). Thus, when the trained subjects with the results portrayed in figure 20.2 carried out maximal isometric contractions of their quadriceps muscles, their mean strength was found to have increased only 15%. More remarkable still, the power output of the same muscles, measured with a cycle ergometer, was unchanged. These discrepancies in performance after training illustrate a general rule, namely: The greatest changes are found in the training exercise itself. This observation has become known as the specificity of the training response.

The possible reasons for this specificity have been considered in some detail by Jones *et al.* (1989); they include differences in optimum muscle length between the training exercise and other tasks, and differences in the relative contributions of synergistic muscles. The main factor, however, appears to be a *neural adaptation* that may be expressed in various ways—for example, by higher initial motoneuron discharge frequencies and by more persistent firing of high-threshold motor units (Grimby *et al.*, 1981). Antagonist muscles, which may initially have been coactivated in the performance of the task, become less responsive as the training performance proceeds (Carolan & Cafarelli, 1992). This occurs presumably through inhibition of their α-*motoneurons* (see figure 20.3).

Invoking "neural adaptation" as the explanation for the discrepancy between improvements in the practiced movement and changes in the maximal isometric force is not entirely satisfactory. For example, one implication is that in the initial condition, full activation of the muscle is not possible; but the *twitch* interpolation technique (chapter 13) shows that most individuals are able to

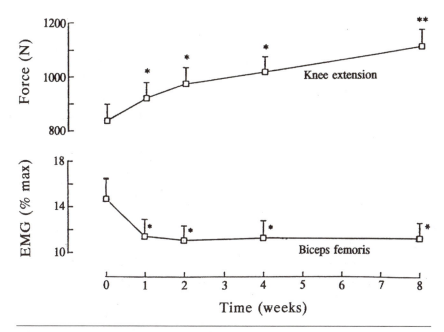

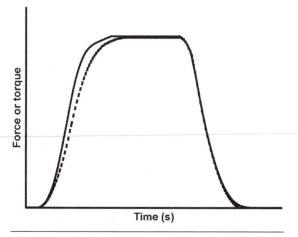

fully activate their muscles (or nearly so). However, it should be kept in mind that the twitch interpolation technique is typically used during a sustained isometric contraction. The remarkable changes in performance that are specific to the practiced task involve generally dynamic contractions, and the duration of these contractions is typically very short.

Peak Rate of Force Generation

Two aspects of neural adaptation could have profound effects on specific shortening contractions without much effect on maximal isometric force. One is a higher frequency of excitation, with the presence of doublet and triplet *action potentials* (see chapter 13). These discharges permit a higher rate of force development (see figure 20.4) for isometric and slow shorten-

Figure 20.3 One of the neural mechanisms involved in strength training. Twenty male subjects trained their knee extensors to perform isometric contractions, and the improvement is seen in the top curve. During the training period, there was a decrease in the associated activity in the biceps femoris, a hamstring muscle that opposes knee extension, as reflected in the EMG (bottom curve). Significant changes are shown by asterisks.

Adapted, by permission, from B. Carolan and E. Carafelli, 1992, "Adaptations in coactivation after isometric resistance training," *Journal of Applied Physiology* 73: 914.

Enhanced peak rate of force generation results from neural adaptation.

ing contractions, and greater shortening velocities in lightly loaded conditions. This aspect of neural adaptation could explain the highly specific improvements that are seen in brief and very fast shortening contractions (Toji *et al.,* 1997). Increased synchronization of motor unit excitation is the other neural factor that could contribute to the improved performance of brief contractions. However, greater synchronization probably does not contribute to increased sustained isometric force (Enoka, 1997).

An increased rate of force development could also result from an increase in tendon stiffness. Kubo *et al.* (2002) used ultrasound to detect *fascicle* shortening and determined stiffness from the force versus tendon length relationship. They found a significant increase in stiffness after 8 weeks of resistance training.

Training with ballistic contractions provides unique adaptations.

Plyometric training, which involves such repeated ballistic motions as hops and drop-jumps, has become a popular form of training for athletes. This type of training results primarily in neural adaptations that permit a higher frequency of firing of recruited motor units early in the effort.

Bodybuilding

Maximal muscle enlargement requires long-term training.

Figure 20.4 An increase in the rate of force development can result in greater force early in a contraction even if maximum force does not increase.

Adapted from Aagaard *et al.,* 2002.

Although a progressive resistance training program can produce quite striking muscle enlargement in dedicated athletes (see figure 20.5, left), scientific studies have resulted in only modest muscle changes. In these studies, the cross-sectional areas

Figure 20.5   Contrasting body shapes of a bodybuilder (left) and an in-line skater (right).

of the muscles were reliably outlined by modern diagnostic imaging machines such as ultrasound, computed transaxial tomography, and magnetic resonance. In a number of training studies lasting 3 to 5 months, the muscle cross-sectional areas increased by only 9% to 23% (Frontera *et al.,* 1988; Ikai & Fukunaga, 1970; MacDougall *et al.,* 1977).

In contrast, MacDougall *et al.* (1984) found that the cross-sectional areas of the biceps brachii muscles in bodybuilders were, on average, 76% greater than in untrained controls. The discrepancy between these results is probably due to the much longer training times undertaken by the bodybuilders in comparison with the experimental subjects, since the former group will have exercised for a period of years rather than months. However, the situation is complicated by the possibility that anabolic steroids may have been used by the bodybuilders, "power" athletes, and professional wrestlers who have such large muscles.

Increased muscle size occurs by fiber hypertrophy.

In the past, there has been argument as to whether the muscle enlargement is entirely due only to fiber hypertrophy or if there is an associated increase in the number of muscle fibers *(hyperplasia).* The latter situation might be expected to arise in one of two ways—by splitting of the muscle fibers or by *satellite cell* proliferation and formation of new muscle cells. In relation to the first possibility, it is certainly true that hypertrophied fibers may split into two or more daughter fibers. In this process, the myonuclei first move into the center of the fibers and there direct the laying down of membranes (Hall-Craggs, 1970). It is unlikely, however, that the daughter fibers become fully separated, for serial sections show that the splits do not usually run the full length of the parent fiber (Isaacs *et al.,* 1973). Further evidence against hyperplasia is that the cross-sectional areas of the muscle fibers, measured in biopsy specimens taken from bodybuilders and control subjects, show that fiber cross-sectional area is proportional to muscle cross-sectional area. Thus, although MacDougall and colleagues (1984) found considerable variation in fiber numbers (172,000-419,000) of the biceps brachii among their subjects, this variation was equally prominent in control subjects and bodybuilders, and the mean numbers of fibers were not significantly different in the two groups. Therefore the number of fibers in a muscle appears to be determined genetically and not to be increased by strength training.

In addition to increases in size and neural adaptations contributing to improved force-generating capabilities, increased angle of pennation has also been reported (Aagaard *et al.,* 2001). When the angle of pennation increases, the physiological cross-sectional area increases to a greater extent than the apparent volume of the muscle. This will permit greater maximal force generation.

Fiber Type Specificity

Type II (fast-twitch) fibers show the greatest hypertrophy.

When needle biopsies of muscles are obtained before and after training and are stained histochemically, it is possible to determine to what extent the different fiber types hypertrophy. The results of four such studies are given in table 20.1 (McComas, 1994). In three of these, the type II (fast-twitch) fibers show much larger increases in cross-sectional area than the type I (slow-twitch) fibers. The greater areas of the fibers, in turn, are due to increased numbers and sizes of *myofibrils*. It is thought that myofibrils increase in size and then undergo splitting to result in the increased numbers. It is not known if myofibrils can be synthesized de novo or whether all are formed by the splitting of thickened myofibrils (Goldspink, 1970). Associated with the new myofibrils are increases in the numbers of mitochondria and in the amounts of T-tubular and sarcoplasmic reticular membranes. The increases in mitochondrial volume appear not to be in proportion to the change in myofibrillar volume, but the usual ratio of *sarcoplasmic reticulum* volume to myofibrillar volume is maintained (Alway *et al.,* 1988).

Fiber type transformation may occur during strength training.

There is evidence that sprinters have a low incidence of slow-twitch (type I) fibers but other types of power athletes such as throwers, weightlifters, and high jumpers do not (Saltin *et al.,* 1977). In this study, it was not known whether the lower proportion of type I fibers in sprinters was due to training or to genetic endowment.

In a subsequent investigation in which sprinting was performed on a cycle ergometer, it was found that very high intensity training for just a few weeks could cause a reduction in type I fibers (Jansson *et al.,* 1990). Although the observed change in fiber type composition was not as large as reported for the difference between sprinters and other athletes, it can be imagined that a larger change could occur if the training were to be continued for a longer time. Furthermore, the simple identification of fibers as type I or type II may mask certain induced changes in the pattern of *myosin heavy chain (MHC)* isoforms. For example, although the proportions of type I and type II fibers are normal in bodybuilders, there may be an almost total absence of fibers with the MHC IIB (IIX) isoform but a much higher than normal incidence of fibers with the IIA isoform (Klitgaard *et al.,* 1990b). Similarly, Staron and colleagues (1990) have found that the incidence of type IIA fibers can increase at the expense of IIB (IIX) fibers; they studied a group of 24 women who underwent strength training over a 20-week period. Other researchers have noted increased proportions of hybrid (IIAX) fibers at the expense of IIX fibers (Ploutz *et al.,* 1994).

Table 20.1 **Effects of Strength Training on Cross-Sectional Areas of Muscle Fibers Belonging to Main Histochemical Types**

Authors	Number of male (m) & female (f) subjects	Muscle	Duration of training (weeks)	Changes in mean fiber area (%)		
				I	IIA	IIB(X)
Hather *et al.* (1991)	8 m	Vastus lateralis	19	+14	------ +32 ------	
Houston *et al.* (1983)	6 m	Vastus lateralis	10	+3	+21	+18
MacDougall *et al.* (1980)	7 m	Triceps brachii	22-26	+15	------ +17 ------	
Staron *et al.* (1990)	24 f	Vastus lateralis	20	+15	+45	+57[a]

[a]Includes type IIAB (AX) fibers.

Reprinted, by permission, from A.J. McComas, 1994, "Neuromuscular adaptations in man associated with alterations in activity," *Medicine and Science in Sports and Exercise* 26: 1499-1509.

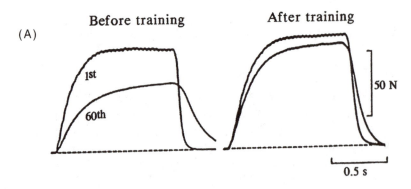

(A)

Before training   After training

1st

60th

50 N

0.5 s

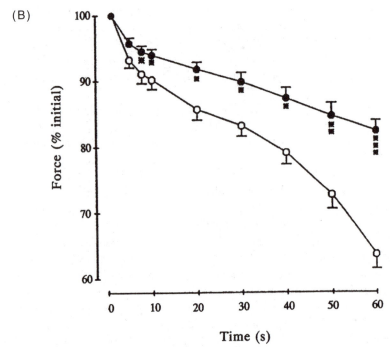

(B)

Figure 20.6 Responses of human thumb adductor muscles to endurance training. (A) Forces generated during 1st and 60th 30-Hz tetani. (B) Comparison of mean (±SE) losses in force during successive 30-Hz tetani in eight subjects before (○) and after (●) training. It can be seen that the trained muscles have become less fatigable. Means differed significantly at $p < 0.05$ (■), < 0.01 (■■), and < 0.001 (■■■).

From "Muscle fatigue, effects of training and disuse," by K. Hainaut and J. Duchateau, 1989, *Muscle and Nerve,* 12, p. 664. Copyright 1989 by John Wiley & Sons, Inc. Reprinted with permission of John Wiley & Sons, Inc.

Human Endurance Training

It is an everyday experience that frequent repetition of a motor task improves endurance, and this is reflected in the long training distances covered by competitive runners, cyclists, and swimmers. The beneficial effects of repetition are not restricted to young adults and to those in good health; impressive gains in endurance can be achieved by the elderly (for review, see Vandervoort *et al.,* 1986) and by patients with a previous myocardial infarction (Todd *et al.,* 1992). Figure 20.6 shows an example of the impressive changes in endurance that can be achieved.

Hainaut and Duchateau (1989) asked their subjects to perform 200 submaximal adductions of the thumb every day over a 3-month period. At the end of this time, *tetanic stimulation* of the adductor pollicis showed that force was maintained better than before when tested over a 1-min sampling time. In more proximal muscles of the limbs, endurance training results in muscles that are not only more effective during sustained activity but also, in the case of long-distance runners, more slender (figure 20.5, right). The explanation for the smaller girth is that the myofibrils, and the fibers themselves, are reduced in cross-sectional area. It is likely that this adaptation allows greater diffusion of metabolites and nutrients between the contractile filaments and the *cytoplasm* and between the cytoplasm and the interstitial fluid. This adaptation would diminish the impact of exercise on disturbance to homeostasis.

Endurance training produces muscles that have greater resistance to fatigue.

Mitochondrial Adaptation

Ingjer (1979) trained seven young women in cross country running for 24 weeks. In muscle biopsies, he observed that the numbers of capillaries had increased significantly around fibers of all histochemical types, while the type I fibers had gained the most mitochondria. During exercise, the more plentiful mitochondria would be able to keep the fibers better supplied with *ATP*, using aerobic metabolism. Similarly, the more extensive capillary bed would be expected to improve the delivery of oxygen and circulating energy sources (glucose, free fatty acids) to the fibers, while the products of muscle activity, especially H^+, K^+, and lactate, would be removed more effectively. Surprisingly, it appears that the increased size of the capillary bed is associated with a correspondingly greater muscle blood flow only during severe exercise (Hudlicka, 1990). During less extreme exercise, however, it is conceivable that oxygen extraction is enhanced by the larger contact area between the capillary blood and the interstitial space, even though the blood flow rate in individual capillaries may be lower than that in untrained muscles.

Fiber Type Transformation

In four studies of endurance athletes, the incidences of type I muscle fibers were higher, and those of type IIB (IIX) fibers lower, in comparison with controls (see table 20.2). These results leave open the question of whether differences in fiber composition are due to transformation from one fiber type to another or whether athletes are able to excel in endurance events because they are genetically endowed with a preponderance of type I and type IIA fibers. To resolve this issue, longitudinal studies have been undertaken in which previously untrained subjects engaged in various types of prolonged physical activity—for example, cross country running, skiing, or cycling.

Table 20.2 Fiber Type Composition (%) in the Vastus Lateralis Muscles of Endurance-Trained Athletes

Type of training	(n)	I	(P <)	IIA	(P <)	IIB(X)	(P <)	Authors
Controls	69	54		32		13		Jansson & Kaijser (1977)
			0.01		n.s.		0.01	
Orienteers	8	68		24		3.3		
Controls	6	51		41		7.1		Howald (1982)
			0.05		0.05		0.05	
Long-distance runners	9	78		19		2.5		
Controls	4	38		31		26		Fridén *et al.* (1984)
			n.s.		n.s.		0.01	
Cross country runners	6	52		35		12		
Sedentary controls	4	51		33		13		Baumann *et al.* (1987)
			0.001		0.001		0.001	
Professional cyclists	13	80		17		0.6		

n = number of subjects., n.s. = not statistically significant.

Pflügers Archiv. European Journal of Physiology, Exercise training induces transitions of myosin isoform subunits within histochemically typed human muscle fibres, H. Baumann, M. Jäggi, F. Soland, H. Howald, & M.C. Schaub, 409, 349-360, 1987, copyright of Springer-Verlag.

Table 20.3   Alterations in Fiber Type Composition (%) in Longitudinal Studies of Endurance-Trained Athletes

Type of training	(n)	% Fiber types							Authors
		I	(P <)	IIA	(P <)	IIB (X)	(P <)		
8 wk high intensity	12	b 41	n.s.	37	0.05	19	0.05		Andersen & Henriksson
		a 43		42		14			(1977)
8 wk high-intensity endurance	4	b 49	n.s.	38.3	n.s.	12.8	0.05		Baumann *et al.* (1987)
		a 48.8		43.3		8			
24 wk cross country running	7	b 58	n.s.	26	0.005	9.2	0.005		Ingjer (1979)
		a 57		32		3.4			
6 wk high-intensity endurance	10	b 50	0.05	37	n.s.	12.4	0.05		Howald *et al.* (1985)
		a 56		34		9.6			
15 wk high-intensity continuous/interval work	24	b 41	0.01	42	n.s.	17	0.01		Simoneau *et al.* (1985)
		a 47		42		11			
5 wk skiing with 80-kg sledge (500 mi)	7	b 29	n.s.	48	n.s.	21	n.s.		Schantz & Henriksson
		a 28		42		14			(1983)

n = number of subjects. b = before training. a = after training. n.s. = not significantly different (P > 0.05).

Some of these results are shown in table 20.3. It can be seen that, in contrast to the rather variable and generally unimpressive results of human strength training regimens (previously discussed), endurance training produces well-defined, consistent changes. Like results of electrical stimulation in animals, the results of human endurance training are to make some of the type II fibers acquire the physiological, biochemical, and structural features of type I fibers; other fast-twitch fibers alter their myosin ATPase activity and so convert from type IIB (IIX) to type IIA.

By employing antibody staining and gel electrophoresis, it has been possible to study in more detail the changes in MHC isoforms that are responsible for the altered histochemical staining. As an example, Schantz and Dhoot (1987) studied the triceps brachii muscles of six subjects who skied and pulled sledges for 800 km in mountainous territory over a 36-day period. At the end of this training period, the authors found that a significant proportion of fibers contained both fast and slow isoforms of the MHCs and hence could be regarded as hybrid fibers. In parallel with the changes in the MHCs was the appearance of slow isoforms of *troponins* I, T, and C. Similar results have been found by Klitgaard *et al.* (1990b) and by Baumann *et al.* (1987); in the latter study, in which subjects trained on a cycle ergometer, there was also a transformation of MHCs from the fast to the slow variety. There have been several reports of fiber type transformation following endurance training (Simoneau *et al.*, 1985; Schantz & Henriksson, 1983; Howald *et al.*, 1985). The general conclusion from the different studies is that if fibers transform as a result of endurance training, they do so in the following sequence:

$$IIX \rightarrow IIXIIA \rightarrow IIA \rightarrow IIAI \rightarrow I$$

It appears that the changes in fiber type composition continue as long as endurance training is sustained. Rusko (1992) obtained muscle biopsies from 9 elite skiers who retired from competition and training and 10 who continued to train. Histochemical analysis revealed an increase from 57% slow-twitch fibers to 68% in the group that continued training for the 8 years between biopsies. The control (detrained) group experienced a 6% decrease in slow-twitch fibers. Apparently the rate of fiber type transformation was less than 1.5% per year. This rather slow change is in contrast with the much more rapid changes reported in several other studies, but it must be remembered that this was a group of elite athletes who may already have undergone substantial change.

Training Studies in Animals

As noted earlier, animal studies have advantages over human investigations in that they can be more comprehensive; in addition, they lend themselves to large numbers of observations with good controls. Four strategies have been employed to alter muscle fiber properties in animals (for review, see Timson, 1990): exercise, weightlifting, internal loading (*tenotomy, ablation*), and chronic stimulation. The first three will be discussed in one section, while chronic stimulation will be discussed in another.

Exercise, Weightlifting, and Internal Loading (Synergist Ablation or Tenotomy)

Increased muscle use is not always an easy thing to achieve in animal models used to study the mechanisms of adaptation in muscle. The ingenuity of an experimenter in devising an animal model of exercise has often been exceeded by that of the rat (the usual choice) in minimizing effort. A rat on a treadmill may prefer to be carried against the backstop and allow the belt to slip underneath its body, while a rat placed in a water tank may elect not to swim but to rest on the bottom and come up for air as required!

- Treadmill running and swimming.

  -Treadmill running was used successfully by Gollnick *et al.* (1981). In combination with synergist ablation (see further on), this program resulted in weight increases in the rat plantaris and soleus muscles of 88% and 44%, respectively. These impressive examples of muscle hypertrophy were not accompanied by any change in muscle fiber number, determined by counting the number of fibers in the entire cross section of the muscle.

  -Several animal experiments using exercise have demonstrated that the changes in the motoneurons that take place during training programs are not limited to impulse firing patterns (see p. 300), for there are also changes in the histochemistry of the cells and in their fast axonal transport. In histochemical studies, Gerchman *et al.* (1975) exercised rats by swimming and demonstrated that acid phosphatase activity was diminished in motoneurons, while the reaction for glucose-6-phosphate dehydrogenase was intensified. In a similar type of study, this time using treadmill running over a period of 8 weeks, there was a pronounced increase in the oxidative enzyme activity of rat soleus motoneurons (Suzuki *et al.,* 1991). Just as an increase in oxidative enzyme activity enables muscle fibers to function more effectively during prolonged contractile efforts by improving the supply of ATP, so would there be a similar benefit to motoneurons in the same circumstances. Treadmill running has also been effective in increasing both the velocity of fast axonal transport and the amount of material carried down the axon (Jasmin *et al.,* 1988). It is possible that the extra protein is needed to repair any damage that may have taken place in the motor nerve terminals during contractile activity.

Muscle enlargement results from removing or tenotomizing synergistic muscles.

- **Synergist ablation or tenotomy.** The term "internal loading" describes a surgical manipulation that results in a muscle's having to carry an increased load whether supporting body weight, moving a limb, or contracting against an antagonist. Two methods have been devised. Tenotomy of synergistic muscles can produce an appreciable weight gain in the remaining agonist muscle during the next 7 days. However, much of the enlargement is due to edema rather than to the formation of myofibrillar proteins (Armstrong *et al.,* 1979). An additional disadvantage is that the tenotomized muscle tends to reattach itself to the residual muscle bellies through the growth of *connective tissue.* A better model is ablation of the synergistic muscles.

  -For example, the medial and lateral gastrocnemii can be removed so that plantarflexion of the foot is necessarily undertaken largely by the soleus and plantaris. With this approach, increases in muscle mass are usually in the 30% to 50% range and are sustained; they correspond to the changes in muscle fiber cross-sectional area (Timson *et al.,* 1985). The increase in tetanic tension in the hypertrophied muscles is associated with prolongation of the isometric twitch and with the conversion of a proportion of fast-twitch fibers to slow-twitch (Noble *et al.,* 1983). In contrast, the activities of glycolytic and oxidative enzymes are mostly unaltered.

Muscle volume per nucleus is maintained in hyper-trophied muscle.

-It is curious that in spite of the increased size of muscles after a strength training program, the muscle volume per nucleus is maintained at a constant value. Since muscle nuclei are considered postmitotic (cannot divide), how can this be? Rosenblatt and Parry (1992) showed in mice that the extra nuclei come from satellite cells. The authors irradiated muscles, then removed synergists. Synergist removal is a standard way of inducing compensatory hypertrophy in an animal model. In the treated animals, the radiation prevented mitosis in satellite cells, just as it would in cancer cells, and the irradiated muscles did not hypertrophy.

Striking enlarge-ments can be induced in avian muscles by weights.

• **Stretch-induced enlargement.** Increasing the load that a muscle must shorten against in animal experiments has met with mixed success, but loading some muscles to impose a chronic stretch has revealed a special kind of hypertrophy.

-Weightlifting models have required rats to maintain their positions in inclined tubes despite having weights attached to their tails (Exner *et al.,* 1973), to climb vertical poles with progressively larger weights strapped to their backs (Gordon, 1967), to lift weighted food baskets (Goldspink, 1964), and to stand upright with weights attached to their waists or to their necks (Ho *et al.,* 1980; Klitgaard, 1988). Some of these experiments were unsuccessful in increasing muscle weight; and in others, the changes, though significant statistically, were modest (8-21%).

-In birds, substantial increases in muscle mass have been observed after a weight has been attached to a wing (figure 20.7A). This treatment imposes a sustained stretch on the anterior latissimus dorsi (ALD) muscle. In one such study, the wet weight of the chicken ALD muscle was found to increase by 180% within 5 weeks (Sola *et al.,* 1973). Such increases are due to the addition of new sarcomeres at the ends of existing muscle fibers as well as to thickening of the fibers; in addition, the number of fibers may increase by 50%, following proliferation of the satellite cells (hyperplasia; Alway *et al.,* 1989; figure 20.7B). There is also a change in the MHC isoforms (Kennedy *et al.,* 1986). The slow tonic ALD muscle exhibits two types of fibers, one specific for the SM-1 heavy myosin isoform and the other for the SM-2 form. Following a 4-week treatment regimen, all the fibers reacted positively for the SM-2 isoform, while the SM-1 form was eliminated in the hypertrophied ALD muscle. Thus, the genetic expression of one particular heavy myosin isoform can be suppressed following such a treatment.

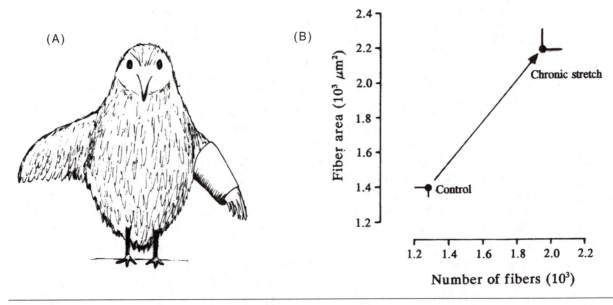

Figure 20.7   (A) Japanese quail with tubular weight attached to one wing. (B) Increased numbers and cross-sectional areas of fibers in stretched ALD muscles, compared with controls; the mean values have been given with their SEs.

Data of Alway *et al.* 1989.

-The situation in human subjects is in contrast to that in birds, in which chronic load bearing induces marked fiber hyperplasia (see p. 302); humans do not demonstrate hyperplasia. Perhaps the difference between the species is related to the fact that some avian muscles are largely composed of slow tonic fibers. These are fibers that give graded electrical and mechanical responses following nerve stimulation, and over which the motor nerve terminals are extensively distributed rather than grouped together in a single *end-plate* region or band (Ginsborg, 1960). While slow tonic fibers are a prominent feature of avian and amphibian muscles, in mammals they are found only in external ocular and inner ear muscles (Hess, 1970) and in *muscle spindles.*

Chronic Stimulation

Chronic stimulation induces all the properties of slow-twitch muscles.

A muscle can be chronically stimulated through electrodes placed around the motor nerve. A pioneering study was that of Salmons and Vrbová (1969), who attempted to change the twitch durations of the rabbit tibialis anterior by stimulating muscles continuously at 10 Hz. In these experiments, the stimulator was powered by a mercury cell and was embedded in epoxy resin; it was implanted in the abdomen and connected to stimulating electrodes around the peroneal nerve. After several weeks of continuous electrical stimulation, the twitch duration was significantly increased (see figure 18.2 in chapter 18).

The same type of experiment was repeated by Al-Amood *et al.* (1973), who used a miniature stimulator mounted on a vertebral spinous process to excite one of the *ventral roots* of the cat; these authors also found that slowing of contraction could be made to occur in a previously fast muscle. In keeping with the altered contractile properties of chronically stimulated muscles has been the demonstration by Streter *et al.* (1973) of a change in the structure of myosin. In the slowed tibialis anterior muscle of the rabbit, these workers were able to show by gel electrophoresis that the myosin had acquired the light chain pattern normally characteristic of slow muscle.

Although the stimulation paradigm has the obvious drawback that the motor units are activated without regard to their normal recruitment pattern, it has the advantage that the temporal pattern and amount of superimposed impulse activity are known precisely. From the experiments of Lømo and colleagues (Gorza *et al.,* 1988; Westgaard & Lømo, 1988), it is evident that the slowing of the twitch is dependent on the low frequency of the applied stimuli, while the resistance to fatigue is a consequence of the total number of stimuli applied (see figure 18.3 in chapter 18).

Sequence of Muscle Change

The changes in the muscles occur in an orderly sequence.

The chronic stimulation model has made it possible to determine the times of occurrence of the changes in the muscle that are characteristic of endurance training. The sequence is as follows for rat muscle stimulated at 10 Hz (Lieber, 1988):

Time	Change
3 hr	Swelling of the sarcoplasmic reticulum
4 days	Increased numbers and sizes of mitochondria, leading to a rise in oxidative enzyme activity
	Capillary formation, leading to a rise in muscle blood flow
14 days	Increased width of *Z-line*
	Decreased Ca^{2+} ATPase activity in the *SR*
28 days	Appearance of slow-twitch isoforms of myosin heavy and light chains and of troponin; decrease in muscle bulk, associated with reduced muscle fiber cross-sectional area

At 28 days, a chronically stimulated muscle, if previously fast-twitch, will have completed its conversion to slow-twitch. The conversion is evident both in the prolongation of the isometric twitch and in the greater incidence of type I fibers as shown by myosin ATPase staining.

Mitochondrial Adaptations

The changes in the mitochondria resulting from chronic stimulation are impressive; for example, in rabbit tibialis anterior muscles stimulated at 10 Hz during alternate hours, there is a sevenfold rise in mitochondrial volume after 28 days (Reichmann *et al.,* 1985). Despite this large increase, the ultrastructure of the organelles is well preserved, and the rise in citric acid enzyme activity is proportional to the new volume. If the stimulation is continued beyond 5 to 6 weeks, however, the mitochondrial volume declines, presumably because the muscle fibers, having been converted to type I, have a lower rate of energy utilization.

Adaptive Changes in DNA and RNA Processing

Regardless of whether the training program has been carried out in human subjects or experimental animals, the resulting structural changes in the muscle fibers must involve increases or decreases in protein formation. In turn, the protein effects must be a consequence of alterations in the processing of *DNA* or *RNA* in the fibers. With the recent application of molecular biology techniques to this area of study, it is now possible to understand some of the DNA and RNA events.

Transcription and Translation

Alterations in both gene transcription and mRNA translation may occur as a result of training.

In those situations in which new types of protein are produced in the muscle fibers (i.e., myosin heavy and light chains and troponin), it is clear that the genes for the respective isoforms must have been transcribed, with the appearance of the corresponding messenger RNAs (mRNAs) in the *sarcoplasm*. In some situations, however, synthesis of a type of protein already present in the muscle fiber may change too rapidly to be explained by an alteration in the amount of mRNA. For example, protein synthesis in the gastrocnemius muscle can be completely halted during the first 10 min of contractile activity (Bylund-Fellenius *et al.*, 1984), a time that is much shorter than the half-life of mRNAs.

Conversely, increases in the rates of synthesis of contractile and mitochondrial proteins may occur too soon to be due to the *transcription* of new mRNAs. In such situations it would appear that the *translation* of mRNA is modified. In many instances, however, it is probable that protein synthesis is regulated at both transcriptional and translational levels, as in the case of the increase in citrate synthase activity that follows 12-hr-per-day stimulation of rat fast-twitch muscle. Figure 20.8 shows that in the first 6 days there is a significant rise in enzyme activity without any change in mRNA level; between 7 and 10 days, however, the mRNA increases six- to sevenfold (Seedorf *et al.*, 1986).

Thomsen and Luco (1944) found that if the ankles of cats were fixed in the dorsiflexed position, the stretched soleus muscles enlarged during the next 7 days, before regressing. Similarly, if one-half of the diaphragm is denervated, it will undergo hypertrophy due

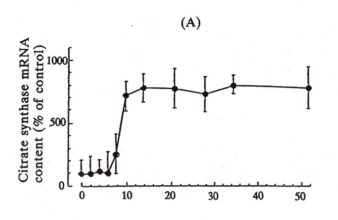

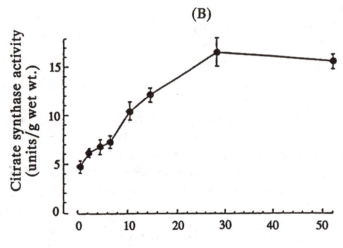

Period of stimulation (days)

Figure 20.8 Effects of chronic electrical stimulation on citrate synthase in rabbit fast-twitch muscle fibers. (A) Delayed rise in mRNA content (transcription). (B) Immediate increase in enzyme activity (translation).

Adapted, by permission, from U. Seedorf et al., 1986, "Neural control of gene expression in skeletal muscle. Effects of chronic stimulation on lactate dehydrogenase isoenzymes and citrate synthase," *Biochemical Journal* 239: 117.

Muscle stretch is an important stimulus for altering DNA transcription and mRNA processing.

Soluble factors, released by stretching the muscle fiber or the extracellular matrix, activate second messenger systems.

Second messengers activate immediate-early genes, which then modify the genes for muscle proteins.

to the stretching produced by the contracting half. More recently, Goldspink and colleagues (1991) showed that lengthening the rabbit tibialis anterior, by fixation of the ankle in a cast, could increase the muscle wet weight by 20% and the RNA content fourfold in as short a time as 4 days. If the lengthened muscle is stimulated, both muscle mass and RNA content increase still further. Associated with these changes is the appearance of slow MHCs, due to expression of the corresponding gene; at the same time, the gene for the fast MHCs appears to be repressed. To account for the rapid growth of the tibialis anterior, at least 30,000 MHC molecules must be synthesized by each *myonucleus* per minute (Goldspink, 1985). The converse experiment, in which the rabbit soleus muscle is fixed in a shortened position, causes expression of the IIB MHC as early as 2 days (Goldspink *et al.,* 1992).

Second Messenger Systems

Experiments performed on cultured fibers derived from skeletal or heart muscle have identified some of the cellular mechanisms that are triggered by stretching. In one system, cells are grown in a culture medium coated onto a silicone sheet; the sheet can then be stretched by known amounts for specified times. It turns out that stretch is a potent stimulus, for a 10% stretch, applied for only 1 min, is able to affect gene transcription 30 min later (Sadoshima & Izumo, 1993). It seems probable that stretch operates by releasing soluble factors from the muscle fibers or the extracellular matrix that operate as *second messengers,* and that some of these factors are prostaglandins (Vandenburgh *et al.,* 1991).

The soluble factors released by stretch act on the nucleus either directly, as in the case of the prostaglandins, or through other second messenger systems, as shown in figure 20.9 (see also Sadoshima & Izumo, 1993). Ca^{2+} will also be available as a second messenger when, as in the intact organism, the stretch of the muscle is the outcome of impulse activity and is associated with a 20- to 100-fold rise in intracellular Ca^{2+} concentration (Allen *et al.,* 1992). The first genes to be affected are immediate-early genes, such as c-fos (Sadoshima & Izumo, 1993). These genes, in turn, control the transcription of other genes in the nucleus, including those that code for all the special proteins needed for the transformation of the muscle fibers (e.g., myosin heavy and light chains, troponins and other Ca^{2+}-binding proteins, and the various muscle enzymes). Master regulatory genes (see chapter 5) may also be involved, to ensure that the genes responsible for the different types of fast or slow protein are expressed together.

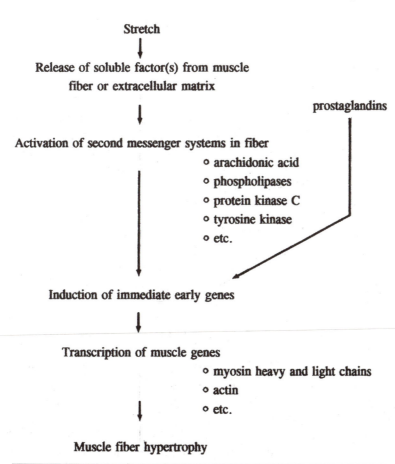

Figure 20.9 Sequence of intracellular events that results in muscle fiber hypertrophy following stretch.

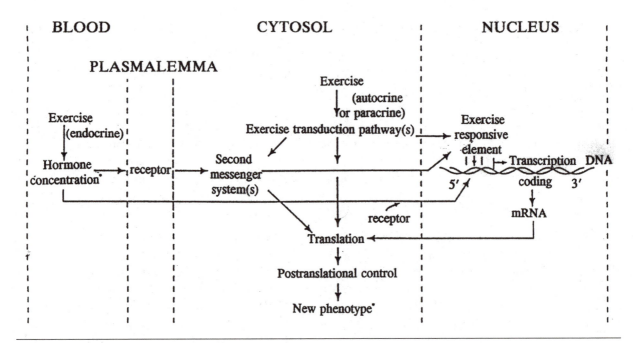

Figure 20.10   Simplified scheme of pathways involved in response of muscle fibers to training.

Adapted, by permission, from F.W. Booth, 1988, "Perspectives on molecular and cellular exercise physiology," *Journal of Applied Physiology* 65: 1462.

Finally, the increase in capillaries that accompanies endurance training may be the result of greater production of *fibroblast growth factor* (Kraus & Williams, 1990). Figure 20.10 summarizes some of the pathways that are likely to be involved in bringing about the adaptive changes in skeletal muscle fibers.

Applied Physiology

Exercise programs can benefit patients with neuro-muscular diseases.

In this section we will briefly consider whether or not exercise training programs are effective in patients with chronic degenerative neuromuscular disorders, such as old polio, muscular dystrophy, and *spinal muscular atrophy.* On the one hand, it might be supposed that an appropriate training program would enlarge those muscle fibers that were still relatively healthy and enable them to develop more force. On the other hand, as will be shown in the next chapter, exercise can damage healthy muscle if it is excessive in intensity or duration. Were this to happen in muscles that were already weak through disease, the consequences could be catastrophic in terms of the activities of daily living. Tasks that could previously have been accomplished with some difficulty might now become impossible. With these considerations in mind, there is an obvious need to carry out studies on the effects of exercise programs in patients. Though this might appear a simple task, it is usually difficult to collect a group of patients having the same diagnosis, with moderate rather than severe weakness, and with the willingness and opportunity to travel to the gym for supervised daily training sessions. The issue of the role of exercise in muscular dystrophies, as an example, remains controversial (Ansved, 2001).

In one small study involving patients with various diagnoses, significant improvements in strength were observed (McCartney *et al.,* 1988). More striking, though, was the ability of training to decrease *fatigability.* This last effect is especially important, for an improvement in stamina may be more useful in many daily tasks than an increase in strength. Another relevant study, this time restricted to persons with old polio, also showed increased muscle performance following an exercise program (Feldman & Soskolne, 1987). It cannot be stressed sufficiently, however, that in any exercise program patients must start with small loads and proceed slowly, for fear of overtaxing the already damaged muscles.

In the next chapter, we will examine the muscle damage that can be brought about by the mechanical stresses of intensive exercise, even in healthy individuals. We will also look at the way the fibers are able to repair themselves.

21

Injury and Repair

There are many ways in which muscle fibers can be damaged (see "Causes of Muscle Fiber Necrosis"). External causes include crushing and laceration injuries to the body and extremes of heat and cold. Internal causes include muscle tears and tendon ruptures following sudden forceful contractions. Muscle tears may be associated with considerable bleeding into the muscle belly due to breaching of the walls of the intramuscular blood vessels. Unaccustomed exercise, especially that involving *lengthening contractions,* may also damage muscles (see the following sections). Finally, degeneration or *necrosis* (death) of muscle fibers is a feature of a number of diseases, particularly those due to inflammation of the muscles *(polymyositis, dermatomyositis)* or to inherited defects (e.g., *Duchenne muscular dystrophy, malignant hyperthermia).* Especially alarming is the muscle necrosis that may accompany local infections with virulent strains of streptococcus A. So rapidly may the necrosis spread that, in the case of an infected limb, amputation may be required as a lifesaving measure. As will be shown, the degenerative and necrotic processes that come into play are similar regardless of the nature of the provocative event. Further, in almost all cases there is an attempt at *regeneration,* such that entirely new fibers, or segments of fibers, can be formed. Especially in young fibers subjected to mechanical injury, the regeneration is rapid and successful.

We will commence our study of muscle injury by examining the changes that may develop after strenuous exercise in healthy untrained individuals.

Causes of Muscle Fiber Necrosis

Alcohol intake
- Bupivacaine
- Rifampicin
- Quinacrine
- Calvacin
- Others

Diseases
- Polymyositis, dermatomyositis
- Muscular dystrophy
- Biochemical disorders (myophosphorylase, carnitine deficiencies)

Exercise (especially eccentric)

Irradiation

Ischemia

Mechanical injuries (stretching, crushing, cutting)

Thermal injuries (heating, freezing)

Biological toxins (venoms, streptococcus A toxin)

Muscle Contraction-Induced Damage

Repeated lengthening contractions can cause surprisingly severe morphological changes in the muscle fibers. From chapter 11 it will be recalled that lengthening contractions occur when the muscle unsuccessfully resists elongation, acting as a brake. Naturally occurring examples are contractions in the biceps brachii as a heavy load is being slowly lowered and the activity in the gluteus maximus that checks the forward swing of the leg during running and walking. In terms of the sliding filament mechanism of muscle contraction, the *myosin cross-bridges* make repeated connections with the *actin* filaments while the muscle is activated. However, during lengthening contractions, the actin filaments, instead of being propelled toward the center of the myosin filament, are pulled in the opposite direction by the external forces on the muscle. The external forces may be a large load, gravity, or the contraction of an antagonist muscle. With reference to the pattern of striations along the muscle fiber (see figures 1.1 in chapter 1 and 11.2 in chapter 11), the *H-zone* in the center of the *sarcomere* will widen as the sarcomere lengthens, as will the *I-band*.

The changes in muscles that result from repeated lengthening contractions have been studied in human subjects by several groups (Fridén *et al.,* 1983; Newham *et al.,* 1983; Clarkson & Sayers, 1999; Gibala *et al.,* 2000, 1995). The changes are barely evident within the first 24 hr of exercise and are then restricted to streaming of the *Z-lines.* By 10 to 15 days, however, a significant proportion of the muscle fibers have undergone necrosis, and there are prominent infiltrations of mononuclear *inflammatory cells* in and around the degenerating fibers (Jones *et al.,* 1986). After an additional 2 to 3 weeks, much of the muscle damage has been repaired by regeneration of fiber segments; nevertheless, many fibers still have central *nuclei* and vary considerably in diameter.

> **Lengthening contractions can produce fiber necrosis.**

The tendency of lengthening contractions to cause severe inflammatory and degenerative changes in muscles has been confirmed in the animal study of McCully and Faulkner (1985). The extensor digitorum longus (EDL) muscles of anesthetized mice were repeatedly tetanized while being stretched by a servomotor. Three days later, 37% of fibers had degenerated, and the maximum tetanic tension was only 22% of the control value (figure 21.1); at this time inflammatory cells were prominent, but 1 day later regeneration processes were already under way (figure 21.2, see also Faulkner *et al.,* 1989).

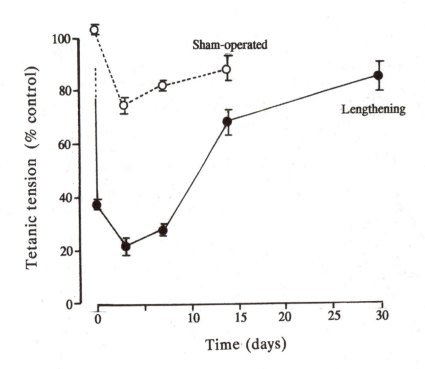

Figure 21.1 Decline in force-generating capacity of mouse extensor digitorum longus muscles at various times after repeated tetani, delivered with the muscle lengthened. The decline was significantly greater and longer lasting than those following shortening and isometric contractions (not shown). Results for sham-operated animals are also given; all values are means ± SEs.

Adapted, by permission, from K.K. McCully and J.A. Faulkner, 1985, "Injury to skeletal muscle fibers of mice following lengthening contractions," *Journal of applied Physiology* 59: 125.

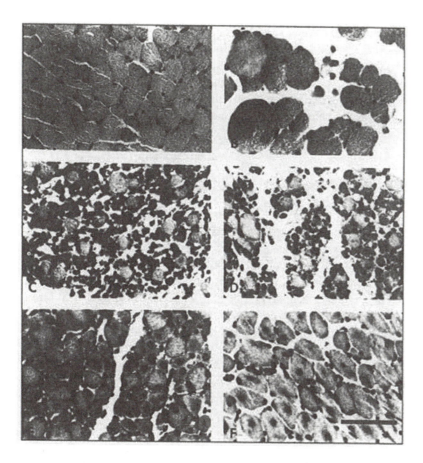

Figure 21.2 Damage caused by lengthening contractions in mouse extensor digitorum longus muscles. The sections are from (A) a control muscle and (B) experimental muscles at 1 day, (C) 3 days, (D) 4 days, (E) 7 days, and (F) 14 days after the stimulated contractions. In B, the fibers are already swollen and rounded, and in C and D, many are undergoing necrosis and have been invaded by inflammatory cells. In E, regeneration is well under way, and in F, the regenerated fibers can be recognized by their central nuclei.

Photographs courtesy of Dr. John Faulkner and Dr. Richard Hinkle; see also McCully and Faulkner 1985.

Biochemical and Inflammatory Events Accompanying Initial Insult

The fiber damage is due to Ca^{2+} and free radicals in the cytoplasm.

As previously noted, during lengthening contractions the sarcomeres, especially in the central regions of the muscle fibers, are being overstretched. Evidence from human and animal studies of structural damage to muscle fibers following lengthening contractions includes sarcolemmal disruption, swelling and disruption of the sarcotubular system, distortion of contractile proteins, damage to the cytoskeleton, and abnormalities in the extracellular matrix of the fibers (Fridén & Lieber, 2001). Jones *et al.* (1984) believe that the entry of excessive amounts of Ca^{2+} into the muscle fiber from the interstitial fluid is a critical step in producing damage. Jones *et al.* reached this conclusion on the basis of experiments in mouse muscles that had been either stimulated to exhaustion, treated with calcium ionophores, or treated with various metabolic inhibitors (cyanide, dinitrophenol, or iodoacetic acid). They found that the muscle fiber damage that would normally occur under these conditions could be prevented by removing Ca^{2+} from the bathing fluid. In muscles that had performed lengthening contractions, it would be reasonable to suppose that Ca^{2+} gained access to the overstretched regions of muscle fibers through minute tears in the *plasmalemma*, but it has been demonstrated that Ca^{2+} channel blockers can attenuate the degradation of *desmin* and Z-band streaming that occurs following a bout of lengthening contractions (Beaton *et al.*, 2002). This observation indicates that some of the relevant increase in cytoplasmic Ca^{2+} results from entry through *Ca^{2+} channels*. Inside the *cytosol*, Ca^{2+} ions activate a calcium-activated *protease* known as *calpain* and a phospholipase, which could digest structural proteins and lipid membranes, respectively. Jackson *et al.* (1984) consider that of the two calcium-activated enzymes, phospholipase is the most likely culprit, and they suggest that additional damage ensues from the liberation of fatty acids and oxidation of the latter with corresponding generation of *free radicals* (see Jones & Round, 1990). An experimental finding consistent with free radical involvement is that intraperitoneal injection of *superoxide dismutase,* an enzyme that scavenges free radicals, significantly reduces the loss of tension in mouse EDL muscles subjected to lengthening contractions (Zerba *et al.,* 1990).

As a result of the damage to the muscle fiber, fluid enters the cells, and neutrophils invade the area, perhaps attracted by some of the products of protein degradation. *Macrophages* invade the tissue after neutrophils and begin producing free oxygen radicals and *cytokines* (interleukin-1, interleukin-6, and *tumor necrosis factor*); enzymes that enhance tissue degradation are activated. The swelling associated with the inflammatory response may not peak until 5 days following the injury (Clarkson & Sayers, 1999).

Appearance of Muscle Enzymes in Plasma, and Delayed-Onset Muscle Soreness

Heavy muscular exercise of any kind can cause significant rises in the plasma levels of muscle enzymes; of these enzymes, the most sensitive indicator is *creatine kinase.* Increases in enzyme titer of two to five times above the resting level can be observed 24 hr after exercise, and elevated CK levels may persist for as long as 6 days (Fridén & Lieber, 2001). Following prolonged or repetitive lengthening contractions, there is another, much larger, rise in plasma enzymes (up to 100 times normal), which reaches a peak at about the fifth day. It is probable that the large amounts of enzyme are being released from regions of muscle fibers undergoing necrosis (see p. 317). The rises in muscle enzyme are prevented if subjects are trained using lengthening contractions, and there is also much less fiber damage (Jones & Round, 1990); the basis for this increased resistance is not known, but increased desmin content has been observed within 72 hr of a single bout of lengthening contractions (Barash *et al.,* 2002). This increased cytoskeletal protein content may contribute to improved tolerance for the strain of lengthening contractions.

Especially in untrained subjects, lengthening contractions may cause appreciable muscle discomfort or even pain. The discomfort is not usually evident during the period of the exercise, but appears on the following day and can last for several more days. In view of the latent period, the discomfort is referred to as *delayed-onset muscle soreness (DOMS).* While the muscles may be somewhat swollen and also tender on palpation, the discomfort is felt most when the muscles are engaged in activities that involve further lengthening contractions. It is thought that DOMS is caused by a sensitization of muscle nociceptors by the tissue breakdown products such that these receptors become more sensitive to various stimuli, including contractile activity (Proske & Morgan, 2001). When the soreness has passed, further bouts of lengthening contractions produce comparatively minor discomfort, in keeping with the reduction in fiber damage observed with the microscope (discussed earlier).

Muscle Injury From External Causes

The previous section explored the muscle changes that result from lengthening contractions in human subjects and experimental animals. We will now consider damage from external sources—usually mechanical injury of one kind or another or the application of a necrotizing chemical such as the local anesthetic *bupivacaine.* The cellular events that take place in the muscle fibers are qualitatively similar to those occurring after lengthening contractions, but they usually result in more necrosis. The nature of the cellular mechanisms that permit the muscle fibers to repair themselves will also be considered.

Muscle Grafting and Pinching As Experimental Approaches

One of the techniques for studying the effects of muscle injury was pioneered by Studitsky (1952) and subsequently exploited by others (e.g., Carlson, 1970); this technique involves the following sequence: remove a muscle from an animal, mince it into small cubes, and insert the pieces into the former bed (see figure 21.3). Alternatively, the muscle can be excised and replaced intact. In both instances, the transplanted muscles are autografts. In contrast, a homograft is a muscle transferred to the normal anatomical position but in another animal.

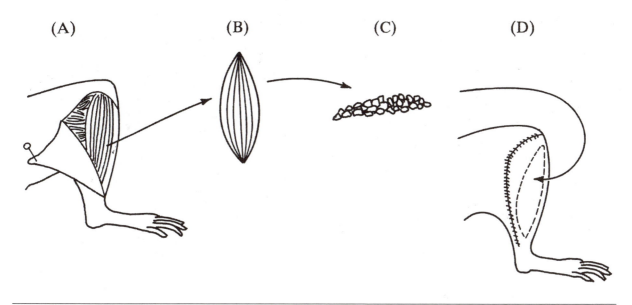

Figure 21.3 The transplantation of minced skeletal (tibialis anterior) muscle using the techniques of Studitsky (1952) and Carlson (1970). The muscle is (A) exposed, (B) removed, (C) cut into small pieces, and (D) packed into the tissue bed left by the same or a different muscle.

Reprinted, by permission, from A.J. McComas, 1977, *Neuromuscular Function and Disorders* (London: Butterworths Publishing Co.), 98.

Less radical procedures than whole-muscle grafting are to divide or pinch a muscle; if a slender muscle with longitudinally running fibers is chosen, the microscopic analysis of subsequent events is simplified. All these methods, and some of those listed in, "Causes of Muscle Fiber Necrosis" on p. 313, have the advantage that the full repertoire of possible cellular events can be examined, including those involving the immune system. In contrast, *in vitro* studies of cultured muscle fibers enable more precise observations to be made by interference or phase-contrast microscopy; in addition, the fates of individual fibers can be followed by time-lapse photography. However, this approach precludes the involvement of systemic responses normally involved in the muscular response to injury.

The influx of Ca^{2+} and complement promotes fiber necrosis.

Local Biochemical Events Following Muscle Injury

Following a sharply localized pinch or incision, a *necrotic zone* develops at the site of damage and extends for a few millimeters on either side (figure 21.4A). The torn plasmalemma allows Ca^{2+} ions to enter the fiber *cytoplasm* from the interstitial fluid; as described in a previous section, Ca^{2+}, once admitted to the fiber, activates proteases that break down proteins in the *myofibrils,* mitochondria, and cytoskeleton. Circulating *complement* components also gain access to the fiber through gaps in the plasmalemma. These assist in the destruction of the myofibrils, organelles, and tubular systems (Arahata & Engel, 1985).

Some of the products formed by the actions of the Ca^{2+}-proteases and complement fixation become molecular signals that attract mononuclear cells from the capillary circulation. A number of these cells, the macrophages, invade the fiber, continue the digestion of its contents, and then remove the debris (figure 21.4A). T-lymphocytes may also arrive at the site of injury, presumably in response to the liberation of cytokines. Figure 1.17, in the section on Duchenne muscular dystrophy in chapter 1, illustrates in schematic form these cellular events.

On both sides of the central necrotic zone, the uninjured myofibrils retract and, with the organelles and tubular systems, form stumps that become covered with newly formed membrane. The membrane evidently has a protective role in that it prevents the entry of Ca^{2+} and complement.

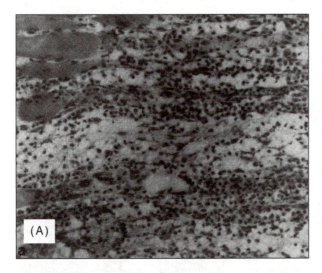

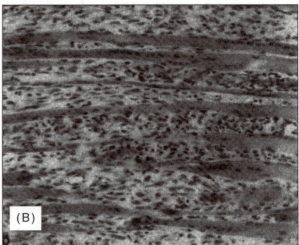

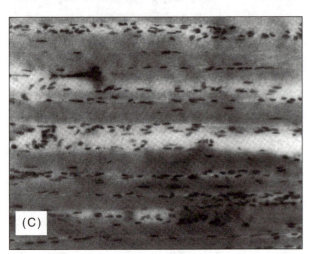

Figure 21.4   Histological changes in rat semitendinosus muscle fibers after crush injury. (A) At 2 days, the damaged fiber segments have undergone necrosis, with digestion and removal by invading macrophages. (B) At 5 days, several newly formed slender *myotubes,* with central nuclei, can be seen in the damaged region. (C) At 10 days, the myotubes have transformed into muscle fibers, many of which have already linked up with the fiber stumps on either side.

Reprinted, by permission, from A. Stuart et al., 1981, "Electrophysiologic features of muscle regeneration," *Experimental Neurology* 74: 157.

Distal to the necrotic region, the fiber is functionally denervated.

Once the plasmalemma is torn and the underlying region of the fiber begins to degenerate, the otherwise intact distal part of the fiber loses communication with the motor *end-plate.* Neither electrical *impulses* nor *trophic* factors are able to span the damaged segment, and hence the distal sarcomeres are functionally denervated (McComas & Mrózek, 1967). Although few electrophysiological studies have been made of this situation, the isolated region would be expected to exhibit all the features of *denervation,* including a fall in resting *membrane potential* and the synthesis of *AChRs.* Further, the release of a diffusible molecular signal appears capable of causing neighboring motor nerve fibers to *sprout* and to form new neuromuscular junctions on the denervated segments (see chapter 17).

Role of Satellite Cells During the Repair Process

Satellite cells are responsible for forming new fiber segments.

At the same time that the central region is undergoing necrosis, this and neighboring parts of the fiber begin the repair process. The first important step is the activation of *satellite cells.* These cells, which can be recognized by electron microscopy, consist of nuclei with very little cytoplasm. Like the *myonuclei,* they lie at the periphery of the muscle fiber, but they are surrounded by their own plasmalemma (see figure 1.14 in chapter 1). Normally there are few of these cells bordering a single muscle fiber; in human muscle, for example, they constitute 4% to 11% of the muscle nuclei (e.g., Wakayama, 1976). In response to an unknown signal from the damaged region of a fiber, the previously dormant satellite cells first proliferate and then migrate into the necrotic area before differentiating into *myoblasts.* As in the embryological development of muscle, the myoblasts fuse to form myotubes (figure 21.4B). With the electron microscope, the myoblasts can be recognized not only by their thin profiles and central nuclei, but also by the abundance of ribosomes. The control of cell proliferation, differentiation, and fusion, and the accompanying synthesis of muscle-specific proteins, are under genetic control. Although molecular biology studies of muscle regeneration do not appear to have been made, it is probable that the interaction of myogenic genes and growth factors is similar to the interaction responsible for muscle development in the embryo (see chapter 5).

The myotubes link the separated muscle fiber stumps together.

As the ends of the myotubes extend, they reach the intact stumps of the damaged fibers, and union occurs, with dissolution of the respective plasmalemma; the fibers are once more in continuity (see figure 21.4C). The time course of the bridging process can be estimated by a simple electrophysiological technique in which the muscle is stimulated on one side of the lesion and *action potentials* are recorded on the other side, using microelectrodes to impale individual fibers (Stuart *et al.,* 1981). In rat semitendinosus muscles crushed with watchmaker forceps, *reinnervation* can be detected at 5 days and is complete by 30 days (see figure 21.5). The architecture of the muscle is best restored if the endomysial sheaths surrounding the damaged fibers have been preserved, since the *endomysium* can act as a scaffold for the containment and alignment of the myotubes. If the endomysium is ruptured, however, satellite cells can escape into the interstitial space and form new fibers.

Applied Physiology

In this section we will see how muscle grafting has been taken out of the animal research laboratory into the patient operating theater and used for the correction of physical deformity.

A possible role for stem cell transplantation in muscle repair?

The possibility exists in the future that stem cells, which are precursor cells that can be induced to become specific tissue-type cells, may be used to help repair damaged or diseased muscles. This approach, which would involve injecting stem cells directly into the tissue, would be used to assist the "endogenous" stem cells, the satellite cells, in the muscle repair and regeneration process, or perhaps replace defective satellite cells such as in the muscular dystrophies. Recent evidence has shown that several different cell types of nonmuscle origin can give rise to muscle cells *in vivo,* including *connective tissue* cells, myofibroblasts, vascular endothelial cells, circulating bone marrow-derived stem cells, dermal *fibroblasts,* and neural stem cells (Grounds *et al.,* 2002). Recently, LaBarge and Blau (2002) discovered that adult bone marrow-derived cell transplants became incorporated into mouse muscle fibers as satellite cells following radiation-induced damage. Furthermore, when these same muscles were subsequently damaged by lengthening contractions, these new satellite

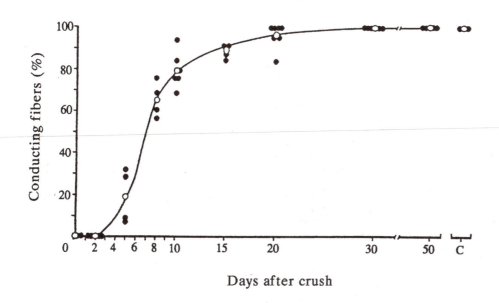

Figure 21.5 Incidence of muscular fibers with restored functional continuity following crush injury. The assessment was determined by recording action potentials in single fibers after stimulating the muscle on the other side of the crush. C = control (intact muscle). Open circles denote mean values.

Reprinted, by permission, from A. Stuart et al., 1981, "Electrophysiologic features of muscle regeneration," *Experimental Neurology* 74: 153.

cells were capable of taking part in the production of muscle fibers. Clearly, the issue of stem cell use to assist in tissue repair and regeneration has very significant implications for muscle repair following exercise-induced damage, injury, and disease.

Muscle grafting can be used in patients with facial palsy.

The technique of muscle grafting has in recent years become more widely used in plastic surgery. However, there might be little benefit to the patient if the fibers in the graft, having been deprived of their blood supply, degenerated and died with no certainty of successful regeneration afterward. For this reason, microvascular anastomoses are performed under the operating microscope, the artery and vein in the cut pedicle (stalk) to the muscle being delicately sutured to corresponding vessels in the tissue bed prepared for the graft. Perhaps nowhere is the challenge of grafting greater than in the face, for success or failure is all too visible. Figure 21.6 shows the design of a grafting procedure for restoring mobility to one side of the face following irreparable damage to the facial nerve on that side. Such damage may be the result of inflammation of the nerve (Bell's palsy) or the inevitable consequence of removing a large tumor from the adjacent eighth cranial nerve within the cranial cavity.

In the restorative operation, a block of gracilis muscle is taken from the inner thigh, complete with its pedicle, and is grafted onto a tissue bed underneath the cheek. Either at the same time or in a later operation, a branch of the "healthy" facial nerve on the normal side of the face is divided and joined to a length of sural nerve taken from the calf. The sural nerve graft is led through the soft subcutaneous tissues between the nose and upper lip and sutured to the nerve stump in the gracilis muscle graft (figure 21.6, see also Harii *et al.,* 1976). The completely severed *axons* in the sural nerve graft will die, but facial nerve axons from the normal side of the face will eventually regenerate through the graft and establish synaptic connections with muscle fibers in the transplanted gracilis. Henceforth, the two sides of the face will contract together under the influence of the single facial nerve. The results of the operation can be very satisfactory to the patient, for both in repose and in smiling, evidence of paralysis may be lost.

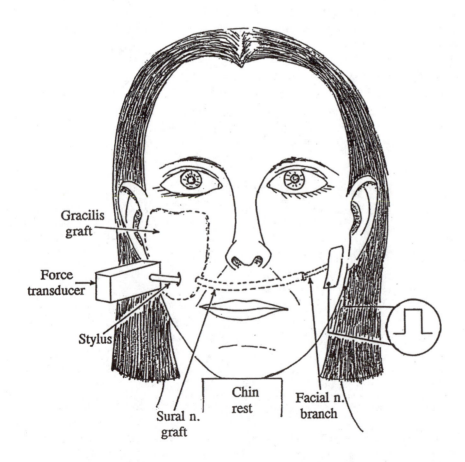

Gracilis graft

Force transducer

Stylus

Sural n. graft

Chin rest

Facial n. branch

Figure 21.6 Double-grafting procedure for facial nerve palsy. A piece of gracilis muscle is transplanted from the thigh to the paralyzed side of the face. The graft receives its nerve supply from the opposite side of the face, through a facial nerve branch that regenerates through a sural nerve graft. The figure also shows the stylus of a force transducer indenting the skin so that the contraction of the stimulated muscle can be recorded.

Grafted human muscles may not fully transform.

From our understanding of the trophic influences exerted by *motoneurons* on muscle fibers (see chapter 18), we would expect the slower-twitch muscle fibers of the gracilis to transform into the fast-twitch fibers characteristic of healthy facial muscle. Curiously, this may not happen, as in the case of the patient whose results are illustrated in figure 21.7. It can be seen that when the healthy facial nerve is stimulated, the *twitch* remains much slower in the grafted gracilis muscle than in the normal facial muscle on the opposite side (Hawrylyshyn *et al.,* 1996). Possibly some human muscles are less "plastic," in response to altered innervation than the muscles that have been studied extensively in animals.

There is one more topic for consideration before our survey of skeletal muscle is complete. This is the appearance and the physiological properties of muscle in old age. As we shall see, there are very marked changes both in the muscle fibers and in their nerve supply. Even so, muscle performance can still be enhanced through *training*.

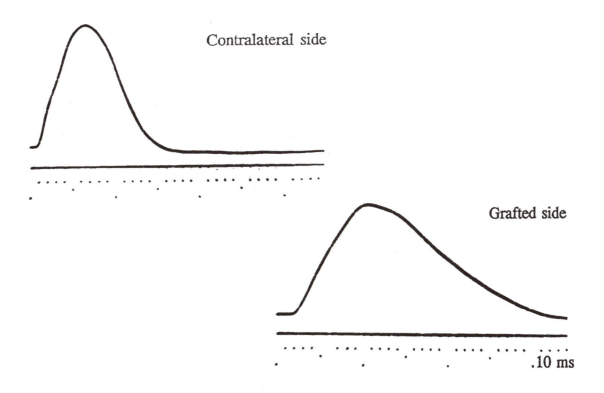

Contralateral side

Grafted side

.10 ms

Figure 21.7 Comparison of muscle contractions on the two sides of the face in a patient who had undergone the nerve and muscle grafting procedure shown in figure 21.6. On stimulation of the facial nerve, the twitch was much slower on the side of the graft.

22

Aging

Not only are men and women living significantly longer than before, but many are able to pursue full and active lives in their later years. Nevertheless, important changes in the motor system accompany aging; muscles become thinner, tendon compliance increases (Kubo *et al.*, 2003), and movements become slower and less precise. The study of world records indicates that peak physical performance is reached in the 20s and is maintained for 5 to 10 years before declining. This decline in performance is evident in figure 22.1, which shows the best speeds achieved for the 100-m and 10-km distances by male and female runners of different ages. It is interesting to note that the apparent age of optimal performance has increased in some sports recently (figure skating, competitive swimming). This development reflects the willingness of successful participants to persist with high-level *training* and performance into adulthood because this has become financially feasible. This shift in the age of peak performance raises an important issue concerned with the mechanism of decline in physical performance. How much of this decline is due to a predetermined aging process, and how much relates to *adaptation* to a decreasing physical demand on the motor system? The various features of the aging motor system will now be considered.

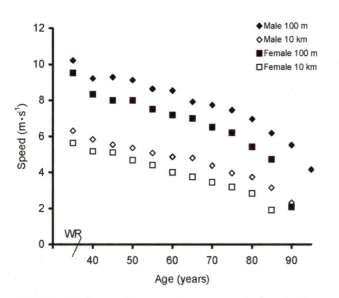

Figure 22.1 Record running speeds for 100-m and 10-km distances at different ages for masters athletes. Initial (left side) values are world records.

Changes in Muscle With Aging

The aging process begins at an early adult age in humans; muscle fiber number begins to decrease after the mid-20s. This development contributes to decreases in muscle size and increased *connective tissue* content. It is difficult to separate changes attributable to aging and changes due to decreasing chronic activity. The changes in muscle that occur with aging can be categorized as changes in strength, structure, size, fiber type, and contractile characteristics. The *motoneurons* also undergo changes that are associated with aging. The apparently downhill slide can be arrested at least temporarily by strength training programs, indicating that at least some of these changes can be attributable to inactivity. In the following sections we will describe the changes that occur in muscle as we get older.

Changes in Muscle Strength and Structure

Comparisons of maximal isometric voluntary contractions (MVCs) in young and elderly subjects have been made for a number of muscle groups, as shown in table 22.1. In this type of study, it is important not only to examine elderly subjects in good health but also to establish whether or not the voluntary contractions are indeed maximal. One way of resolving the latter uncertainty is to employ the twitch interpolation technique (see chapter 13). For muscles of the ankle it appears that with encouragement from an observer, the elderly can recruit *motor units* as effectively as younger subjects (Vandervoort & McComas, 1986). It is therefore likely that the substantial reductions in strength, shown for the various muscle groups in table 22.1, are the result of changes in the muscle fibers and their nerve supply rather than in the motor commands generated in the central nervous system. The loss of strength affects men and women equally and involves all the muscle groups tested. The decline is, as expected, greatest in the oldest subjects, in whom some muscles can develop only half the force observed in younger people.

The ability to generate force at high velocity decreases.

In contrast to the numerous studies of isometric contractions, little attention has been given to the forces developed during dynamic contractions in the elderly. It would appear, however, that strength is compromised to a greater extent at high velocities (e.g., $180° \cdot s^{-1}$) than at low ones, or in

Table 22.1   Summary of Studies on Changes in Isometric Muscle Strength With Aging

Studies (categorized by muscle action)	Sex	Age composition of elderly subject group[a]	n	% Decline in strength of elderly vs. young adults[b]
Ankle plantarflexion				
Davies & White (1983)	M	70 ± 1.3	13	43
Vandervoort & McComas (1986)	M	80-100	13	45
Vandervoort & McComas (1986)	F	80-100	8	55
Knee extension				
Larsson *et al.* (1979)	M	60-69	16	25
Clarkson *et al.* (1981)	M	55-73	15	39
Young *et al.* (1984)	F	71-81	25	35
Young *et al.* (1985)	M	70-79	12	39
Murray *et al.* (1980)	M	70-86	24	45
Murray *et al.* (1985)	F	70-86	24	37
Handgrip				
Fisher & Birren (1947)	M	53-68	20	17
Shephard (1969)	M	60-69	10	18
Mathiowetz *et al.* (1985)	M	75-94	25	53
Mathiowetz *et al.* (1985)	F	75-94	26	59
Kallman *et al.* (1990)	M	60-69	158	12
Kallman *et al.* (1990)	M	80-89	42	66
Elbow extension				
Davies *et al.* (1986)	M	0 = 70 ± 2.8	20	39
Davies *et al.* (1986)	F	0 = 69 ± 2.9	11	28

[a]Age range in years is presented when available; otherwise the mean is age in years ± SD. [b]Compiled from published data, using the mean strength values of young and old subjects.

Adapted from Vandervoort *et al.* 1986.

comparison with isometric conditions (Larsson *et al.*, 1979). A diminution in both force and velocity of shortening would reduce power to an even greater extent (see figure 11.25 in chapter 11).

In fact, maximal velocity of shortening is also decreased in the elderly. Two factors appear to contribute to this impairment. The first is a decrease in *sarcomere* number (Hooper, 1981) such that at rest, muscle fiber lengths are smaller without any compensatory change in average sarcomere length. This development means that the muscle fiber:tendon length ratio decreases, leading to a reduced effective range of motion as well as decreased maximal angular velocity of shortening. It is interesting to note, however, that the usual adaptation of muscle fiber length to increased excursion is maintained in elderly rats (Burkholder, 2001). This would indicate that the reduced length results from reduced use and in particular to reduced excursion in habitual movement. This will need to be confirmed.

The second factor that contributes to a change in maximal velocity appears to be a change at the molecular level. *Myosin isoform* composition is not substantially altered, but *in vitro* motility assay indicates a slower maximal velocity for a given *isoform* of old rat muscle *myosin* (Höök *et al.*, 1999). This observation suggests there may be a posttranslational change in the myosin that alters its functional capabilities. The slower turnover of myosin (and other proteins) in older animals probably contributes to the magnitude of this change.

Muscle Twitches Become Smaller and Slower

We get slower with age.

There is argument as to when loss of strength commences, but agreement that only modest declines can be detected prior to 60 years. The results of Vandervoort and McComas (1986) suggest that ankle dorsiflexor and plantarflexor muscles lose approximately 1.3% of their former strength each year beyond the age of 52 years (figure 22.2), and it is likely that similar conclusions would apply to other muscle groups. Another approach to investigating muscle strength in the elderly is to measure the force evoked by electrical stimulation. Since *tetanic stimulation* is painful, measurements have usually been made of *twitch* tension instead. The twitch results also show a decline in force with aging, though it is proportionately less than the reduction in maximal voluntary strength (Vandervoort & McComas, 1986). The twitch studies are of interest for another reason; for as muscles age, there is a progressive prolongation of the *contraction* and *half-relaxation times*. This feature has been demonstrated not only in small muscles of the hand and foot (Botelho *et al.*, 1954; Campbell *et al.*, 1973; Newton & Yemm, 1986) but also in larger ones, such as the dorsiflexors and plantarflexors of the ankle (Davies & White, 1983; Keh-Evans *et al.*, 1992; Vandervoort & McComas, 1986), and can be seen in figure 22.3.

The cellular mechanisms responsible for the prolongation may involve the *sarcoplasmic reticulum,* since its volume and Ca^{2+}-pumping capacity are reduced in aged rodent muscle (Larsson & Salviati, 1989). Although motoneuron firing rates decrease in elderly human subjects

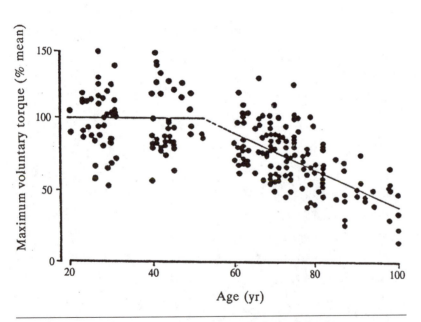

Figure 22.2 Maximal voluntary torques of ankle dorsiflexor and plantarflexor muscles in male and female subjects of different ages. The results have been expressed as percentages of respective mean values for the youngest age group (20-30 years). The horizontal line for subjects aged 52 years or less signifies the lack of an age effect before this time.

Reprinted, by permission, from A.A. Vandervoort and A.J. McComas, 1986, " Contractile changes in opposing muscles of the human ankle joint with aging," *Journal of Applied Physiology* 61: 364.

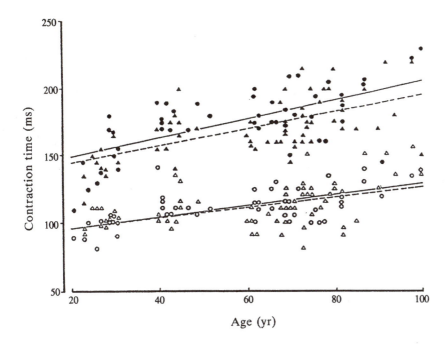

Figure 22.3 Prolongation of muscle twitches with age. The results are expressed as contraction times (see figure 11.8 in chapter 11) of ankle dorsiflexor and plantarflexor muscles. ▲, ● = male and female plantarflexor values, respectively; △, ○ = male and female dorsiflexor values, respectively.

Reprinted, by permission, from A.A. Vandervoort and A.J. McComas, 1986, "Contractile changes in opposing muscles of the human ankle joint with aging," *Journal of Applied Physiology* 61: 365.

Calcium release is impaired as we get older.

(Borg, 1981; Soderberg *et al.,* 1991), the prolongation of the contractile response enables fused contractions to occur at the lower excitation frequencies.

Another change that may contribute to the decreased twitch amplitude in aging is a decrease in density of the *dihydropyridine receptors (DHPRs)*. Loss of the DHPRs leaves more *ryanodine receptors (RYRs)* without association with DHPRs. This could attenuate the release of Ca^{2+} from the sarcoplasmic reticulum. It is easy to see how compromised Ca^{2+} release from the sarcoplasmic reticulum could decrease the twitch contractions, but this mechanism has also been credited with decreasing maximal contractile force. This would be the case if the compromised release per stimulus prevented saturation of *troponin* C. Gonzalez *et al.* (2003) have shown that overexpression of *insulin-like growth factor* I (*IGF*-I) in mice prevents this age-dependent decline in *activation* of skeletal muscle. Not only was maximal specific force maintained better in these mutant mice, but it was shown that peak intracellular $[Ca^{2+}]$ was higher. This observation supports the notion that part of the decrease in *specific tension* is due to impaired activation of the contractile machinery.

Muscle Mass Is Reduced in the Elderly

Muscle mass can be estimated in life from cross-sectional images made by *ultrasonography,* computerized transaxial tomography, or *magnetic resonance imaging*. In the investigation of Klitgaard *et al.* (1990a), however, cross sections were cut in muscles of elderly men who had died from accidents. All such studies provide clear evidence of reduced muscle bulk in elderly subjects. Results of such studies are given in table 22.2.

In some muscles, the reductions in fiber volume may be considerably greater than the percentage declines in cross-sectional area given in table 22.2, due to the replacement of contractile tissue by fat and connective tissue (Inokuchi *et al.,* 1975; Overend *et al.,* 1992). This substitution for contractile tissue explains why the force developed per unit cross-sectional area may be substantially lower in the elderly in comparison with younger subjects (Vandervoort & McComas, 1986). The loss of muscle mass in the aged is known as *sarcopenia,* and this condition is accompanied by several changes in functional capabilities. Of these, the loss of strength is the most profound.

Substantial Losses of Muscle Fibers Occur

A decrease in muscle cross-sectional area could be due to fiber *atrophy* or decreased numbers of fibers. It would be a daunting prospect to count all the muscle fibers in one of the larger human

Table 22.2　Human Studies Demonstrating Loss of Muscle Mass With Age[a]

Study	Sex	Age	n	% Decline
Triceps surae				
Vandervoort & McComas (1986)	M	82-100	5	23
Vandervoort & McComas (1986)	F		5	23
Quadriceps femoris				
Klitgaard *et al.* (1990a)	M	68 ± 0.5	7	24
Young *et al.* (1984)	F	71-81	25	33
Young *et al.* (1985)	M	70-79	12	25
Vastus lateralis				
Lexell *et al.* (1983)	M	70-73	6	18
Biceps brachii, brachialis				
Klitgaard *et al.* (1990c)	M	68 ± 0.5	7	20
Klein *et al.* (2002)	M	86 ± 4	13	32

[a]Values are for cross-sectional area, as determined by imaging.

muscles. Instead Lexell *et al.* (1983) examined every 48th square millimeter of a cross section and estimated that the vastus lateralis muscles of elderly men (70-73 years) contain, on average, 25% fewer fibers than the corresponding muscles of young men (19-37 years). Still-older subjects, in their 80s, have approximately half the number of fibers of young subjects (Lexell *et al.*, 1988; Grimby & Saltin, 1983).

The study by Inokuchi *et al.* (1975) differs from the preceding ones in having been carried out on the rectus abdominis, a trunk muscle, and in the use of an automated scanning microscope to count the fibers. This investigation, performed on muscles from victims of accidental death, revealed that the loss of fibers is considerably greater in the rectus abdominis than in the vastus lateralis. The loss commences in, or before, the third decade in both men and women (figure 22.4). Despite increases in fat cells and connective tissue, the cross-sectional areas of the muscles decrease by 40% in men and 32% in women between the third and ninth decades. As table 22.3 shows, evidence of muscle fiber loss with increasing age has been found in most animal studies as well.

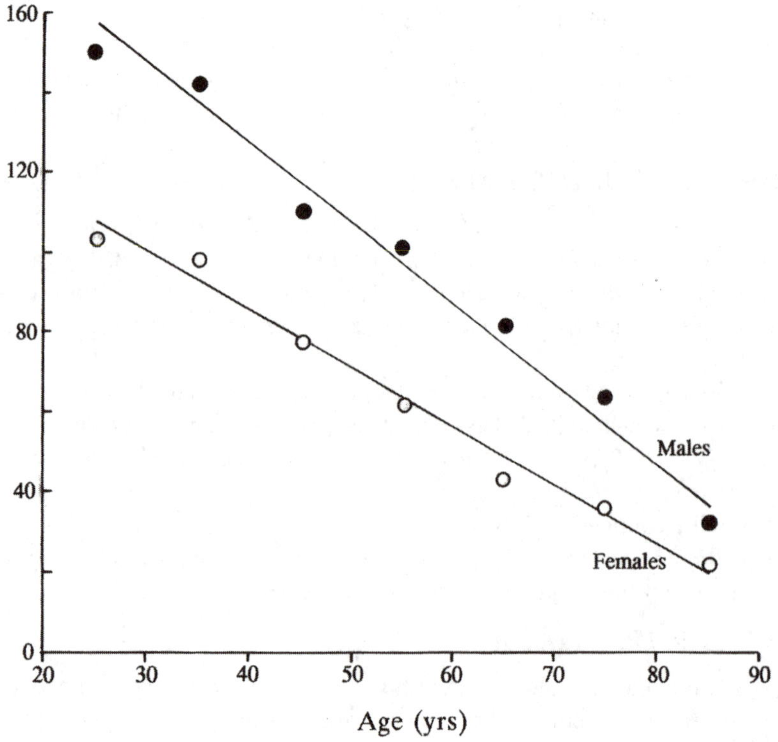

Figure 22.4 Loss of muscle fibers with age. The total numbers of fibers were counted at a single level of the rectus abdominis muscle. The mean values are shown for males (●) and parous females (○) for the midpoint of each decade rather than for mean age. The two straight lines were fitted by eye. In comparison with these results, the values for nulliparous females (not shown) are better maintained until the eighth decade.

Data of Inokuchi *et al.* (1975).

Table 22.3   Animal Studies Demonstrating Loss of Muscle Fibers With Age

Study	Species	Sex	Age (mo)	n	% Decline
Soleus					
Ansved & Edström (1991)	Rat	M	21-25	11	14
Ishihara et al. (1987)	Rat	F	34	15	24
Alnaqeeb & Goldspink (1987)	Rat	M	24	5	13
Gutmann & Hanzlíková (1965)	Rat	–	24	–	25
Caccia et al. (1979)	Rat	F	30	4	27
Rowe (1969)	Mouse	M	25	–	6
	Mouse	F	25	–	21
Extensor digitorum longus					
Alnaqeeb & Goldspink (1987)	Rat	M	24	8	21
Caccia et al. (1979)	Rat	F	30	4	15
Ishihara & Araki (1988)	Rat	F	24	5	22
Rowe (1969)	Mouse	M	25	–	12
	Mouse	F	25	–	16
Sternomastoid					
Rowe (1969)	Mouse	M	25	–	5
	Mouse	F	25	–	7
Hooper (1981)	Mouse	M	18	12	7
Tibialis anterior					
Ishihara et al. (1987)	Rat	F	34	15	13
Rowe (1969)	Mouse	M	25	–	2
	Mouse	F	25	–	(+4)
Hooper (1981)	Mouse	M	18	12	19
Biceps brachii					
Rowe (1969)	Mouse	M	25	–	13
	Mouse	F	25	–	(+4)
Hooper (1981)	Mouse	M	18	12	15
Peroneus digiti quiniti					
Tuffery (1971)	Cat	–	216	1	50
	Cat	–	228	1	63
Plantaris					
Arabadjis et al. (1990)	Rat	F	24	5	5

Arabadjis et al., 1990; Ansved and Edström, 1991; Ishihara et al., 1987; Ishihara and Araki, 1988; Ihemelandu, 1980.

Type II Fibers Undergo the Greatest Atrophy

The cross-sectional areas of human muscle fibers have been measured in the vastus lateralis, biceps brachii, and tibialis anterior. Tissue has been obtained either from needle biopsies during life or from transverse sections of whole muscles obtained post mortem. The picture that emerges from studies of this kind is that muscle fibers maintain their sizes into the seventh decade; beyond that time there is progressive shrinkage of the type II fibers, with lesser changes taking place in the type I fibers (for example, Klitgaard et al., 1990a; Lexell, 1993). A small proportion of fibers will undergo *hypertrophy,* however, and may then start to split longitudinally.

In elderly subjects there is sometimes the appearance of an unusually large incidence of type I fibers in muscle cross sections. Two quantitative investigations, in the vastus lateralis and tibialis anterior, have supported this impression (Larsson, 1983; Jakobsson *et al.,* 1988). The majority of studies, however, have not shown any significant alteration in fiber type proportions with age (Grimby *et al.,* 1982; Lexell *et al.,* 1988).

In contrast to these negative findings, muscle histochemistry in the elderly often shows an unusual pattern of fiber type distribution. There is a loss of the normal mosaic pattern. Instead, fibers of similar types tend to be grouped together, a finding usually indicative of chronic *denervation* and *reinnervation* (see chapter 17). Using monoclonal antibodies to *myosin heavy chains* (*MHCs*), Klitgaard *et al.* (1990c) have been able to demonstrate, in vastus lateralis muscles from elderly men, unusually high incidences of fibers containing more than one MHC isoform; for example, type IIA and IIB (IIX) isoforms may be found together, or type I isoforms with those of type IIA. This association of isoforms would be expected if fibers were changing from one histochemical type to another, as part of the process responsible for fiber type grouping. Such a change in fiber type, as well as grouping of cells of a like fiber type, is consistent with the process of denervation and reinnervation described in chapter 17.

It has been reported that *activity-dependent potentiation* is decreased in elderly humans (Vandervoort & McComas, 1986; Klein *et al.,* 2002) and rats (Carlsen & Walsh, 1987). Since this property of muscle is typically attributed to fast-twitch fibers, and fiber type composition is not consistently altered, the change is not anticipated. In the case of human muscle, it has been observed that substantial change in *posttetanic potentiation* occurs in the 20s and the 40s for ankle dorsiflexors, with little further change (Vandervoort & McComas, 1986). It is not clear what the mechanism of this change might be, but it is consistent with the observed decrease in *staircase* and posttetanic potentiation in disuse atrophy, which occurs due to an absence of *myosin light chain* phosphorylation (Tubman *et al.,* 1996). It is also interesting to note that the age-associated decline in potentiation occurs later in plantarflexors than dorsiflexors, at least for males (Vandervoort & McComas, 1986), and there does not appear to be a change in the triceps brachii muscles (Klein *et al.,* 2002). The decreased potentiation in the elderly is one of the contractile changes that can be reversed with training (Hicks *et al.,* 1991). This is consistent with the notion that this is an atrophy-related issue.

Aged Muscles Show Other Degenerative Features

Some of the histopathological features of muscle in the elderly have been mentioned already; these include muscle fiber atrophy (particularly of type II fibers), fiber type grouping, and the replacement of fibers by fat and connective tissue. To these must be added the following changes in a proportion of fibers (Rubinstein, 1960; Serratrice *et al.,* 1968; Tomlinson *et al.,* 1969):

- *Hyaline* degeneration, *vacuoles* at ends of fibers, *lipofuscin* pigmentation, and loss of *myofibrils* so as to form spirals (Ringbinden) or central masses surrounded by *cytoplasm*
- *Necrosis,* with infiltrations of *macrophage* cells in and around the degenerating fibers
- Groups of very small fibers with dense (pyknotic) *nuclei,* also angulated fibers
- Central nucleation of fibers
- Hypertrophy and splitting of muscle fibers

With the electron microscope, other changes have been observed (Orlander *et al.,* 1978; Shafiq *et al.,* 1978; Tomonaga, 1977):

- Increased endomysial connective tissue
- Disorganization of *sarcomere* spacing, with streaming of the *Z-disks*
- Reductions in the sizes of mitochondria
- Accumulation of reticular material from the sarcoplasmic reticulum and *T-tubules*

The same kinds of abnormality reported for aging human muscle have been reported in the muscles of aged rodents as examined with the light and electron microscopes (Alnaqeeb & Goldspink, 1987; Gutmann & Hanzlíková, 1966). In addition, Cardasis (1983), studying electron micrographs of the rat soleus, has described accumulations of enlarged mitochondria, with indistinct *cristae,* underneath the *plasmalemmae* of the muscle fibers; in these areas of the fibers, there are often deposits of glycogen also. *Lysosomes* are frequently seen at the poles of the *myonuclei,* and *polysomes* are seen in close association with the *myofilaments.* These degenerative changes are consistent with deterioration and dismantling of the precise and functional structures of muscle that were described in chapter 1.

Motoneuron Changes in Aging

A question that now arises is whether the changes in muscle fiber number described thus far are due to intrinsic degeneration of the muscle fibers or whether they result from denervation. Histological study of muscle itself by conventional methods is of limited value, for many of the features previously described can occur as part of the so-called secondary "myopathic" complications of denervation (Drachman *et al.,* 1967), as well as in a primary muscle disorder. Although it is accepted that the presence of clusters of atrophied muscle fibers (Tomlinson *et al.,* 1969), fiber type grouping (Jennekens *et al.,* 1971), or angulated fibers is suggestive of denervation, some of the relevant studies are still subject to criticism over the source of material. Thus, unselected post mortem specimens of muscle are not acceptable since almost any terminal illness may be expected to affect the neuromuscular system—whether as a result of disuse (through extended bed rest), malnutrition, chronic infection, or ischemia. Similarly, the neuromuscular system may be involved as a remote effect of cancer or as a consequence of metabolic disturbances such as renal failure and diabetes. Even elderly subjects killed in accidents may have been suffering serious chronic illness beforehand.

Numbers of Motor Units

Losses of motoneurons can be demonstrated in the elderly.

Although the source of the material may still be problematic, a more direct and satisfactory approach to the investigation of muscle innervation during aging is to count the numbers of motoneurons in the spinal cord or the numbers of large *axons* in the *ventral roots.* The former method is not without difficulty, for care must be taken not to count the same motoneuron twice in successive cross sections or to include cells that are interneurons rather than γ-*motoneurons* or α-*motoneurons.* The method was used successfully by Tomlinson and Irving (1977), who found a reduction in the population of lumbosacral motoneurons that begins in the seventh decade; by the ninth decade, the loss is approximately 29% (see also Kawamura *et al.,* 1977).

One can more easily count ventral root fibers than motoneurons, remembering that those with large diameters are the axons of α-motoneurons and that smaller axons belong to the γ-motoneurons. Unfortunately, criticism can be made of the choice of material used by Gardner (1940); nevertheless, his data strongly suggest that about a quarter of the motor axons are lost in old age (see table 22.4). In a more recent study, directed to the eighth cervical nerve root, Mittal and Logmani (1987) also found a reduction in the number of myelinated nerve fibers, together with diminished nerve fiber diameters.

Motoneuron populations can also be estimated by the electrophysiological

Motor unit counting reveals losses of functioning motoneurons with aging.

Table 22.4   Numbers of Axons in Eighth and Ninth Thoracic Ventral Roots in Post Mortem Specimens From 60 Subjects of Varying Ages

Age (yr)	Mean no. of axons	
	(T8)	(T9)
10-19	5,698	6,081
20-29	6,205	6,204
30-39	6,816	5,648
40-49	5,495	5,506
50-59	5,361	5,293
60-69	4,805	4,845
70-79	4,420	4,628
80-89	4,923	5,086
Max loss (%)	28.8	25.4

Results for each decade have been averaged.

Adapted from "Decrease in Human Neurones With Age," by E. Gardner, 1940, *Anatomical Record,* 77, p. 532.

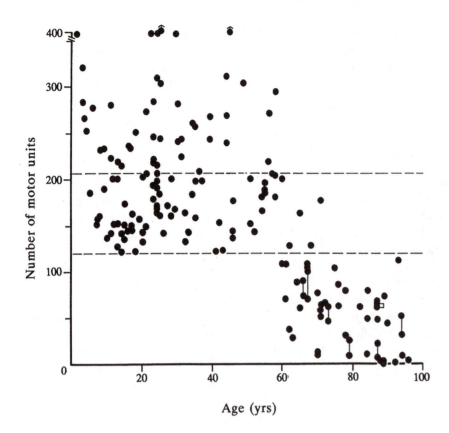

Figure 22.5 Loss of functioning motor units with age. The results were obtained from the extensor digitorum brevis (EDB) muscles of 207 healthy subjects, aged 7 months to 97 years. The upper and lower horizontal lines show, respectively, the mean value (210 units) and the lower limit of the range for subjects below 60 years (120 units). Linked values are bilateral.

Reprinted, by permission, from A.J. McComas, 1977, *Neuromuscular Function and Disorders* (London: Butterworth Publishing Co.), 52.

technique described in "Estimation of Human Motor Units" in chapter 12, in which the size of the maximal evoked muscle *action potential* is compared with that of an average *motor unit action potential.* The advantage of this approach is that, unlike the anatomical ones previously described, it provides information as to the numbers of functioning motoneurons and excludes any axons that might still be present in an aged spinal cord but no longer capable of exciting muscle fibers. The first application of this technique was to the extensor digitorum brevis muscle (EDB) by Campbell and colleagues (1970, 1973), who were careful to study only those subjects considered to be in good physical condition for their age (see figure 22.5). As would be expected of such an approximate technique, the results for any one age exhibit considerable scatter. Nevertheless, when the results for different ages are compared, a striking feature emerges. Until the age of 60, the number of functioning motor units shows little change; beyond the age of 60, there is a progressive fall in the number of functioning units, and by 70, the population of units is reduced to less than half its original size. In some of the oldest subjects, only a very small number of motor units remained, and figure 22.6 shows deterioration of the sole surviving EDB unit in a man of 92.

    The same electrophysiological method has also been applied to the median-innervated thenar muscles and the hypothenar group (Brown, 1972; Sica *et al.,* 1974), as well as to larger muscles such as the soleus (Sica *et al.,* 1976; Vandervoort & McComas, 1986) and the biceps brachialis (Doherty & Brown, 2002). In these muscles also, there are losses of units after the age of 60, although they are less obvious in the biceps and hypothenar group than in the other muscles (Galea, 1996).

    The finding of a reduced number of motor units in old age is not in itself sufficient evidence for denervation. For example, the reduction could arise by a gradual paring away of muscle fibers within motor units as part of a myopathic process until some units cease to exist; a "destitute" motoneuron would be left. Were this the case, one would expect the surviving motor units, because of their reduced fiber populations, to generate smaller potentials than normal. In fact, the mean amplitudes of the motor unit action potentials are significantly larger in the elderly than in controls (Campbell *et al.,* 1973). Considered together, these observations suggest that although some motoneurons (or axons) cease to function, others successfully adopt denervated muscle fibers by sending out new

Aging affects motor units selectively.

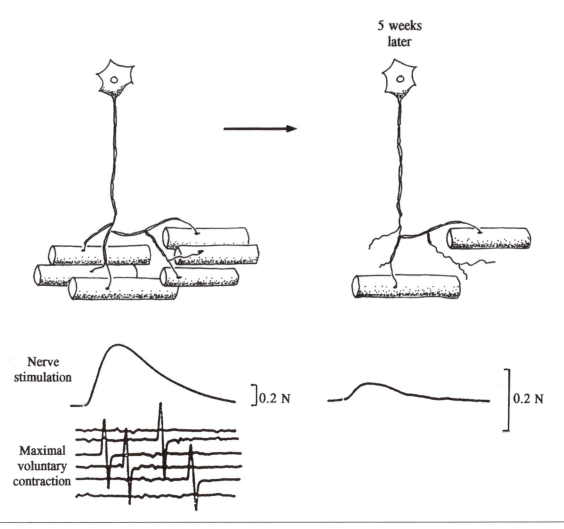

Figure 22.6 Severe denervation in a 92-year-old man. During maximal voluntary contractions, all motor unit action potentials were identical, indicating that only a single motor unit was active. When stimulated through its axon, the same unit developed a large twitch, reflecting substantial *collateral reinnervation.* Five weeks later, the twitch force had fallen to only 5% of its previous value, suggesting the imminent death of the motor unit.

Adapted, by permission, from M.J. Campbell, A.J. McComas, and F. Petito, 1973, "Physiological changes in aging muscles," *Journal of Neurology, Neurosurgery, and Psychiatry* 36: 177.

axonal branches. In old age, the capacity of the surviving motoneurons to reinnervate is limited, however, for it is found that the motor unit potentials are not as large as those in younger subjects with similar amounts of denervation.

Animal Studies Also Demonstrate Motoneuron Depletion

In animals, there is also evidence of a loss of motoneurons during aging. For example, in rats aged between 300 and 800 days, a mean loss of 10% of ventral root axons was demonstrated by Duncan (1934). In still older rats, aged from 900 to 1,100 days, Bari and Andrew (1964, cited in Andrew, 1971) found an 18% to 38% loss of ventral horn neurons. This last study also showed that degenerative changes may be present in a high proportion of the surviving neurons. Cell bodies are smaller and accumulate pigment, the Nissl substance is reduced in amount, and the nuclei are pale and vacuolated.

Neurons can also be counted in squash preparations of single segments of mouse spinal cord. The total number of large ventral horn neurons (presumably mainly motoneurons) is unchanged up to 50 weeks of age but falls by 15% to 20% during the next 60 weeks (Wright & Spink, 1959). Finally, motoneurons can be labeled by *retrograde* transport of *horseradish peroxidase.* Such studies in aged rats also demonstrate reduced numbers of motoneurons; in addition, it appears that the most susceptible motoneurons are those innervating type IIB muscle fibers (Ishihara & Araki, 1988; Ishihara *et al.,* 1987). The findings of Gutmann and Hanzlíková (1966) stand in contrast, for these authors found normal numbers of motor axons in the soleus nerves of aged rats. However, it is possible that some axons were no longer functional, in view of the reduced electrophysiological estimates of motor units for this muscle (Caccia *et al.,* 1979) and for the rat plantaris (Pettigrew & Gardiner, 1987) and medial gastrocnemius (Einsiedel & Luff, 1992).

Size of Aging Motor Units

Motor units are larger in elderly subjects.

Human motor unit potentials are enlarged through collateral reinnervation in old age, regardless of whether the potentials are evoked by electrical stimulation and recorded with a surface electrode (Campbell *et al.,* 1973) or recruited during voluntary contraction and picked up with needle electrodes (Howard *et al.,* 1988; Stålberg & Fawcett, 1982; Stålberg *et al.,* 1989). Although the evidence is indirect, the greater sizes of the motor unit potentials strongly suggest that the motor units are themselves enlarged through collateral reinnervation.

Few studies have been made of motor unit twitches in aging. In one such investigation, carried out on the EDB muscle, Campbell *et al.* (1973) found that the twitch tensions were often increased; similar observations have been made by Galganski *et al.* (1993) in the first dorsal interosseous muscle of the hand and by Doherty *et al.* (1993) in the thenar muscle groups. Another finding in the elderly is that the contraction times of the motor units are prolonged (Newton *et al.,* 1988), presumably due to delay in the pumping of Ca^{2+} back into the sarcoplasmic reticulum.

In aged animals, motor unit twitches are larger and prolonged.

Surprisingly few physiological studies of single motor units have been made in aged animals, but the information gained has been detailed and has enabled correlations to be made between the contractile and histochemical properties of the units (Larsson *et al.,* 1991; Pettigrew & Gardiner, 1987). As in the human investigations, prolongation of the twitch is found, and the twitch and tetanic tensions of the units are increased. That the motor units contain more fibers than in young animals can be shown by the glycogen depletion method; Larsson *et al.* (1991) found the type IIB units to have increased their territories and their muscle fiber complements by 40% and 31%, respectively. These authors also detected an increased proportion of aged muscle fibers containing the IIX MHC, and some hybrid units containing fibers of more than one histochemical type, suggesting that the fibers were undergoing transitions. Although motor units varied in their susceptibility to fatigue, there were no significant differences in *fatigability* between the results in young and old animals.

Other neurons resemble motoneurons by degenerating in later life.

Do the changes in motoneurons reflect a more widespread loss of neural function with advancing age? On the basis of frequently quoted studies on cerebellar Purkinje cells, cerebral cortex, olfactory bulb, and optic nerve, it is generally believed that there is a steady loss of neurons throughout adult life. Wright and Spink (1959) have argued that, in the mouse cord, the loss of motoneurons is delayed until relatively late in life. The ingeniously simple experiments on whole brains of mice by Johnson and Erner (1972) shed new light on this controversy. After fixation in formalin, the brains were mashed in water and then further broken up into a fine suspension by ultrasonic vibration. Using a standard dilution, the suspension was stained with thionine, and samples were examined in a standard hemocytometer chamber. The somata of the neurons were well preserved and could be identified and counted; the total brain neuron content was determined by a multiplication factor appropriate for the dilution of the suspension. Johnson and Erner found that a marked loss of neurons did take place during aging but that this occurred only relatively late (see figure 22.7). Allowing for differences in the life spans of the species, these results are similar to those of Campbell and associates (1973) for human motor units, discussed earlier.

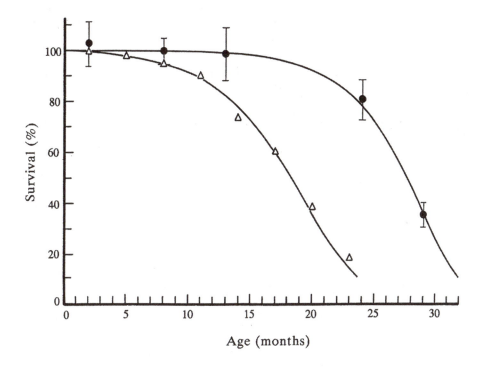

Figure 22.7 Percentages of mice (Δ) and their neurons (●) surviving at different ages. The smooth curves have been drawn according to a "Gompertz" function as explained by Johnson and Erner.

Adapted from *Experimental Gerontology*, Vol. 7, H.A. Johnson and S. Erner "Neuron survival in the aging mouse," pg. 113, Copyright 1972, with permission from Elsevier.

Changes in Axons and Neuromuscular Junctions

The loss of axons in the ventral roots of elderly subjects has already been commented on. In this section, we shall see that the surviving axons may show morphological changes and that the neuromuscular junctions, too, are different in old age. The possibility of diminished function, raised by these modifications, will be explored.

Nerve impulses are conducted more slowly in elderly subjects.

It is easy to measure the maximal *impulse* conduction velocities in human subjects (see chapter 9), and a consistent finding in the elderly is that the values decrease (Falco *et al.*, 1992; Norris *et al.*, 1953). Histological examination of nerve fibers, including teased specimens, reveals certain features that contribute to the lower velocities; these features include segmental *demyelination*, remyelination, decreased internodal length, and dropout of the fibers with largest diameters (Arnold & Harriman, 1970; Dyck, 1975).

Increasing Neuromuscular Complexity

Neuromuscular junctions become more complex with age.

In human muscles, the intramuscular motor axons can be examined by the intravital methylene blue method. The dye is applied to the surface of the muscle over the innervation zone, and the specimen is excised 5 min afterward; the nerve fibers are stained blue (Cöers & Woolf, 1959). This technique has revealed that the neuromuscular junctions are often complex, due to the presence of additional nerve twigs from the original motor axons (Harriman *et al.*, 1970). In the preterminal axons, spherical swellings are evident and there is an increased incidence of axon sprouting, suggesting that denervation of muscle fibers may be occurring with subsequent reinnervation by surviving motor axons.

In cat muscle fibers teased and impregnated with silver, there is also evidence of complex *end-plates* in older animals (Barker & Ip, 1966). As many as five axon *sprouts* may grow from the *nodes of Ranvier* to supply the same end-plate (Tuffery, 1971; see figure 22.8).

As in human muscles, some cat axons have dilatations. The presence of swollen, retracted axon terminals suggests that remodeling of the junctions continues throughout life. It is probable that in aging, the axon sprouts are needed to maintain the safety margin for *synaptic transmission*. In addition, by opening up new areas of contact for the transfer of *trophic* material, the new branches may boost the declining influence of the motoneuron on the muscle fiber and vice versa.

Studies in small, short-lived mammals, such as the mouse and rat, have confirmed that aging neuromuscular junctions have abnormal structures. However, there are no additional axonal branches derived from nodal sprouts, so that although the end-plate area is enlarged, the axonal arborization is less dense than in young animals (figure 22.9). Ultraterminal sprouts are found, though (Rosenheimer, 1990). With the electron microscope, it can be seen that some of the synaptic gutters are empty, indicating that the motor nerve terminals have withdrawn (Cardasis, 1983; Wernig *et al.*, 1984). The remaining nerve endings tend to be smaller than in young animals and may contain fewer synaptic vesicles but more *neurofilaments* and *microtubules* (Banker *et al.*, 1983). More than one ending may lie within the same synaptic gutter. At some junctions, *Schwann cells* are partially or completely interposed between the motor nerve terminals and the synaptic folds, reducing the possibility of effective synaptic transmission. On the postsynaptic side of the junction, the most obvious changes during aging are the

> **In aged rodents, nerve endings may withdraw from their synaptic gutters.**

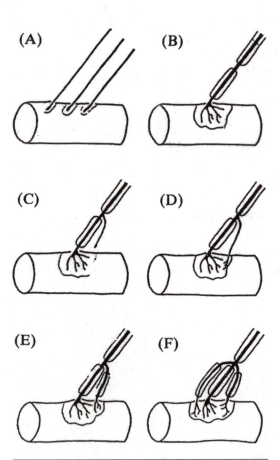

Figure 22.8 Changes in the neuromuscular junction with age, as studied in the cat. At birth, (A) muscle fibers are multiply innervated, but (B) within the next few weeks, the surplus motor axons are lost. In adulthood, (C) a sprout may form at a node of Ranvier, (D) reach the end-plate, and (E) increase the complexity of the end-plate. (F) The new sprout, and others that appear later, become myelinated.

Reprinted, by permission, from A.R.Tuffery, 1971, "Growth and degeneration of motor end-plate in normal cat hind limb muscles," *Journal of Anatomy* 110: 226.

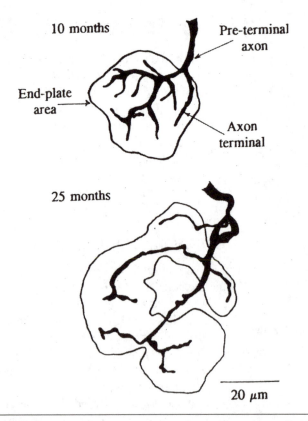

Figure 22.9 Changes in motor nerve endings with age, as studied in the rat. By 25 months, an old age for a rat, the terminals are longer but have lost some of their side branches. The total area occupied by the end-plate is larger.

Reprinted from *International Journal of Developmental Neuroscience*, 8, J.L. Rosenheimer, Factors affecting denervation-like changes at the neuromuscular junction during aging, pg. 646, Copyright 1990, with kind permission from Elsevier Science Ltd.

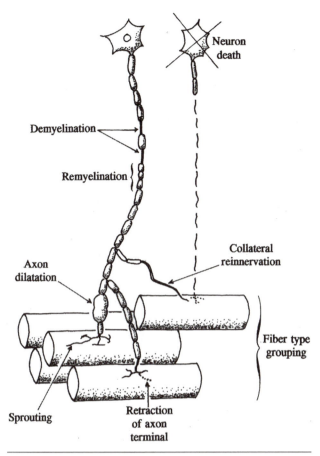

Figure 22.10 Summary of aging changes that may occur in the motor innervation of mammalian muscle fibers.

widening of the secondary synaptic clefts, with the inclusion of *collagen,* and the shallowness of the primary clefts. Surprisingly, although the amount of synaptic folding is decreased, there is no reduction in the numbers of *ACh* receptors (Banker *et al.,* 1983). The various changes in the neuromuscular junction and motor axon with aging are summarized in figure 22.10 and in table 22.5.

Neuromuscular Function in Old Age

How well do the neuromuscular junctions function in old age? Gutmann *et al.* (1971a), in a microelectrode examination of neuromuscular junctions in old rats, reported that the frequency of *MEPPs* was reduced, but other authors have not found this to be true of all muscles (Banker *et al.,* 1983; Smith, 1984). Despite the relative paucity of synaptic vesicles in the nerve terminal (see the preceding section), the numbers that are released with each impulse may be greater than normal (Banker *et al.,* 1983; Smith, 1984). It is likely that the presence of more than one (small) axon terminal over the same *synaptic cleft* is responsible for this overcompensation. Another factor may be an increased influx of Ca^{2+} into the terminal during the impulse (Alshuaib & Fahim, 1990). In human muscles, single-fiber *EMG* recordings have shown that neuromuscular

Table 22.5 Electron Microscopic Morphometric Data of Nerve Terminals in Young and Old Mouse Extensor Digitorum Longus Muscles

Parameter investigated	Young	Old	Percentage change in old vs. young
Number of junctional regions (number of animals)	28 (5)	27 (5)	15 ↓
Junctional length of nerve terminal (μm)	4.6 ± 3.0	2.9 ± 3.0	55 ↓
Nerve terminal area (μm²)	6.3 ± 3.2	2.9 ± 2.9	55 ↓
Number of vesicles per terminal	415.1 ± 180.8	157.6 ± 117.0	52 ↓
Number of vesicles per μm² terminal area free of mitochondria	111.2 ± 42.6	78.5 ± 33.0	29 ↓
Number of SER[a] profiles per terminal	3.6 ± 2.0	5.9 ± 5.2	65 ↑
Number of cisternae per terminal	3.6 ± 2.1	4.4 ± 5.6	22 ↑
Number of coated vesicles per terminal	5.0 ± 3.0	7.5 ± 4.6	49 ↑
Number of mitochondria per terminal	16.6 ± 8.5	3.9 ± 5.1	77 ↓
Mitochondrial area (μm²)	2.2 ± 1.5	0.4 ± 0.5	84 ↓
Mitochondrial area as a percentage of the nerve terminal area	33.3 ± 12.0	11.0 ± 12.0	67 ↓
Synaptic cleft width (nm)	55.7 ± 16.6	62.0 ± 24.1	11 ↑

[a]SER = smooth endoplasmic reticulum. All data are expressed as mean ± SD. All percentage changes are significant at the *p* = <0.05 level.

Reprinted, by permission, from M.A. Fahim and N. Robbins, 1982, "Ultrastructural studies of young and old mouse neuromuscular junctions," *Journal of Neurocytology* 11(4): 641-656. With kind permission of Springer Science and Business Media.

**Aged neuro-
muscular
junctions
appear to
function
well.**

**Nerve
regeneration
is slower
in aged
animals.**

jitter (see chapter 10) increases with age (Stålberg & Trontelj, 1979), although muscle responses to repetitive nerve stimulation are well maintained.

Drahota and Gutmann (1961) crushed rat sciatic nerves and observed that axonal *regeneration* is considerably delayed and much less vigorous in older animals than in younger ones. The most complete study is that of Pestronk *et al.* (1980), who examined terminal sprouting in rat soleus muscles after pharmacological denervation had been produced by the application of *botulinum toxin.* In contrast to the striking increases in motor end-plate length and in terminal arborization that occurred in younger animals, there was no significant sprouting response in the oldest (28-month-old) rats. Axon regeneration after a sciatic nerve crush was also studied using axonal transport as a marker of the outgrowth from the site of the lesion. It was found that some axons in older rats grow back as quickly as those in young animals, but the average rate of regeneration is considerably slower (figure 22.11). Although these results showed the impaired response of aged motoneurons to injury, they do not distinguish between an intrinsic failure of the nerve cells to generate new axonal branches and a lack of stimulatory cues (e.g., growth factors) supplied by the denervated tissues and Schwann cells. The reduced response to injury may be related to the decrease in axon transport that has been shown to occur in aged rats (Frolkis *et al.,* 1990).

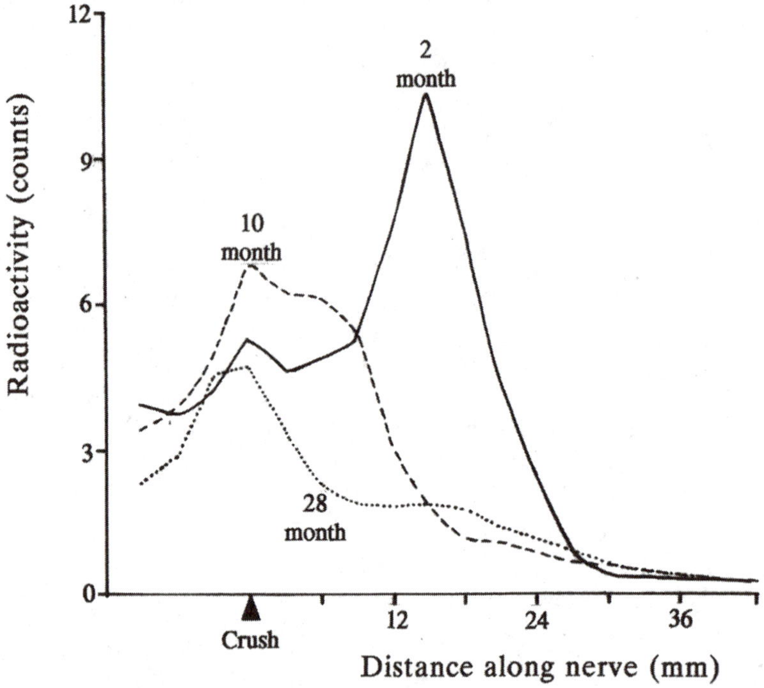

Figure 22.11 Delayed regeneration in motor axons of old animals. Rat sciatic nerves were crushed 9 days previously, and the spinal cords were injected with radioactive amino acid (^{35}S-methionine) so that the radioactivity could be incorporated into the regenerating axons. The majority of axons, as reflected in the peak of radioactivity, grew farthest in the 2-month-old animals and least in the 28-month-old rats. See also figure 8.2 in chapter 8.

Adapted, by permission, from A. Pestronk, D.B. Drachman, and J.W. Griffin, 1980, "Effects of aging on nerve sprouting and regeneration," *Experimental Neurology* 70: 75.

Applied Physiology

**Elderly
subjects
benefit from
training
programs.**

In this section, there are two topics of consideration. The question whether exercise programs are effective in the elderly is an important one. In a society that encourages comfort and passive entertainment, it is all too easy for the elderly to adopt sedentary lifestyles and to be at risk for certain types of injury and illness. We will also learn about a deadly disease in older people in which there is a rapid loss of motoneurons.

Training Programs

Given the various degenerative changes in the muscles and motoneurons of elderly people, can improvements in strength and endurance come from training programs? This question has been addressed by several investigators who have applied heavy resistance and isometric training techniques to a variety of large and small limb muscles. Very substantial improvements in strength and metabolic capabilities were reported in most of these studies, as table 22.6 shows; it is likely that the changes resulted from a combination of *neural adaptation* and muscle hypertrophy (see review in Williams *et al.,* 2002). According to Grimby (1988), the type II fibers show the most

Table 22.6 Summary of Strength Training Studies in Elderly Persons

Study reference	Sex	Age (yr)	n	Type of control	Muscle action	Type of training	Type of test	Training sessions per wk	No. of wk	Increase in strength (%)
Perkins & Kaiser (1961)	M	62-86	5	None	Plantarflexion Knee extension Hip extension	Isometric (n = 10)	Same	3	6	50 57 29
	F		15		Plantarflexion Knee extension Hip extension	Concentric weightlifting (n = 10)	Same	3	6	63 41 64
Chapman et al. (1972)	M	63-88	20	Contralateral limb	Index finger flexion	Concentric weightlifting	Isometric	3	6	33
Aniansson & Gustaffsson (1981)	M	69-74	12	Matched cohort	Knee extension	Calisthenics	Isokinetic 30°/s-180°/s	3	12	14-22
							Isometric			9
Aniansson et al. (1984)	F	63-84	15	Matched cohort	Knee extension	Calisthenics, elastic bands	Isokinetic 30°/s-180°/s	3	26	7-11
							Isometric			13
Kaufman (1985)	F	65-73	10	None	Little finger abduction	Isometric	Isometric	3	6	72

prominent increases in cross-sectional area following strength training in the elderly. Although none of the studies appears to have investigated changes in endurance in a systematic manner, it is an everyday observation that some elderly athletes are able to run considerable distances, and even postmyocardial infarction patients have been trained to run marathons (Todd *et al.,* 1992). Furthermore, the increase in mitochondrial volume that accompanies these training programs is consistent with improved endurance. To provide a sense of the magnitude of change that can be achieved, the results of a specific study will be presented.

Jubrias *et al.* (2001) studied 40 elderly subjects who were randomly assigned to one of three groups: control, *resistance training,* or endurance training. Training for 24 weeks resulted in substantial performance improvement. Subjects in the resistance program increased their leg press by 64%. Endurance training resulted in a substantial increase in the intensity of exercise that could be sustained for 40 min. Endurance-trained subjects had less *phosphocreatine* breakdown and less acidosis during a standard stimulation exercise. These observations relate to a 21% decrease in *ATP* demand. They also had an increased rate of phosphocreatine resynthesis, indicating an aerobic metabolic adaptation. Resistance training resulted in increased muscle cross-sectional area and increased mitochondrial volume density. No change in muscle fiber type composition was seen. However, it should be mentioned that others have observed substantial changes in fiber type distribution in training programs for the elderly (Aniansson & Gustafsson, 1981).

The results of the study by Jubrias *et al.* (2001) indicate that the remarkable adaptability of skeletal muscle is well maintained in elderly muscle. These results and others like these provide encouraging evidence that at least some of the changes seen with aging can be reversed with regular exercise, and indicate that some of the changes seen with aging are a direct consequence of reduced physical demand placed on our muscles as we get older. It should be pointed out, however, that it remains to be demonstrated whether or not such training programs actually have a lasting effect on the age-related deterioration of the neuromuscular system after the training is terminated.

Amyotrophic Lateral Sclerosis

In amyotrophic lateral sclerosis, motoneurons die prematurely.

Amyotrophic lateral sclerosis, or *ALS,* is a fatal disease in which the motoneurons appear to age unusually quickly. The age-adjusted incidence increases in the sixth and seventh decades, when motor unit populations normally start to decline (see figure 22.5), and continues to rise thereafter. Although age is undoubtedly a factor in causation, in some patients other influences are at play. Thus, a small proportion of cases are familial, and in others there are biochemical or immunological abnormalities. Occasionally, there is a striking association with a severe electric shock or with mechanical trauma.

Among the well-known people who died from this condition was Lou Gehrig, who played at first base for the New York Yankees baseball team between 1925 and 1939. Amyotrophic lateral sclerosis is often referred to as Lou Gehrig's disease. Other famous people who have had ALS include Ezzard Charles, the former heavyweight boxing champion of the world, and David Niven, the British film actor. Figure 22.12 shows how rapidly the motoneurons degenerate; once a motoneuron pool is affected, approximately half the cells will cease to function within 6 months. The disease may present in a variety of ways—for example, clumsiness of the hands (as in Lou Gehrig), *weakness* of the legs, difficulty in swallowing, or even a change in voice. As the disease progresses, the muscles become increasingly weak and wasted, and motor units may begin to twitch spontaneously (fasciculation). The mean life expectancy is only 3 years from the time of diagnosis.

Much research is now being devoted to establishing the etiology of this condition, and one success has been the identification of a gene on chromosome 21 in some of the familial cases (Siddique *et al.,* 1991). This gene codes for an enzyme, cytosolic *superoxide dismutase,* that disposes of free oxygen radicals in the cell body. If the radicals are allowed to linger because of the enzyme deficiency, it is easy to imagine that they would damage the metabolic and genetic machinery of the cell and cause its premature death (McNamara & Fridovich, 1993). Attractive though this explanation is, current opinion holds that it may not be the full story and that the enzyme abnormality may exert its effect in some other way. In some of the sporadic cases, there may be similar

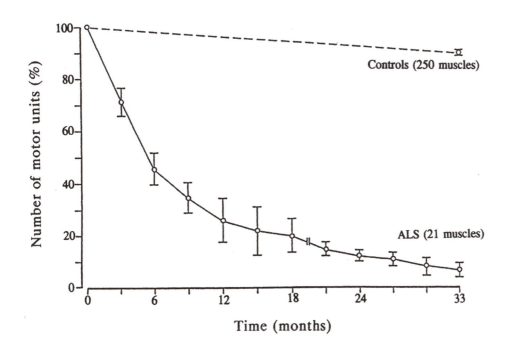

Figure 22.12 Rapid loss of motor units in ALS. The values were obtained from various muscles using the electrophysiological method described in chapter 12.

From "The extent and time course of motoneuron involvement in amyotrophic lateral sclerosis," by M. Dantes and A. J. McComas, 1991, *Muscle and Nerve*, 14, p. 419. Copyright 1991 by John Wiley & Sons, Inc. Reprinted with permission of John Wiley & Sons, Inc.

enzyme dysfunction, but others may have completely different etiologies, for example, viral infection (Turnbull *et al.,* 2003).

This exploration of aging changes brings to an end the study of skeletal muscle and its nerve supply. If parts of the study have appeared complex, it is nonetheless inevitable that yet more detail will emerge about many of the physiological processes and their ultrastructural correlates. At the same time, there can be satisfaction that so much is known about the many facets of muscle and nerve fibers and that this knowledge has extended to the identification of precise abnormalities responsible for an increasing number of neuromuscular disorders.

Glossary of Terms

Note. For terms that appear in only a few chapters, chapter numbers are listed; for terms that appear in many chapters or throughout the book, chapter numbers are not given.

A-band [ch 1, 5, 11]—The central dark (highly refractive, or anisotropic) region of the sarcomere, containing primarily thick filaments, but also containing the overlap region of thin filaments.

ablation [ch 20]—Removal by surgical technique. This approach (removal of specific muscles) is often used to overload the remaining synergists to study the impact of increased chronic activation on muscle properties.

absolute refractory period [ch 9]—Time immediately after an action potential when it is not possible to activate the cell again. *(See also "refractory period".)*

accommodation [ch 9]—The diminished response of a cell or system when a stimulus is applied slowly or continuously.

acetylcholine (ACh)—Chemical transmitter (acetic acid ester of choline) released by motor nerve terminals and also by nerves in other neural systems, like the autonomic nervous system.

acetylcholine receptor (AChR) [ch 1, 3, 5, 6, 7, 16, 18. 19]—Ion channel protein in muscle fiber membrane, usually localized to the neuromuscular junction, that binds acetylcholine to produce excitation.

acetylcholine receptor-inducing activity (ARIA) [ch 3, 6, 18]—42-kD glycoprotein secreted by motor nerve terminals; induces adult-form acetylcholine receptors.

acetylcholinesterase (AChE)—Enzyme located in basement membrane and the perijunctional region of the neuromuscular junction that splits acetylcholine and terminates neuromuscular transmission.

acetyl-CoA [ch 10, 14]—Two-carbon substrate derived from pyruvic acid or β-oxidation of free fatty acids; combines with oxaloacetate to form citric acid in Krebs cycle. Also the substrate from which acetylcholine is synthesized.

ACh—See acetylcholine.

AChR—See acetylcholine receptor.

actin—Globular protein that combines to form two-stranded α-helix (filamentous actin) making up the thin filaments of the sarcomere. Also one of the cytoskeletal proteins.

α-actinin [ch 1]—Cytoskeletal protein associated with Z-disk and subsarcolemmal sites.

action potential—Triggered and patterned exchange of ions across the membrane of an excitable cell that is propagated along the membrane.

activation (of muscle), also "contractile activation" [ch 7, 11, 14, 16, 18, 22]—The process by which muscle is "turned on" or induced to activity.

activation energy [ch 14]—The energy required by a muscle for the processes accomplishing activation of the contractile processes. The total energy required for contraction, minus the energy for myosin ATPase activity.

activin [ch 5]—A peptide growth factor that induces genes during development, specifically involved in patterning of the mesoderm.

activity-dependent potentiation [ch 11, 13, 15, 22]—Enhancement of submaximal contraction resulting from prior activity. See also staircase and posttetanic potentiation.

adaptation [ch 18, 19, 20, 22]—The physiological change in response to a chronic stimulus that permits subsequent exposure to that stimulus to result in a less pronounced disturbance to homeostasis.

adenosine diphosphate (ADP) [ch 11, 14, 15]—Chemical product of removal of one of the three phosphate groups from ATP: ATP → ADP + Pi.

adenosine triphosphate (ATP)—Nucleotide triphosphate that provides energy for muscular contraction and many other forms of biological work when hydrolyzed to ADP and Pi.

adenylate cyclase [ch 1, 7]—Enzyme responsible for synthesizing one of the second messengers, cyclic AMP from ATP, often activated by hormone-receptor complex.

ADP—See adenosine diphosphate.

afferent [ch 4, 18]—A neuron conducting signals from the periphery to the central nervous system. Also an adjective indicating incoming to the neuron.

after-hyperpolarization [ch 9, 12, 17, 18]—The membrane potential after an action potential, when the membrane potential is more negative than the resting membrane potential that existed prior to the action potential.

agrin—Protein released from motor nerve terminals that regulates clustering of acetylcholine receptors at the end-plate.

allosteric regulator [ch 14]—Ion or molecule that binds to a receptor, channel, pump, or enzyme and modifies its function (activates or inhibits).

ALS—See amyotrophic lateral sclerosis.

amorphous ground substance [ch 1]—Amorphous means "without body." A collection of molecules without structure (see glycosaminoglycans).

amyotrophic lateral sclerosis (ALS) [ch 17, 22]—Lou Gehrig's disease. Deterioration of the nervous system, characterized by premature or rapid loss of motoneurons, weakness, and loss of voluntary motor coordination.

ankyrin [ch 1, 2, 3]—Plasma membrane protein that anchors other proteins such as microfilaments and cytoskeletal proteins to plasma membrane.

anterograde (transport) [ch 2, 8, 17, 18]—Movement of substances along the axon away from the soma.

antigen [ch 10, 16]—Chemical compound that triggers an immune response, for example, antibody production. If this response is overzealous, it is considered to be an allergic reaction.

antiport [ch 1, 7]—A transport protein that transfers two solutes in opposite directions across a membrane (e.g., N^+-K^+ pump).

aponeurosis [ch 1]—Tendinous sheet that serves to connect individual muscle fibers to the tendon proper in pinnate muscles.

apoptosis [ch 6]—Cell death, under normal physiological conditions; also called "programmed cell death." This is in contrast with necrotic cell death that results from injury or infection.

apoptosis-inhibiting protein [ch 6]—A protein that prevents the cascade of events associated with apoptosis, possibly by inhibition of caspases that are important in accomplishing apoptosis.

ARIA—See acetylcholine receptor-inducing activity.

ATP—See adenosine triphosphate.

atrophy—Decreased tissue size or wasting, often resulting from lack of use or specific disease process.

autophagic [ch 16]—Literally, self-eating. An autophagic vacuole is a membrane-bound substance that is subsequently dismantled or digested.

axis cylinder—Central part of an axon.

axolemma—Plasma membrane that surrounds the axon.

axon—That part of a neuron that conducts the action potential from source to destination, that is, from axon hillock of soma to nerve terminal. α-Axon refers to axon of α-motoneuron; γ-axon refers to axon of γ-motoneuron; β-axon is derived from a β-motoneuron but innervates an intrafusal fiber.

axon hillock [ch 2, 13, 18]—Site where the neuronal soma forms a junction with the axon. Also known as the initial segment, the most excitable region of the motoneuron, and is the locus from which action potentials are generated.

axoplasmic transport—Energy-requiring process by which chemical substances and organelles are conveyed along the length of the neuronal axon.

axotomy (also axotomize) [ch 16, 18]—Severing of an axon.

bands of Bungner [ch 16, 17]—Columns of dividing Schwann cells contained in an intact basement membrane of a damaged axon.

basal lamina [ch 5, 6]—Portion of basement membrane that lies adjacent to the muscle fiber membrane, made up of the electron-lucent lamina lucida and the electron-dense lamina densa.

basement membrane—Glycoprotein complex that surrounds the muscle and nerve fiber. Lies between endomysium and plasmalemma and consists of different layers.

basophilia [ch 16]—That property of tissue components having a high affinity for basic (high pH) dyes used for histochemical staining of tissue slices.

BDNF—See brain-derived neurotrophic factor.

Becker muscular dystrophy [ch 1]—Form of dystrophy closely resembling pseudohypertrophic muscular dystrophy but with a late onset and slow progression; an X-linked recessive trait.

blastocoel [ch 5]—The cavity that develops in a mass of differentiating, dividing cells.

blastula [ch 6]—A hollow ball of cells that evolves from the fertilized egg during development.

botulinum toxin [ch 16, 17, 18, 22]—Bacterial toxin, from *Clostridium botulinum*. This toxin has specific effects on the motoneuron terminal;

cleavage of synaptobrevin, that impairs neuromuscular transmission.

brain-derived neurotrophic factor (BDNF) [ch 6, 16, 17, 18]—Polypeptide trophic factor, first found in brain and subsequently in other tissues including muscle.

α-bungarotoxin (BTX) [ch 3, 5, 6, 19]—Toxin derived from highly poisonous venom of the banded krait; binds irreversibly to acetylcholine receptors and is frequently used to identify location and quantity of AChR. β-bungarotoxin and γ-bungarotoxin act presynaptically.

bupivacaine [ch 21]—Local anesthetic with long-lasting effects.

cadherin [ch 6, 17]—Membrane protein involved in calcium-dependent cell adhesion.

caffeine [ch 11, 15, 16]—A xanthine derivative (methylxanthine) that inhibits phosphodiesterase and blocks adenosine receptors. Caffeine enhances Ca^{2+} release from sarcoplasmic reticulum and increases Ca^{2+} sensitivity.

calcitonin gene-related peptide (CGRP) [ch 3, 6, 7, 17, 18]—A peptide of 37 amino acids, a product of the calcitonin gene that arises due to alternative processing of mRNA. CGRP is involved in cell signaling processes in several cells of the body. For example, it is secreted at the neuromuscular junction and stimulates synthesis of acetylcholine receptors.

Ca^{2+} ATPase—See sarco-endoplasmic reticulum calcium ATPase.

calcium (release) channels (Ca^{2+} channels)—Ion channels that permit Ca^{2+} to cross a membrane. These can be in an external membrane, as in the dihydropyridine receptor, or in an internal membrane, as in the ryanodine receptor of the sarcoplasmic reticulum.

calcium sensitivity (Ca^{2+} sensitivity) [ch 11, 15]—When the contractile response increases or decreases for a given calcium concentration, then the calcium sensitivity has changed. This is quantified by expression of the calcium concentration at the half-maximal active force. When this concentration is higher, the calcium sensitivity has decreased.

calmodulin [ch 7, 8, 11, 17]—Calcium-binding protein that serves as a signal in activation of a number of cellular responses.

calpain [ch 15, 21]—A calcium-activated proteolytic enzyme.

calsequestrin [ch 7, 11]—A protein found in the terminal cisternae of sarcoplasmic reticulum. Binds or sequesters large amounts of Ca^{2+}.

cAMP—See cyclic AMP.

capacitance (C) [ch 9, 10]—Property of being able to store an electric charge.

carnitine [ch 14]—Naturally occurring amino acid required for mitochondrial oxidation of long-chain fatty acids; acts as a carrier of acyl groups (fatty acids) across mitochondrial membrane to matrix.

carnitine palmityltransferase [ch 14]—An enzyme that combines an acyl group from acyl-CoA (for example, palmityl CoA) with carnitine. This makes transfer across a lipid membrane easier.

CAT—See computerized axial tomography.

catch-like property [ch 13]—The higher force maintained at a given frequency when initial activation is with doublet or triplet activation at higher frequencies.

caveolae [ch 1]—Small in-pocketings of membrane connected to surface membrane by narrow necks. In some cases, these represent the preliminary stage of pinocytosis.

CDF—See choline acetyltransferase development factor.

CGRP—See calcitonin gene-related peptide.

cholesterol [ch 1]—A fatlike steroid alcohol, $C_{27}H_{45}OH$, important precursor for steroid hormones and a possible contributor to arteriosclerosis.

choline [ch 1, 3, 9, 10]—Quaternary amine, a B vitamin that is part of phosphatidylcholine and acetylcholine.

choline acetyltransferase [ch 8, 10, 18]—Enzyme that combines choline with acetate to form acetylcholine.

choline acetyltransferase development factor (CDF) [ch 18]—A 20- to 22-kD polypeptide found in muscle that stimulates choline acetyltransferase synthesis in neurons.

cholinergic [ch 17]—Neural system that relies on acetylcholine for synaptic transmission.

cholinesterase—Enzyme that catalyzes hydrolysis of acyl group from esters of choline.

chromatolysis [ch 16, 18]—Disintegration of the granular components (Nissl bodies) of a nerve cell body following damage to the peripheral parts.

ciliary neurotrophic factor (CNTF) [ch 17, 18]—Neurotrophin derived from Schwann cells, with sustaining effects on motoneurons.

citric acid cycle [ch 14, 16]—Also known as Krebs cycle or tricarboxylic acid cycle. Sequence of chemical reactions that begin with oxaloacetate combining with acetyl-CoA to form citric acid. Following reactions release CO_2 and reducing equivalents for the electron transport chain; end result is formation of oxaloacetate.

clathrin [ch 3, 8]—Protein that encapsulates new vesicles at nerve terminals, the latter becoming "coated vesicles."

CNTF—See ciliary neurotrophic factor.

colchicine [ch 2, 18]—Drug that causes depolymerization of neuronal microtubules by inhibiting the addition of further tubulin molecules.

collagen—Molecules of tropocollagen, forming a protein substance in skin, tendon, bone, cartilage, and other connective tissue.

collateral reinnervation—Capture of otherwise noninnervated muscle fiber by intact neighboring motoneuron, as opposed to regrowth or regeneration of original motoneuron.

complement [ch 10, 12, 21]—A series of at least 20 distinct serum proteins that interact in immuno-response to lyse targeted cells.

computerized axial tomography (CAT) [ch 19]—Technique whereby body organs are scanned with X rays and a computer is used to reconstruct cross-sectional scans along a single axis.

concentric contraction—See shortening contraction.

connectin—See titin.

connective tissue—Lies between muscle fibers offering structure to muscle belly and allowing movements and forces produced in fibers to be transmitted to the tendon; composed of epimysium, perimysium, and endomysium in muscle and epineurium, perineurium, and endoneurium in nerve.

μ-conotoxin [ch 10]—Chemical that blocks skeletal muscle Na^+ channels, thus preventing a propagated action potential following neural stimulation.

contraction time—Time elapsing between the first detectable active force of a contraction and the instant of peak force.

contracture [ch 11, 14, 15, 16, 17]—Condition of fixed high resistance to passive stretch of a muscle, resulting from fibrosis of the tissues supporting muscles or joints or from disorders of muscle fibers.

costameres [ch 1, 11]—Protein filaments at Z-disk that connect myofibrils to plasmalemma and basement membrane, and probably to endomysium.

creatine [ch 14, 15]—N-methyl-guanidinoacetic acid. Is phosphorylated to phosphocreatine, an important energy store that can quickly rephosphorylate adenosine diphosphate to adenosine triphosphate. Creatine participates in the transport (creatine shuttle) of high-energy phosphate from the mitochondria to sites in the muscle where ATP is hydrolyzed.

creatine kinase—An enzyme that removes phosphate from phosphocreatine and gives it to ADP to form ATP, or the reverse of this reaction. It occurs as three major isoforms (specific to brain, cardiac muscle, and skeletal muscle), but skeletal muscle contains two isoforms, cytoplasmic and mitochondrial.

crista, cristae (pl)—Inwardly directed folds of the inner membrane of a mitochondrion. The folds of the mitochondrial inner membrane that separate the intermembrane space from the mitochondrial matrix.

cross-bridge—When the myosin head (S1 segment) interacts with actin to produce force or movement by muscle fibers, it does so by forming a cross-bridge.

curare—See tubocurarine.

cyclic AMP [ch 7, 14]—Cyclic adenosine monophosphate. Messenger molecule in cell interior, produced by activation of adenyl cyclase.

cytochalasin [ch 2]—Class of drugs that cause depolymerization of neuronal microfilaments by inhibiting addition of further actin molecules.

cytokine [ch 18, 21]—Non-antibody proteins secreted by cells of the immune system that serve to regulate the immune system and act as intercellular mediators.

cytoplasm—Cytosol, filaments, and organelles contained within plasmalemma but excluding the nucleus. Sometimes referred to as sarcoplasm in skeletal muscle fiber.

cytosol—Intracellular aqueous solution of inorganic ions, sugars, amino acids, peptides, and proteins.

delayed-onset muscle soreness (DOMS) [ch 15, 19, 21]—Muscle soreness or tenderness that develops over 24 to 48 hr following a bout of unfamiliar exercise.

delayed rectifier [chap 7]—A rectifier is a device that regulates the direction of transfer. The delayed rectifier is a slowly activated ion channel that permits transfer in only one direction.

demyelinate (also demyelination) [ch 2, 9, 16, 22]—To remove or destroy the myelin sheath of axons.

denervate (also denervation or denervated)—To remove neural connection; not having neural connection.

dendrite—Membranous extensions of neuron soma other than the axon, via which incoming signals from other cells are received.

depolarization—Reduction in the potential across a cell membrane.

derecruitment [ch 13]—The consequence of no longer activating.

dermatomyositis [ch 21]—An inflammatory response, often with necrosis, involving the skin and underlying muscle.

desmin [ch 1, 5, 21]—Intermediate filament that acts to anchor membrane proteins to cytoskeletal proteins. This makes movement of proteins somewhat restricted within the membrane.

DGC—See dystrophin-glycoprotein complex.

DHP channel or receptor (DHPR)—See dihydropyridine receptor.

dihydropyridine receptor (DHPR)—L-type Ca^{2+} channel, located primarily in T-tubule of skeletal muscle fibers, that selectively binds a class of drugs referred to as dihydropyridines.

DMD—See Duchenne muscular dystrophy.

DNA—Deoxyribonucleic acid, the genetic material, most of which is located in the cell nucleus.

DOMS—See delayed-onset muscle soreness.

down-regulation [ch 6, 7]—Decrease in quantity of receptor or enzymes in functional location.

Duchenne muscular dystrophy (DMD) [ch 1, 9, 11, 21]—Most common and severe type of pseudohypertrophic muscular dystrophy; begins in early childhood and is chronic and progressive. Is characterized by increasing weakness in pelvic and shoulder girdles, pseudohypertrophy of muscles followed by atrophy, lordosis, and a peculiar swinging gait with legs kept wide apart. The underlying mechanism of the disease is deleterious modification of the gene for the protein dystrophin.

dynamic response [ch 4]—High-frequency burst of impulses fired by Ia axon while muscle spindles undergo stretch.

dynein [ch 8, 17]—ATP-driven motor protein responsible for retrograde axonal transport via its interactions with microtubules.

α-dystrobrevin [ch 3]—Dystrophin-associated protein that tethers intermediate filament proteins to dystrophin-associated protein complex.

dystrophin [ch 1, 3]—One of the cytoplasmic proteins that anchor basement membrane to muscle fiber plasmalemma through attachments to laminin.

dystrophin-glycoprotein complex (DGC) [ch 3]—Complex composed of several components, including dystrophin, dystroglycans, and dystrobrevin, among others. Conveys structural integrity to muscle fiber plasma membrane by connecting subsarcolemmal cytoskeleton and extracellular matrix.

EAN—See experimental allergic neuritis.

eccentric contraction—See lengthening contraction.

E-C coupling—See excitation-contraction coupling.

ectoderm [ch 5, 6]—Layer of cells covering the embryo following gastrulation, and the source of epidermis and its derivatives, as well as the CNS.

efferent [ch 13]—A neuron conducting action potentials from the CNS to the periphery. Also, an adjective indicating emanating from the neuron.

effort—The magnitude of conscious attempt to accomplish a task. In the case of motor unit recruitment, effort should be construed as the cortical activity put forth toward accomplishment of the muscle activation.

E_K, E_{Na}, E_{Cl} [ch 9, 15]—Equilibrium potentials for potassium, sodium, and chloride ions, respectively. See equilibrium potential.

elastin [ch 1]—Yellow scleroprotein, the essential constituent of elastic connective tissue; is flexible and elastic. A connective tissue protein that allows the property of elasticity.

electrochemical gradient—Driving force for motion of ions; considers electromotive forces and concentration gradient for a given ion, usually across a semipermeable membrane.

electromyogram (EMG)—Biological signal detectable from muscle by needle or surface electrodes. Reflects electrical activities of recruited motor units. The dimension of the investigated motor unit pool (i.e., the number of MUAPs contributing to the signal) depends on selectivity of the electrodes.

electromyography (EMG)—Recording of the electrical signal that occurs during muscular activation.

electron transfer chain [ch 14]—Sequence of cytochromes across which an electron is passed, progressing to lower energy states while conserving energy in active transport of H^+ across the inner mitochondrial membrane.

E_m [ch 9]—Membrane potential. Electrical potential difference across the nerve or muscle membrane.

EMG—See electromyogram and electromyography.

endocytosis [ch 10, 18]—Cellular uptake of fluid

and material by invagination of the plasmalemma membrane; also phagocytosis and pinocytosis.

endoderm [ch 5]—The innermost layer of cells in the developing embryo. This layer is destined to become parts of the digestive and respiratory systems and several glands.

endomysium [ch 1, 21]—Thin connective tissue sheath of reticular fibrils that surrounds a muscle fiber, with connections to the basement membrane.

endoneurium [ch 2]—Connective tissue layer surrounding individual nerve fibers.

endoplasmic reticulum [ch 2]—Membrane-bound organelle of broad cellular distribution that accumulates and releases Ca^{2+} in a regulated manner for cell signaling processes.

end-plate—Region of muscle fiber underlying the motor nerve terminals.

end-plate potential [ch 3, 6, 10, 15]—Depolarization of muscle fiber by acetylcholine released at nerve terminal.

ephrin [ch 6]—Axon guidance molecule present during development, which functions as an axon growth repellent.

epimysium [ch 1, 4]—Fibrous connective tissue sheath that surrounds entire muscle and separates it from other muscles.

epineurium [ch 2]—Thick coat of connective tissue surrounding a nerve trunk.

equilibrium potential—Electrical potential required across a membrane to maintain a given concentration gradient for a specific ion if the membrane was freely permeable to that ion.

Evans blue dye [ch 8, 18]—An azo dye that is commonly used to measure blood volume because it binds to albumin.

excitation-contraction coupling (E-C coupling) [ch 7, 11, 13, 15]—Sequence of steps from generation of an action potential on the muscle fiber membrane to release and binding of Ca^{2+} to regulatory proteins and interaction of actin and myosin in contraction.

exocytosis [ch 3, 10]—Removal of cellular contents by pinching off a vesicle from the plasmalemma. The opposite of endocytosis.

experimental allergic neuritis (EAN) [ch 16]—A demyelinating condition that is experimentally induced by injection of parts of a protein contained in myelin. This protein stimulates an allergic reaction that causes demyelination.

extrafusal (muscle fiber)—Voluntary skeletal muscle fiber that does the work of contraction. Located outside the muscle spindle.

extrinsic proteins [ch 1]—One of two types of proteins embedded in the plasmalemma. Extrinsic proteins are exposed only at the internal or at the external surface of the membrane.

F [ch 9]—Faraday constant (in Nernst equation).

facilitation [ch 10]—The effect of a nerve impulse acting across a synapse and bringing the postsynaptic membrane closer to threshold.

FAD—See flavin adenine dinucleotide.

fascicle—Small bundle or cluster, especially of nerve, tendon, or muscle fibers.

fasciclin [ch 6]—A cell adhesion molecule important during development of the embryo for axon pathfinding and adherence of growing axons to one another during growth.

fast Fourier transform (FFT) [ch 13]—Mathematical transform of a signal that appears to be noise (i.e., EMG) but can be resolved into a series of sine waves of known frequency.

fast-oxidative-glycolytic (FOG) [ch 12]—Fast-twitch oxidative glycolytic type of motor unit.

fatigability—The propensity to experience fatigue. The property that allows fatigue to occur.

feedforward [ch 14]—A regulatory mechanism that effects adjustment in the controlled parameter in anticipation of change. Usually this is achieved by parallel transmission of the signal that causes the parameter change to effect adjustment in the control system.

FF [ch 12, 13, 17, 19]—Fast-twitch fatigable type of motor unit.

FFT [ch 13]—See fast Fourier transform.

FG [ch 12]—Fast-twitch glycolytic type of motor unit.

FGF—See fibroblast growth factor.

fibrillation [ch 15, 16, 18, 19]—Spontaneous and repeated activation of muscle membrane. Oscillations of membrane potential that result in passing the threshold of activation.

fibroblast—An undifferentiated connective tissue cell that has the capability to differentiate into any of a number of connective tissue cells.

fibroblast growth factor (FGF)—Peptide that stimulates proliferation of connective tissue fibroblasts and that is also involved in muscle development and neurotrophic mechanisms similar to those with agrin. Also bFGF for basic FGF.

fibronectin [ch 1, 5, 6, 17]—With laminin, this protein links integrin molecules of plasmalemma to collagen fibers of basement membrane.

fibronectin receptor [ch 1]—Intrinsic plasmalemma protein that links the internal cytoskeleton to laminin outside the cell.

filopodia [ch 6]—Thin protrusions, supported by microfilaments.

FInt [ch 12, 13]—Fast-twitch intermediate type of motor unit.

flavin adenine dinucleotide (FAD) [ch 14]—A riboflavin-containing molecule that accepts electrons from the Krebs Cycle and transfers these to the electron transfer chain.

flower spray ending [ch 4]—Endings of the sensory nerve fiber, *Group II axon*. These constitute the secondary endings.

force-frequency relationship [ch 11, 13, 15]—The relationship between the frequency of activation of a muscle and the active force it generates in response to that activation.

force-length relationship [ch 15]—The relationship in skeletal muscle that exists between the maximal, isometric force and the muscle or sarcomere length.

force-velocity relationship [ch 11, 12, 15, 20]—The relationship in skeletal muscle that exists between the force and the speed of contraction of the shortening/lengthening.

FR [ch 12, 13, 19]—Fast-twitch fatigue-resistant type of motor unit.

free radicals [ch 21]—Atom or molecule with an unpaired electron. Strongly oxidizing species, thus causing a wide range of tissue damage. Usually oxygen free radical or nitrogen free radical.

functional electrical stimulation (FES) [ch 13]—The use of electrical stimulation to activate muscles or motoneurons in a pattern that will allow functional movements in an otherwise paralyzed individual.

furaptra [ch 11]—A fluoroprobe (chemical that fluoresces when bound with a target) used for quantification of free Ca^{2+} concentration. Furaptra has a low affinity for Ca^{2+}, so reacts quite quickly, permitting detection of the time-course of a Ca^{2+} transient.

fusimotoneurons [ch 2, 4]—Motoneurons innervating the muscle spindle.

fusion frequency [ch 11]—The frequency of activation above which oscillations in force associated with individual activations are no longer detectable.

g [ch 9]—Symbol for conductance; the ease with which electric current flows across a structure (e.g., ions across a cell membrane).

GABA—See gamma-amino butyric acid.

gamma-amino butyric acid [ch 2]—An inhibitory neurotransmitter.

GAP-43 [ch 17, 18]—A neural protein that is primarily expressed during growth, as in development or nerve regeneration.

gap junction [ch 5, 6]—Connection between two cells consisting of many pores that allow passage of molecules. Electrical synapses are gap junctions.

gap substance [ch 2, 9]—Extracellular material composed of mucopolysaccharides, found in nodes of Ranvier of myelinated axons.

gating current [ch 7, 9]—Electrical current detectable in association with opening of an ion channel but not due to movement of ions through the channel.

GBS—See Guillain-Barré syndrome.

GDP—See guanosine diphosphate.

GDNF—See glial cell-derived neurotrophic factor.

generator potential [ch 4]—Depolarization of sensory nerve endings caused by opening of stretch-activated ion channels.

glia [ch 2, 18]—Supportive cells in the central nervous system.

glial cell-derived neurotrophic factor [ch 17, 18]—A 20-kD neurotrophin identified originally as a product of glial cells.

globular head [ch 8, 11]—A globular protein component of a molecular motor that interacts with actin (in the case of myosin) or tubulin (in the case of dynein or kinesis), permitting physical translocation of the relative position of the molecular motor.

gluconeogenesis [ch 14]—Formation of glucose from molecules that are not carbohydrates, as from amino acids, lactate, and glycerol portion of fats.

glutamate [ch 2, 3, 7]—Amino acid that serves as an excitatory neurotransmitter.

glycerol [ch 1, 14]—A trihydroxy sugar alcohol that is the backbone of many lipids and an important intermediate in carbohydrate and lipid metabolism.

glycine [ch 2]—Amino acid that serves as an inhibitory neurotransmitter.

glycogen supercompensation [ch 14]—Enhanced glycogen storage through exercise and diet procedure. Reduced training and a high-carbohydrate diet permit storage of up to double the normal amount of glycogen in the skeletal muscles.

glycolysis [ch 14, 15]—Enzymatic conversion of glucose to the simpler compounds lactate or pyruvate, resulting in energy stored in the form of adenosine triphosphate.

glycosaminoglycans (GAGs) [ch 1]—Long, unbranched disaccharide polymers that are acidic (negatively charged) due to presence of hydroxyl, carboxyl, and sulfate side groups on disaccharide units. Hyaluronate, chondroitin sulfate, keratan sulfate, heparan sulfate, and heparin are major GAGs that coat proteins in basement membrane and also form amorphous ground substance.

Golgi apparatus [ch 2, 5, 6, 8]—Flattened tube-like membrane-bound reticulum that plays a role in glycosylation of proteins before they are transported out of the soma.

Golgi tendon organs [ch 2, 4, 13, 15]—Receptors found in series with extrafusal fibers at musculotendinous junction; sensitive to contractile force.

G protein [ch 1, 14]—GTP-binding protein on cytoplasmic surface of plasmalemma; involved in hormone-initiated activation of adenylate cyclase.

growth cone [ch 6, 17, 18]—The bulbous tip of a growing axon.

GTP—See guanosine triphosphate.

guanosine diphosphate (GDP) [ch 14]—Nucleoside consisting of guanosine linked to ribose with two phosphate groups.

guanosine triphosphate (GTP) [ch 1, 14]—Energy-rich compound analogous to ATP, involved in several metabolic reactions (Krebs cycle) and an activated precursor in synthesis of RNA. See guanosine diphosphate.

Guillain-Barré syndrome (BBGBS) ch 16, 17]—Peripheral inflammatory response that destroys axons, probably resulting from production of antibodies to myelin.

half-relaxation time [ch 11, 15, 19, 22]—The elapsed time from the peak of a contraction halfway to the passive force, usually measured on a twitch contraction, sometimes measured as a half-time for relaxation later on the relaxation curve.

heat of activation [ch 14]—Energy released as heat, in association with active transport of ions involved in the process of activation of muscle contraction.

hemicholinium [ch 3, 10]—A drug that interferes with synthesis of acetylcholine in cholinergic nerve terminals by inhibition of choline transport.

heparin sulfate proteoglycan [ch 1, 3, 6]—Also heparan sulfate; a sulfated mucopolysaccharide, situated outside many cells of the body, in the basement membrane.

h gate [ch 7]—Part of the molecular structure of the fast sodium channel that inactivates the transmission of Na^+.

homeostasis [ch 1, 7, 15, 20]—Tendency to maintain the normal physiological conditions of the organism, a constant environment.

homonymous motoneuron [ch 4]—Motoneuron activated by afferent neuron that arises in the same muscle.

hormone receptor [ch 1]—Protein embedded in the plasmalemma that binds to specific hormones or ligands and, through direct or indirect pathways, alters the intracellular environment.

horseradish peroxidase (HRP)—Substance used to visualize path and connectivity of neurons by injection into target area or organ and allowing time for transport along length of neuron by axoplasmic transport.

H-reflex [ch 4]—Phenomenon whereby electrical stimulation of large afferent fibers results in excitation of α-motoneuron of the homonymous muscle.

HRP—See horseradish peroxidase.

hyaline [ch 22]—Clear or transparent, granular free.

hybrid fiber [ch 12, 20]—Skeletal muscle fiber containing more than one myosin isoform.

hydrophilic [ch 1, 7]—Readily absorbing or interacting with water.

hydrophobic [ch 1, 7]—Not readily absorbing or interacting with water.

hyperplasia [ch 20]—An abnormal increase in the number of cells in a tissue or organ.

hyperpolarization [ch 9, 15, 18]—Any increase in the amount of electrical charge separated by the cell membrane and hence in the strength of the transmembrane potential.

hypertrophy—Growth of a tissue or organ as a result of increased size of individual cells.

H-zone [ch 1, 11, 21]—Pale area in the center of the muscle fiber A-band.

i [ch 9]—Symbol for electric current.

I-band [ch 1, 5, 11, 21]—Light (poorly refractive, or isotropic) region of muscle fiber, which alternates with A-band and is associated with the presence of only thin, actin filaments (no thick filaments).

idiopathic [ch 16]—Of unknown origin, or started on its own.

IGF—See insulin-like growth factor.

immunohistochemistry [ch 5]—Technique of histochemical localization of immunoreactive substances using labeled antibodies.

impulse—See action potential.

inflammatory cells [ch 10, 21]—Any cells (primarily leukocytes and macrophages) that participate in the inflammatory response, which includes pain, swelling, and heat generation in response to irritation, infection, or injury.

initial segment [ch 2]—That portion of the axon closest to the cell body, also known as the axon hillock, which is the most excitable part of the neuron.

inorganic phosphate (Pi) [ch 14, 15]—Ionized phosphate, carrying a net negative charge.

inositol triphosphate (IP$_3$) [ch 11, 14]—A second messenger, formed by the action of phospholipase c on phosphatidylinositol 4,5-bisphosphate, which is one of the phospholipids that makes up the cell membrane.

inscription(s) [ch 1, 12]—Band(s) of connective tissue that separate(s) bundles of muscle fibers in series.

in situ [ch 11, 15]—Research that is conducted in situ involves evaluation of properties and responses in the natural location (i.e., in the body) with only partial isolation.

insulin-like growth factor (IGF) [ch 5, 7, 16, 17, 18, 22]—A polypeptide with insulin-like features. IGF is released by the liver or fibroblasts and promotes growth in target organs by stimulation of DNA synthesis.

integrin [ch 1, 2]—Class of proteins that span the surface membrane, linking basement membrane and endomysium to plasmalemma and cytoskeletal structures (e.g., fibronectin receptor, dystrophin-associated glycoproteins).

internode [ch 2, 7, 9]—Distance between two successive nodes of Ranvier.

interpolated twitch [ch 13, 19]—To evaluate the relative level of activation achieved by voluntary means, a single or double stimulation is imposed during the voluntary effort. The amplitude of the extra tension observed in response to this stimulation, expressed relative to the response to the same stimulation applied in the absence of voluntary effort, is an estimate of the relative inactivation.

intrafusal muscle fiber [ch 2, 4, 17]—Specialized muscle fiber found in muscle spindle.

intrinsic proteins [ch 1]—One of two types of proteins embedded in plasmalemma. Intrinsic proteins penetrate the full thickness of the membrane.

in vitro—In glass. Experimental procedures conducted in a test tube or isolated tissue chamber are considered to be conducted in vitro.

in vivo—Within the living body. Studies conducted in vivo rely on measurements within and on the body, without disruption or removal of the parts of interest.

inward rectifier [ch 7]—Ion channel that restricts passage of the selective ion to the inward direction only.

ion channels [ch 1, 4, 7, 19]—Proteins embedded in the plasmalemma that selectively permit specific ions to move across the membrane down their electrochemical gradient.

ion transfer capacity [ch 7]—The maximum rate at which ions can pass through a channel or given area of membrane.

IP$_3$—See inositol triphosphate.

isoform(s) [ch 5, 6, 12, 18, 20, 22]—Different forms of a protein resulting from small variations in amino acid sequence that may be produced from different genes, or from the same gene by alternate splicing.

isotonic [ch 11]—Referring to constant force or resistance. This is an experimental control issue. A constant load is often used to study dynamic contractions. Typically an in vivo contraction cannot be isotonic, because a constant resistance or load imposed on a limb will require a changing force as the angle about the axis of rotation changes.

jitter [ch 19, 22]—Abnormal neuromuscular transmission, detected using intramuscular EMG electrodes, in which timing of firing of two muscle fibers in the same motor unit shows high variability.

junctional feet [ch 1, 11]—Structural projections of ryanodine receptor from sarcoplasmic reticulum toward T-tubule, presumably for signaling between DHPR and RYR.

junctional fold [ch 1, 3, 10]—The specialized folding of the muscle fiber membrane to increase surface area for interaction with acetylcholine is called the secondary synaptic cleft and is made up of junctional folds.

kinase [ch 1, 6, 14, 18]—Class of enzyme that catalyzes the transfer of a high-energy phosphate group from a donor (usually ATP) to an acceptor (usually an enzyme that is activated or inhibited by this phosphorylation). This process represents posttranslational control of a protein.

kinesin [ch 8, 17]—Protein motor composed of two pairs of polypeptides, involved in fast anterograde axoplasmic transport.

Krebs cycle—See citric acid cycle.

lactic acid—Derived from pyruvic acid by the action of lactate dehydrogenase. Lactic acid is a weak acid that dissociates into H^+ and lactate.

Lambert-Eaton myasthenic syndrome (LEMS) [ch 10]—Muscular weakness due to inadequate neurotransmitter release, probably due to impaired Ca^{2+} channels of the nerve terminals.

lamina densa [ch 1] (see basal lamina)—Thicker (10-15 nm) of the two basement laminas.

lamina lucida [ch 1] (see basal lamina)—Thinner (2.5 nm) of the two basement laminas. Termed lamina lucida because of its electron translucence.

laminin [ch 1, 3, 6, 17]—With fibronectin, this protein links integrin molecules of plasmalemma to collagen fibers of basement membrane.

lateral sacs [ch 1, 11]—See terminal cisternae.

LEMS—See Lambert-Eaton myasthenic syndrome.

lengthening contraction [ch 5, 11, 13, 21]—A contraction for which the load exceeds the isometric force of the active parts (motor units), resulting in lengthening. Also referred to as an eccentric contraction.

leukemia-inhibiting factor [ch 18]—A cytokine originally named for its effect on blood-forming cells that has, among other effects, positive effects on motor and sensory nerve survival.

ligand [ch 1, 7]—Molecule that binds to a protein (like a receptor or channel) and participates in regulating function of that protein.

lipofuscin granules [ch 2, 22]—A fatty brown pigmentation, resulting from lipid peroxidation.

lysosomes—Membrane-bound organelles, containing hydrolytic enzymes that can be released externally or internally for the purpose of proteolysis.

macrophage [ch 16, 17, 21, 22]—Blood-derived mononuclear cell of immune system origin that participates in removal of debris by phagocytosis.

magnetic resonance imaging (MRI) [ch 11, 19, 22]—Imaging technique based on principles of nuclear magnetic resonance, used to produce high-quality images of the inside of the body. Also known as nuclear magnetic resonance imaging (NMRI) and nuclear magnetic resonance (NMR).

magnetic resonance spectroscopy (MRS) [ch 11, 15]—The use of pulses of magnetic radiation to activate specific elements in tissue and detect the transition to the nonexcited state. This is commonly done to quantify phosphorus in muscle tissue in the forms associated with energy transduction (PCr, ATP, Pi).

malignant hyperthermia [ch 11, 21]—Fever, induced by halogenated anesthetics or depolarizing muscle relaxants, characterized by uncontrolled release of Ca^{2+} into the sarcoplasm, resulting from a genetic abnormality of the ryanodine receptor.

McArdle's syndrome (disease) [ch 11, 14, 15]—Glycogen storage disease. Absence or impairment of myophosphorylase, the enzyme responsible for glycogenolysis.

MEF2—See myocyte enhancer factor.

membrane potential—Difference in electrical charge between inside and outside of plasma membrane in the resting (nonactivated) state.

MEPP—See miniature end-plate potential.

mesoderm [ch 5, 6]—The middle layer of cells in the embryo following gastrulation. This layer is destined to become muscle, bone, cartilage, connective tissue, and blood.

M-filaments [ch 1]—Protein filaments, approximately 5 nm in diameter, that run parallel to the myosin filaments in the M-region in each myofibril.

m gate [ch 7]—Molecular portion of a Na^+ channel that opens when the channel is activated, to permit ion transfer.

MHC—See myosin heavy chain.

microfilaments [ch 2, 5]—Molecules found in neurons composed of actin monomers; can lengthen and shorten.

microtubules—Tubular structures in neurons (and many other cell types) formed by polymerization of tubulin; can lengthen and shorten by protein synthesis (or degradation) carried out at the soma and are about 20 nm wide.

miniature end-plate potential (MEPP)—Transient depolarization of end-plate due to spontaneous release of single vesicles of transmitter (ACh) from a motoneuron terminal.

MLP—See muscle LIM protein.

molecular motors [ch 8, 11, 14]—Proteins that can undergo structural change to accomplish movement; carry particulate matter in axoplasmic transport, allow mobility of cells and organelles.

morphogens [ch 5]—Variety of chemical substances in embryonic tissues that influence the movement and organization of cells by forming a concentration gradient.

motoneurons—Large cells with soma in ventral horn of spinal cord that innervate muscle fibers. α-Motoneuron innervates extrafusal muscle fibers. γ-Motoneuron innervates intrafusal muscle fibers; also called fusimotor neuron.

motor unit—Single α-motoneuron and the colony of muscle fibers (20-2,000) it innervates.

motor unit action potential (MUAP) [ch 13, 22]—Recorded electrical signal corresponding to synchronous activation of muscle cells associated with a single motor unit, detected by indwelling (fine wire or needle) electrodes.

M-region or M-line [ch 1, 11, 12, 14]—Narrow, dark area in center of H-zone of sarcomere. Contains creatine kinase, M-protein, and other cytoskeletal proteins that serve to stabilize the array of myosin filaments.

MRI—See magnetic resonance imaging.

mRNA—See ribonucleic acid.

MRS—See magnetic resonance spectroscopy.

MUAP—See motor unit action potential.

muscle LIM protein (MLP) [ch 5]—A muscle protein essential to myogenic differentiation.

muscle-specific kinase (MuSK) [ch 3, 6]—Phosphorylates tyrosine in β-subunits of AChRs.

muscle spindle—Specialized encapsulated muscle receptor which detects and transmits information concerning length and rate of change of length.

muscle wisdom [ch 13, 15]—Decreasing frequency of firing of individual motor units during sustained maximal effort; corresponds with an increase in contraction and half-relaxation times of the twitch response.

MuSK—See muscle-specific kinase.

MW—Molecular weight.

M-wave—EMG signal typically recorded from the surface (skin) representing a muscle compound action potential, resulting from synchronous activation of a number of motor units.

myasthenia gravis [ch 10]—An autoimmune disease that results in rapid fatigue of muscles on repeated activation. Usually due to immune reaction against acetylcholine receptors. Inadequate numbers of available receptors make it difficult to activate individual motor units.

myelin [ch 2, 3, 9, 16]—Lipid content of Schwann cells, wrapped around myelinated axons.

myelin lamella, myelin lamellae (pl) [ch 2, 17]—The myelin sheath can be divided into layers or lamellae.

myelin sheath [ch 2, 6, 16, 17]—Layer of myelin contained in Schwann cell, wrapped around some mammalian axons, with a gap between cells (node of Ranvier).

myelocele [ch 6]—A tissue protrusion containing nerves. In spina bifida, this includes the spinal cord, which protrudes because of a defect in the vertebral canal.

myelomeningocele [ch 6]—Similar to a myelocele, except that the protrusion includes only the covering of the spinal cord (meninges).

myf-5, myf-6 [ch 5]—Myogenic regulatory factors (MRFs) active during myogenesis. Myf-6 is also known as MRF 4.

myoblasts—Precursor cells destined to become muscle cells by fusion to form myotubes that evolve into muscle cells.

myo D [ch 5]—One of the myogenic regulatory factors (MRFs) active during myogenesis.

myofibril—Cylindrical structure within muscle, containing contractile proteins and separated from other myfibrils by sarcoplasmic reticulum and intermyofibrillar mitochondria.

myofilament—Any of the ultramicroscopic thread-like structures composing myofibrils of striated muscle fibers (e.g., myosin, actin, titin, nebulin, desmin, vimentin).

myogenic enhancer factor [ch 5]—A transcription factor important in regulating the expression of muscle-specific genes, especially during development.

myogenic master regulatory genes [ch 5]—Genes necessary for the expression of the muscle phenotype.

myogenin [ch 3, 5]—One of the myogenic regulatory factors (MRFs) active during myogenesis.

myokinase [ch 14]—An enzyme that uses two ADP to form one ATP and one AMP (adenosine monophosphate).

myonucleus, myonuclei (pl)—Any nucleus in a muscle fiber.

myophosphorylase [ch 11, 14, 15]—Muscle enzyme that hydrolyzes glucosyl-1-phosphate units from glycogen.

myosin—Molecular motor that forms the thick filaments of the sarcomere and interacts with actin to accomplish force development, motion, or both. See myosin heavy chain.

myosin ATPase [ch 4, 12, 14, 17]—Any ATPase has the capability to hydrolyze ATP to ADP, releasing the chemical energy from the terminal phosphate bond. Myosin ATPase is located in the head region (S1 subunit) and is activated by binding with actin.

myosin heavy chain (MHC) [ch 5, 16, 17, 18, 19, 20, 22]—A large elongated protein consisting of a long tail extending to the carboxy terminal and a globular domain at the amino terminal. A myosin molecule contains two myosin heavy chains, in which the tail regions are joined together in α-helix configuration, while the amino terminal portions remain as separate "heads" that contain ATPase and interact with actin during contraction.

myosin isoform—One of several myosin molecules, with slight variation in amino acid sequence that gives the protein unique characteristics.

myosin light chain [ch 11, 22]—Small peptides (light chains) that associate with the myosin S1 segment. Two forms of myosin light chain bind to each S1 segment: essential light chain and regulatory light chain.

myostatin [ch 5]—A member of the transforming growth factor beta (TGF-β) superfamily of molecules. The product of the myostatin gene inhibits growth of skeletal muscles. Inhibition of the myostatin gene during development results in "double-muscled" animals.

myotonia [ch 7]—Increased muscle tone or contractile response due to spontaneous extra activations or delayed relaxation or various causes.

myotrophin [ch 18]—Substance originating from an extra-myocytic source that exerts trophic effects on muscle tissue.

myotube [ch 3, 5, 6, 11, 21]—Premuscle cell derived from the fusion of several myoblasts and destined to become a multinucleated muscle cell.

NAD—See nicotinamide adenine dinucleotide.

N-CAM—See neural cell adhesion molecule.

nebulin [ch 1, 5, 11]—Strengthening protein found in actin filaments. May serve as a template for thin filament formation.

necrosis—Morphological changes that indicate cell disruption and death usually caused by progressive enzymatic degradation.

necrotic zone [ch 21]—An area of tissue in which there is evidence of necrosis, cell deterioration, and death.

negative after-potential [ch 7, 9]—Membrane potential during slow repolarization, late in the action potential.

nerve growth factor (NGF) [ch 16, 17, 18]—The prototype for neurotrophins, molecules that exert trophic effects on nerve cells. NGF has effects primarily on sensory and sympathetic neurons and influences neurite outgrowth and neuron survival.

netrin [ch 6]—Protein that plays a role in guiding axon growth by attraction.

neural adaptation [ch 20, 22]—Training-induced performance enhancement that can be explained by changes in activation properties of the motor system. These include enhanced recruitment, firing frequency, and coordination as well as inhibition of antagonist muscles.

neural cell adhesion molecule (N-CAM) [ch 3, 6, 17]—Molecule involved in a variety of contact-mediated interactions among cells.

neural crest [ch 5, 6]—Embryologic tissue derived from the neural plate that gives rise to adult cells of spinal and autonomic ganglia, peripheral glial cells, and some nonneural cells.

neural tube [ch 5, 6]—Embryologic structure that develops into the central nervous system.

neurapraxia [ch 19]—Nerve injury resulting from compression.

neuregulin (NGR) [ch 6]—A growth factor that activates tyrosine kinase and is essential for muscle development.

neurite-promoting factor [ch 17]—Any substance that stimulates the production and extension of neurites in cultured neurons.

neuroblast [ch 6]—An embryonic cell that has committed to a neural pathway and is expressing a neuronal phenotype.

neurofilament(s)—Very fine linear protein structure(s) inside axon that serve structural or transport functions.

neurotrophic factor or neurotrophin—Small chemical compound that is released from nerve and has a stimulating or inhibiting influence on a target tissue.

neurula [ch 6]—Early embryonic stage at which neurulation (formation of the neural plate, neural folds, and the neural tube) begins.

NGF—See nerve growth factor.

nicotinamide adenine dinucleotide (NAD) [ch 14]—A coenzyme that acts as an electron carrier in cellular metabolism.

Nissl bodies [ch 2]—Organelles found in cytoplasm of neuron soma, composed of tightly packed arrays of endoplasmic reticulum.

nociceptive ending [ch 4]—Receptor for pain, also called nociceptor, triggered by physical or chemical injury.

nodes of Ranvier—Gaps between Schwann cells on axons of myelinated neurons, where action potentials are generated.

NRG—See neuregulin.

nuclear bag (fibers) [ch 4]—Type of intrafusal muscle fiber, possessing a large number of nuclei in the middle portion.

nuclear chain (fibers) [ch 4, 5]—Type of intrafusal muscle fiber, possessing nuclei distributed relatively evenly along their lengths.

nucleolus, nucleoli (pl) [ch 2, 16, 17, 18]—Small dense area of the nucleus where ribosomal RNA is synthesized.

nucleus, nuclei (pl)—Organelle surrounded by a nuclear envelope and containing the genetic material for the cell.

ouabain [ch 7]—Blocking agent that combines with α subunit and possibly with β subunit of the sodium-potassium pump, commonly used to identify distribution of Na^+-K^+ pumps or the impact of removing the pump. A glycoside derived from *Strophanthus gratus*.

β-oxidation [ch 14]—Metabolism of fatty acids, severing two-carbon units in the form of acetate from the long carbon chain.

oxidative phosphorylation [ch 14]—Process in the mitochondria in which ADP is converted to ATP using the energy derived from aerobic metabolism.

paciniform corpuscle [ch 4]—Encapsulated nerve ending ensheathed in thin wrappings of connective tissues. May play minor role in providing information about contraction and relaxation; is sensitive to rapid small deformations; is abundant in skin, periosteum, and abdominal viscera.

paranodal region [ch 2]—That area of the axon that is adjacent to the node of Ranvier.

parvalbumin [ch 7, 11, 15, 16]—Ca^{2+}-binding protein present in sarcoplasm of small mammalian fast-twitch muscle and amphibian muscle. Accelerates relaxation by sequestering Ca^{2+}.

patch clamp [ch 7, 9]—Technique of measuring membrane transport on tiny portion of membrane that may contain only a single channel. Voltage across the piece of membrane can be controlled to evaluate the regulation and conductance properties of the channel.

paxillin [ch 1]—One of several attachment proteins that connect contractile actin with the extracellular matrix.

PCr—See phosphocreatine.

pennate or pennation—See pinnate.

percutaneous [ch 12]—Through the skin, as in percutaneous needle biopsy or percutaneous electrical stimulation.

perimysium [ch 1]—Connective tissue sheath that encloses a bundle (fascicle) of muscle fibers. Also provides a pathway for major blood vessels and nerves to run through the muscle belly.

perineurium [ch 2]—Connective tissue layer around each bundle of axons within a nerve.

permeability [ch 5, 7, 9, 10, 16, 17]—The property of a membrane that allows certain substances to cross the membrane with ease.

phagocytosis [ch 16]—Intake of membrane and adjacent fluid and contents into the cell to form an intracellular membrane-bound vesicle.

phalloidin [ch 2]—Drug that stabilizes actin filaments.

phosphate [ch 1, 14, 15, 18]—An ester of phosphoric acid, often associated with energy content. See adenosine triphosphate.

phosphocreatine (PCr) [ch 1, 14, 15, 22]—Phosphorylated form of creatine, which provides an immediate source of energy for replenishment of ATP. An increase in free creatine concentration stimulates oxidative phosphorylation.

phospholamban [ch 7]—A protein that binds and inhibits SERCA2a. This inhibition is lost with phosphorylation that can be achieved by calmodulin kinase.

phospholipid [ch 1]—Lipid that contains phosphorus, that is, phosphoglycerides or sphingomyelins. The principal lipid in the plasmalemma.

phosphorylation—Process of introducing a phosphate group into an organic molecule. This reaction is typically catalyzed by a kinase and reversed by a phosphatase. Phosphorylation reactions are used to activate an enzyme or alter properties of a binding site or channel.

pinnate [ch 1, 20]—Referring to a muscle with fiber alignment in parallel but at a distinct angle (or angles) relative to the line of action of the muscle.

pinocytosis [ch 18]—See endocytosis.

plasmalemma, plasmalemmae (pl)—Plasma membrane, composed of a bilayer of phospholipids with cholesterol and a multitude of proteins embedded in it. Surrounds all living cells.

plate ending [ch 4]—Type of nerve ending found in polar regions of intrafusal muscle fibers, most commonly nuclear bag fibers.

polymerization [ch 2]—The act of forming a polymer, a larger unit formed by the combination of several small units.

polymyositis [ch 21]—Muscular pain and weakness associated with inflammation.

polyneuronal innervation [ch 6, 12]—Individual muscle fibers innervated by more than one motoneuron.

polyradiculoneuritis [ch 16]—Also known as Guillain-Barré syndrome. A postinfectious, immune-mediated disease that affects specific myelin proteins, causing degeneration of the latter.

polysome [ch 22]—A string of ribosomes participating in protein synthesis.

postsynaptic membrane [ch 3, 6, 10]—Muscle membrane at the neuromuscular junction, equipped to receive chemical signal from nerve terminal.

posttetanic potentiation [ch 11, 22]—The enhancement of submaximal active force or shortening resulting from a prior tetanic contraction.

power stroke [ch 11]—The mechanical consequence when a cross-bridge makes the transition from weak to strong binding, associated with relative movement of the thin filament.

primary endings [ch 4]—Spiral endings of a large afferent fiber within the equatorial region on each of the intrafusal fibers.

primary synaptic cleft [ch 3, 10, 16]—70-nm gap between membrane of axon terminal and muscle fiber plasmalemma. *(See also "synaptic cleft".)*

prostanoid receptors [ch 5]—Membrane receptors for prostaglandins.

protease [ch 3, 6, 15, 19, 21]—An enzyme that dismantles proteins, usually at specific amino acid pairs.

protein kinase—Any protein that catalyzes the transfer of a phosphate group from ATP to form a phosphoprotein.

proteoglycan [ch 3]—A protein that contains at least one, and most often several, covalently linked glycosaminoglycans. There are many proteoglycans that differ in their protein and glycosaminoglycan constituents and that perform myriad functions at the cell surface, intracellularly, and in the extracellular matrix.

proteolysis [ch 6, 7]—Enzymatic breakdown or destruction of protein.

pseudofacilitation [ch 15]—The enlargement of the M-wave or motor unit action potential, independent of the number of muscle fibers activated. This enlargement occurs due to hyperpolarization of the membrane caused by activation of the Na^+-K^+ pump.

quantal size [ch 3, 10]—The number of molecules of transmitter per synaptic vesicle.

r [ch 9]—Symbol for resistance to electrical current.

R [ch 9]—Universal gas constant (in Nernst equation).

rapsyn [ch 3, 6]—One of several proteins found at the top of synaptic folds at the neuromuscular junction; helps maintain structural integrity of folds.

rate-coding [ch 13]—The increase in muscle force that is achieved by increasing the rate of firing on already active motor units, in contrast with motor unit recruitment.

reactive hyperemia [ch 15]—Increased blood flow in response to a period of relative deprivation of blood flow.

recruitment—The process of increasing the number of active motor units.

reducing equivalents [ch 14]—Quantity of molecules containing the electrons made available by specific reactions in glycolysis or Krebs cycle. Transfer of reducing equivalents to the electron transfer chain results in conservation of the energy associated with these electrons.

refractory period [ch 9, 13, 16, 20]—Time following an impulse during which there is a depression of membrane excitability. *(See "absolute refractory period" and "relative refractory period".)*

regeneration—Repair and regrowth of an anatomical part.

reinnervation [ch 3, 16, 17, 18, 21, 22]—Restoration of innervation either by regeneration of the original neural source or by sprouting from an adjacent neuron (collateral reinnervation).

relative refractory period [ch 9]—Time immediately after one action potential when a stronger signal is required to elicit another action potential. *(See also "refractory period".)*

resistance training [ch 20, 22]—Improving strength by repeated movements against a resistance that requires additional motor unit activation.

respiratory chain [ch 14]—Sequence of cytochromes on the inner mitochondrial membrane that accept an electron from NADH or $FADH_2$ and pass this from one to the next, with corresponding transfer of H^+ to the intermembrane space.

reticular lamina [ch 1]—Fibrous portion of the basement membrane, visible by the electron microscope.

retinacula [ch 1]—Connective tissue structures that keep an organ or tissue in place. In the case of a skeletal muscle, hold a tendon close to a joint that the tendon crosses.

retinoic acid [ch 5, 6]—Also known as vitamin A. Probable factor in gene regulation.

retrograde—Backward or in reverse direction. In the motoneuron, retrograde transport is in a direction toward the spinal cord.

ribonucleic acid (RNA)—Ribonucleic acid, present in all cells, transcribed from DNA and used to manufacture proteins in cytoplasm; mRNA is messenger RNA, the form of RNA that carries the code for the sequence of amino acids in a given protein.

rigor [ch 11]—A state of semipermanent attachment of a myosin head to actin that occurs in the absence of ATP.

RNA—See ribonucleic acid.

ryanodine receptor (RYR)—Calcium-releasing channel found in terminal cisternae of sarcoplasmic reticulum of muscle fiber.

S [ch 12, 13, 19]—Slow-twitch type of motor unit.

sag [ch 12]—The fall in peak force observed during a brief incompletely fused tetanic contraction that is typical of fast-twitch motor units.

saltatory conduction [ch 9]—Conduction of action potentials, which proceeds by leaps rather than by gradual transitions.

sarco-endoplasmic reticulum calcium ATPase (SERCA) [ch 7, 12, 14]—The calcium pump in the sarcoplasmic or endoplasmic reticulum. This pump, with at least three isoforms, allows sequestration of Ca^{2+} in the lumen of the reticulum.

sarcolemma—Plasmalemma and basement membrane of a muscle fiber.

sarcomere—The basic contractile structure within a muscle; extends from Z-disk to Z-disk. Contains actin and myosin (contractile proteins) as well as titin, nebulin, and regulatory proteins.

sarcopenia [ch 22]—The decreased muscle mass seen with advancing age due to decreased fiber size and number.

sarcoplasm—Aqueous protoplasm or cytosol of skeletal muscle.

sarcoplasmic reticulum—A system of interconnected vesicular cytoplasmic membranes that function in storage, release, and transport of Ca^{2+} for activation and relaxation of muscle.

satellite cell—Small cell at periphery of muscle fiber, responsible for repair and regeneration of the fiber.

saturating disk [ch 3, 10]—Refers to the arrangement of the active zone, the site of vesicle release, opposite a secondary synaptic cleft and the corresponding acetylcholine receptors.

Schmidt-Lanterman incisures [ch 2, 16]—Clefts in myelin sheath of myelinated axons, wherein a narrow cytoplasmic process is sent from Schwann cell to axis cylinder.

Schwann cell—Type of neural satellite cell that forms the myelin sheath.

sciatin [ch 18]—A myotrophic substance isolated from adult sciatic nerve that promotes survival of embryonic skeletal muscles in culture. After an initial report in 1980, later shown to be identical to the plasma protein transferrin.

secondary ending [ch 4]—Sensory nerve terminal, found in the muscle spindle, where a group II axon terminates in flower spray endings on the intrafusal muscle fibers.

secondary synaptic cleft [ch 3, 7, 10, 17]—Small repeated invaginations (folds) in plasmalemma of sole-plate that increase surface area of end-plate. (See also "synaptic cleft".)

second messenger—Chemical used to disseminate a message from a primary messenger (usually a hormone or similar chemical that binds to a receptor) to the interior of the cell.

selectivity filter [ch 7]—A mechanism whereby passage of ions through a channel is restricted to a single species, based on size, water shell, and charge/binding properties.

semaphorins [ch 6]—A family of cell surface and secreted proteins that provide chemical clues for axon guidance—in this case, chemorepellent.

SERCA—See sarco-endoplasmic reticulum calcium ATPase.

shortening contraction [ch 5, 11, 13, 20]—A contraction that occurs when the muscle is active and shortens against any resistance. Also referred to as a concentric contraction.

size principle [ch 13, 17]—The orderly recruitment of motor units from smallest to largest as demand increases.

Slit [ch 6]—Protein that plays a role in guiding axon growth by preventing growth in the wrong direction.

SMA—See spinal muscular atrophy.

smooth endoplasmic reticulum [ch 3, 8]—An intracellular network of tubules and vesicles that, unlike rough endoplasmic reticulum, contains no ribosomes and is therefore not involved in protein synthesis. Some of the primary functions are intracellular transport, phospholipid synthesis, action and storage of enzymes, and calcium regulation.

SNAP-25 [ch 10]—Synaptosome-associated protein. Protein of the neurilemma that participates in vesicle exocytosis.

SNARE [ch 10]—Soluble N-ethylmaleimide-sensitive factor attachment protein receptor; an essential participant in vesicle docking, fusion, or both.

sole-plate [ch 3, 6, 16]—Region of sarcoplasm and contents beneath motor nerve terminals.

soma—Motoneuron cell body; contains nucleus of the cell, as well as mitochondria and Nissl bodies.

somite [ch 5, 6]—Grouping of mesoderm cells that forms on both sides of the neural groove during embryonic development.

spasticity [ch 4]—Tonic involuntary contraction of striated muscle.

specific tension [ch 19, 22]—Active force of a muscle contraction expressed relative to the physiological cross-sectional area.

spectrin—Plasma membrane protein involved in anchoring of microfilaments to plasma membrane in neurons and skeletal muscle cells.

spike-triggered averaging [ch 12, 13]—A technique allowing visualization of an event that occurs regularly after an identifiable trigger (e.g., motor unit action potential) but is otherwise obscured by noise. The relevant signal becomes evident with averaging of a large number of triggered events.

spinal cord isolation [ch 18, 19]—An experimental technique that eliminates descending and sensory input to a specific motoneuron pool.

spinal muscular atrophy (SMA) [ch 6, 13, 17, 20]—Genetic abnormality leading to degeneration of motoneuron cell bodies and subsequent muscle atrophy.

spindle [ch 4, 17]—Most important muscle sensory organ, responsible for signaling information about length of muscle fibers. Composed of small intrafusal muscle fibers and enclosed in connective tissue sheath; fusiform in shape and embedded in parallel with extrafusal fibers of the muscle.

sprout—The process of forming a new axonal branch. Nodal sprout originates at a node of Ranvier; a terminal sprout originates on a nerve terminal; a preterminal sprout originates after the last myelin sheath.

SR—See sarcoplasmic reticulum.

staircase [ch 11, 22]—The progressive enhancement of active force during repetitive submaximal contractions.

static response [ch 4]—Steady discharge of impulses from the Ia and IIa axons when muscle spindle is passively stretched and held at a new length. The IIa axon has only a static response (see dynamic response).

STX [ch 7, 9]—Saxitonin. Paralytic poison manufactured by certain algae.

substantia gelatinosa [ch 6]—The gelatinous gray matter forming the dorsal aspect of the posterior column of the spinal cord. Primary location for integration of heat and pain stimuli.

succinylcholine [ch 10, 11, 19]—Depolarizing neuromuscular blocking agent; acts as skeletal muscle relaxant; used as chloride salt for surgery.

summation [ch 11, 13, 15]—The additive effect of sequential activation when the contractile response of the first activation is still present as the following contractile response occurs. Summation is nonlinear. The resulting contraction is generally larger than the sum of two twitches.

superoxide dismutase [ch 21, 22]—Free radical scavenger that forms hydrogen peroxide from superoxide, thereby protecting the cell from potential damage from superoxide.

symport [ch 1]—A transport protein that moves two solutes across a membrane in the same direction.

synapse—The structural arrangement between a neuron and the target organ. In skeletal muscle, also the neuromuscular junction.

synaptic cleft—Gap between end of an α-motoneuron and underlying end-plate of a muscle fiber. *(See also "primary synaptic cleft" and "secondary synaptic cleft".)*

synaptic delay [ch 10]—Time elapsed between depolarization of nerve terminal and depolarization of postsynaptic membrane (muscle fiber membrane in the case of the neuromuscular junction).

synaptic transmission—The transfer of a signal from a neuron to a target organ.

synaptobrevin [ch 10]—Small vesicle membrane protein that interacts with SNARE in the process of vesicle emptying in synaptic transmission.

synaptogenesis [ch 16]—Synapse formation.

synaptotagmin [ch 10]—Vesicle membrane protein with Ca^{2+}-binding domain that is essential for vesicle fusion.

synemin [ch 1]—Intermediate filament found at Z-disk; functions to maintain myofibrils in register.

syntaxin [ch 10]—Neurolemma-associated protein; interacts with synaptobrevin for docking of synaptic vesicles in preparation for release of neurotransmitter.

talin [ch 1, 2]—Plasma membrane protein involved in anchoring microfilaments and cytoskeletal proteins to plasma membrane in neurons and skeletal muscle fibers.

TAP$_1$—See terminal anchorage protein.

taxol [ch 2]—Drug that stabilizes microtubules.

tenascin [ch 1]—Glycoprotein contained in the extracellular matrix; links muscle fiber proteins to collagen fibers of the tendon.

tenotomy [ch 19, 20]—Severing of the tendon of a muscle.

tensin [ch 1]—One of several attachment proteins; links contractile actin to extracellular proteins.

terminal anchorage protein (TAP$_1$) [ch 17]—A proteoglycan-like molecule that secures the nerve terminal to the basal lamina.

terminal cisternae—Widened portion of the sarcoplasmic reticulum that occurs adjacent to the transverse tubules. Sequesters Ca^{2+} with calsequestrin and contains ryanodine receptors in the membrane for triggered release of Ca^{2+}.

tetanic contraction or stimulation—Repetitive stimulation with a frequency that is high enough to result in incomplete relaxation between activations is considered tetanic stimulation. The resulting contractile response is considered a tetanic contraction (see also summation).

tetanus [ch 7, 11, 15, 16]—Train of electrical stimuli that will elicit a tetanic contraction if the interval between stimuli does not permit complete relaxation of force.

tetanus toxin [ch 10, 17]—A neurotoxin that binds to synaptobrevin and blocks synaptic transmission at inhibitory neurons of the spinal cord. Results in uncontrolled spasm (tetany) of affected muscles.

tetrad [ch 11]—An entity composed of four parts.

tetrodotoxin (TTX) [ch 7, 9, 16, 17, 18, 19]—A paralytic poison secreted by Japanese puffer fish; binds to and inhibits fast sodium channels.

TGF-β—See transforming growth factor.

titin [ch 1, 5, 11]—Giant protein spanning from Z-disk to M-band, in part through the thick filament, in the sarcomere of striated muscle. Pairs of titin strands help keep the myosin filament in the center of the sarcomere during contraction and relaxation.

Torpedo **[ch 3, 7, 10]**—Electric ray from which acetylcholine receptors were first purified and characterized.

trail ending [ch 4]—Type of motor nerve ending found on intrafusal muscle fibers, most commonly seen in nuclear chain fibers.

training [ch 19, 20, 21, 22]—Practice of a movement requiring complex or extraordinary effort for the purpose of improving subsequent performance of the movement.

transcranial magnetic stimulation [ch 15]—The use of focused magnetic field application for the purpose of activating the motor cortex.

transcription—The process of RNA synthesis using DNA as a template, by the enzyme RNA polymerase.

transferrin [ch 18]—A plasma protein that transports iron. See also sciatin.

transforming growth factor (β) [ch 5]—Family of cytokine growth factors that take part in embryonic development and in tissue remodeling and repair following injury.

translation—Process of synthesis of polypeptide chains and proteins, as determined by the sequence of bases in a messenger RNA.

transverse (T-) tubule—Narrow membranous channel, confluent with the extracellular fluid that conducts impulses into the interior of a muscle fiber.

triad [ch 1, 11]—Group of three entities. In skeletal muscle, the three entities are two terminal cisternae and one transverse tubule.

tricarboxylic acid (TCA) cycle—See citric acid cycle.

trophic—Sustaining action of a molecule or cellular activity.

trophism [ch 17, 18, 19]—Chemically mediated induction to grow or develop.

tropomyosin—Protein associated with actin and troponin T. When troponin C does not have Ca^{2+} bound to it, then tropomyosin prevents interaction of actin and myosin.

troponin—Regulatory protein on the thin myofilament, composed of three subunits: troponin I, which inhibits actin–myosin interaction; troponin C, which binds Ca^{2+} when the concentration increases on muscle activation; and troponin T, which is bound to and interacts with tropomyosin.

T-tubules—See transverse tubules.

TTX—See tetrodotoxin.

TTX-resistant sodium channels [ch 18]—Sodium channels that are not inhibited by TTX, typically expressed following denervation.

tubocurarine [ch 10, 19]—Active part of curare, used to prevent skeletal muscle contraction via inhibition of neuromuscular transmission.

tubulin—Globular protein; polymerizes to form microtubules in neurons.

tumor necrosis factor [ch 21]—A cytokine that is involved in immune responses and leads to preferential destruction of tumor cells.

twitch—Muscle contractile response to a single excitation.

twitch–tetanus ratio [ch 16]—The ratio of the active force for a twitch to the active force of a tetanic contraction.

ubiquitin-proteosome system [ch 19]—Proteolytic system that involves tagging of proteins for degradation by ubiquitin, followed by degradation of the tagged protein by the proteosome complex, which uses ATP in the process.

ultrasonography [ch 19, 22]—Also ultrasound. An imaging technique based on the difference in sound transmission velocity through tissues of different density (different speeds of sound conduction).

uniport [ch 1]—A transport protein that carries only one solute across the membrane.

up-regulate [ch 7, 16, 19]—To increase the quantity of receptor or enzymes in functional location.

utrophin [ch 3]—Protein similar to dystrophin localized near the neuromuscular junction.

vacuole [ch 3, 15, 16, 22]—A space or a cavity. An apparently unfilled vesicle.

VAMP—See vesicle-associated membrane protein.

vasa nervorum [ch 2]—Network of blood vessels surrounding the nerve and supplying it with blood.

ventral root—Site where outgoing (efferent) axons leave the spinal cord.

vesicle-associated membrane protein (VAMP) [ch 6, 10]—Proteins found in synaptic vesicles; play an important role in vesicular docking to presynaptic membrane; also known as synaptobrevins.

vesicular acetylcholine transporter [ch 10]—Transport protein in the membrane of the nerve terminal that transports acetylcholine into the nerve terminal.

vimentin [ch 1]—Protein filaments that wrap around myofibrils at the Z-disk.

vinblastine [ch 2, 8]—Drug that interferes with microtubule formation in neurons by inducing formation of tubulin crystals.

vincristine [ch 2, 8]—Drug that interferes with microtubule formation in neurons by inducing formation of tubulin crystals. One of two alkaloids (the other being vinblastine) produced by the periwinkle plant that is useful in the treatment of cancer.

vinculin [ch 1, 2]—Plasma membrane (attachment) protein involved in anchoring of proteins such as microfilaments and cytoskeletal proteins to the plasma membrane in neurons and skeletal muscle fibers.

voltage clamp [ch 7]—Technique to quantify activation and conducting properties of ion channels. The membrane potential is regulated from one potential (i.e., resting) to another (holding potential), and the current needed to maintain this membrane potential is measured.

voltage sensor [ch 1, 7, 11, 15]—Amino acid sequence within the ion channel protein, where a change in membrane potential can be detected.

Wallerian degeneration [ch 16, 17]—The sequence of events in dismantling an axon following injury, first described by Waller.

weakness—A restricted ability to generate muscular force.

X-ray crystallography—See X-ray diffraction.

X-ray diffraction [ch 11]—Scattering of X rays by atoms of a substance with uniform repeating units so the diffraction pattern allows interpretation of the structure of the substance. Used to identify protein structures and dimensions of filament arrays.

Z-line (disk)—Dark structure in the center of the muscle fiber I-band, to which actin filaments are attached.

References

Aagaard, P., Andersen, J. L., Dyhre-Poulsen, P., Leffers, A. M., Wagner, A., Magnusson, S. P., Halkjaer-Kristensen, J., & Simonsen, E. B. (2001). A mechanism for increased contractile strength of human pennate muscle in response to strength training: changes in muscle architecture. Journal of Physiology 534, 613-623. [20]

Aagaard, P., Simonsen, E. B., Andersen, J. L., Magnusson, P., & Dyhre-Poulsen, P. (2002). Increased rate of force development and neural drive of human skeletal muscle following resistance training. Journal of Applied Physiology 93, 1318-1326. [20]

Abbate, F., Bruton, J. D., de Haan, A., & Westerblad, H. (2002). Prolonged force increase following a high-frequency burst is not due to a sustained elevation of [Ca2+]i. American Journal of Physiology 283, C42-C47. [13]

Adams, G. R., Haddada, F., McCue, S. A., Bodell, P. W., Zeng, M., Qin, A. X., & Baldwin, K. M. (2000a). Effects of spaceflight and thyroid deficiency on rat hindlimb development. II. Expression of MHC isoforms. Journal of Applied Physiology 88, 904-916. [5]

Adams, G. R., McCue, S. A., Bodell, P. W., Zeng, M., & Baldwin, K. M. (2000b). Effects of spaceflight and thyroid deficiency on hindlimb development. I. Muscle mass and IGF-I expression. Journal of Applied Physiology 88, 894-903. [5]

Adams, R. D., Denny-Brown, D., & Pearson, C. (1962). Diseases in Muscle. A Study in Pathology, 2nd ed. Harper and Row, New York. [16]

Adrian, E. D. & Bronk, D. W. (1929). The discharge of impulses in motor nerve fibres. II. The frequency of discharge in reflex and voluntary contractions. Journal of Physiology 67, 119-151. [13]

Adrian, R. H. & Bryant, S. H. (1974). On the repetitive discharge in myotonic muscle fibres. Journal of Physiology 240, 505-515. [7]

Adrian, R. H., Costantin, L. L., & Peachey, L. D. (1969). Radial spread of contraction in frog muscle fibres. Journal of Physiology 204, 231-257. [11]

Aguayo, A., Attiwell, M., Trecarten, J., Perkins, S., & Bray, G. M. (1977). Abnormal myelination in transplanted Trembler mouse Schwann cells. Nature 265, 73-75. [2]

Aguayo, A., Perkins, S., Bray, G., & Duncan, I. (1978). Transplantation of nerves from patients with Charcot-Marie-Tooth (CMT) disease into immune-suppressed mice. Journal of Neuropathology and Experimental Neurology 37, 582. [2]

Aitken, J. T., Sharman, M., & Young, J. Z. (1947). Maturation of regenerating nerve fibres with various peripheral connexions. Journal of Anatomy 81, 1-22. [17]

Akima, H., Kawakami, Y., Kubo, K., Sekiguchi, C., Ohshima, H., Miyamoto, A., & Fukunaga, T. (2000). Effect of short-duration spaceflight on thigh and leg muscle volume. Medicine and Science in Sports and Exercise 32, 1743-1747. [19]

Al Majed, A. A., Neumann, C. M., Brushart, T. M., & Gordon, T. (2000). Brief electrical stimulation promotes the speed and accuracy of motor axonal regeneration. Journal of Neuroscience 20, 2602-2608. [17]

Al-Amood, W. S., Buller, A. J., & Pope, R. (1973). Long-term stimulation of cat fast-twitch skeletal muscle. Nature 244, 225-227. [18, 20]

Albani, A., Lowrie, M. B., & Vrbová, G. (1988). Reorganization of motor units in reinnervated muscles of the rat. Journal of Neurological Sciences 88, 195-206. [6]

Albe-Fessard, D., Liebeskind, J., & Lamarre, Y. (1965). Projection au niveau du cortex somato-moteur du singe d'afférences provemant des recepteurs musclaires. Comptes Rendus de l'Academie des Sciences 261, 3891-3894. [4]

Albers, R. W. (1967). Biochemical aspects of active transport. Annual Review of Biochemistry 36, 727-756. [7]

Alberts, B., Bray, D., Lewis, J., Raff, M., Roberts, K., & Watson, J. D. (1983). Molecular Biology of the Cell, 1st ed. Garland, New York. [14]

Alberts, B., Bray, D., Lewis, J., Raff, M., Roberts, K., & Watson, J. D. (1989). Molecular Biology of the Cell, 2nd ed. Garland, New York. [1]

Albuquerque, E. X., Schuh, F. T., & Kauffman, F. C. (1971). Early membrane depolarization of the fast mammalian muscle after denervation. Pflügers Archiv: European Journal of Physiology 328, 36-50. [16, 18]

Aljure, E. F. & Borrero, L. M. (1968). The influence of muscle length on the development of fatigue in toad sartorius. Journal of Physiology 199, 241-252. [15]

Allen, D. G., Lee, J. A., & Westerblad, H. (1989). Intracellular calcium and tension during fatigue in isolated single muscle fibres from Xenopus laevis. Journal of Physiology 415, 433-458. [11, 15]

Allen, D. G., Westerblad, H., Lee, J. A., & Lännergren, J. (1992). Role of excitation-contraction coupling in muscle fatigue. Sports Medicine 13, 116-126. [20]

Allen, F. & Warner, A. (1991). Gap junction communication during neuromuscular junction formation. Neuron 6, 101-111. [6]

Allen, G. M., Gandevia, S. C., Neering, I. R., Hickie, I., Jones, R., & Middleton, J. (1994). Muscle performance, voluntary activation and perceived effort in normal subjects and patients with prior poliomyelitis. Brain 117, 661-670. [19]

Allt, G. & Cavanagh, J. B. (1969). Ultrastructural changes in the region of the node of Ranvier in the rat caused by diphtheria toxin. Brain 92, 459-468. [9]

Almenar-Queralt, A. & Goldstein, L. S. (2001). Linkers, packages and pathways: new concepts in axonal transport. Current Opinion in Neurobiology 11, 550-557. [8]

Alnaes, E. & Rahamimoff, R. (1975). On the role of mitochondria in transmitter release from motor nerve terminals. Journal of Physiology 248, 285-306. [3]

Alnaqeeb, M. A. & Goldspink, G. (1987). Changes in fibre type, number and diameter in developing and ageing skeletal muscle. Journal of Anatomy 153, 31-45. [22]

Alshuaib, W. B. & Fahim, M. A. (1990). Aging increases calcium influx at motor nerve terminal. International Journal of Developmental Neuroscience 8, 655-666. [22]

Alway, S. E., MacDougall, J. D., Sale, D. G., Sutton, J. R., & McComas, A. J. (1988). Functional and structural adaptations in skeletal muscle of trained athletes. Journal of Applied Physiology 64, 1114-1120. [20]

Alway, S. E., Winchester, P. K., Davis, M. E., & Gonyea, W. J. (1989). Regionalized adaptations and muscle fiber proliferation in stretch-induced enlargement. Journal of Applied Physiology 66, 771-781. [20]

Amos, L. A., Linck, R. W., & Klug, A. (1976). Molecular structure of flagellar microtubules. In Cell Motility. Book C. Microtubules and Related Proteins, eds. Goldman, R., Pollard, T., & Rosenbaum, J., pp. 847-868. Cold Spring Harbor Laboratory, New York. [2]

Andersen, P. & Henriksson, J. (1977). Training induced changes in the subgroups of human type II skeletal muscle fibres. Acta Physiologica Scandinavica 99, 123-125. [20]

Anderson, D. C., King, S. C., & Parsons, S. M. (1982). Proton gradient linkage to active uptake of [3H] acetylcholine by Torpedo electric organ synaptic vesicles. Biochemistry 21, 3037-3043. [10]

Andreassen, S. & Arendt-Neilsen, L. (1987). Muscle fibre conduction velocity in motor units of the human anterior tibial muscle: a new size principle parameter. Journal of Physiology 391, 561-571. [9, 12]

Andrew, W. (1971). The Anatomy of Ageing in Man and Animal. Grune and Stratton, New York. [22]

Aniansson, A. & Gustafsson, E. (1981). Physical training in elderly men with special reference to quadriceps muscle strength and morphology. Clinical Physiology 1, 87-98. [22]

Aniansson, A., Ljungberg, P., Rundgren, A., & Wattequist, H. (1984). Effect of a training programme for pensioners on condition and muscular strength. Archives of Gerontology and Geriatrics 3, 229-241. [22]

Ansved, T. & Edström, L. (1991). Effects of age on fibre structure, ultrastructure and expression of desmin and spectrin in fast- and slow-twitch rat muscles. Journal of Anatomy 174, 61-79. [22]

Ansved, T. (2001). Muscle training in muscular dystrophies. Acta Physiologica Scandinavica 171, 359-366. [20]

Arabadjis, P. G., Heffner, R. R., Jr., & Pendergast, D. R. (1990). Morphologic and functional alterations in aging rat muscle. Journal of Neuropathology and Experimental Neurology 49, 600-609. [22]

Arahata, K. & Engel, A. G. (1985). Endomysial killer/natural killer (K/NK) cells, T cells and macrophages (MF) in polymyositis (PM), inclusion body myositis (IBM), and Duchenne dystrophy (DD). Neurology 35, 205. [21]

Arasaki, K., Tamaki, M., Hosoya, Y., & Kudo, N. (1997). Validity of electromyograms and tension as a means of motor unit number estimation. Muscle & Nerve 20, 552-560. [12]

Armstrong, C. M. & Bezanilla, F. (1973). Currents related to movement of the gating particles of the sodium channels. Nature 242, 459-461. [7]

Armstrong, C. M., Bezanilla, F., & Rojas, E. (1973). Destruction of sodium conductance inactivation in squid axons perfused with pronase. Journal of General Physiology 62, 375-391. [7]

Armstrong, C. M., Bezanilla, F. M., & Horowicz, P. (1972). Twitches in the presence of ethylene glycol bis(β-aminoethyl ether)-N,N1-tetraacetic acid. Biochimica et Biophysica Acta 267, 605-608. [11]

Armstrong, R. B., Marum, P., Tullson, P., & Saubert, C. W. (1979). Acute hypertrophic response of skeletal muscle to removal of synergists. Journal of Applied Physiology 46, 835-842. [20]

Arnáson, B. G. W., Winkler, G. F., & Hadler, N. M. (1969). Cell-mediated demyelination of peripheral nerve in tissue culture. Laboratory Investigation; A Journal of Technical Methods and Pathology 21, 1-10. [16]

Arnold, H. H. & Winter, B. (1998). Muscle differentiation: more complexity to the network of myogenic regulators. Current Opinion in Genetics & Development 8, 539-544. [5]

Arnold, N. & Harriman, D. G. (1970). The incidence of abnormality in control human peripheral nerves studied by single axon dissection. Journal of Neurology, Neurosurgery, and Psychiatry 33, 55-61. [22]

Ashley, C. C. & Ridgway, E. B. (1968). Simultaneous recording of membrane potential, calcium transient and tension in single muscle fibers. Nature 219, 1168-1169. [11]

Ashley, C. C. & Ridgway, E. B. (1970). On the relationships between membrane potential, calcium transient and tension in single barnacle muscle fibres. Journal of Physiology 209, 105-130. [11]

Asmussen, E. (1979). Muscle fatigue. Medicine and Science in Sports 11, 313-321. [15]

Åstrand, P. O. (1967). Diet and athletic performance. Federation Proceedings 26, 1772-1777. [14]

Aström, K. E. & Waksman, B. H. (1962). The passive transfer of experimental allergic encephalomyelitis and neuritis with living lymphoid cells. Journal of Pathology and Bacteriology 83, 89-106. [16]

Auld, V. J., Goldin, A. L., Krafte, D. S., Catterall, W. A., Lester, H. A., Davidson, N., & Dunn, R. J. (1990). A neutral amino acid change in segment 11 S4 dramatically alters the gating properties of the voltage-dependent sodium channel. Proceedings of the National Academy of Sciences of the United States of America 87, 323-327. [7]

Auld, V. J., Goldin, A. L., Krafte, D. S., Marshall, J., Dunn, J. M., Catterall, W. A., Lester, H. A., Davidson, N., & Dunn, R. J. (1988). A rat brain Na+ channel α subunit with novel gating properties. Neuron 1, 449-461. [7]

Axelsson, J. & Thesleff, S. (1959). A study of supersensitivity in denervated mammalian skeletal muscle. Journal of Physiology 147, 178-193. [16]

Bagust, J. & Lewis, D. M. (1974). Isometric contractions of motor units in self-reinnervated fast and slow twitch muscles of the cat. Journal of Physiology 237, 91-102. [13, 17]

Baichwal, R. R., Bigbee, J. W., & DeVríes, G. H. (1988). Macrophage-mediated myelin-related mitogenic factor for cultured Schwann cells. Proceedings of the National Academy of Sciences of the United States of America 85, 1701-1705. [16]

Baker, L. P., Chen, Q., & Peng, H. B. (1992). Induction of acetylcholine receptor clustering by native polystyrene beads. Journal of Cell Science 102, 543-555. [3]

Baker, P. F., Hodgkin, A. L., & Shaw, T. I. (1962a). Replacement of the axoplasm of giant nerve fibres with artificial solutions. Journal of Physiology 164, 330-354. [9]

Baker, P. F., Hodgkin, A. L., & Shaw, T. I. (1962b). The effects of changes in internal ionic concentrations on the electrical properties of perfused giant axons. Journal of Physiology 164, 355-374. [9]

Balice-Gordon, R. J. & Lichtman, J. W. (1994). Long-term synapse loss induced by focal blockade of postsynaptic receptors. Nature 372, 519-524. [6]

Ballantyne, J. P. & Campbell, M. J. (1973). Electrophysiological study after surgical repair of sectioned human peripheral nerves. Journal of Neurology, Neurosurgery, and Psychiatry 36, 797-805. [17]

Bangsbo, J., Graham, T. E., Kiens, B., & Saltin, B. (1992). Elevated muscle glycogen and anaerobic energy production during exhaustive exercise in man. Journal of Physiology 451, 205-227. [14]

Banister, E. W. & Cameron, B. J. C. (1990). Exercise-induced hyperammonemia: peripheral and central effects. International Journal of Sports Medicine 11, 129-142. [14]

Banker, B. Q. (1986). The congenital myopathies. In Myology: Basic and Clinical, eds. Engel, A. G. & Banker, B. Q., pp. 1527-1581. McGraw-Hill, New York. [5]

Banker, B. Q., Kelly, S. S., & Robbins, N. (1983). Neuromuscular transmission and correlative morphology in young and old mice. Journal of Physiology 339, 355-375. [22]

Barash, I. A., Peters, D., Fridén, J., Lutz, G. J., & Lieber, R. L. (2002). Desmin cytoskeletal modifications after a bout of eccentric exercise in the rat. American Journal of Physiology 283, R958-R963. [21]

Barbeau, H., Ladouceur, M., Mirbagheri, M. M., & Kearney, R. E. (2002). The effect of locomotor training combined with functional electrical stimulation in chronic spinal cord injured subjects: walking and reflex studies. Brain Research-Brain Research Reviews 40, 274-291. [13]

Barbet, J. P., Thornell, L. E., & Butler-Browne, G. S. (1991). Immunocytochemical characterisation of two generations of fibers during the development of the human quadriceps muscle. Mechanisms of Development 35, 3-11. [5]

Barchi, R. L. (1988). Probing the molecular structure of the voltage-dependent sodium channel. Annual Review of Neuroscience 11, 455-495. [7]

Barcroft, H. & Millen, J. L. E. (1939). The blood flow through muscle during sustained contraction. Journal of Physiology 97, 17-31. [15]

Barker, D. (1974). The morphology of muscle receptors. In Handbook of Sensory Physiology. Muscle Receptors, ed. Hunt, C. C., pp. 1-190. Springer-Verlag, Berlin. [4]

Barker, D. & Ip, M. C. (1966). Sprouting and degeneration of mammalian motor axons in normal and de-afferentated skeletal muscle. Proceedings of the Royal Society of London. Series B: Biological Sciences 163, 538-554. [22]

Barker, D. & Saito, M. (1981). Autonomic innervation of receptors and muscle fibres in cat skeletal muscle. Proceedings of the Royal Society of London. Series B: Biological Sciences 212, 317-332. [15]

Barnard, E. A., Dolly, J. O., Porter, C. W., & Albuquerque, E. X. (1975). The acetylcholine receptor and the ionic conductance modulation system of skeletal muscle. Experimental Neurology 48, 1-28. [3]

Basmajian, J. V. (1963). Control and training of individual motor units. Science 141, 440-441. [13]

Basmajian, J. V. & DeLuca, C. J. (1985). Muscles Alive: Their Functions Revealed by Electromyography, 5th ed. Williams & Wilkins, Baltimore. [13]

Bastian, J. & Nakajima, S. (1974). Action potential in the transverse tubules and its role in the activation of skeletal muscle. Journal of General Physiology 63, 257-278. [11]

Bateson, D. S. & Parry, D. J. (1983). Motor units in a fast-twitch muscle of normal and dystrophic mice. Journal of Physiology 345, 515-523. [12]

Baumann, H., Jäggi, M., Soland, F., Howald, H., & Schaub, M. C. (1987). Exercise training induces transitions of myosin isoform subunits within histochemically typed human muscle fibres. Pflügers Archiv: European Journal of Physiology 409, 349-360. [20]

Bay, C. M. & Strichartz, G. R. (1980). Saxitoxin binding to sodium channels of rat skeletal muscles. Journal of Physiology 300, 89-103. [7]

Beam, K. G., Caldwell, J. H., & Campbell, D. T. (1985). Na channels in skeletal muscle concentrated near the neuromuscular junction. Nature 313, 588-590. [7]

Beam, K. G., Knudson, C. M., & Powell, J. A. (1986). A lethal mutation in mice eliminates the slow calcium current in skeletal muscle cells. Nature 320, 168-170. [11]

Beaton, L. J., Tarnopolsky, M. A., & Phillips, S. M. (2002). Contraction-induced muscle damage in humans following calcium channel blocker administration. Journal of Physiology 544, 849-859. [21]

Beeson, D. & Barnard, E. (1990). Acetylcholine receptors at the neuromuscular junction. In Neuromuscular Transmission: Basic and Applied Aspects, eds. Vincent, A. & Wray, D., pp. 157-181. Manchester University Press, Manchester. [3]

Bélanger, A. Y. & McComas, A. J. (1981). Extent of motor unit activation during effort. Journal of Applied Physiology 51, 1131-1135. [13, 15]

Bélanger, A. Y. & McComas, A. J. (1983). Contractile properties of muscles in myotonic dystrophy. Journal of Neurology, Neurosurgery, and Psychiatry 46, 625-631. [15]

Bellemare, F. & Bigland-Ritchie, B. (1984). Assessment of human diaphragm strength and activation using phrenic nerve stimulation. Respiration Physiology 58, 263-277. [13]

Bellemare, F., Bigland-Ritchie, B., & Woods, J. J. (1986). Contractile properties of the human diaphragm in vivo. Journal of Applied Physiology 61, 1153-1161. [12]

Bellemare, F., Woods, J. J., Johansson, R., & Bigland-Ritchie, B. (1983). Motor-unit discharge rates in maximal voluntary contractions of three human muscles. Journal of Neurophysiology 50, 1380-1392. [12, 13]

Belmar, J. & Eyzaguirre, C. (1966). Pacemaker site of fibrillation potentials in denervated mammmalian muscle. Journal of Neurophysiology 29, 425-441. [16]

Benezra, R., Davis, R. L., Lassar, A., Tapscott, S., Thayer, M., Lockshon, D., & Weintraub, H. (1990). Id: a negative regulator of helix-loop-helix DNA binding proteins. Control of terminal myogenic differentiation. Annals of the New York Academy of Sciences 599, 1-11. [5]

Bennett, M. K., Calakos, N., & Scheller, R. H. (1992). Syntaxin: a synaptic protein implicated in docking of synaptic vesicles at presynaptic active zones. Science 257, 255-259. [10]

Bennett, M. K. & Scheller, R. H. (1993). The molecular machinery for secretion is conserved from yeast to neurons. Proceedings of the National Academy of Sciences of the United States of America 90, 2559-2563. [10]

Bennett, M. R., Mclachlan, E. M., & Taylor, R. S. (1973a). The formation of synapses in reinnervated mammalian striated muscle. Journal of Physiology 233, 481-500. [17]

Bennett, M. R. & Pettigrew, A. G. (1974). The formation of synapses in striated muscle during development. Journal of Physiology 241, 515-545. [6]

Bennett, M. R., Pettigrew, A. G., & Taylor, R. S. (1973b). The formation of synapses in reinnervated and cross-reinnervated adult avian muscle. Journal of Physiology 230, 331-357. [17]

Benoit, P. & Changeux, J. P. (1978). Consequences of blocking the nerve with a local anaesthetic on the evolution of multiinnervation at the regenerating neuromuscular junction of the rat. Brain Research 149, 89-96. [6]

Berg, D. K. & Hall, Z. W. (1975). Increased extra-junctional acetylcholine sensitivity produced by chronic acetylcholine sensitivity produced by chronic post-synaptic neuromuscular blockade. Journal of Physiology 244, 659-676. [19]

Berg, H. E., Dudley, G. A., Haggmark, T., Ohlsen, H., & Tesch, P. A. (1991). Effects of lower limb unloading on skeletal muscle mass and function in humans. Journal of Applied Physiology 70, 1882-1885. [19]

Berg, H. E., Larsson, L., & Tesch, P. A. (1997). Lower limb skeletal muscle function after 6 wk of bed rest. Journal of Applied Physiology 82, 182-188. [19]

Bergström, J. (1962). Muscle electrolytes in man. Scandinavian Journal of Clinical Laboratory Investigation 68, 1-110. [13]

Bergström, J., Hermansen, L., Hultman, E., & Saltin, B. (1967). Diet, muscle glycogen and physical performance. Acta Physiologica Scandinavica 71, 140-150. [14]

Berliner, E., Young, E. C., Anderson, K., Mahtani, H. K., & Gelles, J. (1995). Failure of a single-headed kinesin to track parallel to microtubule protofilaments. Nature 373, 718-721. [8]

Bessman, S. P. & Fonyo, A. (1966). The possible role of the mitochondrial bound creatine kinase in regulation of mitochondrial respiration. Biochemical and Biophysical Research Communications 22, 597-602. [14]

Bessman, S. P. & Geiger, P. J. (1981). Transport of energy in muscle: the phosphorylcreatine shuttle. Science 211, 448-452. [14]

Bessou, P. & Pagès, B. (1972). Intracellular potentials from intrafusal muscle fibers evoked by stimulation of static and dynamic fusimotor axons in the cat. Journal of Physiology 227, 709-727. [4]

Betz, W. J., Caldwell, J. H., & Ribchester, R. R. (1979). The size of motor units during post-natal development of rat lumbrical muscle. Journal of Physiology 297, 463-478. [12]

Bevan, S., Chiu, S. Y., Gray, P. T. A., & Ritchie, F. R. S. (1985). The presence of voltage-gated sodium, potassium and chloride channels in rat cultured astrocytes. Proceedings of the Royal Society of London. Series B: Biological Sciences 225, 299-313. [2]

Bigland, B. & Lippold, O. C. J. (1954). Motor unit activity in the voluntary contraction of human muscle. Journal of Physiology 125, 322-335. [13]

Bigland-Ritchie, B., Furbush, F., & Woods, J. J. (1986a). Fatigue of intermittent submaximal voluntary contractions: central and peripheral factors. Journal of Applied Physiology 61, 421-429. [15]

Bigland-Ritchie, B., Johansson, R., Lippold, O. C. J., Smith, S., & Woods, J. J. (1983). Changes in motoneurone firing rates during sustained maximal voluntary contractions. Journal of Physiology 340, 335-346. [13, 15]

Bigland-Ritchie, B., Jones, D. A., & Woods, J. J. (1979). Excitation frequency and muscle fatigue: electrical responses during human voluntary and stimulated contractions. Experimental Neurology 64, 414-427. [13]

Bigland-Ritchie, B., Kukulka, C. G., Lippold, O. C., & Woods, J. J. (1982). The absence of neuromuscular transmission failure in sustained maximal voluntary contractions. Journal of Physiology 330, 265-278. [15]

Bigland-Ritchie, B. R., Dawson, N. J., Johansson, R. S., & Lippold, O. C. (1986b). Reflex origin for the slowing of motoneurone firing rates in fatigue of human voluntary contractions. Journal of Physiology 379, 451-459. [4, 15]

Bigland-Ritchie, B. R., Fuglevand, A. J., & Thomas, C. K. (1998). Contractile properties of human motor units: is man a cat? Neuroscientist 4, 240-249. [12]

Billeter, R., Heizmann, C. W., Howald, H., & Jenny, E. (1981). Analysis of myosin light and heavy chain types in single human skeletal muscle fibers. European Journal of Biochemistry 116, 389-395.

Bisby, M. A. (1975). Inhibition of axonal transport in nerves chronically treated with local anesthetics. Experimental Neurology 47, 481-489. [19]

Bischoff, R. & Holtzer, H. (1969). Mitosis and the processes of differentiation of myogenic cells in vitro. Journal of Cell Biology 41, 188-200. [5]

Bischoff, R. & Lowe, M. (1974). Cell surface components and the interaction of myogenic cells. In Exploratory Concepts in Muscular Dystrophy II, ed. Milhorat, A. T., pp. 17-29. Excerpta Medica, Amsterdam. [5]

Black, J. A., Kocsis, J. D., & Waxman, S. G. (1990). Ion channel organization of the myelinated fiber. Trends in Neurosciences 13, 48-54. [7, 9]

Black, M. M. & Lasek, R. J. (1979). Slowing of the rate of axonal regeneration during growth and maturation. Experimental Neurology 63, 108-119. [17]

Blake, D. J., Weir, A., Newey, S. E., & Davies, K. E. (2002). Function and genetics of dystrophin and dystrophin-related proteins in muscle. Physiological Reviews 82, 291-329. [1]

Blasi, J., Chapman, E. R., Link, E., Binz, T., Yamasaki, S., De Camilli, P., Südhof, T. C., Niemann, H., & Jahn, R. (1993). Botulinum neurotoxin A selectively cleaves the synaptic protein SNAP-25. Nature 365, 160-163. [10]

Blewett, C. & Elder, G. C. B. (1993). Quantitative EMG analysis in soleus and plantaris during hindlimb suspension and recovery. Journal of Applied Physiology 74, 2057-2066. [19]

Blinks, J. R., Rüdel, R., & Taylor, S. R. (1978). Calcium transients in isolated amphibian skeletal muscle fibres: detection with aequorin. Journal of Physiology 277, 291-323. [15]

Blumcke, S. & Niedorf, H. R. (1965). [Electron optical studies on the growth end-bulbs of regenerating peripheral nerve fibers]. [German]. Virchows Archiv fur Pathologische Anatomie und Physiologie und fur Klinische Medizin 340, 93-104. [3]

Bodian, D. (1964). An electromicroscopic study of the monkey spinal cord. I. Fine structure of normal motor column. II. Effects of retrograde chromatolysis. III. Cytologic effects of mild and virulent poliovirus infection. Bulletin of Johns Hopkins Hospital 114, 13-119. [2]

Bodine-Fowler, S., Garfinkel, A., Roy, R. R., & Edgerton, V. R. (1990). Spatial distribution of muscle fibers within the territory of a motor unit. Muscle & Nerve 13, 1133-1145. [12]

Bolitho-Donaldson, S. K. & Hermansen, L. (1978). Differential, direct effects of H+ on Ca2+ -activated force of skinned fibres from the soleus, cardiac and adductor magnus muscles of rabbits. Pflügers Archiv: European Journal of Physiology 376, 55-65. [15]

Booth, F. W. (1977). Time course of muscular atrophy during immobilization of hindlimbs in rats. Journal of Applied Physiology 43, 656-661. [19]

Booth, F. W. (1988). Perspectives on molecular and cellular exercise physiology. Journal of Applied Physiology 65, 1461-1471. [20]

Booth, F. W., Lou, W., Hamilton, M. T., & Yan, Z. (1996). Cytochrome c mRNA in skeletal muscles of immobilized limbs. Journal of Applied Physiology 81, 1941-1945. [19]

Borg, J. (1981). Properties of single motor units of the extensor digitorum brevis in elderly humans. Muscle & Nerve 4, 429-434. [22]

Borg, T. K. & Caulfield, J. B. (1980). Morphology of connective tissue in skeletal muscle. Tissue Cell 12, 197-207. [1]

Bostock, H. & Sears, T. A. (1978). The internodal axon membrane: electrical excitability and continuous conduction in segmental demyelination. Journal of Physiology 280, 273-301. [9]

Botelho, S. Y., Cander, L., & Guiti, N. (1954). Passive and active tension-length diagrams of intact skeletal muscle in normal women of different ages. Journal of Neurobiology 6, 379-394. [22]

Bottinelli, R., Schiaffino, S., & Reggiani, C. (1991). Force-velocity relations and myosin heavy chain isoform compositions of skinned fibres from rat skeletal muscle. Journal of Physiology 437, 655-672. [12]

Bowden, R. E. M. & Gutmann, E. (1944). Denervation and re-innervation of human voluntary muscle. Brain 67, 273-313. [16]

Boyd, I. A. (1966). The behaviour of isolated mammalian muscle spindles with intact innervation. Journal of Physiology 186, 109P-110P. [4]

Boyd, I. A. & Davey, M. R. (1968). Composition of Peripheral Nerves Livingstone, Edinburgh. [12]

Brady, S. T. (1985). A novel brain ATPase with properties expected for the fast axonal transport motor. Nature 317, 73-75. [8]

Brandstater, M. E. & Dinsdale, S. M. (1976). Electrophysiological studies in the assessment of spinal cord lesions. Archives of Physical Medicine & Rehabilitation 57, 70-74. [18]

Brandstater, M. E. & Lambert, E. H. (1969). A histochemical study of the spatial arrangement of muscle fibres in single motor units within rat tibialis anterior muscle. Bulletin of the American Association of Electromyography and Electrodiagnosis 82, 15-16. [12]

Brandstater, M. E. & Lambert, E. H. (1973). Motor unit anatomy. Type and spatial arrangement of muscle fibres. In New Developments in Electromyography and Clinical Neurophysiology, ed. Desmedt, J. E., pp. 14-22. Karger, Basel. [12]

Brannstrom, T. & Kellerth, J. O. (1998). Changes in synaptology of adult cat spinal alpha-motoneurons after axotomy. Experimental Brain Research 118, 1-13. [16]

Braun, T., Rudnicki, M. A., Arnold, H. H., & Jaenisch, R. (1992). Targeted inactivation of the muscle regulatory gene Myf-5 results in abnormal rib development and perinatal death. Cell 71, 369-382. [5]

Bray, J. J., Hawken, M. J., Hubbard, J. I., Pockett, S., & Wilson, L. (1976). The membrane potential of rat diaphragm muscle fibres and the effect of denervation. Journal of Physiology 255, 651-667. [16]

Brimijoin, S. (1975). Stop-flow: a new technique for measuring axonal transport, and its application to the transport of dopamine-β-hydroxylase. Journal of Neurobiology 6, 379-394. [8]

Broman, H., De Luca, C. J., & Mambrito, B. (1985). Motor unit recruitment and firing rates interaction in the control of human muscles. Brain Research 337, 311-319. [13]

Brooke, M. H. & Engel, W. K. (1969). The histographic analysis of human muscle biopsies with regard to fiber types. 1. Adult male and female. Neurology 19, 221-233. [5]

Brooke, M. H. & Kaiser, K. K. (1970). Muscle fiber types: how many and what kind? Archives of Neurology 23, 369-379. [12, 13]

Brooke, M. H. & Kaiser, K. K. (1974). The use and abuse of muscle histochemistry. Annals of the New York Academy of Sciences 228, 121-144. [12]

Brooks, G. A. (2000). Intra- and extra-cellular lactate shuttles. Medicine and Science in Sports and Exercise 32, 790-799. [14]

Brooks, G. A., Dubouchaud, H., Brown, M., Sicurello, J. P., & Butz, C. E. (1999). Role of mitochondrial lactate dehydrogenase and lactate oxidation in the intracellular lactate shuttle. Proceedings of the National Academy of Sciences of the United States of America 96, 1129-1134. [14]

Brooks, J. E. (1969). Hyperkalemic periodic paralysis. Intracellular electromyographic studies. Archives of Neurology 20, 13-18. [7]

Brostoff, S., Burnett, P., Lampert, P., & Eylar, E. H. (1972). Isolation and characterization of a protein from sciatic nerve myelin responsible for experimental allergic neuritis. Nature New Biology 235, 210-212. [16]

Brostoff, S. W. & Eylar, E. H. (1972). The proposed amino acid sequence of the P1 protein of rabbit sciatic nerve myelin. Archives of Biochemistry and Biophysics 153, 590-598. [16]

Brown, A. G. & Fyffe, R. E. (1981). Direct observations on the contacts made between Ia afferent fibres and alpha-motoneurones in the cat's lumbosacral spinal cord. Journal of Physiology 313, 121-140. [4]

Brown, G. L. & Harvey, A. M. (1939). Congenital myotonia in the goat. Brain 62, 341-363. [7]

Brown, M. C. & Butler, R. G. (1974). Evidence for innervation of muscle spindle intrafusal fibres by branches of α−motoneurones following nerve injury. Journal of Physiology 238, 41P-43P. [17]

Brown, M. C., Holland, R. L., & Ironton, R. (1978). Degenerating nerve products affect innervated muscle fibres. Nature 275, 652-654. [17]

Brown, M. C. & Ironton, R. (1977). Motor neurone sprouting induced by prolonged tetrodotoxin block of nerve action potentials. Nature 265, 459-461. [19]

Brown, M. C., Jansen, J. K., & Van Essen, D. (1976). Polyneuronal innervation of skeletal muscle in new-born rats and its elimination during maturation. Journal of Physiology 261, 387-422. [6]

Brown, M. C. & Lunn, E. R. (1988). Mechanism of interaction between motoneurons and muscles. Ciba Foundation Symposium 138, 78-96. [17]

Brown, W. F. (1972). A method for estimating the number of motor units in thenar muscles and the changes in motor unit count with ageing. Journal of Neurology, Neurosurgery, and Psychiatry 35, 845-852. [22]

Browne, K., Lee, J., & Ring, P. A. (1954). The sensation of passive movement at the metatarsophalangeal joint of the great toe in man. Journal of Physiology 126, 448-458. [4]

Brushart, T. M., Hoffman, P. N., Royall, R. M., Murinson, B. B., Witzel, C., & Gordon, T. (2002). Electrical stimulation promotes motoneuron regeneration without increasing its speed or conditioning the neuron. Journal of Neuroscience 22, 6631-6638. [17]

Bryant, S. H. (1969). Cable properties of external intercostal muscle fibres from myotonic and nonmyotonic goats. Journal of Physiology 204, 539-550. [7]

Bryant, S. H. & Morales-Aguilera, A. (1971). Chloride conductance in normal and myotonic muscle fibres and the action of monocarboxylic aromatic acids. Journal of Physiology 219, 367-383. [7, 9]

Buchthal, F. & Clemmesen, S. (1941). On the differentiation of muscle atrophy by electromyography. Acta Psychiatrica Neurologica 16, 143-181. [13]

Buchthal, F., Guld, C., & Rosenfalck, P. (1955). Propagation velocity in electrically activated muscle fibres in man. Acta Physiologica Scandinavica 34, 75-89. [9]

Buchthal, F., Guld, C., & Rosenfalck, P. (1957). Multielectrode study of the territory of a motor unit. Acta Physiologica Scandinavica 39, 83-104. [12]

Buchthal, F., Rosenfalck, A., & Behse, F. (1975). Sensory potentials of normal and diseased nerves. In Peripheral Neuropathy, eds. Dyck, P. J., Thomas, P. K., & Lambert, E. H., pp. 442-464. Saunders, Philadelphia. [2]

Buchthal, F. & Rosenfalck, P. (1958). Rate of impulse conduction in denervated human muscle. Electroencephalography and Clinical Neurophysiology 10, 521-526. [9]

Buchthal, F., Rosenfalck, P., & Erminio, F. (1960). Motor unit territory and fiber density in myopathies. Neurology 10, 398-408. [9]

Buchthal, F. & Schmalbruch, H. (1970). Contraction times and fibre types in intact human muscle. Acta Physiologica Scandinavica 79, 435-452. [12]

Buckingham, M. (2001). Skeletal muscle formation in vertebrates. Current Opinion in Genetics & Development 11, 440-448. [5]

Buffelli, M., Pasino, E., & Cangiano, A. (1997). Paralysis of rat skeletal muscle equally affects contractile properties as does permanent denervation. Journal of Muscle Research and Cell Motility 18, 683-695. [16]

Buller, A. J., Dornhorst, A. C., Edwards, R., Kerr, D., & Whelan, R. F. (1959). Fast and slow muscles in mammals. Nature 183, 1516-1517. [12]

Buller, A. J., Eccles, J. C., & Eccles, R. M. (1960a). Differentiation of fast and slow muscles in the cat hindlimb. Journal of Physiology 150, 399-416. [5, 19]

Buller, A. J., Eccles, J. C., & Eccles, R. M. (1960b). Interaction between motoneurones and muscles in respect of the characteristic speeds of their responses. Journal of Physiology 150, 417-439. [17, 18]

Burden, S. J. (2002). Building the vertebrate neuromuscular synapse. Journal of Neurobiology 53, 501-511. [3]

Burke, R. E. (1967). Motor unit types of cat triceps surae muscle. Journal of Physiology 193, 141-160. [12, 13]

Burke, R. E. (1975). A comment on the existence of motor unit "types." In The Nervous System. The Basic Neurosciences, ed. Tower, D. B., pp. 611-619. Raven Press, New York. [12]

Burke, R. E. (1986). Physiology of motor units. In Myology. Basic and Clinical, eds. Engel, A. G. & Banker, B. Q., pp. 419-443. McGraw-Hill, New York. [13]

Burke, R. E. (1999). The significance of supraspinal control of reflex actions. Brain Research Bulletin 50, 325. [4]

Burke, R. E., Dunn, R. P., Fleshman, J. W., Glenn, L. L., Lev-Tov, A., O'Donovan, M. J., & Pinter, M. J. (1982). A HRP study of the relation between cell size and motor unit type in cat ankle extensor motoneurons. Journal of Comparative Neurology 209, 17-28. [12]

Burke, R. E., Levine, D. N., Salcman, M., & Tsairis, P. (1974). Motor units in cat soleus muscle: physiological, histochemical and morphological characteristics. Journal of Physiology 238, 503-514. [12]

Burke, R. E., Levine, D. N., Tsairis, P., & Zajac, F. E. (1973). Physiological types and histochemical profiles in motor units of the cat gastrocnemius. Journal of Physiology 234, 723-748. [12, 15]

Burke, R. E., Levine, D. N., & Zajac, F. E. (1971). Mammalian motor units: physiological-histochemical correlation in three types in cat gastrocnemius. Science 174, 709-712. [12, 13]

Burke, R. E. & Tsairis, P. (1973). Anatomy and innervation ratios in motor units of cat gastrocnemius. Journal of Physiology 234, 749-765. [12]

Burkholder, T. J. (2001). Age does not influence muscle fiber length adaptation to increased excursion. Journal of Applied Physiology 91, 2466-2470. [22]

Butler, J., Cauwenbergs, P., & Cosmos, E. (1986). Fate of brachial muscles of the chick embryo innervated by inappropriate nerves: structural, functional and histochemical analyses. Journal of Embryology and Experimental Morphology 95, 147-168. [6]

Butler, J., Cosmos, E., & Brierley, J. (1982). Differentiation of muscle fiber types in aneurogenic brachial muscles of the chick embryo. Journal of Experimental Zoology 224, 65-80. [5]

Butler, J., Cosmos, E., & Cauwenbergs, P. (1988). Positional signals: evidence for a possible role in muscle fibre-type patterning of the embryonic avian limb. Development 102, 763-772. [5]

Butler-Browne, G. S., Eriksson, P. O., Laurent, C., & Thornell, L. E. (1988). Adult human masseter muscle fibers express myosin isozymes characteristic of development. Muscle & Nerve 11, 610-620. [12]

Byers, M. R., Fink, B. R., Kennedy, R. D., Middaugh, M. E., & Hendrickson, A. E. (1973). Effects of lidocaine on axonal morphology, microtubules, and rapid transport in rabbit vagus nerve in vitro. Journal of Neurobiology 4, 125-143. [19]

Bylund-Fellenius, A.-C., Ojamaa, K. M., Flaim, K. E., Li, J. B., Wassner, S. J., & Jefferson, L. S. (1984). Protein synthesis versus energy state in contracting muscles of perfused rat hindlimb. American Journal of Physiology 246, E297-E305. [20]

Caccia, M. R., Harris, J. B., & Johnson, M. A. (1979). Morphology and physiology of skeletal muscle in aging rodents. Muscle & Nerve 2, 202-212. [22]

Cady, E. B., Jones, D. A., Lynn, J., & Newham, D. J. (1989). Changes in force and intracellular metabolites during fatigue of human skeletal muscle. Journal of Physiology 418, 311-325. [15]

Cajal, S. R. (1909). Histologie du Systeme Nerveux de L'homme et des Vertebrés, pp. 485-489. Maloine, Paris. [4]

Cajal, S. R. (1928). Degeneration and Regeneration of the Nervous System. Oxford University Press, London. [17]

Calancie, B. & Bawa, P. (1985). Voluntary and reflexive recruitment of flexor carpi radialis motor units in humans. Journal of Neurophysiology 53, 1194-1200. [12, 13]

Campbell, L. R., Dayton, D. H., & Sohal, G. S. (1986). Neural tube defects: a review of human and animal studies on the etiology of neural tube defects. Teratology 34, 171-187. [6]

Campbell, M. J. & McComas, A. J. (1970). The effects of ageing on muscle function. In 5th Symposium on Current Research on Muscular Dystrophy and Related Disease Presention. Abstract No. 6. Muscular Dystrophy Group of Great Britain, London. [22]

Campbell, M. J., McComas, A. J., & Petito, F. (1973). Physiological changes in ageing muscles. Journal of Neurology, Neurosurgery, and Psychiatry 36, 174-182. [22]

Cangiano, A. (1973). Acetylcholine supersensitivity: the role of neurotrophic factors. Brain Research 58, 255-259. [18]

Carafoli, E. & Penniston, J. T. (1985). The calcium signal. Scientific American 253, 70-78. [7]

Cardasis, C. A. (1983). Ultrastructural evidence of continued reorganization at the aging (11-26 months) rat soleus neuromuscular junction. Anatomical Record 207, 399-415. [22]

Carlsen, R. C. & Walsh, D. A. (1987). Decrease in force potentiation and appearance of alpha-adrenergic mediated contracture in aging rat skeletal muscle. Pflügers Archiv: European Journal of Physiology 408, 224-230. [22]

Carlson, B. M. (1970). Regeneration of the rat gastrocnemius muscle from sibling and non-sibling muscle fragments. The American Journal of Anatomy 128, 21-31. [21]

Carlsoo, S. (1958). Motor units and action potentials in masticatory muscles; an electromyographic study of the form and duration of the action potentials and an anatomic study of the size of the motor units. Acta Morphologica Neerlando-Scandinavica 2, 13-19. [12]

Carolan, B. & Cafarelli, E. (1992). Adaptations in coactivation after isometric resistance training. Journal of Applied Physiology 73, 911-917. [20]

Caroni, P. (1993). Activity-sensitive signaling by muscle-derived insulin-like growth factors in the developing and regenerating neuromuscular system. Annals of the New York Academy of Sciences 692, 209-222. [16]

Carrington, J. L. & Fallon, J. F. (1988). Initial limb budding is independent of apical ectodermal ridge activity; evidence from a limbless mutant. Development 104, 361-367. [5]

Carson, J. A. & Wei, L. (2000). Integrin signaling's potential for mediating gene expression in hypertrophying skeletal muscle. Journal of Applied Physiology 88, 337-343. [1]

Casey, E. B., Jellife, A. M., Le Quesne, P. M., & Millett, Y. L. (1973). Vincristine neuropathy: clinical and electrophysiological observations. Brain 96, 69-86. [8]

Cashman, N. R., Maselli, R., Wollmann, R. L., Roos, R., Simon, R., & Antel, J. P. (1987). Late denervation in patients with antecedent paralytic poliomyelitis. The New England Journal of Medicine 317, 7-12. [17]

Castle, N. A., Haylett, D. G., & Jenkinson, D. H. (1989). Toxins in the characterization of potassium channels. Trends in Neurosciences 12, 59-65. [7]

Castro, M. J., Apple, D. F., Jr., Staron, R. S., Campos, G. E. R., & Dudley, G. A. (1999). Influence of complete spinal cord injury on skeletal muscle within 6 mo of injury. Journal of Applied Physiology 86, 350-358. [19]

Catterall, W. A. (1988). Structure and function of voltage-sensitive ion channels. Science 242, 50-61. [7, 10]

Chandler, W. K., Rakowski, R. F., & Schneider, M. F. (1976). Effects of glycerol treatment and maintained depolarization on charge movement in skeletal muscle. Journal of Physiology 254, 285-316. [11]

Changeux, J. P. & Danchin, A. (1976). Selective stabilisation of developing synapses as a mechanism for the specification of neuronal networks. Nature 264, 705-712. [6]

Chapman, E. A., DeVries, H. A., & Swezey, R. (1972). Joint stiffness: effects of exercise on young and old men. Journal of Gerontology 27, 218-221. [22]

Chapman, R. A. & Tunstall, J. (1987). The calcium paradox of the heart. Progress in Biophysics and Molecular Biology 50, 67-96. [15]

Charlton, M. P., Silverman, H., & Atwood, H. L. (1981). Intracellular potassium activities in muscles of normal and dystrophic mice: an in vivo electrometric study. Experimental Neurology 71, 203-219. [9]

Charlton, M. P., Smith, S. J., & Zucker, R. S. (1982). Role of presynaptic calcium ions and channels in synaptic facilitation and depression at the squid giant synapse. Journal of Physiology 323, 173-193. [10]

Chen, Z. Y., Chai, Y. F., Cao, L., Lu, C. L., & He, C. (2001). Glial cell line-derived neurotrophic factor enhances axonal regeneration following sciatic nerve transection in adult rats. Brain Research 902, 272-276. [17]

Chevallier, A., Kieny, M., & Mauger, A. (1977). Limb-somite relationship: origin of the limb musculature. Journal of Embryology & Experimental Morphology 41, 245-258. [5]

Chow, I. & Cohen, M. W. (1983). Developmental changes in the distribution of acetylcholine receptors in the myotomes of Xenopus laevis. Journal of Physiology 339, 553-571. [6]

Christensen, E. H. & Hansen, O. (1939). Arbeitsfahigkeit und ernahrung. Skandinavische Archiv fur Physiologie 81, 160-171. [14]

Chu, L.-W. (1954). A cytological study of anterior horn cells isolated from human spinal cord. Journal of Comparative Neurology 100, 381-413. [2]

Clark, A. W., Mauro, A., Longenecker, H. E., Jr., & Hurlbut, W. P. (1970). Effects of black widow spider venom on the frog neuromuscular junction. Effects on the fine structure of the frog neuromuscular junction. Nature 225, 703-705. [3]

Clarke, D. M., Loo, T. W., Inesi, G., & MacLennan, D. H. (1989). Location of high affinity Ca2+-binding sites within the predicted transmembrane domain of the sarcoplasmic reticulum Ca2+-ATPase. Nature 339, 476-478. [7]

Clarkson, P. M., Kroll, W., & Melchionda, A. M. (1981). Age, isometric strength, rate of tension development and fiber type composition. Journal of Gerontology 36, 648-653. [22]

Clarkson, P. M. & Sayers, S. P. (1999). Etiology of exercise-induced muscle damage. Canadian Journal of Applied Physiology 24, 234-248. [21]

Clausen, T. (1986). Regulation of active Na+-K+ transport in skeletal muscle. Physiological Reviews 66, 542-580. [7]

Clausen, T. (1990). Significance of Na+-K+ pump regulation in skeletal muscle. News in Physiological Sciences 5, 148-151. [15]

Clausen, T. (1996). Long- and short-term regulation of the Na+-K+ pump in skeletal muscle. News in Physiological Sciences 11, 24-30. [7]

Clausen, T. & Everts, M. E. (1989). Regulation of the Na,K-pump in skeletal muscle. Kidney International 35, 1-13. [7]

Close, R. (1967). Properties of motor units in fast and slow skeletal muscles of the rat. Journal of Physiology 193, 45-55. [12]

Cöers, C. & Woolf, A. L. (1959). The Innervation of Muscle: a Biopsy Study. Blackwell Scientific Publications, Oxford. [17, 22]

Cole, K. S. & Curtis, H. J. (1939). Electric impedance of the squid giant axon during activity. Journal of General Physiology 22, 649-670. [9]

Cole, K. S. & Moore, J. W. (1960). Ionic current measurements in the squid giant axon membrane. Journal of General Physiology 44, 123-167. [7]

Colling-Saltin, A. S. (1978). Enzyme histochemistry on skeletal muscle of the human foetus. Journal of Neurological Sciences 39, 169-185. [5]

Colquhoun, D., Rang, H. P., & Ritchie, J. M. (1974). The binding of tetrodotoxin and α-bungarotoxin to normal and denervated mammalian muscle. Journal of Physiology 240, 199-226. [9]

Condon, K., Silberstein, L., Blau, H. M., & Thompson, W. J. (1990). Differentiation of fiber types in aneural musculature of the prenatal rat hindlimb. Developmental Biology 138, 275-295. [5]

Connor, E. A. & Smith, M. A. (1994). Retrograde signaling in the formation and maintenance of the neuromuscular junction. Journal of Neurobiology 25, 722-739. [6]

Cook, J. A. (1777). A Voyage Towards the South Pole and Around the World, pp. 112-113. Straham and Cadell, London. [7]

Cook, S. D., Murray, M. N., Whitaker, J. N., & Dowling, P. (1969). Myelinotoxic antibody in the Guillain-Barré syndrome. Neurology 19, 284. [16]

Cook, W. H., Walker, J. H., & Barr, M. L. (1951). A cytological study of transneuronal atrophy in the cat and the rabbit. Journal of Comparative Neurology 94, 267-291. [18]

Cooke, R., Franks, K., Luciani, G. B., & Pate, E. (1988). The inhibition of rabbit skeletal muscle contraction by hydrogen ions and phosphate. Journal of Physiology 395, 77-97. [15]

Cooke, R. & Pate, E. (1985). The effects of ADP and phosphate on the contraction of muscle fibers. Biophysical Journal 48, 789-798. [15]

Cooper, R. G., Stokes, M. J., & Edwards, R. H. (1989). Myofibrillar activation failure in McArdle's disease. Journal of the Neurological Sciences 93, 1-10. [14]

Corley, K., Kowalchuk, N., & McComas, A. J. (1984). Contrasting effects of suspension on hind limb muscles in the hamster. Experimental Neurology 85, 30-40. [19]

Costantin, L. L. (1970). The role of sodium current in the radial spread of contraction in frog muscle fibers. Journal of General Physiology 55, 703-715. [11]

Coupland, M. E., Puchert, E., & Ranatunga, K. W. (2001). Temperature dependence of active tension in mammalian (rabbit psoas) muscle fibres: effect of inorganic phosphate. Journal of Physiology 536, 879-891. [15]

Courbin, P., Koenig, J., Ressouches, A., Beam, K. G., & Powell, J. A. (1989). Rescue of excitation-contraction coupling in dysgenic muscle by addition of fibroblasts in vitro. Neuron 2, 1341-1350. [11]

Creese, R., Hashish, S. E. E., & Scholes, N. W. (1958). Potassium movements in contracting diaphragm muscle. Journal of Physiology 143, 307-324. [15]

Creese, R., Head, S. D., & Jenkinson, D. F. (1987). The role of the sodium pump during prolonged end-plate currents in guinea-pig diaphragm. Journal of Physiology 384, 377-403. [9]

Crick, F. H. C. (1970). Diffusion in embryogenesis. Nature 225, 420-422. [5]

Crowe, A. & Matthews, P. B. C. (1964). The effects of stimulation of static and dynamic fusiform fibres on the response to stretching of the primary endings of muscle receptors. Journal of Physiology 174, 109-131. [4]

Cummins, K. L., Dorfman, L. J., & Perkel, D. H. (1979). Nerve fiber conduction-velocity distributions. II. Estimation based on two compound action potentials. Electroencephalography and Clinical Neurophysiology 46, 647-658. [9]

Curtis, J. H. & Cole, K. S. (1938). Transverse electric impedance of the squid giant axon. Journal of General Physiology 21, 757-765. [9]

Czeh, G., Gallego, R., Kudo, N., & Kuno, M. (1978). Evidence for the maintenance of motoneurone properties by muscle activity. Journal of Physiology 281, 239-252. [18]

Dan, Y. & Poo, M.-M. (1992). Hebbian depression of isolated neuromuscular synapses in vitro. Science 256, 1570-1573. [6]

Dantes, M. & McComas, A. J. (1991). The extent and time course of motoneuron involvement in amyotrophic lateral sclerosis. Muscle & Nerve 14, 416-421. [17, 22]

Dasgupta, A. & Simpson, J. A. (1962). Relation between firing frequency of motor units and muscle tension in the human. Electromyography II, 117-128. [13]

David, J. D., See, W. M., & Higginbotham, C. A. (1981). Fusion of chick embryo skeletal myoblasts: role of calcium influx preceding membrane union. Developmental Biology 82, 297-307. [5]

Davies, C. T., Thomas, D. O., & White, M. J. (1986). Mechanical properties of young and elderly human muscle. Acta Medica Scandinavica. Supplementum 711, 219-226. [22]

Davies, C. T. & White, M. J. (1983). Contractile properties of elderly human triceps surae. Gerontology 29, 19-25. [22]

Davis, C. J. & Montgomery, A. (1977). The effect of prolonged inactivity upon the contraction characteristics of fast and slow mammalian twitch muscle. Journal of Physiology 270, 581-594. [19]

Dawson, M. J., Gadian, D. G., & Wilkie, D. R. (1978). Muscular fatigue investigated by phosphorus nuclear magnetic resonance. Nature 274, 861-866. [15]

de Vries, J. I. (1987). Development of Specific Movement Patterns in the Human Fetus. Drukkerij Van Denderen BV, Groningen. [5]

de Vries, J. I., Visser, G. H., & Prechtl, H. F. (1982). The emergence of fetal behaviour. I. Qualitative aspects. Early Human Development 7, 301-322. [5]

Dean, R. B. (1941). Theories of electrolyte equilibrium in muscle. Biological Symposia 3, 331-339. [7]

DeBarsy, T. & Hers, H.-G. (1990). Normal metabolism and disorders of carbohydrate metabolism. Bailliere's Clinical Endocrinology and Metabolism 4, 499-522. [14]

DeLorme, T. L. (1945). Restoration of muscle power by heavy resistance exercises. Journal of Bone and Joint Surgery 27, 645-667. [20]

DeLuca, C. J., LeFever, R. S., McCue, M. P., & Xenakis, A. P. (1982). Behaviour of human motor units in different muscles during linearly varying contractions. Journal of Physiology 329, 113-128. [13]

Denborough, M. A. & Lovell, R. R. H. (1960). Anaesthetic deaths in a family. Lancet ii, 45. [11]

Dengler, R., Konstanzer, A., Hesse, S., Schubert, M., & Wolf, W. (1989). Collateral nerve sprouting and twitch forces of single motor units in conditions with partial denervation in man. Neuroscience Letters 97, 118-122. [17]

Dengler, R., Stein, R. B., & Thomas, C. K. (1988). Axonal conduction velocity and force of single human motor units. Muscle & Nerve 11, 136-145. [12]

Dennis, M. J. & Miledi, R. (1974). Non-transmitting neuromuscular junctions during an early stage of end-plate reinnervation. Journal of Physiology 239, 553-570. [17]

Dennis, M. J., Ziskind-Conhaim, L., & Harris, A. J. (1981). Development of neuromuscular junctions in rat embryos. Developmental Biology 81, 266-279. [6]

Denny-Brown, D. (1929). The histological features of striped muscle in relation to its functional activity. Proceedings of the Royal Society of London. Series B: Biological Sciences 104, 371-411. [12]

Denny-Brown, D. & Brenner, C. (1944). Paralysis of nerve induced by direct pressure and by tourniquet. Archives of Neurology and Psychiatry (Chicago) 51, 1-26. [19]

Denslow, J. S. (1948). Double discharge in human motor units. Journal of Neurophysiology 11, 209-215. [13]

Denzer, A. J., Hauser, D. M., Gesemann, M., & Ruegg, M. A. (1997). Synaptic differentiation: the role of agrin in the formation and maintenance of the neuromuscular junction. Cell and Tissue Research 290, 357-365. [3]

Desmedt, J. E. & Hainaut, K. (1977). Inhibition of the intracellular release of calcium by dantrolene in barnacle giant muscle fibers. Journal of Physiology 265, 565-585. [11]

Detwiler, S. R. (1920). On the hyperplasia of nerve centers resulting from excessive peripheral loading. Proceedings of the National Academy of Sciences of the United States of America 6, 96-101. [6]

Devreotes, P. N. & Fambrough, D. M. (1975). Acetylcholine receptor turnover in membranes of developing muscle fibers. Journal of Cell Biology 65, 335-358. [3]

Di Prampero, P. E. & Narici, M. V. (2003). Muscles in microgravity: from fibres to human motion. Journal of Biomechanics 36, 403-412. [19]

Dickson, B. J. (2002). Molecular mechanisms of axon guidance. Science 298, 1959-1964. [6]

DiMauro, S. & DiMauro, P. M. (1973). Muscle carnitine palmityltransferase deficiency and myoglobinuria. Science 182, 929-931. [14]

Dodge, F. A., Jr. & Rahamimoff, R. (1967). Co-operative action a calcium ions in transmitter release at the neuromuscular junction. Journal of Physiology 193, 419-432. [10]

Doherty, T. J. & Brown, W. F. (1997). Age-related changes in the twitch contractile properties of human thenar motor units. Journal of Applied Physiology 82, 93-101. [12]

Doherty, T. J. & Brown, W. F. (2002). Motor unit number estimation—methods and application. In Clinical Neurophysiology and Neuromuscular Disease, eds. Brown, W. F., Bolton, C. F., & Aminoff, M., pp. 274-290. Saunders, Philadelphia. [22]

Doherty, T. J., Komori, T., Stashuk, D. W., Kassam, A., & Brown, W. F. (1994). Physiological properties of single thenar motor units in the F-response of younger and older adults. Muscle & Nerve 17, 860-872. [9]

Doherty, T. J., Vandervoort, A. A., Taylor, A. W., & Brown, W. F. (1993). Effects of motor unit losses on strength in older men and women. Journal of Applied Physiology 74, 868-874. [22]

Donoghue, M. J., Morris-Valero, R., Johnson, Y. R., Merlie, J. P., & Sanes, J. R. (1992). Mammalian muscle cells bear a cell-autonomous, heritable memory of their rostrocaudal position. Cell 69, 67-77. [5]

Drachman, D. B., Murphy, S. R., Nigam, M. P., & Hills, J. R. (1967). "Myopathic" changes in chronically denervated muscle. Archives of Neurology 16, 14-24. [22]

Draeger, A., Weeds, A. G., & Fitzsimons, R. B. (1987). Primary, secondary and tertiary myotubes in developing skeletal muscle: a new approach to the analysis of human myogenesis. Journal of the Neurological Sciences 81, 19-43. [5, 12]

Drahota, Z. & Gutmann, E. (1961). The influence of age on the course of reinnervation of muscle. Gerontologica 5, 88-109. [22]

Drahota, Z. & Gutmann, E. (1963). Long-term regulatory influence of the nervous system on some metabolic differences in muscles of different function. Physiologica Bohemoslovaca 12, 339-348. [17]

Dubois-Dalcq, M., Buyse, M., Buyse, G., & Gorce, F. (1971). The action of Guillain-Barré syndrome serum on myelin. A tissue culture and electron microscopic analysis. Journal of Neurological Sciences 13, 67-83. [16]

Dubowitz, V. (1967). Cross-innervated mammalian skeletal muscle: histochemical, physiological and biochemical observations. Journal of Physiology 193, 481-496. [17, 18]

Dubowitz, V. (1985). Muscle Biopsy. A Practical Approach. Baillière Tindall, London. [12]

Duchateau, J. & Hainaut, K. (1987). Electrical and mechanical changes in immobilized human muscle. Journal of Applied Physiology 62, 2168-2173. [19]

Duchen, L. W. (1970a). Changes in motor innervation and cholinesterase localization induced by botulinum toxin in skeletal muscle of the mouse: differences between fast and slow muscles. Journal of Neurology, Neurosurgery, and Psychiatry 33, 40-54. [17]

Duchen, L. W. (1970b). The effects in the mouse of nerve crush and regeneration on the innervation of skeletal muscles paralysed by Clostridium botulinum toxin. Journal of Pathology 102, 9-14. [17]

Duchen, L. W. & Stefani, E. (1971). Electrophysiological studies of neuromuscular transmission in hereditary "motor end-plate disease" of the mouse. Journal of Physiology 212, 535-548. [15]

Duchen, L. W., Stolkin, C., & Tonge, D. A. (1972). Light and electron microscopic changes in slow and fast skeletal muscle fibres and their motor end-plates in the mouse after the local injection of tetanus toxin. Journal of Physiology 222, 136P-137P. [17]

Duchenne, G. B. A. (1861). De L'électrisation Localisée et de Son Application à la Physiologie à la Pathologie et à Thérapeutique, 2nd ed. Baillière et fils, Paris. [1]

Dudley, G. A., Duvoisin, M. R., Convertino, V. A., & Buchanan, P. (1989). Alterations of the in vivo torque-velocity relationship of human skeletal muscle following 30 days exposure to simulated microgravity. Aviation, Space and Environmental Medicine 60, 659-663. [19]

Dulhunty, A. F. & Franzini-Armstrong, C. (1975). The relative contributions of the folds and caveolae to the surface membrane of frog skeletal muscle fibres at different sarcomere lengths. Journal of Physiology 250, 513-539. [1]

Duncan, D. (1934). A determination of the number of nerve fibres in the eight thoracic and the largest lumbar ventral roots of the albino rat. Journal of Comparative Neurology 59, 47-60. [22]

Dyck, P. J. (1975). Pathological alterations of the peripheral nervous system of man. In Peripheral Neuropathy, eds. Dyck, P. J., Thomas, P. K., & Lambert, E. H., pp. 296-336. Saunders, Philadelphia. [22]

Dyck, P. J., Lais, A. C., & Low, P. A. (1978). Nerve xenografts to assess cellular expression of the abnormality of myelination in inherited neuropathy and Friedreich ataxia. Neurology 28, 261-265. [2]

Dyck, P. J., Nukada, H., Lais, A. C., & Karnes, J. L. (1984). Permanent axotomy: a model of chronic neuronal degeneration preceded by axonal atrophy, myelin remodelling and degeneration. In Peripheral Neuropathy, eds. Dyck, P. J., Thomas, P. K., Lambert, E. H., & Bunge, R., pp. 666-690. Saunders, Philadelphia. [16]

Eaton, L. M. & Lambert, E. H. (1957). Electromyography and electric stimulation of nerves in diseases of motor unit. Journal of the American Medical Association 163, 1117-1124. [10]

Eberstein, A. & Sandow, A. (1963). Fatigue mechanisms in muscle fibres. In Effects of Use and Disuse on Neuromuscular Functions, ed. Guthmann, E., pp. 515-526. Prague Publication House, Czechoslovakia. [15]

Eccles, J. C., Eccles, R. M., & Lundberg, A. (1958a). Action potentials of alpha motoneurones supplying fast and slow muscles. Journal of Physiology 142, 275-291. [12]

Eccles, J. C., Libet, B., & Young, R. R. (1958b). The behaviour of chromatolysed motoneurones studied by intracellular recording. Journal of Physiology 143, 11-40. [18]

Eccles, J. C. & Liley, A. W. (1959). Factors controlling the liberation of acetylcholine at the neuromuscular junction. American Journal of Physical Medicine 38, 96-103. [10]

Eccles, J. C. & Sherrington, C. S. (1930). Numbers and contraction-values of individual motor-units examined in some muscles of the limb. Proceedings of the Royal Society of London. Series B: Biological Sciences 106, 326-357. [12]

Edgerton, V. R., Bodine-Fowler, S., Roy, R., Ishihara, A., & Hodgson, J. A. (1996). Neuromuscular adaptation. In Exercise: Regulation and Integration of Multiple Systems, eds. Rowell, L. B. & Shepherd, J. T., pp. 54-88. Oxford University Press, New York. [5, 6, 18]

Edgerton, V. R., Zhou, M. Y., Ohira, Y., Klitgaard, H., Jiang, B., Bell, G., Harris, B., Saltin, B., Gollnick, P. D., Roy, R. R., Day, M. K., & Greenisen, M. (1995). Human fiber size and enzymatic properties after 5 and 11 days of spaceflight. Journal of Applied Physiology 78, 1733-1739. [19]

Edman, K. A. P. (1988). Double-hyperbolic nature of the force-velocity relation in frog skeletal muscle. Advances in Experimental Medicine 226, 643-652.

Edman, K. A. P. & Lou, F. (1991). Changes in force and stiffness during fatigue of frog isolated muscle fibres. Journal of Physiology. [15]

Edman, K. A. P. & Lou, F. (1992). Myofibrillar fatigue versus failure of activation during repetitive stimulation of frog muscle fibres. Journal of Physiology 457, 655-673. [15]

Edström, L. & Kugelberg, E. (1968). Histochemical composition, distribution of fibres and fatiguability of single motor units. anterior tibial muscle of the rat. Journal of Neurology, Neurosurgery, and Psychiatry 31, 424-433. [12]

Edwards, R. H. T. (1981). Human muscle function and fatigue. In Human Muscle Fatigue: Physiological Mechanisms Ciba Found. Symp. 82, eds. Porter, R. & Whelan, J., pp. 1-18. Pitman, London. [15]

Edwards, R. H. T., Hill, D. K., & Jones, D. A. (1975). Heat production and chemical changes during isometric contractions of the human quadriceps muscle. Journal of Physiology 251, 303-315. [13]

Edwards, R. H. T., Hill, D. K., Jones, D. A., & Merton, P. A. (1977). Fatigue of long duration in human skeletal muscle after exercise. Journal of Physiology 272, 769-778. [15]

Einsiedel, L. J. & Luff, A. R. (1992). Alterations in the contractile properties of motor units within the ageing rat medial gastrocnemius. Journal of Neurological Sciences 112, 170-177. [22]

Eisenberg, B. R. (1974). Quantitative ultrastructural analysis of adult mammalian skeletal muscle fibers. In Exploratory Concepts in Muscular Dystrophy II, ed. Milhorat, A. T., pp. 258-269. Excerpta Medica, Amsterdam. [12]

Eisenberg, B. R. & Kuda, A. M. (1976). Discrimination between fiber populations in mammalian skeletal muscle by using ultrastructural parameters. Journal of Ultrastructure Research 54, 76-88. [12]

Eisenberg, R. S. & Gage, P. W. (1969). Ionic conductances of the surface and transverse tubular membranes of frog sartorius fibers. Journal of General Physiology 53, 279-297. [9]

Ekstedt, J. (1964). Human single muscle fibre action potentials. Acta Physiologica Scandinavica 61, 1-96. [12]

Ekstedt, J. & Stålberg, E. (1967). Myasthenia gravis. Diagnostic aspects by a new electrophysiological method. Opuscula Medica 12, 73-76. [10]

Elder, G. B., Dean, D., McComas, A. J., Paes, B., & DeSa, D. (1983). Infantile centronuclear myopathy. Evidence suggesting incomplete innervation. Journal of the Neurological Sciences 60, 79-88. [5]

Elder, G. C., Bardbury, K., & Roberts, R. (1982). Variability of fiber type distributions within human muscles. Journal of Applied Physiology 53, 1473-1480. [19]

Elder, G. C. & McComas, A. J. (1987). Development of rat muscle during short- and long-term hindlimb suspension. Journal of Applied Physiology 62, 1917-1923. [19]

Elder, G. C. B. & Kakulas, B. A. (1993). Histochemical and contractile property changes during human muscle development. Muscle & Nerve 16, 1246-1253. [5, 20]

Eldridge, L. (1984). Lumbosacral spinal isolation in cat: surgical preparation and health maintenance. Experimental Neurology 83, 318-327. [19]

Elek, J. M., Kossev, A., Dengler, R., Schubert, M., Wohlfahrt, K., & Wolf, W. (1992). Parameters of human motor unit twitches obtained by intramuscular microstimulation. Neuromuscular Disorders: NMD 2, 261-267. [12]

Elmqvist, D., Hofmann, W. W., Kugelberg, J., & Quastel, D. M. J. (1964). An electrophysiological investigation of neuromuscular transmission in myasthenia gravis. Journal of Physiology 174, 417-434. [10]

Elsberg, C. A. (1917). Experiments on motor nerve regeneration and the direct neurotization of paralysed muscles by their own and foreign nerves. Science 45, 318-320. [17]

Elul, R., Miledi, R., & Stefani, E. (1968). Neurotrophic control of contracture in slow muscle fibres. Nature 217, 1274-1275. [17]

Emerson, C. P., Jr. (1993). Skeletal myogenesis: genetics and embryology to the fore. Current Opinion in Genetics & Development 3, 265-274. [5]

Endo, M. (1966). Entry of fluorescent dyes into the sarcotubular system of the frog muscle. Journal of Physiology 185, 224-238. [11]

Endo, M. (1973). Length-dependence of activation of skinned muscle fibres by calcium. Cold Spring Harbor Symposia on Quantitative Biology 37, 505-510. [11]

Engel, A. G. (1987). Molecular biology of end-plate diseases. In The Vertebrate Neuromuscular Junction, ed. Salpeter, M. M., pp. 361-424. Alan R. Liss, New York. [10]

Engel, A. G. & Angelini, C. (1973). Carnitine deficiency of human skeletal muscle with associated lipid storage myopathy: a new syndrome. Science 179, 899-902. [14]

Engel, A.G. & Banker, B.Q. (1986). Myology: Basic and Clinical. McGraw-Hill, New York. [14]

Engel, A. G., Tsujihata, M., Lindstrom, J. M., & Lennon, V. A. (1976). The motor end-plate in myasthenia gravis and in experimental auto-immune myasthenia gravis. A quantitative ultrastructural study. Annals of the New York Academy of Sciences 274, 60-79. [10]

Engel, A. G., Walls, T. J., Nagel, A., & Uchitel, O. (1990). Newly recognized congenital myasthenic syndromes: I. Congenital paucity of synaptic vesicles and reduced quantal release. II. High-conductance fast-channel syndrome. III. Abnormal acetylcholine receptor (AChR) interaction with acetylcholine. IV. AChR deficiency and short channel-open time. Progress in Brain Research 84, 125-137. [10]

Engel, W. K. (1962). The essentiality of histo- and cytochemical studies of skeletal muscle in the investigation of neuromuscular disease. Neurology 12, 778-784. [12]

England, P. J. (1986). Intracellular calcium receptor mechanisms. British Medical Bulletin 42, 375-383. [7]

Enoka, R. M. (1988). Muscle strength and its development. New perspectives. Sports Medicine 6, 146-168. [20]

Enoka, R. M. (1995). Morphological features and activation patterns of motor units. Journal of Clinical Neurophysiology 12, 538-559. [12]

Enoka, R. M. (1997). Neural adaptations with chronic physical activity. Journal of Biomechanics 30, 447-455. [20]

Entwistle, A., Zalin, R. J., Bevan, S., & Warner, A. E. (1988a). The control of chick myoblast fusion by ion channels operated by prostaglandins and acetylcholine. Journal of Cell Biology 106, 1693-1702. [5]

Entwistle, A., Zalin, R. J., Warner, A. E., & Bevan, S. (1988b). A role for acetylcholine receptors in the fusion of chick myoblasts. Journal of Cell Biology 106, 1703-1712. [5]

Erlanger, J. & Schloepfle, G. M. (1946). A study of nerve degeneration and regeneration. American Journal of Physiology 147, 550-581. [16]

Everts, M. E., Retterstøl, K., & Clausen, T. (1988). Effects of adrenaline on excitation-induced stimulation of the sodium-potassium pump in rat skeletal muscle. Acta Physiologica Scandinavica 134, 189-198. [7]

Exner, G. U., Staudte, H. W., & Pette, D. (1973). Isometric training of rats—effects upon fast and slow muscle and modification by an anabolic hormone (nandrolone decanoate). II. Male rats. Pflügers Archiv: European Journal of Physiology 345, 15-22. [20]

Exner, S. (1885). Noitz zu der Fage von der Faserverthelung mehreren nerven in einem Muskeln. Pflügers Archiv: European Journal of Physiology 36, 572-576. [17]

Faaborg-Andersen, K. (1957). Electromyographic investigation of intrinsic laryngeal muscles in humans. Acta Physiologica Scandinavica 41 (Suppl 140), 1-149. [12]

Fahim, M. A., Holley, J. A., & Robbins, N. (1983). Scanning and light microscopic study of age changes at a neuromuscular junction in the mouse. Journal of Neurocytology 12, 13-25. [3]

Fahim, M. A. & Robbins, N. (1982). Ultrastructural studies of young and old mouse neuromuscular junctions. Journal of Neurocytology 11, 641-656. [22]

Fahim, M. A. & Robbins, N. (1986). Remodelling of the neuromuscular junction after subtotal disuse. Brain Research 383, 353-356. [19]

Falco, F. J., Hennessey, W. J., Braddom, R. L., & Goldberg, G. (1992). Standardized nerve conduction studies in the upper limb of the healthy elderly. American Journal of Physical Medicine & Rehabilitation 71, 263-271. [22]

Fallentin, N., Jorgensen, K., & Simonsen, E. B. (1993). Motor unit recruitment during prolonged isometric contractions. European Journal of Applied Physiology & Occupational Physiology 67, 335-341. [13]

Fambrough, D., Hartzell, H. C., Rash, J. E., & Ritchie, A. K. (1974). Receptor properties of developing muscle. Annals of the New York Academy of Sciences 228, 47-61. [1, 5]

Fambrough, D. M. (1979). Control of acetylcholine receptors in skeletal muscle. Physiological Reviews 59, 165-227. [3]

Fambrough, D. M., Wolitzky, B. A., Tamkun, M. M., & Takeyasu, K. (1987). Regulation of the sodium pump in excitable cells. Kidney International 32, S97-116. [7, 15]

Farmer, T. W., Buchthal, F., & Rosenfalck, P. (1960). Refractory period of human muscle after the passage of a propagated action potential. Electroencephalography and Clinical Neurophysiology 12, 455-466. [9]

Fatt, P. & Katz, B. (1951). An analysis of the end-plate potential recorded with an intracellular electrode. Journal of Physiology 115, 320-370. [10]

Fatt, P. & Katz, B. (1952). Spontaneous subthreshold activity at motor nerve endings. Journal of Physiology 117, 109-128. [3]

Faulkner, J. A. (2003). Terminology for contractions of muscles during shortening, while isometric, and during lengthening. Journal of Applied Physiology 95, 455-459. [11]

Faulkner, J. A., Jones, D. A., & Round, J. M. (1989). Injury to skeletal muscles of mice by forced lengthening during contractions. Quarterly Journal of Experimental Physiology 74, 661-670. [21]

Faulkner, J. A., Maxwell, L. C., Ruff, G. L., & White, T. P. (1979). The diaphragm as a muscle. Contractile properties. American Review of Respiratory Disease 119, 89-92. [12]

Feasby, T. E., Hahn, A. F., & Gilbert, J. J. (1982). Passive transfer studies in Guillain-Barré polyneuropathy. Neurology 32, 1159-1167. [16]

Feinstein, B., Lindegård, B., Nyman, E., & Wohlfart, G. (1955). Morphologic studies of motor units in normal human muscles. Acta Anatomica 23, 127-142. [1, 12]

Feldman, R. M. & Soskolne, C. L. (1987). The use of non-fatiguing strengthening exercises in post-polio syndrome. Birth Defects Original Article Series 23, 335-341. [20]

Fell, R. D., Gladden, L. B., Steffen, J. M., & Musacchia, X. J. (1985). Fatigue and contraction of slow and fast muscle in hypokinetic/hypodynamic rats. Journal of Applied Physiology 58, 65-69. [19]

Fenn, W. O. (1936). Electrolytes in muscle. Cold Spring Harbor Symposia on Quantitative Biology IV, 252-259. [15]

Fenton, J., Garner, S., & McComas, A. J. (1991). Abnormal M-wave responses during exercise in myotonic muscular dystrophy: a Na(+)–K+ pump defect? Muscle & Nerve 14, 79-84. [7]

Ferrando, A. A., Lane, H. W., Stuart, C. A., Davis-Street, J., & Wolfe, R. R. (1996). Prolonged bed rest decreases skeletal muscle and whole body protein synthesis. American Journal of Physiology 270, E627-E633. [19]

Fertuck, H. C. & Salpeter, M. M. (1976). Quantification of junctional and extrajunctional acetylcholine receptors by electron microscope autoradiography after 125I-alpha-bungarotoxin binding at mouse neuromuscular junctions. Journal of Cell Biology 69, 144-158. [10]

Fewings, J. D., Harris, J. B., Johnson, M. A., & Bradley, W. G. (1977). Progressive denervation of skeletal muscle induced by spinal irradiation in rats. Brain 100, 157-183. [16]

Fex, S., Sonesson, B., Thesleff, S., & Zelená, J. (1966). Nerve implants in botulinum poisoned mammalian muscle. Journal of Physiology 184, 872-882. [17]

Fill, M., Coronado, R., Mickelson, J. R., Vilven, J., Ma, J. J., Jacobson, B. A., & Louis, C. F. (1990). Abnormal ryanodine receptor channels in malignant hyperthermia. Biophysical Journal 57, 471-475. [11]

Fine, E. G., Decosterd, I., Papaloizos, M., Zurn, A. D., & Aebischer, P. (2002). GDNF and NGF released by synthetic guidance channels support sciatic nerve regeneration across a long gap. European Journal of Neuroscience 15, 589-601. [17]

Finer, J. T., Simmons, R. M., & Spudich, J. A. (1994). Single myosin molecule mechanics: piconewton forces and nanometre steps. Nature 368, 113-118. [11]

Fischbach, G. D. (1972). Synapse formation between dissociated nerve and muscle cells in low density cell cultures. Developmental Biology 28, 407-429. [6]

Fischbach, G. D., Harris, D. A., Falls, D. L., Dubinsky, J. M., English, K. L., & Johnson, F. A. (1989). The accumulation of acetylcholine receptors at developing chick nerve-muscle synapses. In Neuromuscular Junction, eds. Sellin, L. C., Libelius, R., & Thesleff, S., pp. 515-532. Elsevier Science Publishers, Amsterdam. [18]

Fischbach, G. D., Nameroff, M., & Nelson, P. G. (1971). Electrical properties of chick skeletal muscle fibers developing in cell culture. Journal of Cellular Physiology 78, 289-299. [5]

Fischbach, G. D. & Robbins, N. (1969). Changes in contractile properties of disused soleus muscles. Journal of Physiology 201, 305-320. [19]

Fischbach, G. D. & Rosen, K. M. (1997). ARIA: a neuromuscular junction neuregulin. Annual Review of Neuroscience 20, 429-458. [6]

Fisher, M. B. & Birren, J. E. (1947). Age and strength. Journal of Applied Physiology 31, 490-497. [22]

Fitch, S. & McComas, A. (1985). Influence of human muscle length on fatigue. Journal of Physiology 362, 205-213. [15]

Fitzsimonds, R. M. & Poo, M.-M. (1998). Retrograde signaling in the development and modification of synapses. Physiological Reviews 78, 143-170. [6]

Flucher, B. E. & Daniels, M. P. (1989). Distribution of Na+ channels and ankyrin in neuromuscular junctions is complementary to that of acetylcholine receptors and the 43 kd protein. Neuron 3, 163-175. [10]

Flück, M., Ziemiecki, A., Billeter, R., & Müntener, M. (2002). Fibre-type specific concentration of focal adhesion kinase at the sarcolemma: influence of fibre innervation and regeneration. Journal of Experimental Biology 205, 2337-2348. [1]

Foehring, R. C., Sypert, G. W., & Munson, J. B. (1987). Motor-unit properties following cross-reinnervation of cat lateral gastrocnemius and soleus muscle with medial gastrocnemius nerve. II. Influence of muscle on motoneurons. Journal of Neurophysiology 57(4), 1227-1245. [17]

Fong, C. N., Atwood, H. L., & Charlton, M. P. (1986). Intracellular sodium-activity at rest and after tetanic stimulation in muscles of normal and dystrophic (dy2J/dy2J) C57BL/6J mice. Experimental Neurology 93, 359-368. [9]

Fontaine, B., Khurana, T. S., Hoffman, E. P., Bruns, G. A., Hains, J. L., Trofatter, J. A., Hanson, M. P., Rich, J., McFarlane, H., McKenna Yasek, D., Romano, D., Gusella, J. F., & Brown, R. H. (1990). Hyperkalemic periodic paralysis and the adult muscle sodium channel alpha-subunit gene. Science 250, 1000-1002. [7]

Fontaine, B., Klarsfeld, A., Hökfelt, T., & Changeux, J.-P. (1986). Calcitonin gene-related peptide, a peptide present in spinal cord motoneurons, increases the number of acetylcholine receptors in primary cultures of chick embryo myotubes. Neuroscience Letters 71, 59-65. [18]

Fosset, M., Jaimovich, E., Delpont, E., & Lazdunski, M. (1983). Nitrendipine receptors in skeletal muscle. Properties and preferential localization in transverse tubules. Journal of Biological Chemistry 258, 6086-6092. [11]

Frank, E., Jansen, J. K., Lømo, T., & Westgaard, R. (1974). Maintained function of foreign synapses on hyperinnervated skeletal muscle fibres of the rat. Nature 247, 375-376. [17]

Frank, E., Jansen, J. K., Lømo, T., & Westgaard, R. H. (1975). The interaction between foreign and original motor nerves innervating the soleus muscle of rats. Journal of Physiology 247, 725-743. [17]

Frank, G. B. (1958). Inward movement of calcium as a link between electrical and mechanical events in contraction. Nature 182, 1800-1801. [11]

Franzini-Armstrong, C. (1970). Studies of the triad. I. Structure of the junction in frog twitch fibers. Journal of Cell Biology 47, 488-499. [11]

Franzini-Armstrong, C. (1980). Structure of sarcoplasmic reticulum. Federation Proceedings 39, 2403-2409. [11]

Franzini-Armstrong, C. & Porter, K. R. (1964). Sarcolemmal invaginations constituting the T system in fish muscle fibers. Journal of Cell Biology 22, 675-696. [1]

Freimann, R. (1954). Untersuchungen über Zahl und Anordnung der Muskelspindeln in den Kaumuskeln des Menschen. Anatomischer Anzeiger 100, 258-264.

Freund, H.-J. (1983). Motor unit and muscle activity in voluntary motor control. Physiological Reviews 63, 387-436. [13]

Freund, H.-J., Budingen, H.-J., & Dietz, V. (1975). Activity of single motor units from human forearm muscles during voluntary isometric contractions. Journal of Neurophysiology 38, 933-946. [13]

Friday, B. B., Horsley, V., & Pavlath, G. K. (2000). Calcineurin activity is required for the initiation of skeletal muscle differentiation. Journal of Cell Biology 149, 657-666. [5]

Fridén, J. & Lieber, R. L. (2001). Eccentric exercise-induced injuries to contractile and cytoskeletal muscle fibre components. Acta Physiologica Scandinavica 171, 321-326. [21]

Fridén, J., Sjöström, M., & Ekblom, B. (1983). Myofibrillar damage following intense eccentric exercise in man. International Journal of Sports Medicine 4, 170-176. [20, 21]

Fridén, J., Sjöström, M., & Ekblom, B. (1984). Muscle fibre type characteristics in endurance trained and untrained individuals. European Journal of Applied Physiology & Occupational Physiology 52, 266-271. [20, 21]

Frolkis, V. V., Tanin, S. A., Marcinko, V. I., & Muradian, K. K. (1990). Effects of hormones on the fast axoplasmic transport of substances in ventral horns of the spinal cord in rats of different ages. Archives of Gerontology and Geriatrics 11, 33-41. [22]

Frontera, W. R., Meredith, C. N., O'Reilly, K. P., Knuttgen, H. G., & Evans, W. J. (1988). Strength conditioning in older men: skeletal muscle hypertrophy and improved function. Journal of Applied Physiology 64, 1038-1044. [20]

Fu, A. K., Ip, F. C., Lai, K. O., Tsim, K. W., & Ip, N. Y. (1997). Muscle-derived neurotrophin-3 increases the aggregation of acetylcholine receptors in neuron-muscle co-cultures. Neuroreport 8, 3895-3900. [6]

Fu, S. Y. & Gordon, T. (1995). Contributing factors to poor functional recovery after delayed nerve repair: prolonged denervation. Journal of Neuroscience 15, 3886-3895. [17]

Fu, Y.-H., Pizzuti, A., Fenwick, R. G., Jr., King, J., Rajnarayan, S., Dunne, P. W., Dubel, J., Nasser, G. A., Ashizawa, T., DeJong, P., Wieringa, B., Korneliuk, R., Perryman, M. P., Epstein, H. F., & Caskey, C. T. (1992). An unstable triplet repeat in a gene related to myotonic muscular dystrophy. Science 255, 1256-1258. [7]

Fuglevand, A. J. & Keen, D. A. (2003). Re-evaluation of muscle wisdom in the human adductor pollicis using physiological rates of stimulation. Journal of Physiology 549, 865-875. [13, 15]

Fuglsang-Frederiksen, A. & Scheel, U. (1978). Transient decrease in number of motor units after immobilization in man. Journal of Neurology, Neurosurgery, and Psychiatry 41, 924-929. [19]

Fujii, J., Otsu, K., Zorzato, F., de Leon, S., Khanna, V. K., Weiler, J. E., O'Brien, P. J., & MacLennan, D. H. (1991). Identification of a mutation in porcine ryanodine receptor associated with malignant hyperthermia. Science 253, 448-451. [11]

Fukunaga, H., Engel, A. G., Osame, M., & Lambert, E. H. (1982). Paucity and disorganization of presynaptic membrane active zones in the Lambert-Eaton myasthenic syndrome. Muscle & Nerve 5, 686-697. [10]

Fulks, R. M., Li, J. B., & Goldberg, A. L. (1975). Effects of insulin, glucose, and amino acids on protein turnover in rat diaphragm. Journal of Biological Chemistry 250, 290-298. [5]

Funakoshi, H., Belluardo, N., Arenas, E., Yamamoto, Y., Casabona, A., Persson, H., & Ibáñez, C. F. (1995). Muscle-derived neurotrophin-4 as an activity-dependent trophic signal for adult motor neurons. Science 268, 1495-1499. [16]

Funakoshi, H., Frisen, J., Barbany, G., Timmusk, T., Zachrisson, O., Verge, V. M., & Persson, H. (1993). Differential expression of mRNAs for neurotrophins and their receptors after axotomy of the sciatic nerve. Journal of Cell Biology 123, 455-465. [16]

Furst, D. O., Osborn, M., & Weber, K. (1989). Myogenesis in the mouse embryo: differential onset of expression of myogenic proteins and the involvement of titin in myofibril assembly. Journal of Cell Biology 109, 517-527. [5]

Gage, P. W. & Eisenberg, R. S. (1969a). Action potentials, afterpotentials, and excitation-contraction coupling in frog sartorius fibers without transverse tubules. Journal of General Physiology 53, 298-310. [11]

Gage, P. W. & Eisenberg, R. S. (1969b). Capacitance of the surface and transverse tubular membrane of frog sartorius muscle fibers. Journal of General Physiology 53, 265-278. [9]

Galavazi, G. & Szirmai, J. A. (1971). Cytomorphometry of skeletal muscle: the influence of age and testosterone on the rat m. levator ani. Zeitschrift fur Zellforschung und mikroskopische Anatomie 121, 507-530. [5]

Galea, V. (1996). Changes in motor unit estimates with aging. Journal of Clinical Neurophysiology 13, 253-260. [22]

Galea, V., de Bruin, H., Cavasin, R., & McComas, A. J. (1991). The numbers and relative sizes of motor units estimated by computer. Muscle & Nerve 14, 1123-1130. [12]

Galea, V. & McComas, A. (1991). Effects of ischaemia on M-wave potentiation in human biceps brachii muscles. Journal of Physiology 438, 212P. [7]

Galea, V., McFadden, L., Cupido, C., & McComas, A. J. (1993). Delayed depression of human muscle excitability following fatigue. Canadian Journal of Neurological Sciences 20, 359. [15]

Galganski, M. E., Fuglevand, A. J., & Enoka, R. M. (1993). Reduced control of motor output in a human hand muscle of elderly subjects during submaximal contractions. Journal of Neurophysiology 69, 2108-2115. [22]

Galkin, A. V., Giniatullin, R. A., Mukhtarov, M. R., Svandova, I., Grishin, S. N., & Vyskocil, F. (2001). ATP but not adenosine inhibits nonquantal acetylcholine release at the mouse neuromuscular junction. European Journal of Neuroscience 13, 2047-2053. [10]

Gallego, R., Kuno, M., Nunez, R., & Snider, W. D. (1979). Dependence of motoneurone properties on the length of immobilized muscle. Journal of Physiology 291, 179-189. [18]

Gandevia, S. C. (2001). Spinal and supraspinal factors in human muscle fatigue. Physiological Reviews 81, 1725-1789. [15]

Gandevia, S. C. & McCloskey, D. I. (1976). Joint sense, muscle sense, and their combination as position sense, measured at the distal interphalangeal joint of the middle finger. Journal of Physiology 260, 387-407. [4]

Gandevia, S. C., McCloskey, D. I., & Burke, D. (1992). Kinaesthetic signals and muscle contraction. Trends in Neurosciences 15, 62-65. [4]

Gans, C. & Gaunt, A. S. (1991). Muscle architecture in relation to function. Journal of Biochemistry 24, 53-65. [1]

Gardner, E. (1940). Decrease in human neurones with age. Anatomical Record 77, 529-536. [22]

Garland, S. J., Garner, S. H., & McComas, A. J. (1988a). Reduced voluntary electromyographic activity after fatiguing stimulation of human muscle. Journal of Physiology 401, 547-556. [15]

Garland, S. J., Garner, S. H., & McComas, A. J. (1988b). Relationship between numbers and frequencies of stimuli in human muscle fatigue. Journal of Applied Physiology 65, 89-93. [15]

Garner, S.H., Hicks, A.L. & McComas, A.J. (1989). Prolongation of twitch potentiating mechanism throughout human muscle fatigue and recovery. Experimental Neurology 103, 277-281. [15]

Garnett, R. & Stephens, J. A. (1980). The reflex responses of single motor units in human first dorsal interosseous muscle following cutaneous afferent stimulation. Journal of Physiology 303, 351-364. [13]

Garnett, R. A., O'Donovan, M. J., Stephens, J. A., & Taylor, A. (1979). Motor unit organization of human medial gastrocnemius. Journal of Physiology 287, 33-43. [12, 13]

Gatev, V., Stamatova, L., & Angelova, B. (1977). Contraction time in skeletal muscles of normal children. Electromyography and Clinical Neurophysiology 17, 441-452. [5]

Gauthier, G. F. (1986). Skeletal muscle fiber types. In Myology: Basic and Clinical, eds. Engel, A. G. & Banker, B. Q., pp. 255-286. McGraw-Hill, New York. [12]

Gaviria, M. & Ohanna, F. (1999). Variability of the fatigue response of paralyzed skeletal muscle in relation to the time after spinal cord injury: mechanical and electrophysiological characteristics. European Journal of Applied Physiology & Occupational Physiology 80, 145-153. [19]

Gelfan, S. & Carter, S. (1967). Muscle sense in man. Experimental Neurology 18, 469-473. [4]

Gerchman, L. B., Edgerton, V. R., & Carrow, R. E. (1975). Effects of physical training on the histochemistry and morphology of ventral motor neurons. Experimental Neurology 49, 790-801. [20]

Geren, B. B. (1954). The formation from the Schwann cell surface of myelin in the peripheral nerves of chick embryos. Experimental Cell Research 7, 558-562. [2]

Gerrits, H. L., de Haan, A., Hopman, M. T., Der Woude, L. H., Jones, D. A., & Sargeant, A. J. (1999). Contractile properties of the quadriceps muscle in individuals with spinal cord injury. Muscle & Nerve 22, 1249-1256. [19]

Ghosh, S. & Dhoot, G. K. (1998). Evidence for distinct fast and slow myogenic cell lineages in human foetal skeletal muscle. Journal of Muscle Research and Cell Motility 19, 431-441. [5]

Gibala, M. J., Interisano, S. A., Tarnopolsky, M. A., Roy, B. D., MacDonald, J. R., Yarasheski, K. E., & MacDougall, J. D. (2000). Myofibrillar disruption following acute concentric and eccentric resistance exercise in strength-trained men. Canadian Journal of Physiology and Pharmacology 78, 656-661. [21]

Gibala, M. J., MacDougall, J. D., Tarnopolsky, M. A., Stauber, W. T., & Elorriaga, A. (1995). Changes in human skeletal muscle ultrastructure and force production after acute resistance exercise. Journal of Applied Physiology 78, 702-708. [21]

Gielen, C. C. A. M. & van der Gon, J. J. D. (1990). The activation of motor units in coordinated arm movements in humans. News in Physiological Sciences 5, 159-163. [13]

Gillespie, C. A., Simpson, D. R., & Edgerton, V. R. (1974). Motor unit recruitment as reflected by muscle fibre glycogen loss in a prosimian (bushbaby) after running and jumping. Journal of Neurology, Neurosurgery, and Psychiatry 37, 817-824. [13]

Gillespie, M. J., Gordon, T., & Murphy, P. R. (1986). Reinnervation of the lateral gastrocnemius and soleus muscles in the rat by their common nerve. Journal of Physiology 372, 485-500. [12]

Gillespie, M. J. & Stein, R. B. (1983). The relationship between axon diameter, myelin thickness and conduction velocity during atrophy of mammalian peripheral nerves. Brain Research 259, 41-56. [16]

Gilliatt, R. W. & Hjorth, R. J. (1972). Nerve conduction during Wallerian degeneration in the baboon. Journal of Neurology, Neurosurgery, and Psychiatry 35, 335-341. [16]

Gilliatt, R. W. & Taylor, J. C. (1959). Electrical changes following section of the facial nerve. Proceedings of the Royal Society of Medicine 52, 1080-1083. [16]

Gilliatt, R. W., Westgaard, R. H., & Williams, I. R. (1978). Extrajunctional acetylcholine sensitivity of inactive muscle fibres in the baboon during prolonged nerve pressure block. Journal of Physiology 280, 499-514. [19]

Ginsborg, B. L. (1960). Some properties of avian skeletal muscle fibres with multiple neuromuscular junctions. Journal of Physiology 193, 581-598. [20]

Gitlin, G. & Singer, M. (1974). Myelin movements in mature mammalian peripheral nerve fibers. Journal of Morphology 143, 167-186. [2]

Gladden, L. B. (1996). Lactate transport and exchange during exercise. Section 12: Exercise: regulation and integration of multiple systems. In Handbook of Physiology, eds. Rowell, L. B. & Shepherd, J. T., pp. 614-648. Oxford University Press, Bethsda, MD. [14]

Glatt, H. R. & Honegger, C. G. (1973). Retrograde axonal transport for cartography of neurones. Experientia 29, 1515-1517. [8, 18]

Glazner, G. W. & Ishii, D. N. (1995). Insulinlike growth factor gene expression in rat muscle during reinnervation. Muscle & Nerve 18, 1433-1442. [17]

Glover, J. C., Petursdottir, G., & Jansen, J. K. (1986). Fluorescent dextran-amines used as axonal tracers in the nervous system of the chicken embryo. Journal of Neuroscience Methods 18, 243-254. [12]

Godfrey, E. W., Nitkin, R. M., Wallace, B. G., Rubin, L. L., & McMahan, U. J. (1984). Components of Torpedo electric organ and muscle that cause aggregation of acetylcholine receptors on cultured muscle cells. Journal of Cell Biology 99, 615-627. [18]

Goldberg, A. L. (1968). Role of insulin in work-induced growth of skeletal muscle. Endocrinology 83, 1071-1073. [5]

Goldberg, A. L. & Goodman, H. M. (1969). Relationship between growth hormone and muscular work in determining muscle size. Journal of Physiology 200, 655-666. [5]

Goldberg, L. J. & Derfler, B. (1977). Relationship among recruitment order, spike amplitude, and twitch tension of single motor units in human masseter muscle. Journal of Neurophysiology 40, 879-890. [13]

Goldkamp, O. (1967). Electromyography and nerve conduction studies in 116 patients with hemiplegia. Archives of Physical Medicine and Rehabilitation 48, 59-63. [18]

Goldman, D. & Sapru, M. K. (1998). Molecular mechanisms mediating synapse-specific gene expression during development of the neuromuscular junction. Canadian Journal of Applied Physiology 23, 390-395. [6]

Goldman, D. E. (1943). Potential, impedance, and rectification in membranes. Journal of General Physiology 27, 37-60. [9]

Goldspink, G. (1964). The combined effects of exercise and reduced food intake on skeletal muscle fibres. Journal of Cellular and Comparative Physiology 63, 209-216. [20]

Goldspink, G. (1965). Cytological basis of decrease in muscle strength during starvation. American Journal of Physiology 209, 100-114. [5, 20]

Goldspink, G. (1970). The proliferation of myofibrils during muscle fibre growth. Journal of Cell Science 6, 593-603. [5]

Goldspink, G. (1985). Malleability of the motor system: a comparative approach. Journal of Experimental Biology 115, 375-391. [20]

Goldspink, G. (1999). Changes in muscle mass and phenotype and the expression of autocrine and systemic growth factors by muscle in response to stretch and overload. Journal of Anatomy 194 (Pt 3), 323-334. [5]

Goldspink, G., Scutt, A., Loughna, P. T., Wells, D. J., Jaenicke, T., & Gerlach, G. F. (1992). Gene expression in skeletal muscle in response to stretch and force generation. American Journal of Physiology 262, R356-R363. [20]

Goldspink, G., Scutt, A., Martindale, J., Jaenicke, T., Turay, L., & Gerlach, G. F. (1991). Stretch and force generation induce rapid hypertrophy and myosin isoform gene switching in adult skeletal muscle. Biochemical Society Transactions 19, 368-373. [20]

Golgi, C. (1903). Sui nervi dei tendini dell'uomo e di altri vertebrati e diun nuovo organo nervoso terminale musculo-tendineo. Reprinted in his Opera omnia. In Opera Omnia, pp. 171-198. Ultrico Hoepli, Milan. [4]

Gollnick, P. D., Karlsson, J., Piehl, K., & Saltin, B. (1974a). Selective glycogen depletion in skeletal muscle fibres of man following sustained contractions. Journal of Physiology 241, 59-67. [13]

Gollnick, P. D., Piehl, K., & Saltin, B. (1974b). Selective glycogen depletion pattern in human muscle fibres after exercise of varying intensity and at varying pedalling rates. Journal of Physiology 241, 45-57. [13]

Gollnick, P. D., Sjodin, B., Karlsson, J., Jansson, E., & Saltin, B. (1974c). Human soleus muscle: a comparison of fiber composition and enzyme activities with other leg muscles. Pflügers Archiv: European Journal of Physiology 348, 247-255. [12]

Gollnick, P. D., Timson, B. F., Moore, R. L., & Riedy, M. (1981). Muscular enlargement and number of fibers in skeletal muscles of rats. Journal of Applied Physiology 50, 936-943. [20]

Gomez-Pinilla, F., Ying, Z., Opazo, P., Roy, R. R., & Edgerton, V. R. (2001). Differential regulation by exercise of BDNF and NT-3 in rat spinal cord and skeletal muscle. European Journal of Neuroscience 13, 1078-1084. [18]

Gomez-Pinilla, F., Ying, Z., Roy, R. R., Molteni, R., & Edgerton, V. R. (2002). Voluntary exercise induces a BDNF-mediated mechanism that promotes neuroplasticity. Journal of Neurophysiology 88, 2187-2195. [18]

Gonzalez, E., Messi, L., Zheng, Z., & Delbono, O. (2003). IGF-1 prevents age-related decrease in specific-force and intracellular Ca2+ in single intact muscle fibres from transgenic mice. Journal of Physiology 552, 833-844. [22]

González-Serratos, H. (1971). Inward spread of activation in vertebrate muscle fibres. Journal of Physiology 212, 777-799. [11]

González-Serratos, H., Somlyo, A. V., McClellan, G., Shuman, H., Borrero, L. M., & Somlyo, A. P. (1978). Composition of vacuoles and sarcoplasmic reticulum in fatigued muscle: electron probe analysis. Proceedings of the National Academy of Sciences of the United States of America 75, 1329-1333. [15]

Good, P. J., Richter, K., & David, I. B. (1990). Studies on embryonic induction: establishing molecular markers for neural development. In Genetics of Pattern Formation and Growth Control, ed. Mahowald, A. P., pp. 125-135. Wiley-Liss, New York. [6]

Goodwin, G. M., McCloskey, D. I., & Matthews, P. B. (1972). Proprioceptive illusions induced by muscle vibration: contribution by muscle spindles to perception? Science 175, 1382-1384. [4]

Gordon, A. M., Huxley, A. F., & Julian, F. J. (1966a). Tension development in highly stretched vertebrate muscle fibres. Journal of Physiology 184, 143-169. [11]

Gordon, A. M., Huxley, A. F., & Julian, F. J. (1966b). The variation in isometric tension with sarcomere length in vertebrate muscle fibres. Journal of Physiology 184, 170-192. [11]

Gordon, E. E. (1967). Anatomical and biochemical adaptations of muscle to different exercises. Journal of the American Medical Association 201, 755-758. [20]

Gordon, T., Gillespie, J., Orozco, R., & Davis, L. (1991). Axotomy-induced changes in rabbit hindlimb nerves and the effects of chronic electrical stimulation. Journal of Neuroscience 11, 2157-2169. [16]

Gordon, T. & Stein, R. B. (1982). Reorganization of motor-unit properties in reinnervated muscles of the cat. Journal of Neurophysiology 48, 1175-1190. [17]

Gordon, T., Yang, J. F., Ayer, K., Stein, R. B., & Tyreman, N. (1993). Recovery potential of muscle after partial denervation: a comparison between rats and humans. Brain Research Bulletin 30, 477-482. [17]

Gordon-Weeks, P. R. (1989). GAP-43—what does it do in the growth cone? Trends in Neurosciences 12, 363-365. [17]

Gorza, L., Gundersen, K., Lømo, T., Schiaffino, S., & Westgaard, R. H. (1988). Slow-to-fast transformation of denervated soleus muscles by chronic high-frequency stimulation in the rat. Journal of Physiology 402, 627-649. [18, 20]

Grabowski, W., Lobsiger, E. A., & Luttgau, H. CH. (1972). The effect of repetitive stimulation at low frequencies upon the electrical and mechanical activity of single muscle fibres. Pflügers Archiv: European Journal of Physiology 334, 222-239. [15]

Grafstein, B. & Forman, D. S. (1980). Intracellular transport in neurons. Physiological Reviews 60, 1167-1283. [8]

Grafstein, B. & McQuarrie, I. G. (1978). Role of the nerve cell body in axonal regeneration. In Neuronal Plasticity, ed. Cotman, C. W., pp. 155-195. Raven, New York. [18]

Grana, E. A., Chiou-Tan, F., & Jaweed, M. M. (1996). End-plate dysfunction in healthy muscle following a period of disuse. Muscle & Nerve 19, 989-993. [19]

Green, H. J., Daub, B., Houston, M. E., Thomson, J. A., Fraser, I., & Ranney, D. (1981). Human vastus lateralis and gastrocnemius muscles. A comparative histochemical and biochemical analysis. Journal of Neurological Sciences 52, 201-210. [12]

Green, L. S., Donoso, J. A., Heller-Bettinger, I. E., & Samson, F. E. (1977). Axonal transport disturbances in vincristine-induced peripheral neuropathy. Annals of Neurology 1, 255-262. [8]

Greenhaff, P. L. (2001). The creatine-phosphocreatine system: there's more than one song in its repertoire. Journal of Physiology 537, 657. [14]

Gregorio, C. C., Granzier, H., Sorimachi, H., & Labeit, S. (1999). Muscle assembly: a titanic achievement? Current Opinion in Cell Biology 11, 18-25. [1]

Grieshammer, U., Sassoon, D., & Rosenthal, N. (1992). A transgene target for positional regulators marks early rostrocaudal specification of myogenic lineages. Cell 69, 79-93. [5]

Grimby, G. (1988). Physical activity and effects of muscle training in the elderly. Annals of Clinical Research 20, 62-66. [22]

Grimby, G., Danneskiold-Samsøe, B., Hvid, K., & Saltin, B. (1982). Morphology and enzymatic capacity in arm and leg muscles in 78-81 year old men and women. Acta Physiologica Scandinavica 115, 125-134. [22]

Grimby, G. & Saltin, B. (1983). The ageing muscle. Clinical Physiology 3, 209-218. [22]

Grimby, L. (1984). Firing properties of single human motor units during locomotion. Journal of Physiology 346, 195-202. [13]

Grimby, L. & Hannerz, J. (1968). Recruitment order of motor units on voluntary contraction: changes induced by proprioceptive afferent activity. Journal of Neurology, Neurosurgery, and Psychiatry 31, 565-573. [13]

Grimby, L., Hannerz, J., & Hedman, B. (1981). The fatigue and voluntary discharge properties of single motor units in man. Journal of Physiology 316, 545-554. [15, 20]

Grossman, E. J., Roy, R. R., Talmadge, R. J., Zhong, H., & Edgerton, V. R. (1998). Effects of inactivity on myosin heavy chain composition and size of rat soleus fibers. Muscle & Nerve 21, 375-389. [19]

Grounds, M. D., White, J. D., Rosenthal, N., & Bogoyevitch, M. A. (2002). The role of stem cells in skeletal and cardiac muscle repair. Journal of Histochemistry and Cytochemistry: Official Journal of the Histochemistry Society 50, 589-610. [21]

Gundersen, K., Leberer, E., Lømo, T., Pette, D., & Staron, R. S. (1988). Fibre types, calcium-sequestering proteins and metabolic enzymes in denervated and chronically stimulated muscles of the rat. Journal of Physiology 398, 177-189. [16, 18]

Gurdon, J. B., Harger, P., Mitchell, A., & Lemaire, P. (1994). Activin signalling and response to a morphogen gradient. Nature 371, 487-492. [5]

Gurdon, J. B., Mohun, T. J., Sharpe, C. R., & Taylor, M. V. (1989). Embryonic induction and muscle gene activation. Trends in Genetics: TIG 5, 51-56. [5]

Gustafsson, B. (1979). Changes in motoneurone electrical properties following axotomy. Journal of Physiology 293, 197-215. [16]

Guth, L., Albers, R. W., & Brown, W. C. (1964). Quantitative changes in cholinesterase activity of denervated fibers and sole plates. Experimental Neurology 10, 236-250. [16]

Guthrie, S. (2002). Neuronal development: sorting out motor neurons. Current Biology 12, R488-R490. [6]

Gutmann, E. (1971). Histology of degeneration and regeneration. In Electrodiagnosis and Electromyography, ed. Licht, S., pp. 113-133. Waverly Press, Baltimore. [17, 18]

Gutmann, E., Gutmann, L., Medawar, P. B., & Young, J. Z. (1942). The rate of regeneration of a nerve. Journal of Experimental Biology 19, 14-44. [17]

Gutmann, E. & Hanzlíková, V. (1965). Age changes of motor end-plates in muscle fibres of the rat. Gerontologica 11, 12-24. [22]

Gutmann, E. & Hanzlíková, V. (1966). Motor unit in old age. Nature 209, 921-922. [22]

Gutmann, E., Hanzlíková, V., & Vyskocil, F. (1971a). Age changes in cross striated muscle of the rat. Journal of Physiology 216, 331-343. [22]

Gutmann, E. & Holubár, J. (1950). The degeneration of peripheral nerve fibres. Journal of Neurology, Neurosurgery, and Psychiatry 13, 89-105. [16]

Gutmann, E. & Holubár, J. (1952). Degenerace terminálních orgánu v príene pruhovaném a hladkém svalstvu. Physiologica Bohemoslovenica 1, 168-175. [16]

Gutmann, E. & Sandow, A. (1965). Caffeine-induced contracture and potentiation of contraction in normal and denervated rat muscle. Life Sciences 4, 1149-1156. [16]

Gutmann, E., Schiaffino, S., & Hanzlíková, V. (1971b). Mechanism of compensatory hypertrophy in skeletal muscle of the rat. Experimental Neurology 31, 451-464. [19]

Haggar, R. A. & Barr, M. L. (1950). Quantitative data on the size of synaptic end-bulbs in the cat's spinal cord. Journal of Comparative Neurology 93, 17-35. [2]

Häggmark, T. & Eriksson, E. (1979). Hypotrophy of the soleus muscle in man after Achilles tendon rupture. Discussion of findings obtained by computed tomography and morphologic studies. American Journal of Sports Medicine 7, 121-126. [19]

Haida, N., Fowler, W. M., Abresch, R. T., Larson, D. B., Sharman, R. B., Taylor, R. G., & Entrikin, R. K. (1989). Effect of hind-limb suspension on young and adult skeletal muscle. I. Normal mice. Experimental Neurology 103, 68-76. [19]

Hainaut, K. & Duchateau, J. (1989). Muscle fatigue, effects of training and disuse. Muscle & Nerve 12, 660-669. [20]

Hakkinen, K., Komi, P. V., & Alen, M. (1985). Effect of explosive type strength training on isometric force- and relaxation-time, electromyographic and muscle fibre characteristics of leg extensor muscles. Acta Physiologica Scandinavica 125, 587-600. [20]

Halkjaer-Kristensen, J. & Ingemann-Hansen, T. (1985). Wasting of the human quadriceps muscle after knee ligament injuries. II. Muscle fiber morphology. Scandinavian Journal of Rehabilitation Medicine. Supplement 13, 5-55. [19]

Hall, Z. W. & Sanes, J. R. (1993). Synaptic structure and development: the neuromuscular junction. Cell 72 Suppl, 99-121. [3, 6]

Hall-Craggs, E. C. (1970). The longitudinal division of fibres in overloaded rat skeletal muscle. Journal of Anatomy 107, 459-470. [20]

Hamburger, V. (1947). A Manual of Experimental Embryology. University of Chicago Press, Chicago. [6]

Hamburger, V. (1958). Regression versus peripheral control of differentiation in motor hypoplasia. The American Journal of Anatomy 102, 365-409. [6]

Hanson, J. (1974). The effects of repetitive stimulation on the action potential and the twitch of rat muscle. Acta Physiologica Scandinavica 90, 387-400. [15]

Harii, K., Ohmori, K., & Torii, S. (1976). Free gracilis muscle transplantation, with microneurovascular anastomoses for the treatment of facial paralysis. A preliminary report. Plastic & Reconstructive Surgery 57, 133-143. [21]

Harriman, D. G., Taverner, D., & Woolf, A. L. (1970). Ekbom's syndrome and burning paraesthesiae. A biopsy study by vital staining and electron microscopy of the intramuscular innervation with a note on age changes in motor nerve endings in distal muscles. Brain 93, 393-406. [22]

Harris, A. J. & Miledi, R. (1971). The effect of type D botulinum toxin on frog neuromuscular junctions. Journal of Physiology 217, 497-515. [10]

Harris, J. B. & Luff, A. R. (1970). The resting membrane potentials of fast and slow skeletal muscle fibres in the developing mouse. Comparative Biochemistry and Physiology 33, 923-931. [5]

Harris, J. B. & Thesleff, S. (1972). Nerve stump length and membrane changes in denervated skeletal muscle. Nature New Biology 236, 60-61. [16, 18]

Harris, J. B. & Wilson, P. (1971). Mechanical properties of dystrophic mouse muscle. Journal of Neurology, Neurosurgery, and Psychiatry 34, 512-520. [12]

Hartshorne, R. P., Keller, B. U., Talvenheimo, J. A., Catterall, W. A., & Montal, M. (1985). Functional reconstitution of the purified brain sodium channel in planar lipid bilayers. Proceedings of the National Academy of Sciences of the United States of America 82, 240-244. [7]

Hather, B. M., Tesch, P. A., Buchanan, P., & Dudley, G. A. (1991). Influence of eccentric actions on skeletal muscle adaptations to resistance training. Acta Physiologica Scandinavica 143, 177-185. [20]

Hawrylyshyn, T., McComas, A. J., & Heddle, S. B. (1996). Limited plasticity of human muscle. Muscle & Nerve 19, 103-105. [21]

Hayward, L., Wesselmann, U., & Rymer, W. Z. (1991). Effects of muscle fatigue on mechanically sensitive afferents of slow conduction velocity in the cat triceps surae. Journal of Neurophysiology 65, 360-370. [15]

Heckman, C. J. & Binder, M. D. (1991). Computer simulation of the steady-state input-output function of the cat medial gastrocnemius motoneuron pool. Journal of Neurophysiology 65, 952-967. [13]

Heilbrunn, L. V. & Wiercinski, F. J. (1947). The action of various cations on muscle protoplasm. Journal of Cellular and Comparative Physiology 29, 15-32. [11]

Henderson, C. E., Camu, W., Mettling, C., Gouin, A., Poulsen, K., Karihaloo, M., Ruilamas, J., Evans, T., McMahon, S. B., Armanini, M. P., Berkemeier, L., Phillips, H. S., & Rosenthal, A. (1993). Neurotrophins promote motor neuron survival and are present in embryonic limb bud. Nature 363, 266-270. [18]

Henderson, C. E., Huchet, M., & Changeux, J.-P. (1983). Denervation increases a neurite-promoting activity in extracts of skeletal muscle. Nature 302, 609-611. [17]

Henderson, C. E., Phillips, H. S., Pollock, R. A., Davies, A. M., Lemeulle, C., Armanini, M., Simpson, L. C., Moffet, B., Vandlen, R. A., Koliatsos, V. E., & Rosenthal, A. (1994). GDNF: a potent survival factor for motoneurons present in peripheral nerve and muscle. Science 266, 1062-1064. [18]

Henneman, E. (1957). Relation between size of neurons and their susceptibility to discharge. Science 126, 1345-1347. [13]

Henneman, E., Shahani, B. T., & Young, R. R. (1976). Voluntary control of human motor units. In The Motor System: Neurophysiology and Muscle Mechanisms, ed. Shahani, M., pp. 73-78. Elsevier, Amsterdam. [13]

Henneman, E., Somjen, G., & Carpenter, D. O. (1965). Functional significance of cell size in spinal motoneurons. Journal of Neurophysiology 28, 560-580. [13]

Hennig, R. & Lømo, T. (1985). Firing patterns of motor units in normal rats. Nature 314, 164-166. [18]

Hennig, R. & Lømo, T. (1987). Effects of chronic stimulation on the size and speed of long-term denervated and innervated rat fast and slow skeletal muscles. Acta Physiologica Scandinavica 130, 115-131. [18]

Herbison, G. J., Jaweed, M. M., & Ditunno, J. F. (1978). Muscle fiber atrophy after cast immobilization in the rat. Archives of Physical Medicine and Rehabilitation 59, 301-305. [19]

Hess, A. (1970). Vertebrate slow muscle fibers. Physiological Reviews 50, 40-62. [20]

Heuser, J. E. & Reese, T. S. (1973). Evidence for recycling of synaptic vesicle membrane during transmitter release at the frog neuromuscular junction. Journal of Cell Biology 57, 315-344. [3, 10]

Heuser, J. E. & Reese, T. S. (1981). Structural changes after transmitter release at the frog neuromuscular junction. Journal of Cell Biology 88, 564-580. [3]

Heuser, J. E., Reese, T. S., Dennis, M. J., Jan, Y., Jan, L., & Evans, L. (1979). Synaptic vesicle exocytosis captured by quick freezing and correlated with quantal transmitter release. Journal of Cell Biology 81, 275-300. [10]

Heuser, J. E., Reese, T. S., & Landis, D. M. (1974). Functional changes in frog neuromuscular junctions studied with freeze-fracture. Journal of Neurocytology 3, 109-131. [10]

Hicks, A. & McComas, A. J. (1989). Increased sodium pump activity following repetitive stimulation of rat soleus muscles. Journal of Physiology 414, 337-349. [7, 9, 15]

Hicks, A. L., Cupido, C. M., Martin, J., & Dent, J. (1991). Twitch potentiation during fatiguing exercise in the elderly: the effects of training. European Journal of Applied Physiology & Occupational Physiology 63, 278-281. [22]

Hikida, R. S., Gollnick, P. D., Dudley, G. A., Convertino, V. A., & Buchanan, P. (1989). Structural and metabolic characteristics of human skeletal muscle following 30 days of simulated microgravity. Aviation, Space and Environmental Medicine 60, 664-670. [19]

Hill, A. V. (1938). The heat of shortening and the dynamic constants of muscle. Proceedings of the Royal Society of London. Series B: Biological Sciences 126, 136-195. [11]

Hill, A. V. & Lupton, H. (1923). Muscular exercise and the supply and utilization of oxygen. Quarterly Journal of Medicine 16, 135-171. [14]

Hille, B. (1992). Ionic Channels of Excitable Membranes, 2nd ed. Sinauer Associates, Sunderland, MA. [7]

Hiraoki, T. & Vogel, H. J. (1987). Structure and function of calcium-binding proteins. Journal of Cardiovascular Pharmacology 10 Suppl 1, S14-S31. [7]

Hník, P. (1972). Changes in function of muscle afferents during muscle atrophy (de-efferentation or tenotomy).

In Symposium on Structure and Function of Normal and Diseased Muscle and Peripheral Nerve, (Programme abstract no. 20). Polish Academy of Science, Warsaw. [19]

Ho, K. W., Roy, R. R., Tweedle, C. D., Heusner, W. W., Van Huss, W. D., & Carrow, R. E. (1980). Skeletal muscle fiber splitting with weight-lifting exercise in rats. The American Journal of Anatomy 157, 433-440. [20]

Hochachka, P. W. (2003). Intracellular convection, homeostasis and metabolic regulation. Journal of Experimental Biology 206, 2001-2009. [14]

Hodes, R., Larrabee, M. G., & German, W. (1948). The human electromyogram in response to nerve stimulation and the conduction velocity of motor axons. Study on normal and on injured peripheral nerves. Archives of Neurology and Psychiatry (Chicago) 60, 340-365. [17]

Hodgkin, A. L. (1977). Chance and design in electrophysiology: an informal account of certain experiments on nerve carried out between 1934 and 1952. In The Pursuit of Nature. Informal Essays on the History of Physiology, eds. Hodgkin, A. L., Huxley, A. F., Feldberg, W., Ruston, W. A. H., Gregory, R. A., & McCance, R. A., pp. 1-9. Cambridge University Press, Cambridge. [9]

Hodgkin, A. L. & Horowicz, P. (1959). The influence of potassium and chloride ions on the membrane potential of single muscle fibres. Journal of Physiology 148, 127-160. [9, 15]

Hodgkin, A. L. & Huxley, A. F. (1952a). A quantitative description of membrane current and its application to conduction and excitation in nerve. Journal of Physiology 117, 500-554. [7, 9]

Hodgkin, A. L. & Huxley, A. F. (1952b). Currents carried by sodium and potassium ions through the membrane of the giant axon of Loligo. Journal of Physiology 116, 449-472. [9]

Hodgkin, A. L. & Katz, B. (1949). The effect of sodium ions on the electrical activity of the giant axon of the squid. Journal of Physiology 108, 37-77. [9]

Hoffman, E. P. & Kunkel, L. M. (1989). Dystrophin abnormalities in Duchenne/Becker muscular dystrophy. Neuron 2, 1019-1029. [1]

Hoffman, H. (1950). Local reinnervation in partially denervated muscle: a histophysiological study. Journal of Experimental Biology 28, 383-397. [17]

Hoffman, H. (1951). Fate of interrupted nerve fibres regenerating into partially denervated muscles. Australian Journal of Experimental Biology and Medical Science 29, 211-219. [17]

Hoffman, P. N. & Lasek, R. J. (1980). Axonal transport of the cytoskeleton in regenerating motor neurons: constancy and change. Brain Research 202, 317-333. [18]

Hoffmann, P. (1918). Ueber die Beeinflussang der Sehnenreflexe durch die willkürliche Contraction. Medizinische Klinik XIV, 203. [4]

Hofmann, F., Flockerzi, V., Nastainczyk, W., Ruth, P., & Schneider, T. (1990). The molecular structure and regulation of muscular calcium channels. Current Topics in Cellular Regulation 31, 223-239. [7]

Hofmann, W. W., Kundin, J. E., & Farrell, D. F. (1967). The pseudomyasthenic syndrome of Eaton and Lambert: an electrophysiological study. Electroencephalography and Clinical Neurophysiology 23, 214-224. [10]

Hofmann, W. W. & Thesleff, S. (1972). Studies on the trophic influence of nerve on skeletal muscle. European Journal of Pharmacology 20, 256-260. [18, 19]

Hogan, E. L., Dawson, D. M., & Romanul, F. C. (1965). Enzymatic changes in denervated muscle. II. Biochemical studies. Archives of Neurology 13, 274-282. [16]

Höök, P., Li, X., Sleep, J., Hughes, S., & Larsson, L. (1999). In vitro motility speed of slow myosin extracted from single soleus fibres from young and old rats. Journal of Physiology 520 (Pt 2), 463-471. [22]

Hooper, A. C. (1981). Length, diameter and number of ageing skeletal muscle fibres. Gerontology 27, 121-126. [22]

Hopf, H. C. (1962). Untersuchungen uber die unterschiede in der leitgeschwindigkeit motorischer nervenfasern beim menschen. Deutsches Zeitschrift fur Nervenheikunde 183, 579-588. [9]

Hopf, H. C. (1963). Electromyographic study on so-called mononeuritis. Archives of Neurology 9, 307-312. [9]

Horisberger, J. D., Lemas, V., Kraehenbuhl, J. P., & Rossier, B. C. (1991). Structure-function relationship of Na,K-ATPase. Annual Review of Physiology 53, 565-584. [7]

Hortobagyi, T., Dempsey, L., Fraser, D., Zheng, D., Hamilton, G., Lambert, J., & Dohm, L. (2000). Changes in muscle strength, muscle fibre size and myofibrillar gene expression after immobilization and retraining in humans. Journal of Physiology 524 (Pt 1), 293-304. [19]

Houk, J. & Henneman, E. (1967). Responses of Golgi tendon organs to active contractions of the soleus muscle of the cat. Journal of Neurophysiology 30, 466-481. [4]

Houston, M. E. (2001). Biochemistry Primer for Exercise Science, 2nd ed. Human Kinetics, Champaign, IL. [14]

Houston, M. E., Froese, E. A., Valeriote, St. P., Green, H. J., & Ranney, D. A. (1983). Muscle performance, morphology and metabolic capacity during strength training and detraining: a one leg model. European Journal of Applied Physiology & Occupational Physiology 51, 25-35. [20]

Howald, H. (1982). Training-induced morphological and functional changes in skeletal muscle. International Journal of Sports Medicine 3, 1-12. [20]

Howald, H., Hoppeler, H., Claassen, H., Mathieu, O., & Straub, R. (1985). Influences of endurance training on the ultrastructural composition of the different muscle fiber types in humans. Pflügers Archiv: European Journal of Physiology 403, 369-376. [20]

Howard, J. E., McGill, K. C., & Dorfman, L. J. (1988). Age effects on properties of motor unit action potentials: ADEMG analysis. Annals of Neurology 24, 207-213. [22]

Hubbard, J. I. & Schmidt, R. F. (1963). An electrophysiological investigation of mammalian motor nerve terminals. Journal of Physiology 166, 145-167. [10]

Hubbard, J. I. & Wilson, D. F. (1973). Neuromuscular transmission in a mammalian preparation in the absence of blocking drugs and the effect of D-tubocurarine. Journal of Physiology 228, 307-325. [10]

Hudlicka, O. (1990). The response of muscle to enhanced and reduced activity. Bailliere's Clinical Endocrinology and Metabolism 4, 417-439. [20]

Huey, K. A. & Bodine, S. C. (1998). Changes in myosin mRNA and protein expression in denervated rat soleus and tibialis anterior. European Journal of Biochemistry 256, 45-50. [16]

Hultborn, H. (2001). State-dependent modulation of sensory feedback. Journal of Physiology 533, 5-13. [4]

Hunt, C. C., Wilkinson, R. S., & Fukami, Y. (1978). Ionic basis of the receptor potential in primary endings of mammalian muscle spindles. Journal of General Physiology 71, 683-698. [4]

Hursch, J. B. (1939). Conduction velocity and diameter of nerve fibers. American Journal of Physiology 127, 131-139. [9, 12]

Hutter, O. F. & Noble, D. (1960). The chloride conductance of frog skeletal muscle. Journal of Physiology 151, 89-102. [7, 9]

Huxley, A. F. & Niedergerke, R. (1954). Structural changes in muscle during contraction; interference microscopy of living muscle fibres. Nature 173, 971-976. [11]

Huxley, A. F. & Simmons, R. M. (1971). Proposed mechanism of force generation in striated muscle. Nature 233, 533-538. [11]

Huxley, A. F. & Taylor, R. E. (1958). Local activation of striated muscle fibres. Journal of Physiology 144, 426-441. [11]

Huxley, H. E. (1958). The contraction of muscle. Scientific American 199, 67-82. [11]

Huxley, H. E. (1964). Evidence for continuity between the central elements of the triads and extracellular space in frog sartorius muscle. Nature 202, 1067-1071. [11]

Huxley, H. E. (1972). Molecular basis of contraction in cross-striated muscles. In The Structure and Function of Muscle, ed. Bourne, G. H., pp. 302-387. Academic Press, New York. [1, 11]

Huxley, H. E. (1990). Sliding filaments and molecular motile systems. Journal of Biological Chemistry 265, 8347-8350. [11]

Huxley, H. E. & Hanson, J. (1954). Changes in the cross-striations of muscle during contraction and stretch and their structural interpretation. Nature 173, 971-973. [11]

Ianniruberto, A. & Tajani, E. (1981). Ultrasonographic study of fetal movements. Seminars in Perinatology 5, 175-181. [5]

Ibraghimov-Beskrovnaya, O., Ervasti, J. M., Leveille, C. J., Slaughter, C. A., Sernett, S. W., & Campbell, K. P. (1992). Primary structure of dystrophin-associated glycoproteins linking dystrophin to the extracellular matrix. Nature 355, 696-702. [1, 3]

Ide, C. (1996). Peripheral nerve regeneration. Neuroscience Research 25, 101-121. [17]

Ihemelandu, E. C. (1980). Decrease in fibre numbers of dog pectineus muscle with age. Journal of Anatomy 130, 69-73. [22]

Ikai, M. & Fukunaga, T. (1970). A study on training effect on strength per unit cross-sectional area of muscle by means of ultrasonic measurement. Internationale Zeitschrift für angewandte Physiologie, einschliesslich Arbeitsphysiologie 28, 173-180. [20]

Inesi, G. & Kirtley, M. E. (1990). Coupling of catalytic and channel function in the Ca2+ transport ATPase. Journal of Membrane Biology 116, 1-8. [7]

Inestrosa, N. C. (1982). Differentiation of skeletal muscle cells in culture. Cell Structure and Function 7, 91-109. [5]

Ingemann-Hansen, T. & Halkjaer-Kristensen, J. (1983). Progressive resistance exercise training of the hypotrophic quadriceps muscle in man. The effects on morphology, size and function as well as the influence of duration of effort. Scandinavian Journal of Rehabilitation Medicine 15, 29-35. [19]

Ingjer, F. (1979). Effects of endurance training on muscle fibre ATP-ase activity, capillary supply and mitochondrial content in man. Journal of Physiology 294, 419-432. [20]

Inokuchi, S., Ishikawa, H., Iwamoto, S., & Kimura, T. (1975). Age-related changes in the histological composition of the rectus abdominis muscle of the adult human. Human Biology: An International Record of Research 47, 231-249. [22]

Isaacs, E. R., Bradley, W. G., & Henderson, G. (1973). Longitudinal fibre splitting in muscular dystrophy: a serial cinematographic study. Journal of Neurology, Neurosurgery, and Psychiatry 36, 813-819. [20]

Ishihara, A. & Araki, H. (1988). Effects of age on the number and histochemical properties of muscle fibers and motoneurons in the rat extensor digitorum longus muscle. Mechanisms of Ageing and Development 45, 213-221. [22]

Ishihara, A., Naitoh, H., & Katsuta, S. (1987). Effects of ageing on the total number of muscle fibers and motoneurons of the tibialis anterior and soleus muscles in the rat. Brain Research 435, 355-358. [22]

Ito, Y., Miledi, R., Molenaar, P. C., Vincent, A., Polak, R. L., van Gelder, M., & Davis, J. N. (1976). Acetylcholine in human muscle. Proceedings of the Royal Society of London. Series B: Biological Sciences 192, 475-480. [10]

Jablecki, C. & Brimijoin, S. (1974). Reduced axoplasmic transport of choline acetyltransferase activity in dystrophic mice. Nature 250, 151-154. [10]

Jackson, M. J., Jones, D. A., & Edwards, R. H. (1984). Experimental skeletal muscle damage: the nature of the calcium-activated degenerative processes. European Journal of Clinical Investigation 14, 369-374. [21]

Jacobsen, M. & Rutishauser, U. (1986). Induction of neural cell adhesion moledule (N-CAM) in Xenopus embryos. Developmental Biology 116, 524-531. [6]

Jacobsen, S. & Guth, L. (1965). An electrophysiological study of the early stages of peripheral nerve regeneration. Experimental Neurology 11, 48-60. [17]

Jakobsson, F., Borg, K., Edström, L., & Grimby, L. (1988). Use of motor units in relation to muscle fiber type and size in man. Muscle & Nerve 11, 1211-1218. [22]

Jakubiec-Puka, A., Ciechomska, I., Morga, J., & Matusiak, A. (1999). Contents of myosin heavy chains in denervated slow and fast rat leg muscles. Comparative Biochemistry and Physiology. Part B, Biochemistry & Molecular Biology 122, 355-362. [16]

Jan, L. Y. & Jan, Y. N. (1989). Voltage-sensitive ion channels. Cell 56, 13-25. [7]

Jan, Y. N., Jan, L. Y., & Dennis, M. J. (1977). Two mutations of synaptic transmission in Drosophila. Proceedings of the Royal Society of London. Series B: Biological Sciences 198, 87-108. [7]

Jankala, H., Harjola, V. P., Petersen, N. E., & Harkonen, M. (1997). Myosin heavy chain mRNA transform to faster isoforms in immobilized skeletal muscle: a quantitative PCR study. Journal of Applied Physiology 82, 977-982. [19]

Jansen, J. K., Lømo, T., Nicolaysen, K., & Westgaard, R. H. (1973). Hyperinnervation of skeletal muscle fibers: dependence on muscle activity. Science 181, 559-561. [19]

Jansson, E., Esbjörnsson, M., Holm, I., & Jacobs, I. (1990). Increase in the proportion of fast-twitch muscle fibres by sprint training in males. Acta Physiologica Scandinavica 140, 359-363. [20]

Jansson, E. & Kaijser, L. (1977). Muscle adaptation to extreme endurance training in man. Acta Physiologica Scandinavica 100, 315-324. [20]

Jasmin, B. J., Lavoie, P. A., & Gardiner, P. F. (1988). Fast axonal transport of labeled proteins in motoneurons of exercise-trained rats. American Journal of Physiology 255, C731-C736. [20]

Jennekens, F. G., Tomlinson, B. E., & Walton, J. N. (1971). Data on the distribution of fibre types in five human limb muscles. An autopsy study. Journal of Neurological Sciences 14, 245-257. [22]

Jentsch, T. J., Steinmeyer, K., & Schwarz, G. (1990). Primary structure of Torpedo marmorata chloride channel isolated by expression cloning in Xenopus oocytes. Nature 348, 510-514. [7]

Jobsis, F. F. & Stainsby, W. N. (1968). Oxidation of NADH during contractions of circulated mammalian skeletal muscle. Respiration Physiology 4, 292-300. [14]

Johns, T. R. & Thesleff, S. (1961). Effects of motor inactivation on the chemical sensitivity of skeletal muscle. Acta Physiologica Scandinavica 51, 136-141. [19]

Johnson, H. A. & Erner, S. (1972). Neuron survival in the aging mouse. Experimental Gerontology 7, 111-117. [22]

Johnson, M. A., Polgar, J., Weightman, D., & Appleton, D. (1973). Data on distribution of fibre types in thirty-six human muscles: an autopsy study. Journal of Neurological Sciences 18, 111-129. [12]

Jokl, P. & Konstadt, S. (1983). The effect of limb immobilization on muscle function and protein composition. Clinical Orthopaedics and Related Research 174, 222-229. [19]

Jones, D. A., Jackson, M. J., McPhail, G., & Edwards, R. H. (1984). Experimental mouse muscle damage: the importance of external calcium. Clinical Science (London, England: 1979) 66, 317-322. [21]

Jones, D. A., Newham, D. J., Round, J. M., & Tolfree, S. E. (1986). Experimental human muscle damage: morphological changes in relation to other indices of damage. Journal of Physiology 375, 435-448. [21]

Jones, D. A. & Round, J. M. (1990). Skeletal Muscle in Health and Disease. A Textbook of Muscle Physiology. Manchester University Press, Manchester. [15, 21]

Jones, D. A., Rutherford, O. M., & Parker, D. F. (1989). Physiological changes in skeletal muscle as a result of strength training. Quarterly Journal of Experimental Physiology 74, 233-256. [13, 20]

Jones, S. F. & Kwanbunbumpen, S. (1970). The effects of nerve stimulation and hemicholinium on synaptic vesicles at the mammalian euromuscular junction. Journal of Physiology 207, 31-50. [3]

Jubrias, S. A., Esselman, P. C., Price, L. B., Cress, M. E., & Conley, K. E. (2001). Large energetic adaptations of elderly muscle to resistance and endurance training. Journal of Applied Physiology 90, 1663-1670. [22]

Juel, C. (1986). Potassium and sodium shifts during in vivo isometric muscle contraction and the time course of the ion-gradient recovery. Pflügers Archiv: European Journal of Physiology 406, 458-463. [9]

Juntunen, J. & Teravainen, H. (1972). Structural development of myoneural junctions in the human embryo. Histochemie 32, 107-112. [6]

Kabbara, A. A. & Allen, D. G. (1999). The role of calcium stores in fatigue of isolated single muscle fibres from the cane toad. Journal of Physiology 519, 169-176. [7]

Kaeser, H. E. & Lambert, E. H. (1962). Nerve function studies in experimental polyneuritis. Electroencephalography and Clinical Neurophysiology 22, 29-35. [16]

Kalderon, N. & Gilula, N. B. (1979). Membrane events involved in myoblast fusion. Journal of Cell Biology 81, 411-425. [5]

Kallman, D. A., Plato, C. C., & Tobin, J. D. (1990). The role of muscle loss in the age-related decline of grip strength: cross-sectional and longitudinal perspectives. Journal of Gerontology 45, M82-M88. [22]

Kalow, W., Britt, B. A., Terrau, M. E., & Haist, C. (1970). Metabolic error of muscle metabolism after recovery from malignant hyperthermia. Lancet 2[7679], 895-898. [11]

Kamb, A., Iverson, L. E., & Tanouye, M. A. (1987). Molecular characterization of Shaker, a Drosophila gene that encodes a potassium channel. Cell 50, 405-413. [7]

Kao, C. Y. (1966). Tetrodotoxin, saxitoxin and their significance in the study of excitation phenomena. Pharmacological Reviews 18, 997-1049. [7]

Karpati, G., Carpenter, S., & Eisen, A. A. (1972). Experimental core-like lesions and nemaline rods: a correlative morphological and physiological study. Archives of Neurology 27, 237-251. [19]

Karpati, G. & Engel, W. K. (1968). Correlative histochemical study of skeletal muscle after suprasegmental denervation, peripheral nerve section, and skeletal fixation. Neurology 18, 681-692. [19]

Kasper, C. E. & Xun, L. (1996a). Cytoplasm-to-myonucleus ratios following microgravity. Journal of Muscle Research and Cell Motility 17, 595-602. [19]

Kasper, C. E. & Xun, L. (1996b). Cytoplasm-to-myonucleus ratios in plantaris and soleus muscle fibres following hindlimb suspension. Journal of Muscle Research and Cell Motility 17, 603-610. [19]

Kasprzak, H. & Salpeter, M. M. (1985). Recovery of acetylcholinesterase at intact neuromuscular junctions after in vivo inactivation with di-isopropylfluorophosphate. Journal of Neuroscience 5, 951-955. [3]

Katz, B. (1966). Nerve, Muscle and Synapse. McGraw-Hill, New York. [9]

Katz, B. & Miledi, R. (1965). Propagation of electric activity in motor nerve terminals. Proceedings of the Royal Society of London. Series B: Biological Sciences 161, 453-482. [10]

Katz, B. & Miledi, R. (1968). The role of calcium in neuromuscular facilitation. Journal of Physiology 195, 481-492. [10]

Katz, B. & Miledi, R. (1972). The statistical nature of the acetycholine potential and its molecular components. Journal of Physiology 224, 665-699. [3]

Kaufman, T. L. (1985). Strength training effect in young and aged women. Archives of Physical Medicine and Rehabilitation 65, 223-226. [22]

Kawamura, Y., Okazaki, H., O'Brien, P. C., & Dych, P. J. (1977). Lumbar motoneurons of man. I. Number and diameter histogram of alpha and gamma axons of ventral root. Journal of Neuropathology and Experimental Neurology 36, 853-860. [22]

Keh-Evans, L., Rice, C. L., Noble, E. G., Paterson, D. H., Cunningham, D. A., & Taylor, A. W. (1992). Comparison of histochemical, biochemical and contractile properties of trained aged subjects. Canadian Journal of Aging 11, 412-425. [22]

Kellerth, J.-O. (1973). Intracellular staining of cat spinal motoneurons with Procion Yellow for ultrastructural studies. Brain Research 50, 415-418. [2]

Kelly, A. M. & Schotland, D. L. (1972). The evolution of the "checkerboard" in rat muscle. In Research in Muscle Development and the Muscle Spindle, eds. Banker, B. Q., Przybylski, R. J., Van der Meulen, J. P., & Victor, M., pp. 32-48. Excerpta Medica, Amsterdam. [12]

Kelly, A. M. & Zacks, S. I. (1969). The fine structure of motor endplate morphogenesis. Journal of Cell Biology 42, 154-169. [6]

Kennedy, J. M., Kamel, S., Tambone, W. W., Vrbová, G., & Zak, R. (1986). The expression of myosin heavy chain isoforms in normal and hypertrophied chicken slow muscle. Journal of Cell Biology 103, 977-983. [20]

Kennedy, W. R. (1970). Innervation of normal human muscle spindles. Neurology 20, 463-475. [4]

Kenny-Mobbs, T. (1985). Myogenic differentiation in early chick wing mesenchyme in the absence of the brachial somites. Journal of Embryology and Experimental Morphology 90, 415-436. [5]

Kereshi, S., Manzano, G., & McComas, A. J. (1983). Impulse conduction velocities in human biceps brachii muscles. Experimental Neurology 80, 652-662. [9]

Kernell, D. (1966). Input resistance, electrical excitability, and size of ventral horn cells in cat spinal cord. Science 152, 1637-1640. [12]

Keynes, R. D., Ritchie, J. M., & Rojas, E. (1971). The binding of tetrodotoxin to nerve membranes. Journal of Physiology 213, 235-254. [9]

Kirkwood, S. P., Munn, E. A., & Brooks, G. A. (1986). Mitochondrial reticulum in limb skeletal muscle. American Journal of Physiology 251, C395-C402. [14]

Kjeldsen, K., Everts, M. E., & Clausen, T. (1986). The effects of thyroid hormones on 3H-ouabain binding site concentration, Na,K-contents and 86Rb-efflux in rat skeletal muscle. Pflügers Archiv: European Journal of Physiology 406, 529-535. [7]

Klaus, W., Lüllmann, H., & Muscholl, E. (1960). The binding of tetrodotoxin to nerve membranes. Journal of Physiology 213, 235-254. [16]

Klein, C. S., Rice, C. L., Ivanova, T. D., & Garland, S. J. (2002). Changes in motor unit discharge rate are not associated with the amount of twitch potentiation in

old men. Journal of Applied Physiology 93, 1616-1621. [22]

Klein, C. S., Allman, B. L., Marsh, G. D., & Rice, C. L. (2002). Muscle size, strength, and bone geometry in the upper limbs of young and old men. Journals of Gerontology Series A-Biological Sciences and Medical Sciences 57, 455-459. [22]

Klinkerfuss, G. H. & Haugh, M. J. (1970). Disuse atrophy of muscle. Histochemistry and electron microscopy. Archives of Neurology 22, 309-320. [19]

Klitgaard, H. (1988). A model for quantitative strength training of hindlimb muscle of the rat. Journal of Applied Physiology 64, 1740-1745. [20]

Klitgaard, H., Mantoni, M., Schiaffino, S., Ausoni, S., Gorza, L., Laurent-Winter, C., Schnohr, P., & Saltin, B. (1990a). Function, morphology and protein expression of ageing skeletal muscle: a cross-sectional study of elderly men with different training backgrounds. Acta Physiologica Scandinavica 140, 41-54. [22]

Klitgaard, H., Zhou, M., & Richter, E. A. (1990b). Myosin heavy chain composition of single fibres from m. biceps brachii of male body builders. Acta Physiologica Scandinavica 140, 175-180. [20]

Klitgaard, H., Zhou, M., Schiaffino, S., Betto, R., Salviati, G., & Saltin, B. (1990c). Ageing alters the myosin heavy chain composition of single fibres from human skeletal muscle. Acta Physiologica Scandinavica 140, 55-62. [22]

Knochel, J. P., Blachley, J. D., Johnson, J. H., & Carter, N. W. (1985). Muscle cell electrical hyperpolarization and reduced exercise hyperkalemia in physically conditioned dogs. Journal of Clinical Investigation 75, 740-745. [7]

Knowles, M., Currie, S., Saunders, M., Walton, J. N., & Field, E. J. (1969). Lymphocyte transformation in the Guillain-Barré syndrome. Lancet 2, 1168-1170. [16]

Kobayashi, J., Mackinnon, S. E., Watanabe, O., Ball, D. J., Gu, X. M., Hunter, D. A., & Kuzon, W. M., Jr. (1997). The effect of duration of muscle denervation on functional recovery in the rat model. Muscle & Nerve 20, 858-866. [17]

Koch, H. P. (1990). Comments on "Proposed mechanisms of action in thalidomide embryopathy." Teratology 41, 243-246. [5]

Koch, M. C., Steinmeyer, K., Lorenz, C., Ricker, K., Wolf, F., Otto, M., Zoll, B., Lehmann-Horn, F., Grzeschik, K. H., & Jentsch, T. J. (1992). The skeletal muscle chloride channel in dominant and recessive human myotonia. Science 257, 797-800. [7]

Koliatsos, V. E., Clatterbuck, R. E., Winslow, J. W., Cayouette, M. H., & Price, D. L. (1993). Evidence that brain-derived neurotrophic factor is a trophic factor for motor neurons in vivo. Neuron 10, 359-367. [16]

Körner, G. (1960). Untersuchungen über Zahl, Anordnung und Länge der Muskelspindeln in einegen Schulterei, den Oberarmmunskein und im Muskulus sternalis des Menschen. Anatomischer Anzeiger 108, 99-103.

Koryak, Y. (1995). Contractile properties of the human tricepts surae muscle during simulated weightlessness. European Journal of Applied Physiology & Occupational Physiology 70, 344-350. [19]

Koryak, Y. (1999). Electrical and contractile parameters of muscle in man: effects of 7-day "dry" water immersion. Aviation, Space and Environmental Medicine 70, 459-464. [19]

Kossev, A., Elek, J. M., Wohlfarth, K., Schubert, M., Dengler, R., & Wolf, W. (1994). Assessment of human motor unit twitches—a comparison of spike-triggered averaging and intramuscular microstimulation. Electroencephalography and Clinical Neurophysiology 93, 100-105. [12]

Kowalchuk, N. & McComas, A. (1987). Effects of impulse blockade on the contractile properties of rat skeletal muscle. Journal of Physiology 382, 255-266. [19]

Kraus, W. E. & Williams, R. S. (1990). Intracellular signals mediating contraction-induced changes in the oxidative capacity of skeletal muscle. In The Dynamic State of Muscle Fibres, ed. Pette, D., pp. 601-615. Walter de Gruyter, Berlin. [20]

Krause, U. & Wegener, G. (1996). Control of glycolysis in vertebrate skeletal muscle during exercise. American Journal of Physiology 270, R821-R829. [14]

Krishnan, S., Lowrie, M. B., & Vrbová, G. (1985). The effect of reducing the peripheral field on motoneurone development in the rat. Brain Research 351, 11-20. [6]

Kristensson, K. & Olsson, Y. (1971). Retrograde axonal transport of protein. Brain Research 29, 363-365. [8, 18]

Krnjevic, K. & Miledi, R. (1958). Failure of neuromuscular propagation in rats. Journal of Physiology 140, 440-461. [15]

Kubo, K., Kanehisa, H., & Fukunaga, T. (2002). Effects of resistance and stretching training programmes on the viscoelastic properties of human tendon structures in vivo. Journal of Physiology 538, 219-226. [20]

Kubo, K., Kanehisa, H., Miyatani, M., Tachi, M., & Fukunaga, T. (2003). Effect of low-load resistance training on the tendon properties in middle-aged and elderly women. Acta Physiologica Scandinavica 178, 25-32. [22]

Kucera, J. (1985). Characteristics of motor innervation of muscle spindles in the monkey. American Journal of Physiology 173, 113-125. [4]

Kudina, L. P. & Alexeeva, N. L. (1992). After-potentials and control of repetitive firing in human motoneurones. Electroencephalography and Clinical Neurophysiology 85, 345-353. [12]

Kuffler, S. W. & Yoshikami, D. (1975). The number of transmitter molecules in a quantum: an estimate from iontophoretic application of acetylcholine at the neuromuscular synapse. Journal of Physiology 251, 465-482. [3]

Kugelberg, E. (1949). Electromyography in muscular dystrophies. Differentiation between dystrophies and chronic lower motor neurone lesions. Journal of Neurology, Neurosurgery, and Psychiatry 12, 129-136. [13]

Kugelberg, E., Edström, L., & Abbruzzese, M. (1970). Mapping of motor units in experimentally reinnervated rat muscle. Interpretation of histochemical and atrophic fibre patterns in neurogenic lesions. Journal of Neurology, Neurosurgery, and Psychiatry 33, 319-329. [17]

Kuiack, S. & McComas, A. (1992). Transient hyperpolarization of non-contracting muscle fibres in anaesthetized rats. Journal of Physiology 454, 609-618. [7, 15]

Kukulka, C. G. & Clamann, H. P. (1981). Comparison of the recruitment and discharge properties of motor units in human brachial biceps and adductor pollicis during isometric contractions. Brain Research 219, 45-55. [13]

Kuno, M., Miyata, Y., & Muñoz-Martinez, E. J. (1974a). Differential reaction of fast and slow α-motoneurones to axotomy. Journal of Physiology 240, 725-739. [16, 18]

Kuno, M., Miyata, Y., & Muñoz-Martinez, E. J. (1974b). Properties of fast and slow α-motoneurones following motor reinnervation. Journal of Physiology 242, 273-288. [18]

Kutay, U., Hartmann, E., & Rapoport, T. A. (1993). A class of membrane proteins with a C-terminal anchor. Trends in Cell Biology 3, 72-75. [10]

LaBarge, M. A. & Blau, H. M. (2002). Biological progression from adult bone marrow to mononucleate muscle stem cell to multinucleate muscle fiber in response to injury. Cell 111, 589-601. [21]

Laing, N. G. & Prestige, M. C. (1978). Prevention of spontaneous motoneurone death in chick embryos [proceedings]. Journal of Physiology 282, 33P-34P. [6]

Lambert, E. H. & Elmqvist, D. (1971). Quantal components of end-plate potentials in the myasthenic syndrome. Annals of the New York Academy of Sciences 183, 183-199. [10]

Lampert, P. W. (1969). Mechanism of demyelination in experimental allergic neuritis. Electron microscopic studies. Laboratory Investigation; A Journal of Technical Methods and Pathology 20, 127-138. [16]

Lance-Jones, C. & Landmesser, L. (1980a). Motoneurone projection patterns in embryonic chick limbs following partial deletions of the spinal cord. Journal of Physiology 302, 559-580. [6]

Lance-Jones, C. & Landmesser, L. (1980b). Motoneurone projection patterns in the chick hind limb following early partial reversals of the spinal cord. Journal of Physiology 302, 581-602. [6]

Landau, W. M. (1953). The duration of neuromuscular function after nerve section in man. Journal of Neurosurgery 10, 64-68. [16]

Landmesser, L. (1971). Contractile and electrical responses of vagus-innervated frog sartorius muscles. Journal of Physiology 213, 707-725. [17]

Landmesser, L. (1972). Pharmacological properties, cholinesterase activity and anatomy of nerve-muscle junctions in vagus-innervated frog sartorius. Journal of Physiology 220, 243-256. [17]

Landon, D. N. (1982). The excitable apparatus of skeletal muscle. In Abnormal Nerves and Muscles as Impulse Generators, eds. Culp, W. J. & Achoa, J., pp. 607-631. Oxford University Press, New York. [15]

Landon, D. N. & Langley, O. K. (1971). The local chemical environment of nodes of Ranvier: a study of cation binding. Journal of Anatomy 108, 419-432. [2]

Lang, B., Newsom-Davis, J., Prior, C., & Wray, D. (1983). Antibodies to motor nerve terminals: an electrophysiological study of a human myasthenic syndrome transferred to mouse. Journal of Physiology 344, 335-345. [10]

Langeland, J. A. & Carroll, S. B. (1993). Conservation of regulatory elements controlling hairy pair-rule stripe formation. Development 117, 585-596. [5]

Lännergren, J., Larsson, L., & Westerblad, H. (1989). A novel type of delayed tension reduction observed in rat motor units after intense activity. Journal of Physiology 412, 267-276. [15]

Lännergren, J. & Westerblad, H. (1991). Force decline due to fatigue and intracellular acidification in isolated fibres from mouse skeletal muscle. Journal of Physiology 434, 307-322. [15]

Larsson, L. (1983). Histochemical characteristics of human skeletal muscle during ageing. Acta Physiologica Scandinavica 117, 469-471. [22]

Larsson, L., Ansved, T., Edström, L., Gorza, L., & Schiaffino, S. (1991). Effects of age on physiological, immunohistochemical and biochemical properties of fast-twitch single motor units in the rat. Journal of Physiology 443, 257-275. [22]

Larsson, L., Grimby, G., & Karlsson, J. (1979). Muscle strength and speed of movement in relation to age and muscle morphology. Journal of Applied Physiology 46, 451-456. [22]

Larsson, L., Li, X., Berg, H. E., & Frontera, W. R. (1996). Effects of removal of weight-bearing function on contractility and myosin isoform composition in single human skeletal muscle cells. Pflügers Archiv: European Journal of Physiology 432, 320-328. [19]

Larsson, L. & Salviati, G. (1989). Effects of age on calcium transport activity of sarcoplasmic reticulum in fast- and slow-twitch rat muscle fibres. Journal of Physiology 419, 253-264. [22]

Lasek, R. J., Garner, J. A., & Brady, S. T. (1984). Axonal transport of the cytoplasmic matrix. Journal of Cell Biology 99, 212s-221s. [8]

Lazarides, E. & Capetanaki, Y. G. (1986). The striated muscle cytoskeleton: expression and assembly in development. In Molecular Biology of Muscle Development, eds. Emerson, C., Fischman, D., Nadal-Ginard, B., & Siddiqui, M. A. Q., pp. 749-772. Alan R. Liss, New York. [1]

Lee, J. A., Westerblad, H., & Allen, D. G. (1991). Changes in tetanic and resting [Ca2+]i during fatigue and recovery of single muscle fibres from Xenopus laevis. Journal of Physiology 433, 307-326. [15]

Lee, R. G., Ashby, P., White, D. G., & Aguayo, A. J. (1975). Analysis of motor conduction velocity in the human median nerve by computer simulation of compound muscle action potentials. Electroencephalography and Clinical Neurophysiology 39, 225-237. [9, 12]

Lehmann-Horn, F., Küther, G., Ricker, K., Grafe, P., Ballanyi, K., & Rüdel, R. (1987). Adynamia episodica hereditaria with myotonia: a non-inactivating sodium current and the effect of extracellular pH. Muscle & Nerve 10, 363-374. [7]

Leibrock, J., Lottspeich, F., Hohn, A., Hofer, M., Hengerer, B., Masiakowski, P., Thoenen, H., & Barde, Y. A. (1989). Molecular cloning and expression of brain-derived neurotrophic factor. Nature 341, 149-152. [18]

Leksell, L. (1945). The action potential and excitatory effects of the small ventral root fibres to skeletal muscle. Acta Physiologica Scandinavica 10, 1-84. [12]

Lemire, R. J. (1969). Variations in development of the caudal neural tube in human embryos (Horizons XIV-XXI). Teratology 2, 361-369. [6]

Lenman, A. J., Tulley, F. M., Vrbová, G., Dimitrijevic, M. R., & Towle, J. A. (1989). Muscle fatigue in some neurological disorders. Muscle & Nerve 12, 938-942. [19]

Lennon, V. A. & Carnegie, P. R. (1971). Immunopharmacological disease: a break in tolerance to receptor sites. Lancet 1, 630-633. [10]

Lennon, V. A., Lindstrom, J. M., & Seybold, M. E. (1976). Experimental autoimmune myasthenia gravis: cellular and humoral immune responses. Annals of the New York Academy of Sciences 274, 283-299. [10]

Leon, J. & McComas, A. J. (1984). Effects of vincristine sulfate on touch dome function in the rat. Experimental Neurology 84, 283-291. [8]

Levi-Montalcini, R. (1987). The nerve growth factor 35 years later. Science 237, 1154-1162. [18]

Levine, R. J. C., Kensler, R. W., Yang, Z. H., Stull, J. T., & Sweeney, H. L. (1996). Myosin light chain phosphorylation affects the structure of rabbit skeletal muscle thick filaments. Biophysical Journal 71, 898-907. [11]

Lewis, D. M. (1972). The effect of denervation on the mechanical and electrical responses of fast and slow mammalian twitch muscle. Journal of Physiology 222, 51-75. [16]

Lewis, D. M. & Parry, D. J. (1979). Properties of motor units in mouse soleus. Journal of Physiology 295, 90P. [12]

Lexell, J. (1993). Ageing and human muscle: observations from Sweden. Canadian Journal of Applied Physiology 18, 2-18. [22]

Lexell, J., Henriksson-Larsén, K., Winblad, B., & Sjöström, M. (1983). Distribution of different fiber types in human skeletal muscles: effects of aging studied in whole muscle cross sections. Muscle & Nerve 6, 588-595. [22]

Lexell, J., Taylor, C. C., & Sjöström, M. (1988). What is the cause of the ageing atrophy? Total number, size and proportion of different fiber types studied in whole vastus lateralis muscle from 15- to 83-year-old men. Journal of Neurological Sciences 84, 275-294. [22]

Libelius, R. & Tågerud, S. (1984). Uptake of horseradish peroxidase in denervated skeletal muscle occurs primarily at the endplate region. Journal of Neurological Sciences 66, 273-281. [18]

Lichtman, J. W. & Balice-Gordon, R. J. (1990). Understanding synaptic competition in theory and in practice. Journal of Neurobiology 21, 99-106. [6]

Lie, D. C. & Weis, J. (1998). GDNF expression is increased in denervated human skeletal muscle. Neuroscience Letters 250, 87-90. [16, 17]

Lieber, R. L. (1988). Time course and cellular control of muscle fiber transformation following chronic stimulation. ISI Atlas of Science: Plants and Animals 1, 189-194. [20]

Lin, W. C., Sanchez, H. B., Deerinck, T., Morris, J. K., Ellisman, M., & Lee, K. F. (2000). Aberrant development of motor axons and neuromuscular synapses in erbB2-deficient mice. Proceedings of the National Academy of Sciences of the United States of America 97, 1299-1304. [6]

Lindboe, C. F. & Platou, C. S. (1984). Effect of immobilization of short duration on the muscle fibre size. Clinical Physiology 4, 183-188. [19]

Lindsley, D. B. (1935). Electrical activity of human motor units during voluntary contraction. American Journal of Physiology 114, 90-99. [13]

Lipicky, R. J. & Bryant, S. H. (1973). A biophysical study of the human myotonias. In New Developments in Electromyography and Clinical Neurophysiology, ed. Desmedt, J. E., pp. 451-463. Karger, Basel. [7]

Lipicky, R. J., Bryant, S. H., & Salmon, J. H. (1971). Cable parameters, sodium, potassium, chloride, and water content, and potassium efflux in isolated external intercostal muscle of normal volunteers and patients with myotonia congenita. Journal of Clinical Investigation 50, 2091-2103. [9]

Lissák, K., Dempsey, E. W., & Rosenblueth, A. (1939). The failure of transmission of motor nerve impulses in the course of Wallerian degeneration. American Journal of Physiology 128, 45-56. [16]

Loeb, G. E. (1981). Somatosensory unit input to the spinal cord during normal walking. Canadian Journal of Physiology and Pharmacology 59, 627-635. [4]

Loeb, G. E., Pratt, C. A., Chanaud, C. M., & Richmond, F. J. R. (1987). Distribution and innervation of short, interdigitated muscle fibers in parallel-fibered muscles of the cat hindlimb. Journal of Morphology 191, 1-15. [1, 12]

Loeb, J. A., Hmadcha, A., Fischbach, G. D., Land, S. J., & Zakarian, V. L. (2002). Neuregulin expression at neuromuscular synapses is modulated by synaptic activity and neurotrophic factors. Journal of Neuroscience 22, 2206-2214. [6]

Lohof, A. M., Ip, N. Y., & Poo, M.-M. (1993). Potentiation of developing neuromuscular synapses by the neurotrophins NT-3 and BDNF. Nature 363, 350-353. [18]

Lømo, T. & Gundersen, K. (1988). Trophic control of skeletal muscle membrane properties. In Nerve-Muscle Cell Trophic Communication, eds. Fernandez, H. L. & Donoso, J. A., pp. 61-79. CRC Press, Inc., Boca Raton, FL. [18]

Lømo, T. & Rosenthal, J. (1972). Control of ACh sensitivity by muscle activity in the rat. Journal of Physiology 221, 493-513. [6, 18, 19]

Loughna, P. T., Izumo, S., Goldspink, G., & Nadal-Ginard, B. (1990). Disuse and passive stretch cause rapid alterations in expression of developmental and adult contractile protein genes in skeletal muscle. Development 109, 217-223. [19]

Loughna, P. T. & Morgan, M. J. (1999). Passive stretch modulates denervation induced alterations in skeletal muscle myosin heavy chain mRNA levels. Pflügers Archiv: European Journal of Physiology 439, 52-55. [16]

Lu, D. X., Huang, S. K., & Carlson, B. M. (1997). Electron microscopic study of long-term denervated rat skeletal muscle. Anatomical Record 248, 355-365. [16]

Lubinska, L. & Niemierko, S. (1970). Velocity and intenstiy of bidirectional migration of acetylcholinesterase in transected nerves. Brain Research 27, 329-342. [8]

Luco, J. V. & Eyzaguirre, C. (1955). Fibrillation and hypersensitivity to ACh in denervated muscle: effect of length of degenerating nerve fibers. Journal of Neurophysiology 18, 65-73. [18]

Ludin, H. P. (1970). Microelectrode study of dystrophic human skeletal muscle. European Neurology 3, 116-121. [9]

Luff, A. R., Hatcher, D. D., & Torkko, K. (1988). Enlarged motor units resulting from partial denervation of cat hindlimb muscles. Journal of Neurophysiology 59, 1377-1394. [17]

Lunn, E. R., Perry, V. H., Brown, M. C., Rosen, H., & Gordon, S. (1989). Absence of Wallerian degeneration does not hinder regeneration in peripheral nerve. European Journal of Neuroscience 1, 27-33. [16]

Luther, P. & Squire, J. (1978). Three-dimensional structure of the vertebrate muscle M-region. Journal of Molecular Biology 125, 313-324. [1]

Lüttgau, H. C. (1965). The effect of metabolic inhibitors on the fatigue of the action potential in single muscle fibres. Journal of Physiology 178, 45-67. [15]

Lux, H. D., Schubert, P., Kreutzberg, G. W., & Globus, A. (1970). Excitation and axonal flow: autoradiographic study on motoneurons intracellularly injected with a 3H-amino acid. Experimental Brain Research 10, 197-204. [8]

Lymn, R. W. & Taylor, E. W. (1971). Mechanism of adenosine triphosphate hydrolysis by actomyosin. Biochemistry 10, 4617-4624. [11]

MacCallum, J. B. (1898). On the histogenesis of the striated muscle fibre, and the growth of the human sartorius muscle. Bulletin of Johns Hopkins Hospital 9, 208-215. [1, 5]

MacDougall, J. D. (1986). Morphological changes in human skeletal muscle following strength training and immobilization. In Human Muscle Power, eds. Jones, N. L., McCartney, N., & McComas, A. J., pp. 269-288. Human Kinetics, Champaign, IL. [5]

MacDougall, J. D., Elder, G. C., Sale, D. G., Moroz, J. R., & Sutton, J. R. (1980). Effects of strength training and immobilization on human muscle fibres. European Journal of Applied Physiology & Occupational Physiology 43, 25-34. [19, 20]

MacDougall, J. D., Sale, D. G., Alway, S. E., & Sutton, J. R. (1984). Muscle fiber number in biceps brachii in bodybuilders and control subjects. Journal of Applied Physiology 57, 1399-1403. [20]

MacDougall, J. D., Ward, G. R., Sale, D. G., & Sutton, J. R. (1977). Biochemical adaptation of human skeletal muscle to heavy resistance training and immobilization. Journal of Applied Physiology 43, 700-703. [20]

Macefield, G., Gandevia, S. C., & Burke, D. (1990). Perceptual responses to microstimulation of single afferents innervating joints, muscles and skin of the human hand. Journal of Physiology 429, 113-129. [4]

Macefield, V. G., Fuglevand, A. J., & Bigland-Ritchie, B. (1996). Contractile properties of single motor units in human toe extensors assessed by intraneural motor axon stimulation. Journal of Neurophysiology 75, 2509-2519. [12]

Macefield, V. G., Gandevia, S. C., Bigland-Ritchie, B., Gorman, R. B., & Burke, D. (1993). The firing rates of human motoneurons voluntarily activated in the absence of muscle afferent feedback. Journal of Physiology 471, 429-443. [15]

MacIntosh, B. R. (2003). Role of calcium sensitivity modulation in skeletal muscle performance. News in Physiological Sciences 18, 222-225. [11]

MacIntosh, B. R., Grange, R. W., Cory, C. R., & Houston, M. E. (1993). Myosin light chain phosphorylation during staircase in fatigued skeletal muscle. Pflügers Archiv: European Journal of Physiology 425, 9-15. [15]

MacIntosh, B. R., Grange, R. W., Cory, C. R., & Houston, M. E. (1994). Contractile properties of rat gastrocnemius muscle during staircase, fatigue and recovery. Experimental Physiology 79, 59-70. [15]

MacIntosh, B. R. & MacNaughton, M. B. (2005). The length dependence of muscle active force: Considerations for parallel elastic properties. Journal of Applied Physiology 92: 369-375. [11]

MacIntosh, B. R. & Rassier, D. E. (2002). What is fatigue? Canadian Journal of Applied Physiology 27, 42-55. [11, 15]

MacIntosh, B. R. & Willis, J. C. (2000). Force-frequency relationship and potentiation in mammalian skeletal muscle. Journal of Applied Physiology 88, 2088-2096. [11]

MacLennan, D. H. & Phillips, M. S. (1992). Malignant hyperthermia. Science 256, 789-794. [11]

MacLennan, D. H. & Wong, P. T. (1971). Isolation of a calcium-sequestering protein from sarcoplasmic reticulum. Proceedings of the National Academy of Sciences of the United States of America 68, 1231-1235. [7]

Magill-Solc, C. & McMahan, U. J. (1990a). Agrin-like molecules in motor neurons. Journal de Physiologie 84, 78-81. [6]

Magill-Solc, C. & McMahan, U. J. (1990b). Synthesis and transport of agrin-like molecules in motor neurons. Journal of Experimental Biology 153, 1-10. [3]

Magleby, K. L. & Zengel, J. E. (1982). A quantitative description of stimulation-induced changes in transmitter release at the frog neuromuscular junction. Journal of General Physiology 80, 613-638. [10]

Manzano, G. & McComas, A. J. (1988). Longitudinal structure and innervation of two mammalian hindlimb muscles. Muscle & Nerve 11, 1115-1122. [12]

Markelonis, G. & Tae Hwan, O. H. (1979). A sciatic nerve protein has a trophic effect on development and maintenance of skeletal muscle cells in culture. Proceedings of the National Academy of Sciences of the United States of America 76, 2470-2474. [18]

Marotte, L. R. & Mark, R. F. (1970). The mechanism of selective reinnervation of fish eye muscle. I. Evidence from muscle function during recovery. Brain Research 19, 41-51. [17]

Marsden, C. D. & Meadows, J. C. (1970). The effect of adrenaline on the contraction of human muscle. Journal of Physiology 207, 429-448. [11]

Marsden, C. D., Meadows, J. C., & Merton, P. A. (1969). Muscle wisdom. Journal of Physiology 200, 15P. [15]

Marsden, C. D., Meadows, J. C., & Merton, P. A. (1971). Isolated single motor units in human muscle and their rate of discharge during maximal voluntary effort. Journal of Physiology 217, 12P-13P. [13]

Marsh, E., Sale, D., McComas, A. J., & Quinlan, J. (1981). Influence of joint position on ankle dorsiflexion in humans. Journal of Applied Physiology 51, 160-167. [12]

Martin, T. P., Edgerton, V. R., & Grindeland, R. E. (1988). Influence of spaceflight on rat skeletal muscle. Journal of Applied Physiology 65, 2318-2325. [19]

Martin, T. P., Stein, R. B., Hoeppner, P. H., & Reid, D. C. (1992). Influence of electrical stimulation on the morphological and metabolic properties of paralyzed muscle. Journal of Applied Physiology 72, 1401-1406. [19]

Mathiowetz, V., Kashman, N., Volland, G., Weber, K., Dowe, M., & Rogers, S. (1985). Grip and pinch strength: normative data for adults. Archives of Physical Medicine and Rehabilitation 66, 69-74. [22]

Matsumura, K. & Campbell, K. P. (1994). Dystrophin-glycoprotein complex: its role in the molecular pathogenesis of muscular dystrophies. Muscle & Nerve 17, 2-15. [1]

Matthews, P. B. (1996). Relationship of firing intervals of human motor units to the trajectory of post-spike after-hyperpolarization and synaptic noise. Journal of Physiology 492 (Pt 2), 597-628. [12]

Matthews, P. B. C. (1962). The differentiation of two types of fusiform fibre by their effects on the dynamic response of muscle spindle primary endings. Quarterly Journal of Experimental Physiology 47, 324-333. [4]

Matthews, P. B. C. (1964). Muscle spindles and their motor control. Physiological Reviews 44, 219-288. [4]

Matthews, P. B. C. (1972). Mammalian Muscle Receptors and Their Central Actions. Edward Arnold, London. [4]

Matzuk, M. M., Kumar, T. R., Vassalli, A., Bickenbach, J. R., Roop, D. R., Jaenisch, R., & Bradley, A. (1995). Functional analysis of activins during mammalian development. Nature 374, 354-356. [5]

Mauro, A. (1961). Satellite cell of skeletal muscle fibers. Journal of Biophysics and Biochemical Cytology 9, 493-495. [5]

Maxwell, M. H. & Kleeman, C. R. (1962). Clinical Disorders of Fluid and Electrolyte Metabolism. McGraw-Hill, New York. [9]

Mayer, R. F., Burke, R. E., Toop, J., Hodgson, J. A., Kanda, K., & Walmsley, B. W. (1981). The effect of long-term immobilization on the motor unit population of the cat medial gastrocnemius muscle. Neuroscience 6, 725-739. [19]

Mayer, R. F. & Doyle, A. M. (1970). Studies of the motor unit in the cat. Histochemistry and topology of anterior tibial and extensor digitorum longus muscles. In Muscle Diseases, eds. Walton, J. N., Canal, N., & Scarlato, G., pp. 159-163. Excerpta Medica, Amsterdam. [12]

McArdle, B. (1951). Myopathy due to a deficit in muscle glycogen breakdown. Clinical Science 10, 13-35. [14]

McArdle, J. J. & Albuquerque, E. X. (1973). A study of the reinnervation of fast and slow mammalian muscles. Journal of General Physiology 61, 1-23. [17]

McBride, W. G. (1961). Thalidomide and congenital abnormalities. Lancet 2, 1358. [5]

McBride, W. G. (1978). Role of neural crest and peripheral nerves in limb development. Lancet 2, 792-793. [5]

McCartney, N., Moroz, D., Garner, S. H., & McComas, A. J. (1988). The effects of strength training in patients with selected neuromuscular disorders. Medicine and Science in Sports and Exercise 20, 362-368. [20]

McComas, A. J. (1994). Human neuromuscular adaptations that accompany changes in activity. Medicine and Science in Sports and Exercise 26, 1498-1509. [19, 20]

McComas, A. J. (1977). Neuromuscular Function and Disorders. Butterworths Publishing Co., London. [3, 5, 7, 8, 9, 10, 11, 12, 16, 17, 18, 19, 21, 22]

McComas, A. J., Fawcett, P. R. W., Campbell, M. J., & Sica, R. E. P. (1971a). Electrophysiological estimation of the number of motor units within a human muscle. Journal of Neurology, Neurosurgery, and Psychiatry 34, 121-131. [12]

McComas, A. J., Galea, V., Einhorn, R. W., Hicks, A. L., & Kuiack, S. (1993). The role of the Na+, K+-pump in delaying muscle fatigue. In Neuromuscular Fatigue, eds. Sargeant, A. J. & Kernell, D., pp. 35-43. North-Holland, Amsterdam. [15]

McComas, A. J., Jorgensen, P. B., & Upton, A. R. M. (1974a). The neurapraxic lesion: a clinical contribution to the study of trophic mechanisms. Canadian Journal of Neurological Sciences 1, 170-179. [16]

McComas, A. J. & Mrózek, K. (1967). Denervated muscle fibres in hereditary mouse dystrophy. Journal of Neurology, Neurosurgery, and Psychiatry 30, 526-530. [21]

McComas, A. J., Mrózek, K., Gardner-Medwin, D., & Stanton, W. H. (1968). Electrical properties of muscle fibre membranes in man. Journal of Neurology, Neurosurgery, and Psychiatry 31, 434-440. [9]

McComas, A. J., Sica, R., & Banerjee, S. (1978). Central nervous system effects of limb amputation in man. Nature 271, 73-74. [16]

McComas, A. J., Sica, R. E., Campbell, M. J., & Upton, A. R. (1971b). Functional compensation in partially denervated muscles. Journal of Neurology, Neurosurgery, and Psychiatry 34, 453-460. [6, 17]

McComas, A. J., Sica, R. E. P., McNabb, A. R., Goldberg, W. M., & Upton, A. R. M. (1974b). Evidence for reversible motoneurone dysfunction in thyrotoxicosis. Journal of Neurology, Neurosurgery, and Psychiatry 37, 548-558. [16]

McComas, A. J., Sica, R. E. P., & Petito, F. (1973a). Muscle strength in boys of different ages. Journal of Neurology, Neurosurgery, and Psychiatry 36, 171-173. [5]

McComas, A. J., Sica, R. E. P., Upton, A. R. M., & Aguilera, N. (1973b). Functional changes in motoneurones of hemiparetic patients. Journal of Neurology, Neurosurgery, and Psychiatry 36, 183-193. [18]

McComas, A. J. & Thomas, H. C. (1968). Fast and slow twitch muscles in man. Journal of Neurological Sciences 7, 301-307. [12]

McCrea, D. A. (2001). Spinal circuitry of sensorimotor control of locomotion. Journal of Physiology 533, 41-50. [4]

McCredie, J. (1975). The pathogenesis of congenital malformations. Australasian Radiology 19, 348-355. [5]

McCully, K. K. & Faulkner, J. A. (1985). Injury to skeletal muscle fibres of mice following lengthening contractions. Journal of Applied Physiology 59, 119-126. [21]

McDonagh, J. C., Binder, M. D., Reinking, R. M., & Stuart, D. G. (1980). Tetrapartite classification of motor units of cat tibialis posterior. Journal of Neurophysiology 44, 696-712. [12]

McKenzie, D. K., Bigland-Ritchie, B., Gorman, R. B., & Gandevia, S. C. (1992). Central and peripheral fatigue of human diaphragm and limb muscles assessed by twitch interpolation. Journal of Physiology 454, 643-656. [15]

McLeod, J. G. & Wray, S. H. (1967). Conduction velocity and fibre diameter of the median and ulnar nerves of the baboon. Journal of Neurology, Neurosurgery, and Psychiatry 30, 240-247. [9]

McMahan, U. J. & Slater, C. R. (1984). The influence of basal lamina on the accumulation of acetylcholine receptors at synaptic sites in regenerating muscle. Journal of Cell Biology 98, 1453-1473. [17]

McManaman, J. L., Oppenheim, R. W., Prevette, D., & Marchetti, D. (1990). Rescue of motoneurons from cell death by a purified skeletal muscle polypeptide: effects of the ChAT development factor, CDF. Neuron 4, 891-898. [18]

McNamara, J. O. & Fridovich, I. (1993). Human genetics. Did radicals strike Lou Gehrig? Nature 362, 20-21. [22]

McPhedran, A. M., Wuerker, R. B., & Henneman, E. (1965). Properties of motor units in a homogeneous red muscle (soleus) of the cat. Journal of Neurophysiology 28, 71-84. [12]

McPherron, A. C., Lawler, A. M., & Lee, S. J. (1997). Regulation of skeletal muscle mass in mice by a new TGF-beta superfamily member. Nature 387, 83-90. [5]

Meakin, S. O. & Shooter, E. M. (1992). The nerve growth factor family of receptors. Trends in Neurosciences 15, 323-331. [18]

Medbø, J. I. & Sejersted, O. M. (1990). Plasma potassium changes with high intensity exercise. Journal of Physiology 421, 105-122. [15]

Melki, J., Abdelhak, S., Sheth, P., Bachelot, M. F., Burlet, P., Marcadet, A., Aicardi, J., Barois, A., Carriere, J. P., Fardeau, M., Fontan, D., Ponsot, G., Billsette, T., Angelini, C., Barbosa, C., Ferriere, G., Lanzi, G., Ottolini, A., Babron, M. C., Cohen, D., Hanauer, A., Colerget-Darpox, F., Lathrop, M., Munnich, A., & Frezal, J. (1990). Gene for chronic proximal spinal muscular atrophies maps to chromosome 5q. Nature 344, 767-768. [6]

Melnick, S. C. (1963). 38 cases of the Guillain-Barré syndrome: an immunological study. British Medical Journal 1, 368-373. [16]

Mendell, L. M., Collins, W. F., III, & Munson, J. B. (1994). Retrograde determination of motoneuron properties and their synaptic input. Journal of Neurobiology 25, 707-721. [17]

Mendell, L. M. & Henneman, E. (1971). Terminals of single Ia fibers: location, density, and distribution within a pool of 300 homonymous motoneurons. Journal of Neurophysiology 34, 171-187. [4]

Mense, S. (1986). Slowly conducting afferent fibers from deep tissues—neurobiological properties and central nervous actions. In Progress in Sensory Physiology, ed. Ottoson, D., pp. 139-219. Springer-Verlag, Berlin. [15]

Mense, S. & Meyer, H. (1985). Different types of slowly conducting afferent units in cat skeletal muscle and tendon. Journal of Physiology 363, 403-417. [4]

Mense, S. & Stahnke, M. (1983). Responses in muscle afferent fibres of slow conduction velocity to contractions and ischaemia in the cat. Journal of Physiology 342, 383-397. [4]

Merton, P. A. (1954). Voluntary strength and fatigue. Journal of Physiology 123, 553-564. [13, 15]

Meryon, E. (1852). On granular and fatty degeneration of the voluntary muscles. Medico-Chirurgical Transactions 35, 73-84. [1]

Mesires, N. T. & Doumit, M. E. (2002). Satellite cell proliferation and differentiation during postnatal growth of porcine skeletal muscle. American Journal of Physiology 282, C899-C906. [5]

Mesulam, M.-M. (1982). Principles of horseradish peroxidase neurohistochemistry and their application for tracing neural pathways-axonal transport, enzyme histochemistry and light microscopic analysis. In Tracing Neural Connections with Horseradish Peroxidase, ed. Mesulam, M.-M., pp. 1-151. John Wiley & Sons, New York. [12]

Metzger, J. M. & Moss, R. L. (1992). Myosin light chain 2 modulates calcium-sensitive cross-bridge transitions in vertebrate skeletal muscle. Biophysical Journal 63, 460-468. [11]

Miledi, R. (1960a). Properties of regenerating neuromuscular synapses in the frog. Journal of Physiology 154, 190-205. [17]

Miledi, R. (1960b). The acetylcholine sensitivity of frog muscle fibres after complete or partial denervation. Journal of Physiology 151, 1-23. [18]

Miledi, R., Molenaar, P. C., & Polak, R. L. (1983). Electrophysiological and chemical determination of acetylcholine release at the frog neuromuscular junction. Journal of Physiology 334, 245-254. [3]

Miledi, R. & Parker, I. (1981). Calcium transients recorded with arsenazo III in the presynaptic terminal of the squid giant synapse. Proceedings of the Royal Society of London. Series B: Biological Sciences 212, 197-211. [10]

Miledi, R. & Slater, C. R. (1969). Electron-microscopic structure of denervated skeletal muscle. Proceedings of the Royal Society of London. Series B: Biological Sciences 174, 253-269. [16]

Miledi, R. & Slater, C. R. (1970). On the degeneration of rat neuromuscular junctions after nerve section. Journal of Physiology 207, 507-528. [8, 16]

Miller, R. G., Boska, M. D., Moussavi, R. S., Carson, P. J., & Weiner, M. W. (1988). 31P nuclear magnetic resonance studies of high energy phosphates and pH in human muscle fatigue. Comparison of aerobic and anaerobic exercise. Journal of Clinical Investigation 81, 1190-1196. [15]

Miller, R. G., Giannini, D., Milner-Brown, H. S., Layzer, R. B., Koretsky, A. P., Hooper, D., & Weiner, M. W. (1987). Effects of fatiguing exercise on high-energy phosphates, force, and EMG: evidence for three phases of recovery. Muscle & Nerve 10, 810-821. [15]

Miller, T. M. & Heuser, J. E. (1984). Endocytosis of synaptic vesicle membrane at the frog neuromuscular junction. Journal of Cell Biology 98, 685-698. [10]

Milner, T. E. & Stein, R. B. (1981). The effects of axotomy on the conduction of action potentials in peripheral sensory and motor nerve fibres. Journal of Neurology, Neurosurgery, and Psychiatry 44, 485-496. [16]

Milner-Brown, H. S., Stein, R. B., & Lee, R. G. (1974a). Contractile and electrical properties of human motor units in neuropathies and motor neurone disease. Journal of Neurology, Neurosurgery, and Psychiatry 37, 670-676. [17]

Milner-Brown, H. S., Stein, R. B., & Lee, R. G. (1974b). Pattern of recruiting human motor units in neuropathies and motor neurone disease. Journal of Neurology, Neurosurgery, and Psychiatry 37, 665-669. [17]

Milner-Brown, H. S., Stein, R. B., & Yemm, R. (1973a). Changes in firing rate of human motor units during linearly changing voluntary contractions. Journal of Physiology 230, 371-390. [13]

Milner-Brown, H. S., Stein, R. B., & Yemm, R. (1973b). The orderly recruitment of human motor units during voluntary isometric contractions. Journal of Physiology 230, 359-370. [12, 13]

Mines, G. R. (1913). On functional analysis of the action of electrolytes. Journal of Physiology 46, 188-235. [11]

Mitchell, J. H., Kaufman, M. P., & Iwamoto, G. A. (1983). The exercise pressor reflex: its cardiovascular effects, afferent mechanisms, and central pathways. Annual Review of Physiology 45, 229-242. [4]

Mitchell, P. (1961). Coupling of phosphorylation to electron and hydrogen transfer by a chemi-osmotic type of mechanism. Nature 191, 144-148. [14]

Mitchell, S. W., Morehouse, G. R., & Keen, W. W. (1864). Gunshot Wounds and Other Injuries of Nerves. Lippincott, Philadelphia, PA. [19]

Mittal, K. R. & Logmani, F. H. (1987). Age-related reduction in 8th cervical ventral nerve root myelinated fiber diameters and numbers in man. Journal of Gerontology 42, 8-10. [22]

Mokri, B. & Engel, A. G. (1975). Duchenne dystrophy: electron microscopic findings pointing to a basic or early abnormality in the plasma membrane of the muscle fiber. Neurology 25, 1111-1120. [1]

Molenaar, P. C., Newsom-Davis, J., Polak, R. L., & Vincent, A. (1982). Eaton-Lambert syndrome: acetylcholine and choline acetyltransferase in skeletal muscle. Neurology 32, 1061-1065. [10]

Monaco, A. P., Bertelson, C. J., Liechti-Gallati, S., Moser, H., & Kunkel, L. M. (1988). An explanation for the phenotypic differences between patients bearing partial deletions of the DMD locus. Genomics 2, 90-95. [1]

Monster, A. W. & Chan, H. (1977). Isometric force production by motor units of extensor digitorum communis muscle in man. Journal of Neurophysiology 40, 1432-1443. [12, 13]

Monti, R. J., Roy, R. R., & Edgerton, V. R. (2001). Role of motor unit structure in defining function. Muscle & Nerve 24, 848-866. [1, 12]

Moore, J. W., Narahashi, T., & Shaw, T. I. (1967). An upper limit to the number of sodium channels in nerve membrane? Journal of Physiology 188, 99-105. [9]

Moore, R. L. & Stull, J. T. (1984). Myosin light chain phosphorylation in fast and slow skeletal muscles in situ. American Journal of Physiology 247, C462-C471. [11]

Morais-Cabral, J. H., Zhou, Y., & Mackinnon, R. (2001). Energetic optimization of ion conduction rate by the K+ selectivity filter. Nature 414, 37-42. [7]

Morey-Holton, E.R. & Globus, R.K. (2002). Hindlimb unloading rodent model: technical aspects. Journal of Applied Physiology 92, 1367-1377. [19]

Mozdziak, P. E., Pulvermacher, P. M., & Schultz, E. (2000). Unloading of juvenile muscle results in a reduced muscle size 9 wk after reloading. Journal of Applied Physiology 88, 158-164. [5]

Mudge, A. W. (1993). Neurobiology. Motor neurons find their factors. Nature 363, 213-214. [18]

Mullins, L. J. & Noda, K. (1963). The influence of sodium-free solutions on the membrane potential of frog muscle fibers. Journal of General Physiology 47, 117-132. [9]

Munson, J. B., Foehring, R. C., Mendell, L. M., & Gordon, T. (1997). Fast-to-slow conversion following chronic low-frequency activation of medial gastrocnemius muscle in cats. II. Motoneuron properties. Journal of Neurophysiology 77, 2605-2615. [18]

Murray, M. P., Duthie, E. H., Jr., Gambert, S. R., Sepic, S. B., & Mollinger, L. A. (1985). Age-related differences in knee muscle strength in normal women. Journal of Gerontology 40, 275-280. [22]

Murray, M. P., Gardner, G. M., Mollinger, L. A., & Sepic, S. B. (1980). Strength of isometric and isokinetic contractions: knee muscles of men aged 20 to 86. Physical Therapy 60, 412-419. [22]

Musacchia, X. J., Deavers, D. R., Meininger, G. A., & Davis, T. P. (1980). A model for hypokinesia: effects on muscle atrophy in the rat. Journal of Applied Physiology 48, 479-486. [19]

Musick, J. & Hubbard, J. I. (1972). Release of protein from mouse motor nerve terminals. Nature 237, 279-281. [8, 18]

Nabeshima, Y., Hanaoka, K., Hayasaka, M., Esumi, E., Li, S., Nonaka, I., & Nabeshima, Y. (1993). Myogenin gene disruption results in perinatal lethality because of severe muscle defect. Nature 364, 532-535. [5]

Nardone, A., Romano, C., & Schieppati, M. (1989). Selective recruitment of high-threshold human motor units during voluntary isotonic lengthening of active muscles. Journal of Physiology 409, 451-471. [13]

Nardone, A. & Schieppati, M. (1988). Shift of activity from slow to fast muscle during voluntary lengthening contractions of the triceps surae muscles in humans. Journal of Physiology 395, 363-381. [13]

Natori, R. (1975). The electric potential change of internal membrane during propagation of contraction in skinned fibre of toad skeletal muscle. Japanese Journal of Physiology 25, 51-63. [11]

Needham, D. M. (1971). Machina Carnis. The Biochemistry of Muscular Contraction in its Historical Development. Cambridge University Press, Cambridge. [14, 15]

Neher, E. & Sakmann, B. (1976a). Noise analysis of drug induced voltage clamp currents in denervated frog muscle fibres. Journal of Physiology 258, 705-729. [10]

Neher, E. & Sakmann, B. (1976b). Single-channel currents recorded from membrane of denervated frog muscle fibres. Nature 260, 799-802. [7, 10]

New, H. V. & Mudge, A. W. (1986). Calcitonin gene-related peptide regulates muscle acetylcholine receptor synthesis. Nature 323, 809-811. [3]

Newham, D. J., McPhail, G., Mills, K. R., & Edwards, R. H. (1983). Ultrastructural changes after concentric and eccentric contractions of human muscle. Journal of Neurological Sciences 61, 109-122. [21]

Newton, J. P. & Yemm, R. (1986). Changes in the contractile properties of the human first dorsal interosseous muscle with age. Gerontology 32, 98-104. [22]

Newton, J. P., Yemm, R., & McDonagh, M. J. (1988). Study of age changes in the motor units of the first dorsal interosseous muscle in man. Gerontology 34, 115-119. [22]

Nielsen, O. B., de Paoli, F., & Overgaard, K. (2001). Protective effects of lactic acid on force production in rat skeletal muscle. Journal of Physiology 536, 161-166. [15]

Nieuwkoop, P. D. (1973). The organization center of the amphibian embryo: its origin, spatial organization, and morphogenetic action. Advances in Morphogenesis 10, 1-39. [5]

Nishimune, H., Uyeda, A., Nogawa, M., Fujimori, K., & Taguchi, T. (1997). Neurocrescin: a novel neurite-outgrowth factor secreted by muscle after denervation. Neuroreport 8, 3649-3654. [16, 17]

Nishizono, H., Saito, Y., & Miyashita, M. (1979). The estimation of conduction velocity in human skeletal muscle in situ with surface electrodes. Electroencephalography and Clinical Neurophysiology 46, 659-664. [9]

Nissl, F. (1892). Uber die veränderungen der Ganglienzellen am Facialiskern des kaninchens nach Aureissung der nerven. Allgemeine Zeitschrift für Psychiatrie 48, 197-198. [18]

Niswander, L., Jeffrey, S., Martin, G. R., & Tickle, C. (1994). A positive feedback loop coordinates growth and patterning in the vertebrate limb. Nature 371, 609-612. [5]

Nixon, R. A. & Sihag, R. K. (1991). Neurofilament phosphorylation: a new look at regulation and function. Trends in Neurosciences 14, 501-506. [2, 8]

Noble, E. G., Dabrowski, B. L., & Ianuzzo, C. D. (1983). Myosin transformation in hypertrophied rat muscle. Pflügers Archiv: European Journal of Physiology 396, 260-262. [20]

Noda, M., Shimizu, S., Tanabe, T., Takai, T., Tayano, T., Ikeda, T., Takahashi, H., Nakayama, H., Kanaoka, Y., Minamino, N., Kangawa, K., Matsu, H., Raferty, M. A., Hirose, T., Inayama, S., Hayashida, H., Miyata, T., & Numa, S. (1984). Primary structure of Electrophorus electricus sodium channel deduced from cDNA sequence. Nature 312, 121-127. [3, 7, 9]

Nordstrom, M. A. & Miles, T. S. (1990). Fatigue of single motor units in human masseter. Journal of Applied Physiology 68, 26-34. [12]

Nornes, H. O. & Das, G. D. (1974). Temporal pattern of neurogenesis in spinal cord of rat. I. An autoradiographic study—times and sites of origin and migration and settling patterns of neuroblasts. Brain Research 73, 121-138. [6]

Norris, A. H., Shock, N. W., & Wagman, I. H. (1953). Age changes in the maximum conduction velocity of motor fibers of the human ulnar nerve. Journal of Applied Physiology 5, 589-593. [22]

Norris, F. H. & Gasteiger, E. L. (1955). Action potentials of single motor units in normal muscle. Electroencephalography and Clinical Neurophysiology 7, 115-126. [13]

Nosek, T. M., Fender, K. Y., & Godt, R. E. (1987). It is diprotonated inorganic phosphate that depresses force in skinned skeletal muscle fibers. Science 236, 191-193. [15]

Nosek, T. M., Guo, N., Ginsburg, J. M., & Kolbeck, R. C. (1990). Inositol (1,4,5)triphosphate (IP3) within diaphragm muscle increases upon depolarization. Biophysical Journal 57, 401a. [11]

Novikova, L., Novikov, L., & Kellerth, J. O. (1997). Effects of neurotransplants and BDNF on the survival and regeneration of injured adult spinal motoneurons. European Journal of Neuroscience 9, 2774-2777. [17]

Nusslein-Volhard, C. (1991). Determination of the embryonic axes of Drosophila. Development. Supplement 1, 1-10. [5]

O'Brien, K., Muskiewicz, K., & Gussoni, E. (2002). Recent advances in and therapeutic potential of muscle-derived stem cells. Journal of Cellular Biochemistry. Supplement. 38, 80-87. [5]

O'Brien, R. A., Ostberg, A. J., & Vrbová, G. (1978). Observations on the elimination of polyneuronal innervation in developing mammalian skeletal muscle. Journal of Physiology 282, 571-582. [6]

Ochoa, J., Danta, G., Fowler, T. J., & Gilliatt, R. W. (1971). Nature of the nerve lesion caused by a pneumatic tourniquet. Nature 233, 265-266. [19]

Ochs, S. (1972). Fast transport of materials in mammalian nerve fibers. Science 176, 252-260. [8]

Ochs, S. & Ranish, N. (1969). Characteristics of the fast transport system in mammalian nerve fibers. Journal of Neurobiology 1, 247-261. [8]

Olson, E. N. (1990). MyoD family: a paradigm for development? Genes & Development 4, 1454-1461. [5]

Olsson, Y. & Sjöstrand, J. (1969). Origin of macrophages in Wallerian degeneration of peripheral nerves demonstrated autoradiographically. Experimental Neurology 23, 102-112. [16]

Ontko, J. A. (1986). Lipid metabolism in muscle. In Myology: Basic and Clinical, eds. Engel, A. G. & Banker, B. Q., pp. 697-720. McGraw-Hill, New York. [14]

Oppenheim, R. W. (1989). The neurotrophic theory and naturally occurring motoneuron death. Trends in Neurosciences 12, 252-255. [6]

Oppenheim, R. W. & Nunez, R. (1982). Electrical stimulation of hindlimb increases neuronal cell death in chick embryo. Nature 295, 57-59. [6]

Oppenheim, R. W., Prevette, D., Yin, Q. W., Collins, F., & MacDonald, J. (1991). Control of embryonic motoneuron survival in vivo by ciliary neurotrophic factor. Science 251, 1616-1618. [18]

Orlander, J., Kiessling, K.-H., Larsson, L., Karlsson, J., & Aniansson, A. (1978). Skeletal muscle metabolism and ultrastructure in relation to age in sedentary men. Acta Physiologica Scandinavica 104, 249-261. [22]

Overend, T. J., Cunningham, D. A., Paterson, D. H., & Lefcoe, M. S. (1992). Thigh composition in young and elderly men determined by computed tomography. Clinical Physiology 12, 629-640. [22]

Overgaard, K. & Nielsen, O. B. (2001). Activity-induced recovery of excitability in K+-depressed rat soleus muscle. American Journal of Physiology 280, R48-R55. [13]

Oyesiku, N. M. & Wigston, D. J. (1996). Ciliary neurotrophic factor stimulates neurite outgrowth from spinal cord neurons. Journal of Comparative Neurology 364, 68-77. [17]

Pachter, B. R. & Eberstein, A. (1984). Neuromuscular plasticity following limb immobilization. Journal of Neurocytology 13, 1013-1025. [19]

Pachter, B. R. & Eberstein, A. (1992). Long-term effects of partial denervation on sprouting and muscle fiber area in rat plantaris. Experimental Neurology 116, 246-255. [17]

Padykula, H. A. & Gauthier, G. F. (1967). Ultrastructural features of three fiber types in the rat diaphragm. Anatomical Record 157, 296-297. [12]

Papazian, D. M., Schwarz, T. L., Tempel, B. L., Jan, Y. N., & Jan, L. Y. (1987). Cloning of genomic and complementary DNA from Shaker, a putative potassium channel gene from Drosophila. Science 237, 749-753. [7]

Parcell, A. C., Trappe, S. W., Godard, M. P., Williamson, D. L., Fink, W. J., & Costill, D. L. (2000). An upper arm model for simulated weightlessness. Acta Physiologica Scandinavica 169, 47-54. [19]

Park, C. H., Pruitt, J. H., & Bennett, D. (1989). A mouse model for neural tube defects: the curtailed (Tc) mutation produces spina bifida occulta in Tc/+ animals and spina bifida with meningomyelocele in Tc/t. Teratology 39, 303-312. [6]

Paschal, B. M., Shpetner, H. S., & Vallee, R. B. (1987). MAP 1C is a microtubule-activated ATPase which translocates microtubules in vitro and has dynein-like properties. Journal of Cell Biology 105, 1273-1282. [8]

Pasquet, B., Carpentier, A., Duchateau, J., & Hainaut, K. (2000). Muscle fatigue during concentric and eccentric contractions. Muscle & Nerve 23, 1727-1735. [15]

Pate, E., Bhimani, M., Franks-Skiba, K., & Cooke, R. (1995). Reduced effect of pH on skinned rabbit psoas muscle mechanics at high temperatures: implications for fatigue. Journal of Physiology 486, 689-694. [15]

Pate, E., Lin, M., Franks-Skiba, K., & Cooke, R. (1992). Contraction of glycerinated rabbit slow-twitch muscle fibers as a function of MgATP concentration. American Journal of Physiology 262, C1039-C1046. [15]

Patrick, J. & Lindstrom, J. (1973). Autoimmune response to acetylcholine receptor. Science 180, 871-872. [10]

Peachey, L. D. (1965). The sarcoplasmic reticulum and transverse tubules of the frog's sartorius. Journal of Cell Biology 25, 209-231. [1, 11]

Pearson, C. M., Rimer, D. G., & Mommaerts, W. F. H. M. (1961). A metabolic myopathy due to absence of muscle phosphorylase. American Journal of Medicine 30, 502-515. [14]

Pekiner, C., Dent, E. W., Roberts, R. E., Meiri, K. F., & McLean, W. G. (1996). Altered GAP-43 immunoreactivity in regenerating sciatic nerve of diabetic rats. Diabetes 45, 199-204. [17]

Perkins, L. C. & Kaiser, H. L. (1961). Results of short-term isotonic and isometric exercise programs in persons over sixty. Physical Therapy Review 41, 633-635. [22]

Pernuš, F. & Erzen, I. (1991). Arrangement of fiber types within fascicles of human vastus lateralis muscle. Muscle & Nerve 14, 304-309. [12]

Person, R. S. (1974). Rhythmic activity of a group of human motoneurones during voluntary contraction of a muscle. Electroencephalography and Clinical Neurophysiology 36, 585-595. [13]

Person, R. S. & Kudina, L. P. (1972). Discharge frequency and discharge pattern of human motor units during voluntary contraction of muscle. Electroencephalography and Clinical Neurophysiology 32, 471-483. [13]

Pestronk, A., Drachman, D. B., & Griffin, J. W. (1976). Effect of botulinum toxin on trophic regulation of acetylcholine receptors. Nature 264, 787-789. [19]

Pestronk, A., Drachman, D. B., & Griffin, J. W. (1980). Effects of aging on nerve sprouting and regeneration. Experimental Neurology 70, 65-82. [22]

Petajan, J. H. & Philip, B. A. (1969). Frequency control of motor unit action potentials. Electroencephalography and Clinical Neurophysiology 27, 66-72. [13]

Peter, J. B., Barnard, R. J., Edgerton, V. R., Gillespie, C. A., & Stempel, K. E. (1972). Metabolic profiles of three fiber types of skeletal muscle in guinea pigs and rabbits. Biochemistry 11, 2627-2633. [12, 13]

Pette, D., Ramirez, B. U., Müller, W., Simon, R., Exner, G. U., & Hildebrand, R. (1975). Influence of intermittent long-term stimulation on contractile, histochemical and metabolic properties of fibre populations in fast and slow rabbit muscles. Pflügers Archiv: European Journal of Physiology 361, 1-7. [18]

Pette, D. & Staron, R. S. (2000). Myosin isoforms, muscle fiber types, and transitions. Microscopy Research and Technique 50, 500-509. [12]

Pettigrew, F. P. & Gardiner, P. F. (1987). Changes in rat plantaris motor unit profiles with advanced age. Mechanisms of Ageing and Development 40, 243-259. [22]

Pinçon-Raymond, M., Rieger, F., Fosset, M., & Lazdunski, M. (1985). Abnormal transverse tubule system and abnormal amount of receptors for Ca2+ channel inhibitors of the dihydropyridine family in skeletal muscle from mice with embryonic muscular dysgenesis. Developmental Biology 112, 458-466. [11]

Pinter, M. J., Vanden Noven, S., Muccio, D., & Wallace, N. (1991). Axotomy-like changes in cat motoneuron electrical properties elicited by botulinum toxin depend on the complete elimination of neuromuscular transmission. Journal of Neuroscience 11, 657-666. [16, 19]

Ploutz, L. L., Tesch, P. A., Biro, R. L., & Dudley, G. A. (1994). Effect of resistance training on muscle use during exercise. Journal of Applied Physiology 76, 1675-1681. [20]

Pockett, S. & Slack, J. R. (1982). Source of the stimulus for nerve terminal sprouting in partially denervated muscle. Neuroscience 7, 3173-3176. [17]

Podolsky, R. J. (1964). The maximum sarcomere length for contraction of isolated myofibrils. Journal of Physiology 170, 110-123. [11]

Pollack, G. H. (1995). Muscle contraction mechanism: are alternative engines gathering steam? Cardiovascular Research 29, 737-746. [11]

Pollard, J. D., King, R. H., & Thomas, P. K. (1975). Recurrent experimental allergic neuritis. An electron microscope study. Journal of Neurological Sciences 24, 365-383. [16]

Popham, P., Band, D., & Linton, R. (1990). Potassium infusions cause release of adrenaline in anaesthetized cats. Journal of Physiology 427, 43P. [15]

Popovic, M. B. (2003). Control of neural prostheses for grasping and reaching. Medical Engineering & Physics 25, 41-50. [13]

Porayko, O. & Smith, R. S. (1968). Morphology of muscle spindles in the rat. Experientia 24, 588-589. [4]

Post, R. L., Kume, S., Tobin, T., Orcutt, B., & Sen, A. K. (1969). Flexibility of an active center in sodium-plus-potassium adenosine triphosphatase. Journal of General Physiology 54, 306s-326s. [7]

Potter, L. T. (1970). Synthesis, storage and release of [14C]acetylcholine in isolated rat diaphragm muscles. Journal of Physiology 206, 145-166. [3]

Powell, J. A. & Fambrough, D. M. (1973). Electrical properties of normal and dysgenic mouse skeletal muscle in culture. Journal of Cellular Physiology 82, 21-38. [11]

Prahlad, K. V., Skala, G., Jones, D. G., & Briles, W. E. (1979). Limbless: a new genetic mutant in the chick. Journal of Experimental Zoology 209, 427-434. [5]

Price, S. R., Marco Garcia, N. V., Ranscht, B., & Jessell, T. M. (2002). Regulation of motor neuron pool sorting by differential expression of type II cadherins. Cell 109, 205-216. [6]

Proske, U. & Morgan, D. L. (2001). Muscle damage from eccentric exercise: mechanism, mechanical signs, adaptation and clinical applications. Journal of Physiology 537, 333-345. [21]

Provins, K. A. (1958). The effect of peripheral nerve block on the appreciation and execution of finger movements. Journal of Physiology 143, 55-67. [4]

Pumplin, D. W., Reese, T. S., & Llinas, R. (1981). Are the presynaptic membrane particles the calcium channels? Proceedings of the National Academy of Sciences of the United States of America 78, 7210-7213. [10]

Purves, D. & Lichtman, J. W. (1980). Elimination of synapses in the developing nervous system. Science 210, 153-157. [6]

Purves, D. & Lichtman, J. W. (1985). Principles of Neural Development, p. 13; 271-272. Sinauer Associates, Sunderland, MA. [6]

Purves, D. & Sakmann, B. (1974). Membrane properties underlying spontaneous activity of denervated muscle fibres. Journal of Physiology 239, 125-153. [16]

Rack, P. M. H. & Westbury, D. R. (1969). The effects of length and stimulus rate on tension in the isometric cat soleus muscle. Journal of Physiology 204, 443-460. [11]

Raftery, M. A., Hunkapiller, M. W., Strader, C. D., & Hood, L. E. (1980). Acetylcholine receptor: complex of homologous subunits. Science 208, 1454-1456. [3]

Ranvier, L. (1873). Propriétés et structures différents des muscles rouges et des muscles blancs, chez les lapins et chez les raies. Comptes Rendus Hebdomadaires des Seances de l'Academie des Sciences: D. Sciences Naturelles (Paris) 77, 1030-1034. [12]

Rash, J. E. & Fambrough, D. (1973). Ultrastructural and electrophysiological correlates of cell coupling and cytoplasmic fusion during myogenesis in vitro. Developmental Biology 30, 166-186. [5]

Rasminsky, M. & Sears, T. A. (1972). Internodal conduction in undissected demyelinated nerve fibres. Journal of Physiology 227, 323-350. [9]

Rasmussen, H. N., Van Hall, G., & Rasmussen, U. F. (2002). Lactate dehydrogenase is not a mitochondrial enzyme in human and mouse vastus lateralis muscle. Journal of Physiology 541, 575-580. [14]

Rassier, D. E., MacIntosh, B. R., & Herzog, W. (1999). Length dependence of active force production in skeletal muscle. Journal of Applied Physiology 86, 1445-1457. [11]

Rassier, D.E. & MacIntosh, B.R. (2002). Sarcomere length-dependence of activity-dependent twitch potentiation in mouse skeletal muscle. BioMed Central Physiology. www.biomedcentral.com/content/pdf/1472-6793-2-19.pdf. [11]

Ray, P. M., Belfall, B., Duff, C., Logan, C., Kean, V., Thompson, M. W., Sylvester, J. E., Gorski, J. L., Schmickel, R. D., & Worton, R. G. (1985). Cloning of the breakpoint of an X;21 translocation associated with Duchenne muscular dystrophy. Nature 318, 672-675. [1]

Raymackers, J. M., Gailly, P., Colson-Van Schoor, M., Schwaller, B., Hunziker, W., Celio, M. R., & Gillis, J. M. (2000). Tetanus relaxation of fast skeletal muscles of the mouse made parvalbumin deficient by gene inactivation. Journal of Physiology 527, 355-364. [7]

Rayment, I., Holden, H. M., Whittaker, M., Yohn, C. B., Lorenz, M., Holmes, K. C., & Milligan, R. A. (1993a). Structure of the actin-myosin complex and its implications for muscle contraction. Science 261, 58-65. [11]

Rayment, I., Rypniewski, W. R., Schmidt-Base, K., Smith, R., Tomchick, D. R., Benning, M. M., Winkelmann, D. A., Wesenberg, G., & Holden, H. M. (1993b). Three-dimensional structure of myosin subfragment-1: a molecular motor. Science 261, 50-58. [11]

Redfern, P. A. (1970). Neuromuscular transmission in newborn rats. Journal of Physiology 209, 701-709. [6]

Reichmann, H., Hoppeler, H., Mathieu-Costello, O., von Bergen, F., & Pette, D. (1985). Biochemical and ultrastructural changes of skeletal muscle mitochondria after chronic electrical stimulation in rabbits. Pflügers Archiv: European Journal of Physiology 404, 1-9. [20]

Reicke, H. & Nelson, K. R. (1990). Duchenne de Boulogne: electrodiagnosis of poliomyelitis. Muscle & Nerve 13, 56-62. [1]

Reiser, P. J., Moss, R. L., Giulian, G. G., & Greaser, M. L. (1985). Shortening velocity in single fibers from adult rabbit soleus muscles is correlated with myosin heavy chain composition. Journal of Biological Chemistry 260, 9077-9080. [12]

Renaud, J.-M. (2002). Modulation of force development by Na+, K+, Na+K+ pump and KATP channel during muscular activity. Canadian Journal of Applied Physiology 27, 296-315. [7, 15]

Rich, M. & Lichtman, J. W. (1989). Motor nerve terminal loss from degenerating muscle fibers. Neuron 3, 677-688. [6]

Rieger, F., Bournaud, R., Shimahara, T., Garcia, L., Pinçon-Raymond, M., Romey, G., & Lazdunski, M. (1987). Restoration of dysgenic muscle contraction and calcium channel function by co-culture with normal spinal cord neurons. Nature 330, 563-566. [11]

Rifenberick, D. H., Gamble, J. G., & Max, S. R. (1973). Response of mitochondrial enzymes to decreased muscular activity. American Journal of Physiology 225, 1295-1299. [19]

Riley, D. A., Bain, J. L., Thompson, J. L., Fitts, R. H., Widrick, J. J., Trappe, S. W., Trappe, T. A., & Costill, D. L. (1998). Disproportionate loss of thin filaments in human soleus muscle after 17-day bed rest. Muscle & Nerve 21, 1280-1289. [19]

Riley, D. A., Bain, J. L. W., Thompson, J. L., Fitts, R. H., Widrick, J. J., Trappe, S. W., Trappe, T. A., & Costill, D. L. (2000). Decreased thin filament density and length in human atrophic soleus muscle fibers after spaceflight. Journal of Applied Physiology 88, 567-572. [19]

Ringer, S. (1883). A further contribution regarding the influence of different constituents of the blood on the contraction of the heart. Journal of Physiology 4, 29-42. [11]

Ríos, E., Ma, J., & González, A. (1991). The mechanical hypothesis of excitation-contraction (EC) coupling in skeletal muscle. Journal of Muscle Research and Cell Motility 12, 127-135. [11]

Ritchie, J. M. & Rogart, R. B. (1977). The binding of saxitoxin and tetrodotoxin to excitable tissue. Reviews of Physiology Biochemistry & Pharmacology 79, 1-50. [7]

Robbins, N. & Yonezawa, T. (1971). Physiological studies during formation and development of rat neuromuscular junctions in tissue culture. Journal of General Physiology 58, 467-481. [6]

Robert, E. D. & Oester, Y. T. (1970). Absence of supersensitivity to acetylcholine in innervated muscle subjected to a prolonged pharmacologic nerve block. Journal of Pharmacology and Experimental Therapeutics 174, 133-140. [19]

Robinson, G. A., Enoka, R. M., & Stuart, D. G. (1991). Immobilization-induced changes in motor unit force and fatigability in the cat. Muscle & Nerve 14, 563-573. [19]

Romanes, G. J. (1941). The development and significances of the cell columns in the ventral horn of the cervical and upper thoracic spinal cord of the rabbit. Journal of Anatomy 76, 112-130. [6, 16]

Romanes, G. J. (1951). The motor cell columns of the lumbosacral spinal cord of the cat. Journal of Comparative Neurology 94, 313-364. [2, 6]

Romanul, F. C. & Hogan, E. L. (1965). Enzymatic changes in denervated muscle. I. Histochemical studies. Archives of Neurology 13, 263-273. [16]

Romanul, F. C. & Van der Meulen, J. P. (1966). Reversal of the enzyme profiles of muscle fibres in fast and slow muscles by cross-innervation. Nature 212, 1369-1370. [17]

Romanul, F. C. & Van der Meulen, J. P. (1967). Slow and fast muscles after cross-innervation. Enzymatic and physiological changes. Archives of Neurology 17, 387-402. [18]

Rome, L. C., Syme, D. A., Hollingworth, S., Lindstedt, S. L., & Baylor, S. M. (1996). The whistle and the rattle: the design of sound producing muscles. Proceedings of the National Academy of Sciences of the United States of America 93, 8095-8100. [11]

Rose, P. K. & Tourond, J. (1993). Structural remodelling of the dendritic trees of cat spinal motoneurons following permanent axotomy. Physiology Canada 24, 135. [18]

Rosenblatt, J. D. & Parry, D. J. (1992). Gamma irradiation prevents compensatory hypertrophy of overloaded mouse extensor digitorum longus muscle. Journal of Applied Physiology 73, 2538-2543. [20]

Rosenbleuth, A. & Dempsey, E. W. (1939). A study of Wallerian degeneration. American Journal of Physiology 128, 19-30. [16]

Rosenbleuth, J. (1974). Structure of amphibian motor endplate. Evidence for a granular component projecting from the outer surface of the receptive membrane. Journal of Cell Biology 62, 755-766. [3]

Rosenheimer, J. L. (1990). Factors affecting denervation-like changes at the neuromuscular junction during aging. International Journal of Developmental Neuroscience 8, 643-654. [22]

Rostami, A. M. (1993). Pathogenesis of immune-mediated neuropathies. Pediatric Research 33, S90-S94. [16]

Rotto, D. M. & Kaufman, M. P. (1988). Effect of metabolic products of muscular contraction on discharge of group III and IV afferents. Journal of Applied Physiology 64, 2306-2313. [15]

Round, J. M., Jones, D. A., Chapman, S. J., Edwards, R. H., Ward, P. S., & Fodden, D. L. (1984). The anatomy and fibre type composition of the human adductor pollicis in relation to its contractile properties. Journal of the Neurological Sciences 66, 263-272. [12]

Rowe, R. W. (1969). The effect of senility on skeletal muscles in the mouse. Experimental Gerontology 4, 119-126. [22]

Rowe, R. W. D. (1981). Morphology of perimysial and endomysial connective tissue in skeletal muscle. Tissue Cell 13, 681-690. [1]

Roy, N., Mahadevan, M. S., McLean, M., Shutler, G., Yaraghi, Z., Farahani, R., Baird, S., Besner-Johnston, A., Lefebvre, C., Kang, X., Salih, M., Aubry, H., Tamai, K., Guan, X., Ioannou, P., Crawford, T. O., de Jong, P. J., Surh, L., Ikeda, J.-E., Korneluk, R. G., & MacKenzie, A. (1995). The gene for neuronal apoptosis inhibitory protein is partially deleted in individuals with spinal muscular atrophy. Cell 80, 167-178. [6]

Roy, R. R., Eldridge, L., Baldwin, K. M., & Edgerton, V. R. (1996). Neural influence on slow muscle properties: inactivity with and without cross-reinnervation. Muscle & Nerve 19, 707-714. [18]

Roy, R. R., Kim, J. A., Grossman, E. J., Bekmezian, A., Talmadge, R. J., Zhong, H., & Edgerton, V. R. (2000). Persistence of myosin heavy chain-based fiber types in innervated but silenced rat fast muscle. Muscle & Nerve 23, 735-747. [19]

Rubinstein, L. J. (1960). Ageing changes in muscles. In The Structure and Function of Muscle, ed. Bourne, G. H., pp. 209-226. Academic Press, New York. [22]

Rüdel, R., Ricker, K., & Lehmann-Horn, F. (1993). Genotype-phenotype correlations in human skeletal muscle sodium channel diseases. Archives of Neurology 50, 1241-1248. [7]

Rudnicki, M. A., Braun, T., Hinuma, S., & Jaenisch, R. (1992). Inactivation of MyoD in mice leads to up-regulation of the myogenic HLH gene myf-5 and results in apparently normal muscle development. Cell 71, 383-390. [5]

Rudy, B. (1988). Diversity and ubiquity of K channels. Neuroscience 25, 729-749. [7, 15]

Ruiz i Altaba, A. & Melton, D. A. (1990). Axial patterning and the establishment of polarity in the frog embryo. Trends in Genetics: TIG 6, 57-64. [5]

Rusko, H. K. (1992). Development of aerobic power in relation to age and training in cross-country skiers. Medicine and Science in Sports and Exercise 24, 1040-1047. [20]

Russ, D. W. & Binder-Macleod, S. A. (1999). Variable-frequency trains offset low-frequency fatigue in human skeletal muscle. Muscle & Nerve 22, 874-882. [13]

Rutherford, O. M. & Jones, D. A. (1986). The role of learning and coordination in strength training. European Journal of Applied Physiology & Occupational Physiology 55, 100-105. [20]

Sadoshima, J. & Izumo, S. (1993). Mechanical stretch rapidly activates multiple signal transduction pathways in cardiac myocytes: potential involvement of an autocrine/paracrine mechanism. EMBO J 12, 1681-1692. [20]

Sahlin, K., Fernström, M., Svensson, M., & Tonkonogi, M. (2002). No evidence of an intracellular lactate shuttle in rat skeletal muscle. Journal of Physiology 541, 569-574. [14]

Sale, D., Quinlan, J., Marsh, E., McComas, A. J., & Bélanger, A. Y. (1982a). Influence of joint position on ankle plantarflexion in humans. Journal of Applied Physiology 52, 1636-1642. [11, 12]

Sale, D. G. (1988). Neural adaptation to resistance training. Medicine and Science in Sports and Exercise 20, S135-S145. [20]

Sale, D. G. (1992). Neural adaptation to strength training. In Strength and Power in Sport, ed. Komi, P. V., pp. 249-265. Blackwell Scientific Publications, Oxford. [13]

Sale, D. G., McComas, A. J., MacDougall, J. D., & Upton, A. R. M. (1982b). Neuromuscular adaptation in human thenar muscles following strength training and immobilization. Journal of Applied Physiology 53, 419-424. [19]

Sale, D., Quinlan, J., Marsh, E., McComas, A. J., & Bélanger, A. Y. (1982). Influence of joint position on ankle plantarflexion in humans. Journal of Applied Physiology 52, 1636-1642.

Salkoff, L. (1983). Genetic and voltage-clamp analysis of a Drosophila potassium channel. Cold Spring Harbor Symposia on Quantitative Biology 48 (Pt 1), 221-231. [7]

Salmons, S. & Vrbová, G. (1969). The influence of activity on some contractile characteristics of mammalian fast and slow muscles. Journal of Physiology 201, 535-549. [18, 20]

Saltin, B., Henriksson, J., Nygaard, E., Andersen, P., & Jansson, E. (1977). Fiber types and metabolic potentials of skeletal muscles in sedentary man and endurance runners. Annals of the New York Academy of Sciences 301, 3-29. [20]

Sanders, F. K. & Young, J. Z. (1946). The influence of peripheral connexion on the diameter of regenerating nerve fibres. Experimental Biology 22, 203-212. [18]

Sandow, A. (1965). Excitation-contraction coupling in skeletal muscle. Pharmacological Reviews 17, 265-320. [11]

Sanes, J. R. & Lichtman, J. W. (1999). Development of the vertebrate neuromuscular junction. Annual Review of Neuroscience 22, 389-442. [3, 6]

Sanes, J. R., Marshall, L. M., & McMahan, U. J. (1978). Reinnervation of muscle fiber basal lamina after removal of myofibers. Differentiation of regenerating axons at original synaptic sites. Journal of Cell Biology 78, 176-198. [3, 17]

Santo Neto, H., de Carvalho, V. C., & Marques, M. J. (1998). Estimation of the number and size of human flexor digiti minimi muscle motor units using histological methods. Muscle & Nerve 21, 112-114. [12]

Santos, D. A., Salgado, A. I., & Cunha, R. A. (2003). ATP is released from nerve terminals and from activated muscle fibres on stimulation of the rat phrenic nerve. Neuroscience Letters 338, 225-228. [10]

Sargeant, A. J., Davies, C. T. M., Edwards, R. H. T., Maunder, C., & Young, A. (1977). Functional and structural changes after disuse of human muscle. Clinical Science and Molecular Medicine 52, 337-342. [19]

Schachat, F. H., Bronson, D. D., & Mcdonald, O. B. (1985). Heterogeneity of contractile proteins. A continuum of troponin-tropomyosin expression in mammalian skeletal muscle. Journal of Biological Chemistry 260, 1108-1113. [12]

Schantz, P. & Henriksson, J. (1983). Increases in myofibrillar ATPase intermediate human skeletal muscle fibers in response to endurance training. Muscle & Nerve 6, 553-556. [20]

Schantz, P. G. & Dhoot, G. K. (1987). Coexistence of slow and fast isoforms of contractile and regulatory proteins in human skeletal muscle fibres induced by endurance training. Acta Physiologica Scandinavica 131, 147-154. [20]

Schatzmann, H. J. (1989). The calcium pump of the surface membrane and of the sarcoplasmic reticulum. Annual Review of Physiology 51, 473-485. [7]

Schiaffino, S., Hanzlíková, V., & Pierobon, S. (1970). Relations between structure and function in rat skeletal muscle fibers. Journal of Cell Biology 47, 107-119. [12]

Schiaffino, S. & Reggiani, C. (1996). Molecular diversity of myofibrillar proteins: gene regulation and functional significance. Physiological Reviews 76, 371-423. [12]

Schiavo, G., Benfenati, F., Poulain, B., Rossetto, O., Polverino, D. L., DasGupta, B. R., & Montecucco, C. (1992). Tetanus and botulinum-B neurotoxins block neurotransmitter release by proteolytic cleavage of synaptobrevin. Nature 359, 832-835. [10]

Schmalbruch, H., Al-Amood, W. S., & Lewis, D. M. (1991). Morphology of long-term denervated rat soleus muscle and the effect of chronic electrical stimulation. Journal of Physiology 441, 233-241. [16]

Schmid, S. L., Braell, W. A., Schlossman, D. M., & Rothman, J. E. (1984). A role for clathrin light chains in the recognition of clathrin cages by "uncoating ATPase." Nature 311, 228-231. [3]

Schmidt, E. M. & Thomas, J. S. (1981). Motor unit recruitment order: modification under volitional control. In Progress in Neurophysiology, ed. Desmedt, J. E., pp. 145-148. Karger, Basel. [13]

Schmied, A., Morin, D., Vedel, J. P., & Pagni, S. (1997). The "size principle" and synaptic effectiveness of muscle afferent projections to human extensor carpi radialis motoneurones during wrist extension. Experimental Brain Research 113, 214-229. [12]

Schonberger, L. B., Bregman, D. J., Sullivan-Bolyai, J. Z., Keenlyside, R. A., Ziegler, D. W., Retailliau, H. F., Eddins, D. L., & Bryan, J. A. (1979). Guillain-Barré syndrome following vaccination in the national influenza immunization program, United States, 1976–1977. American Journal of Epidemiology 110, 105-123. [16]

Schröder, J. M. (1972). Altered ratio between axon diameter and myelin sheath thickness in regenerated nerve fibers. Brain Research 45, 49-65. [17]

Schulze, M. L. (1955). Die absolute und relative Zahl der Muskelspindeln in den Kurzen Daumenmuskeln des Menschen. Anatomischer Anzeiger 102, 290-291.

Scott, S. A. (1977). Maintained function of foreign and appropriate junctions on reinnervated goldfish extraocular muscles. Journal of Physiology 268, 87-109. [17]

Seale, P. & Rudnicki, M. A. (2000). A new look at the origin, function, and "stem-cell" status of muscle satellite cells. Developmental Biology 218, 115-124. [5]

Seedorf, U., Leberer, E., Kirschbaum, B. J., & Pette, D. (1986). Neural control of gene expression in skeletal muscle. Effects of chronic stimulation on lactate dehydrogenase isoenzymes and citrate synthase. The Biochemical Journal 239, 115-120. [20]

Sejersted, O. M., Vøllestad, N. K., & Medbo, J. I. (1986). Muscle fluid and electrolyte balance during and following exercise. Acta Physiologica Scandinavica 128 (Supplementum. 556), 119-127. [15]

Sendtner, M., Kreutzberg, G. W., & Thoenen, H. (1990). Ciliary neurotrophic factor prevents the degeneration of motor neurons after axotomy. Nature 345, 440-441. [18]

Sendtner, M., Schmalbruch, H., Stöckli, K. A., Carroll, P., Kreutzberg, G. W., & Thoenen, H. (1992). Ciliary neurotrophic factor prevents degeneration of motor neurons in mouse mutant progressive motor neuronopathy. Nature 358, 502-504. [18]

Serratrice, G., Roux, H., & Aquaron, R. (1968). Proximal muscular weakness in elderly subjects. Report of 12 cases. Journal of Neurological Sciences 7, 275-299. [22]

Shafiq, S. A., Lewis, S. G., Dimino, L. C., & Schutta, H. S. (1978). Electron microscopic study of skeletal muscle in elderly subjects. In Aging in Muscle, eds. Kaldor, G. & Battista, W. J., pp. 68-85. Raven, New York. [22]

Shainberg, A., Yagil, G., & Yaffe, D. (1969). Control of myogenesis in vitro by Ca2+ concentration in nutritional medium. Experimental Cell Research 58, 163-167. [5]

Sheard, P. W. (2000). Tension delivery from short fibers in long muscles. Exercise and Sport Sciences Reviews 28, 51-56. [1]

Sheetz, M. P. & Spudich, J. A. (1983). Movement of myosin-coated fluorescent beads on actin cables in vitro. Nature 303, 31-35. [11]

Sheetz, M. P., Steuer, E. R., & Schroer, T. A. (1989). The mechanism and regulation of fast axonal transport. Trends in Neurosciences 12, 474-479. [8]

Shelanski, M. L. & Wísniewskí, H. M. (1969). Neurofibrillary degeneration: induced by vincristine therapy. Archives of Neurology 20, 199-206. [8]

Shephard, R. J. (1969). The working capacity of the older employee. Archives of Environmental Health 18, 982-986. [22]

Sherrington, C. (1929). Some functional problems attaching to convergence. Proceedings of the Royal Society of London. Series B: Biological Sciences 105, 332-362. [2, 8, 12]

Sherrington, C. S. (1900). The muscular sense. In Textbook of Physiology, ed. Schrafer, E. A., pp. 1002-1025. Pentland, Edinburgh. [4]

Shorey, M. L. (1909). The effect of the destruction of peripheral areas on the differentiation of the neuroblasts. Journal of Experimental Zoology 7, 25-63. [6]

Sica, R. E., Sanz, O. P., & Colombi, A. (1976). The effects of ageing upon the human soleus muscle. An electrophysiological study. Medicina (B Aires) 36, 443-446. [22]

Sica, R. E. P. & McComas, A. J. (1971). Fast and slow twitch units in a human muscle. Journal of Neurology, Neurosurgery, and Psychiatry 34, 113-120. [12]

Sica, R. E. P., McComas, A. J., Upton, A. R. M., & Longmire, D. (1974). Motor unit estimations in small muscles of the hand. Journal of Neurology, Neurosurgery, and Psychiatry 37, 55-67. [22]

Siddique, T., Figlewicz, D. A., Pericak-Vance, M. A., Haines, J. L., Rouleau, G., Jeffers, A. J., Sapp, P., Hung, W. Y., Bebout, J., McKenna-Yasek, D., & et al. (1991). Linkage of a gene causing familial amyotrophic lateral sclerosis to chromosome 21 and evidence of genetic-locus heterogeneity. The New England Journal of Medicine 324, 1381-1384. [22]

Sigworth, F. J. & Neher, E. (1980). Single Na+ channel currents observed in cultured rat muscle cells. Nature 287, 447-449. [7]

Silinsky, E. M. & Hubbard, J. I. (1973). Release of ATP from rat motor nerve terminals. Nature 243, 404-405. [10]

Simmons, R. M., Finer, J. T., Chu, S., & Spudich, J. A. (1996). Quantitative measurements of force and displacement using an optical trap. Biophysical Journal 70, 1813-1822. [11]

Simoneau, J. A., Lortie, G., Boulay, M. R., Marcotte, M., Thibault, M. C., & Bouchard, C. (1985). Human skeletal muscle fiber type alteration with high-intensity intermittent training. European Journal of Applied Physiology & Occupational Physiology 54, 250-253. [20]

Sjøgaard, G. (1986). Water and electrolyte fluxes during exercise and their relation to muscle fatigue. Acta Physiologica Scandinavica 128, 129-136. [15]

Sjøgaard, G., Adams, R. P., & Saltin, B. (1985). Water and ion shifts in skeletal muscle of humans with intense dynamic knee extension. American Journal of Physiology 248, R190-R196. [15]

Skene, J. H. (1989). Axonal growth-associated proteins. Annual Review of Neuroscience 12, 127-156. [18]

Skup, M., Dwornik, A., Macias, M., Sulejczak, D., Wiater, M., & Czarkowska-Bauch, J. (2002). Long-term locomotor training up-regulates TrkB(FL) receptor-like proteins, brain-derived neurotrophic factor, and neurotrophin 4 with different topographies of expression in oligodendroglia and neurons in the spinal cord. Experimental Neurology 176, 289-307. [18]

Slack, J. R. & Pockett, S. (1982). Motor neurotrophic factor in denervated adult skeletal muscle. Brain Research 247, 138-140. [17]

Slater, C. R., Lyons, P. R., Walls, T. J., Fawcett, P. R. W., & Young, C. (1992). Structure and function of neuromuscular junctions in the vastus lateralis of man. A motor point biopsy study of two groups of patients. Brain 115 (Pt 2), 451-478. [10]

Slomic, A., Rosenfalck, A., & Buchthal, F. (1968). Electrical and mechanical responses of normal and myasthenic muscle with particular reference to the staircase phenomenon. Brain Research 10, 1-78. [11, 12]

Small, D. H. (1990). Non-cholinergic actions of acetylcholinesterases: proteases regulating cell growth and development? Trends in Biochemical Sciences 15, 213-216. [3]

Smerdu, V., Karsch-Mizrachi, I., Campione, M., Leinwand, L., & Schiaffino, S. (1994). Type IIx myosin heavy chain transcripts are expressed in type IIb fibers of human skeletal muscle. American Journal of Physiology 267, C1723-C1728. [12]

Smith, D. O. (1984). Acetylcholine storage, release and leakage at the neuromuscular junction of mature adult and aged rats. Journal of Physiology 347, 161-176. [22]

Smith, J. C., Price, B. M., Van Nimmen, K., & Huylebroeck, D. (1990). Identification of a potent Xenopus mesoderm-inducing factor as a homologue of activin A. Nature 345, 729-731. [5]

Smith, J. L., Betts, B., Edgerton, V. R., & Zernicke, R. F. (1980). Rapid ankle extension during paw shakes: selective recruitment of fast ankle extensors. Journal of Neurophysiology 43, 612-620. [13]

Smith, R. S. (1966). Properties of intrafusal muscle fibres. In Muscular Afferents and Motor Control, ed. Granit, R., pp. 69-80. Almqvist & Wiksell, Stockholm. [4]

Soderberg, G. L., Minor, S. D., & Nelson, R. M. (1991). A comparison of motor unit behaviour in young and aged subjects. Age Ageing 20, 8-15. [22]

Sola, O. M., Christensen, D. L., & Martin, A. W. (1973). Hypertrophy and hyperplasia of adult chicken anterior latissimus dorsi muscles following stretch with and without denervation. Experimental Neurology 41, 76-100. [20]

Solandt, D. Y. (1936). The effect of potassium on the excitability and resting metabolism of frog's muscle. Journal of Physiology 86, 162-170. [15]

Söllner, T., Whiteheart, S. W., Brunner, M., Erdjument-Bromage, H., Geromanos, S., Tempst, P., & Rothman, J. E. (1993). SNAP receptors implicated in vesicle targeting and fusion. Nature 362, 318-324. [10]

Solomon, V., Baracos, V., Sarraf, P., & Goldberg, A. L. (1998). Rates of ubiquitin conjugation increase when muscles atrophy, largely through activation of the N-end rule pathway. Proceedings of the National Academy of Sciences of the United States of America 95, 12602-12607. [19]

Son, Y. J., Trachtenberg, J. T., & Thompson, W. J. (1996). Schwann cells induce and guide sprouting and reinnervation of neuromuscular junctions. Trends in Neurosciences 19, 280-285. [17]

Spemann, H. & Mangold, H. (1924). Uber induction von embryonalagen durch implantation artfremder organis atoren. Roux's Archives of Developmental Biology 100, 599-638. [6]

Spudich, J. A. (1994). How molecular motors work. Nature 372, 515-518. [8, 11]

Spudich, J. A., Kron, S. J., & Sheetz, M. P. (1985). Movement of myosin-coated beads on oriented filaments reconstituted from purified actin. Nature 315, 584-586. [11]

St.-Pierre, D. & Gardiner, P. F. (1985). Effect of "disuse" on mammalian fast-twitch muscle: joint fixation compared with neurally applied tetrodotoxin. Experimental Neurology 90, 635-651. [19]

St.-Pierre, D. & Gardiner, P. F. (1987). The effect of immobilization and exercise on muscle function: a review. Physiotherapy Canada 39, 24-36. [19]

St. Johnston, D. & Nusslein-Volhard, C. (1992). The origin of pattern and polarity in the Drosophila embryo. Cell 68, 201-219. [5]

Stacey, M. J. (1969). Free nerve endings in skeletal muscle of the cat. Journal of Anatomy 105, 231-254. [4]

Stainsby, W. N. & Lambert, C. R. (1979). Determinants of oxygen uptake in skeletal muscle. In Exercise and Sport Sciences Reviews, eds. Hutton, R. S. & Miller, D. I., pp. 125-152. Franklin Institute Press. [14]

Stålberg, E. (1966). Propagation velocity in human muscle fibers in situ. Acta Physiologica Scandinavica 70 Suppl. 287, 1-112. [9]

Stålberg, E. & Antoni, L. (1980). Electrophysiological cross section of the motor unit. Journal of Neurology, Neurosurgery, and Psychiatry 43, 469-474. [12]

Stålberg, E., Borges, O., Ericsson, M., Essén-Gustavsson, B., Fawcett, P. R. W., Nordesjö, L.-O., Nordgren, B., & Uhlin, R. (1989). The quadriceps femoris muscle in 20-70-year-old subjects: relationship between knee extension torque, electrophysiological parameters, and muscle fiber characteristics. Muscle & Nerve 12, 382-289. [22]

Stålberg, E. & Fawcett, P. R. (1982). Macro EMG in healthy subjects of different ages. Journal of Neurology, Neurosurgery, and Psychiatry 45, 870-878. [22]

Stålberg, E. & Trontelj, J. V. (1979). Single Fibre Electromyography. Mirvalle Press, Surrey. [22]

Stanley, E. F. & Drachman, D. B. (1978). Effect of myasthenic immunoglobulin on acetylcholine receptors of intact mammalian neuromuscular junctions. Science 200, 1285-1287. [10]

Staron, R. S. (1997). Human skeletal muscle fiber types: delineation, development, and distribution. Canadian Journal of Applied Physiology 22, 307-327. [5]

Staron, R. S. & Johnson, P. (1993). Myosin polymorphism and differential expression in adult human skeletal muscle. Comparative Biochemistry and Physiology. B, Comparative Biochemistry 106, 463-475. [12]

Staron, R. S., Malicky, E. S., Leonardi, M. J., Falkel, J. E., Hagerman, F. C., & Dudley, G. A. (1990). Muscle hypertrophy and fast fiber type conversions in heavy resistance-trained women. European Journal of Applied Physiology 60, 71-79. [20]

Staron, R. S. & Pette, D. (1990). The multiplicity of myosin light and heavy chain combinations in muscle fibers. In The Dynamic State of Muscle Fibres, ed. Pette, D., pp. 315-328. Walter de Gruyter, Berlin. [12]

Stein, R. B., Gordon, T., Jefferson, J., Sharfenberger, A., Yang, J. F., De Zepetnek, J. T., & Belanger, M. (1992). Optimal stimulation of paralyzed muscle after human spinal cord injury. Journal of Applied Physiology 72, 1393-1400. [19]

Steinbach, J. H., Schubert, D., & Eldridge, L. (1980). Changes in cat muscle contractile proteins after prolonged muscle inactivity. Experimental Neurology 67, 655-669. [19]

Steinmeyer, K., Lorenz, C., Pusch, M., Koch, M. C., & Jentsch, T. J. (1994). Multimeric structure of ClC-1 chloride channel revealed by mutations in dominant myotonia congenita (Thomsen). EMBO J 13, 737-743. [7]

Steinmeyer, K., Ortland, C., & Jentsch, T. J. (1991). Primary structure and functional expression of a developmentally regulated skeletal muscle chloride channel. Nature 354, 301-304. [7]

Stephens, J. A. & Stuart, D. G. (1975). The motor units of cat medial gastrocnemius: speed-size relations and their significance for the recruitment order of motor units. Brain Research 91, 177-195. [13]

Stephens, J. A. & Taylor, A. (1972). Fatigue of maintained voluntary muscle contraction in man. Journal of Physiology 220, 1-18. [15]

Stephens, J. A. & Usherwood, T. P. (1977). The mechanical properties of human motor units with special reference to their fatiguability and recruitment threshold. Brain Research 125, 91-97. [12]

Stephenson, G. M. (2001). Hybrid skeletal muscle fibres: a rare or common phenomenon? Clinical and Experimental Pharmacology & Physiology 28, 692-702. [12]

Sterne, G. D., Brown, R. A., Green, C. J., & Terenghi, G. (1997). Neurotrophin-3 delivered locally via fibronectin mats enhances peripheral nerve regeneration. European Journal of Neuroscience 9, 1388-1396. [17]

Stevens, C. F. (1991). Ion channels. Making a submicroscopic hole in one. Nature 349, 657-658. [7]

Stickland, N. C. (1981). Muscle development in the human fetus as exemplified by m. sartorius: a quantitative study. Journal of Anatomy 132, 557-579. [5]

Stonnington, H. H. & Engel, A. G. (1973). Normal and denervated muscle. A morphometric study of fine structure. Neurology 23, 714-724. [16]

Strecker, T. R. & Stephens, T. D. (1983). Peripheral nerves do not play a trophic role in limb skeletal morphogenesis. Teratology 27, 159-167. [5]

Street, S. F. & Ramsey, R. W. (1965). Sarcolemma: transmitter of active tension in frog skeletal muscle. Science 149, 1379-1380. [1]

Strehler, E. E., Carlsson, E., Eppenberger, H. M., & Thornell, L.-E. (1983). Ultrastructural localization of M-band proteins in chicken breast muscle as revealed by combined immunocytochemistry and ultramicrotomy. Journal of Molecular Biology 166, 141-158. [1]

Streter, F. A., Gergely, J., Salmons, S., & Romanul, F. (1973). Synthesis by fast muscle of myosin light chains characteristic of slow muscle in response to long-term stimulation. Nature New Biology 241, 17-19. [20]

Strickholm, A. (1974). Intracellular generated potentials during excitation-contraction coupling in muscle. Journal of Neurobiology 5, 161-187. [11]

Stuart, A., McComas, A. J., Goldspink, G., & Elder, G. (1981). Electrophysiologic features of muscle regeneration. Experimental Neurology 74, 148-159. [21]

Studitsky, A. N. (1952). [The restoration of muscle by means of transportation of minced muscle tissue] (Russian). Doklady Akademii Nauk SSSR 84, 389-392. [21]

Su, Q. N., Namikawa, K., Toki, H., & Kiyama, H. (1997). Differential display reveals transcriptional up-regulation of the motor molecules for both anterograde and retrograde axonal transport during nerve regeneration. European Journal of Neuroscience 9, 1542-1547. [17]

Suarez-Isla, B. A., Orozco, C., Heller, P. F., & Froehlich, J. P. (1986). Single calcium channels in native sarcoplasmic reticulum membranes from skeletal muscle. Proceedings of the National Academy of Sciences of the United States of America 83, 7741-7745. [11]

Sulik, K. K. & Dehart, D. B. (1988). Retinoic-acid-induced limb malformations resulting from apical ectodermal ridge cell death. Teratology 37, 527-537. [5]

Summerbell, D. & Maden, M. (1990). Retinoic acid, a developmental signalling molecule. Trends in Neurosciences 13, 142-147. [5]

Sunderland, S. (1947). Rate of regeneration in human, peripheral nerves. Analysis of the interval between injury and onset of recovery. Archives of Neurology and Psychiatry (Chicago) 58, 251-295. [17]

Sunderland, S. & Ray, L. J. (1950). Denervation changes in mammalian striated muscle. Journal of Neurology, Neurosurgery, and Psychiatry 13, 159-177. [16]

Suter, U., Welcher, A. A., & Snipes, G. J. (1993). Progress in the molecular understanding of hereditary peripheral neuropathies reveals insights into the biology of the peripheral nervous system. Trends in Neurosciences 13, 142-147. [2]

Sutherland, H., Jarvis, J. C., Kwende, M. M., Gilroy, S. J., & Salmons, S. (1998). The dose-related response of rabbit fast muscle to long-term low-frequency stimulation. Muscle & Nerve 21, 1632-1646. [18]

Suzuki, H., Tsuzimoto, H., Ishiko, T., Kasuga, N., Taguchi, S., & Ishihara, A. (1991). Effect of endurance training on the oxidative enzyme activity of soleus motoneurons in rats. Acta Physiologica Scandinavica 143, 127-128. [20]

Svedahl, K. & MacIntosh, B. R. (2003). Anaerobic threshold: the concept and methods of measurement. Canadian Journal of Applied Physiology 28, 299-323. [14]

Svoboda, K., Schmidt, C. F., Schnapp, B. J., & Block, S. M. (1993). Direct observation of kinesin stepping by optical trapping interferometry. Nature 365, 721-727. [8]

Swett, J. E. & Schoultz, T. W. (1975). Mechanical transduction in the Golgi tendon organ: a hypothesis. Archives of Italian Biology 113, 374-382. [4]

Tabary, J. C., Tabary, C., Tardieu, C., Tardieu, G., & Goldspink, G. (1972). Physiological and structural changes in the cat's soleus muscle due to immobilization at different lengths by plaster casts. Journal of Physiology 224, 231-244. [5, 11, 19]

Tanabe, T., Beam, K. G., Powell, J. A., & Numa, S. (1988). Restoration of excitation-contraction coupling and slow calcium current in dysgenic muscle by dihydropyridine receptor complementary DNA. Nature 336, 134-139. [11]

Tanji, J. & Kato, M. (1972). Discharge of single motor units at voluntary contraction of abductor digiti minimi muscle in man. Brain Research 45, 590-593. [13]

Tanji, J. & Kato, M. (1973). Firing rate of individual motor units in voluntary contraction of abductor digiti minimi muscle in man. Experimental Neurology 40, 771-783. [13]

Tanouye, M. A. & Ferrus, A. (1985). Action potentials in normal and Shaker mutant Drosophila. Journal of Neurogenetics 2, 253-271. [7]

Taormino, J. P. & Fambrough, D. M. (1990). Pre-translational regulation of the (Na++K+)-ATPase in response to demand for ion transport in cultured chicken skeletal muscle. Journal of Biological Chemistry 265, 4116-4123. [7]

Taylor, J. L., Butler, J. E., Allen, G. M., & Gandevia, S. C. (1996). Changes in motor cortical excitability during human muscle fatigue. Journal of Physiology 490, 519-528. [15]

Terenghi, G. (1999). Peripheral nerve regeneration and neurotrophic factors. Journal of Anatomy 194 (Pt 1), 1-14. [17]

Thesleff, S. (1960). Supersensitivity of skeletal muscle produced by botulinum toxin. Journal of Physiology 151, 598-607. [19]

Thesleff, S. (1963). Spontaneous electrical activity in denervated rat skeletal muscle. In The Effect of Use and Disuse on Neuromuscular Function, eds. Gutmann, E. & Hník, P., pp. 41-51. Czechoslovak Academy of Sciences, Prague. [16]

Thesleff, S., Molgó, J., & Tågerud, S. (1990). Trophic interrelations at the neuromuscular junction as revealed by the use of botulinum neurotoxins. Journal de Physiologie 84, 167-173. [18, 19]

Thesleff, S. & Ward, M. R. (1975). Studies on the mechanism of fibrillation potentials in denervated muscle. Journal of Physiology 244, 313-323. [16]

Thimm, F. & Baum, K. (1987). Response of chemosensitive nerve fibers of group III and IV to metabolic changes in rat muscles. Pflügers Archiv: European Journal of Physiology 410, 143-152. [4]

Thoenen, H. (1991). The changing scene of neurotrophic factors. Trends in Neurosciences 14, 165-170. [18]

Thomas, C. K., Bigland-Ritchie, B., Westling, G., & Johansson, R. S. (1990a). A comparison of human thenar motor-unit properties studied by intraneural motor-axon stimulation and spike-triggered averaging. Journal of Neurophysiology 64, 1347-1351. [12]

Thomas, C. K., Johansson, R. S., Westling, G., & Bigland-Ritchie, B. (1990b). Twitch properties of human thenar motor units measured in response to intraneural motor-axon stimulation. Journal of Neurophysiology 64, 1339-1346. [12]

Thomas, C. K., Ross, B. H., & Calancie, B. (1987a). Human motor-unit recruitment during isometric contractions and repeated dynamic movements. Journal of Neurophysiology 57, 311-324. [12]

Thomas, C. K., Ross, B. H., & Stein, R. B. (1986). Motor-unit recruitment in human first dorsal interosseous muscle for static contractions in three different directions. Journal of Neurophysiology 55, 1017-1029. [12]

Thomas, C. K., Stein, R. B., Gordon, T., Lee, R. G., & Elleker, M. G. (1987b). Patterns of reinnervation and motor unit recruitment in human hand muscles after complete ulnar and median nerve section and resuture. Journal of Neurology, Neurosurgery, and Psychiatry 50, 259-268. [17]

Thomas, C. K., Woods, J. J., & Bigland-Ritchie, B. (1989). Impulse propagation and muscle activation in long maximal voluntary contractions. Journal of Applied Physiology 67, 1835-1842. [15]

Thomas, P. K. (1970). The cellular response to nerve injury. 3. The effect of repeated crush injuries. Journal of Anatomy 106, 463-470. [17]

Thomas, R. C. (1972). Intracellular sodium activity and the sodium pump in snail neurons. Journal of Physiology 220, 55-71. [15]

Thomsen, G., Woolf, T., Whitman, M., Sokol, S., Vaughan, J., Vale, W., & Melton, D. A. (1990). Activins are expressed early in Xenopus embryogenesis and can induce axial mesoderm and anterior structures. Cell 63, 485-493. [5]

Thomsen, P. & Luco, J. V. (1944). Changes in weight and neuromuscular transmission in muscles of immobilized joints. Journal of Neurophysiology 7, 246-251. [20]

Tibes, U., Hemmer, B., Schweigart, U., Bóning, D., & Fotescu, D. (1974). Exercise acidosis as cause of electrolyte changes in femoral venous blood of trained and untrained man. Pflügers Archiv: European Journal of Physiology 347, 145-158. [7]

Timpe, L. C., Schwarz, T. L., Tempel, B. L., Papazian, D. M., Jan, Y. N., & Jan, L. Y. (1988). Expression of functional potassium channels from Shaker cDNA in Xenopus oocytes. Nature 331, 143-145. [7]

Timson, B. F. (1990). Evaluation of animal models for the study of exercise-induced muscle enlargement. Journal of Applied Physiology 69, 1935-1945. [20]

Timson, B. F., Bowlin, B. K., Dudenhoeffer, G. A., & George, J. B. (1985). Fiber number, area, and composition of mouse soleus muscle following enlargement. Journal of Applied Physiology 58, 619-624. [20]

Titmus, M. J. & Faber, D. S. (1990). Axotomy-induced alterations in the electrophysiological characteristics of neurons. Progress in Neurobiology 35, 1-51. [18]

Todd, I. C., Wosornu, D., Stewart, I., & Wild, T. (1992). Cardiac rehabilitation following myocardial infarction. A practical approach. Sports Medicine 14, 243-259. [20, 22]

Toji, H., Suei, K., & Kaneko, M. (1997). Effects of combined training loads on relations among force, velocity and power development. Canadian Journal of Applied Physiology 22, 328-336. [20]

Tomanek, R. J. & Lund, D. D. (1974). Degeneration of different types of skeletal muscle fibres. II. Immobilization. Journal of Anatomy 118, 531-541. [19]

Tomlinson, B. E. & Irving, D. (1977). The numbers of limb motor neurons in the human lumbosacral cord throughout life. Journal of Neurological Sciences 34, 213-219. [22]

Tomlinson, B. E., Walton, J. N., & Rebeiz, J. J. (1969). The effects of ageing and of cachexia upon skeletal muscle. A histopathological study. Journal of Neurological Sciences 9, 321-346. [22]

Tomonaga, M. (1977). Histochemical and ultrastructural changes in senile human skeletal muscle. Journal of the American Geriatrics Society 25, 125-131. [22]

Tonge, D. A. (1974). Physiological characteristics of re-innervation of skeletal muscle in the mouse. Journal of Physiology 241, 141-153. [17]

Toop, J. (1975). The histochemical development of human skeletal muscle and its motor innervation. In Recent Advances in Myology, eds. Bradley, W. G., Gardner-Medwin, D., & Walton, J. N., pp. 322-329. Excerpta Medica, Amsterdam. [5, 6]

Torre, M. (1953). [Number and dimensions of the motor units of the extrinsic eye muscles and, in general, of skeletal muscles connected with the sensory organs.]. Schweizer Archiv fur Neurologie und Psychiatrie 72, 362-376. [12]

Tower, S. (1939). The reaction of muscle to denervation. Physiological Reviews 19, 1-48. [16]

Tower, S. S. (1937). Trophic control of non-nervous tissues by the nervous system: a study of muscle and bone innervated from an isolated and quiescent region of spinal cord. Journal of Comparative Neurology 67, 241-267. [19]

Toyoshima, C., Nakasako, M., Nomura, H., & Ogawa, H. (2000). Crystal structure of the calcium pump of sarcoplasmic reticulum at 2.6 A resolution. Nature 405, 647-655. [7]

Toyoshima, C. & Unwin, N. (1988). Ion channel of acetyl-choline receptor reconstructed from images of postsynaptic membranes. Nature 336, 247-250. [3, 5]

Trimmer, J. S., Cooperman, S. S., Tomiko, S. A., Zhou, J., Crean, S. M., Boyle, M. B., Kallen, R. G., Sheng, Z., Barchi, R. L., Sigworth, F. J., Goodman, R. H., Agnew, W. S., & Mandel, G. (1989). Primary structure and functional expression of a mammalian skeletal muscle sodium channel. Neuron 3, 33-49. [7]

Trinick, J. (1991). Elastic filaments and giant proteins in muscle. Current Opinion in Cell Biology 3, 112-119. [1]

Tubman, L. A., MacIntosh, B. R., & Rassier, D. E. (1996). Absence of myosin light chain phosphorylation and twitch potentiation in atrophied skeletal muscle. Canadian Journal of Physiology and Pharmacology 74, 723-728. [22]

Tuffery, A. R. (1971). Growth and degeneration of motor end-plates in normal cat hind limb muscles. Journal of Anatomy 110, 221-247. [22]

Turnbull, J., Martin, J., Butler, J., Galea, V., & McComas, A. (2003). MUNE in ALS: natural history and implications in motor unit number estimation. Bromberg, M. B. Supplements to Clinical Neurophysiology 55, 167-176. Amsterdam; Boston, Elsevier Science. Motor unit number estimation (MUNE): Proceedings of the First International Symposium on MUNE, Snowbird, UT, USA, 14-15 July 2001. [22]

Turner, P. R., Westwood, T., Regen, C. M., & Steinhardt, R. A. (1988). Increased protein degradation results from elevated free calcium levels found in muscle from mdx mice. Nature 335, 735-738. [15]

Tweedle, C. D. & Kabara, J. (1977). Lipophilic nerve sprouting factor(s) isolated from denervated muscle. Neuroscience Letters 6, 41-46. [17]

Tzartos, S., Langeberg, L., Hochschwender, S., & Lindstrom, J. (1983). Demonstration of a main immunogenic region on acetylcholine receptors from human muscle using monoclonal antibodies to human receptor. FEBS Letters 158, 116-118. [10]

Uchida, S., Yamamoto, H., Iio, S., Matsumoto, N., Wang, X. B., Yonehara, N., Imai, Y., Inoki, R., & Yoshida, H. (1990). Release of calcitonin gene-related peptide-like immunoreactive substance from neuromuscular junction by nerve excitation and its action on striated muscle. Journal of Neurochemistry 54, 1000-1003. [6]

Ulfhake, B. & Kellerth, J.-O. (1982). Does alpha-motoneurone size correlate with motor unit type in cat triceps surae? Brain Research 251, 201-209. [12]

Ungewickell, E. (1984). First clue to biological role of clathrin light chains. Nature 311, 213. [3]

Urbano, F. J., Rosato-Siri, M. D., & Uchitel, O. D. (2002). Calcium channels involved in neurotransmitter release at adult, neonatal and P/Q-type deficient neuromuscular junctions. Molecular Membrane Biology 19, 293-300. [10]

Usdin, T. B. & Fischbach, G. D. (1986). Purification and characterization of a polypeptide from chick brain that promotes the accumulation of acetylcholine receptors in chick myotubes. Journal of Cell Biology 103, 493-507. [3]

Vale, R. D., Reese, T. S., & Sheetz, M. P. (1985a). Identification of a novel force-generating protein, kinesin, involved in microtubule-based motility. Cell 42, 39-50. [8]

Vale, R. D., Schnapp, B. J., Mitchison, T., Steuer, E., Reese, T. S., & Sheetz, M. P. (1985b). Different axoplasmic proteins generate movement in opposite directions along microtubules in vitro. Cell 43, 623-632. [8]

Vallee, R. B. & Bloom, G. S. (1991). Mechanisms of fast and slow axonal transport. Annual Review of Neuroscience 14, 59-92. [8]

Vallee, R. B., Shpetner, H. S., & Paschal, B. M. (1989). The role of dynein in retrograde axonal transport. Trends in Neurosciences 12, 66-70. [8]

Van Cutsem, M., Duchateau, J., & Hainaut, K. (1998). Changes in single motor unit behaviour contribute to the increase in contraction speed after dynamic training in humans. Journal of Physiology 513, 295-305. [20]

Van Cutsem, M., Feiereisen, P., Duchateau, J., & Hainaut, K. (1997). Mechanical properties and behaviour of motor units in the tibialis anterior during voluntary contractions. Canadian Journal of Applied Physiology 22, 585-597. [12]

Van der Kloot, W. (2003). A chloride channel blocker reduces acetylcholine uptake into synaptic vesicles at the frog neuromuscular junction. Brain Research 961, 287-289. [10]

Van Essen, D. & Jansen, J. K. (1974). Reinnervation of the rat diaphragm during perfusion with α-bungarotoxin. Acta Physiologica Scandinavica 91, 571-573. [17]

Van Hall, G. (2000). Lactate as a fuel for mitochondrial respiration. Acta Physiologica Scandinavica 168, 643-656. [14]

Van Harreveld, A. (1947). On the mechanism of the spontaneous reinnervation in paretic muscles. American Journal of Physiology 150, 670-676. [17]

VanBuren, P., Waller, G. S., Harris, D. E., Trybus, K. M., Warshaw, D. M., & Lowey, S. (1994). The essential light chain is required for full force production by skeletal muscle myosin. Proceedings of the National Academy of Sciences of the United States of America 91, 12403-12407. [11]

Vanden Noven, S., Gardiner, P. F., & Seburn, K. L. (1994). Motoneurons innervating two regions of rat medial gastrocnemius muscle with differing contractile and histochemical properties. Acta Anatomica 150, 282-293. [12]

Vandenboom, R. & Houston, M. E. (1996). Phosphorylation of myosin and twitch potentiation in fatigued skeletal muscle. Canadian Journal of Physiology and Pharmacology 74, 1315-1321. [15]

Vandenburgh, H. H., Hatfaludy, S., Karlisch, P., & Shansky, J. (1991). Mechanically induced alterations in cultured skeletal muscle growth. Journal of Biomechanics 24 Suppl 1, 91-99. [20]

Vandenburgh, H. H., Shansky, J., Solerssi, R., & Chromiak, J. (1995). Mechanical stimulation of skeletal muscle increases prostaglandin F2 alpha production, cyclooxygenase activity, and cell growth by a pertussis toxin sensitive mechanism. Journal of Cellular Physiology 163, 285-294. [5]

Vandervoort, A. A., Hayes, K. C., & Bélanger, A. Y. (1986). Strength and endurance of skeletal muscle in the elderly. Physiotherapy Canada 38, 167-173. [20, 22]

Vandervoort, A. A. & McComas, A. J. (1986). Contractile changes in opposing muscles of the human ankle joint with aging. Journal of Applied Physiology 61, 361-367. [22]

Vandervoort, A. A., Quinlan, J., & McComas, A. J. (1983). Twitch potentiation after voluntary contraction. Experimental Neurology 81, 141-152. [11]

Vassilev, P. M., Scheuer, T., & Catterall, W. A. (1988). Identification of an intracellular peptide segment involved in sodium channel inactivation. Science 241, 1658-1661. [7]

Veratti, E. (1902). Richerche sulla fine struttura della fibra muscolare striata. Memorie Reale Istituto Lombardi 19, 87-133. [1]

Vibert, P. & Cohen, C. (1988). Domains, motions and regulation in the myosin head. Journal of Muscle Research and Cell Motility 9, 296-305. [11]

Viguie, C. A., Lu, D. X., Huang, S. K., Rengen, H., & Carlson, B. M. (1997). Quantitative study of the effects of long-term denervation on the extensor digitorum longus muscle of the rat. Anatomical Record 248, 346-354. [16]

Vincent, A., Lang, B., & Newsom-Davis, J. (1989). Autoimmunity to the voltage-gated calcium channel underlies the Lambert-Eaton myasthenic syndrome, a paraneoplastic disorder. Trends in Neurosciences 12, 496-502. [10]

Vizoso, A. D. (1950). The relationship between internodal length and growth in human nerves. Journal of Anatomy 84, 342-353. [2]

Vogt, T., Nix, W. A., & Pfeifer, B. (1990). Relationship between electrical and mechanical properties of motor units. Journal of Neurology, Neurosurgery, and Psychiatry 53, 331-334. [12]

Vøllestad, N. K. & Blom, P. C. S. (1985). Effect of varying exercise intensity on glycogen depletion in human muscle. Acta Physiologica Scandinavica 125, 395-405. [13]

Vøllestad, N. K., Sejersted, O. M., Bahr, R., Woods, J. J., & Bigland-Ritchie, B. (1988). Motor drive and metabolic responses during repeated submaximal contractions in humans. Journal of Applied Physiology 64, 1421-1427. [15]

Vøllestad, N. K., Tabata, I., & Medbo, J. I. (1992). Glycogen breakdown in different human muscle fibre types during exhaustive exercise of short duration. Acta Physiologica Scandinavica 144, 135-141. [13]

Vøllestad, N. K., Vaage, O., & Hermansen, L. (1984). Muscle glycogen depletion patterns in type I and subgroups of type II fibres during prolonged severe exercise in man. Acta Physiologica Scandinavica 122, 433-441. [13]

Voss, H. (1937). Untersuchungen über Zahl, Anordnung und Länge der Muskelspindeln in den Lumbricalmuskeln des Menschen und einiger. Zeitschrift fur mikroskopisch-anatomische Forschung 42, 509-524.

Voss, H. (1956). Zahl und Anordnung der Muskelspindeln in den oberen Zungennbeinmuskeln, im M. trapezius und M. Latissimus dorsi. Anatomischer Anzeiger 103, 443-446.

Voss, H. (1958). Zahl und Anordung der Muskelspindeln in den unteren Zungenbeinmuskeln, dem M. sternocleidomastoideus unde den brauch- und tiefen Nackenmuskeln. Anatomischer Anzeiger 104, 345-355.

Vrbová, G. (1963). The effect of motoneurone activity on the speed of contraction of striated muscle. Journal of Physiology 169, 513-526. [19]

Vrbová, G. & Lowrie, M. (1989). Role of activity in developing synapses, search for molecular mechanisms. News in Physiological Sciences 4, 75-78. [6]

Vyskocil, F., Hník, P., Rehfeldt, H., Vejsada, R., & Ujec, E. (1983). The measurement of K+ concentration changes in human muscles during volitional contractions. Pflügers Archiv: European Journal of Physiology 399, 235-237. [15]

Wagenknecht, T., Grassucci, R., Frank, J., Saito, A., Inui, M., & Fleischer, S. (1989). Three-dimensional architecture of the calcium channel/foot structure of sarcoplasmic reticulum. Nature 338, 167-170. [11]

Wainman, P. & Shipounoff, G. C. (1941). The effects of castration and testosterone propionate on the striated peroneal musculature in the rat. Endocrinology 29, 975-978. [5]

Wakayama, Y. (1976). Electron microscopic study on the satellite cell in the muscle of Duchenne muscular dystrophy. Journal of Neuropathology and Experimental Neurology 35, 532-540. [21]

Wakelam, M. J. (1985). The fusion of myoblasts. The Biochemical Journal 228, 1-12. [5]

Wakeling, J. M., Pascual, S. A., Nigg, B. M., & von Tscharner, V. (2001). Surface EMG shows distinct populations of muscle activity when measured during sustained sub-maximal exercise. European Journal of Applied Physiology 86, 40-47. [13]

Waksman, B. H. & Adams, R. D. (1955). Allergic neuritis: an experimental disease of rabbits induced by the injection of peripheral nervous tissue and adjuvants. Journal of Experimental Medicine 102, 213-236. [16]

Wallace, B. G., Qu, Z., & Huganir, R. L. (1991). Agrin induces phosphorylation of the nicotinic acetylcholine receptor. Neuron 6, 869-878. [6]

Waller, A. (1850). Experiments on the section of the glossopharyngeal and hypoglossal nerves of the frog, and observation on the alteration produced thereby in the structure of their primitive fibres. Philosophical Transactions of the Royal Society of London 140, 423. [16]

Walsh, B., Tonkonogi, M., Söderlund, K., Hultman, E., Saks, V., & Sahlin, K. (2001). The role of phosphorylcreatine and creatine in the regulation of mitochondrial respiration in human skeletal muscle. Journal of Physiology 537, 971-978. [14]

Walter, C. B. (1988). The influence of agonist premotor silence and the stretch-shortening cycle on contractile rate in active skeletal muscle. European Journal of Applied Physiology & Occupational Physiology 57, 577-582. [20]

Wang, J. & Best, P. M. (1992). Inactivation of the sarcoplasmic reticulum calcium channel by protein kinase. Nature 359, 739-741. [7]

Warren, G. (1993). Cell biology. Bridging the gap. Nature 362, 297-298. [10]

Wasserschaff, M. (1990). Coordination of reinnervated muscle and reorganization of spinal cord motoneurons after nerve transection in mice. Brain Research 515, 241-246. [12]

Watson, W. E. (1968a). Centripetal passage of labelled molecules among mammalian motor axons. Journal of Physiology 196, 122P-123P. [18]

Watson, W. E. (1968b). Observations on the nucleolar and total cell body nucleic acid of injured nerve cells. Journal of Physiology 196, 655-676. [18]

Watson, W. E. (1969). The response of motor neurones to intramuscular injection of botulinum toxin. Journal of Physiology 202, 611-630. [18]

Watson, W. E. (1970). Some metabolic responses of axotomized neurones to contact between their axons and denervated muscle. Journal of Physiology 210, 321-343. [18]

Watson, W. E. (1972). Some quantitative observations upon the responses of neuroglial cells which follow axotomy of adjacent neurones. Journal of Physiology 225, 415-435. [3]

Watson, W. E. (1974). Cellular responses to axotomy and to related procedures. British Medical Bulletin 30, 112-115. [16]

Weintraub, H., Tapscott, S. J., Davis, R. L., Thayer, M. J., Adam, M. A., Lassar, A. B., & Miller, A. D. (1989). Activation of muscle-specific genes in pigment, nerve, fat, liver, and fibroblast cell lines by forced expression of MyoD. Proceedings of the National Academy of Sciences of the United States of America 86, 5434-5438. [5]

Weiss, P. (1944). Damming of axoplasm in constricted nerve: a sign of perpetual growth in nerve fibers. Anatomical Record 88, 464. [8]

Weiss, P. (1969). Neuronal dynamics. In Neurosciences Research Symposium Summaries, eds. Schmitt, F. O. & et al., pp. 255-299. MIT Press, Cambridge, MA. [2, 8]

Weiss, P. & Davis, H. (1943). Pressure block in nerves provided with arterial sleeves. Journal of Neurophysiology 6, 269-286. [8]

Weiss, P. & Edds, M. V., Jr. (1945). Spontaneous recovery of muscle following partial denervation. American Journal of Physiology 145, 587-607. [17]

Weiss, P. & Hiscoe, H. B. (1948). Experiments on the mechanism of nerve growth. Journal of Experimental Zoology 107, 315-395. [8]

Wernig, A., Carmody, J. J., Anzil, A. P., Hansert, E., Marciniak, M., & Zucker, H. (1984). Persistence of nerve sprouting with features of synapse remodelling in soleus muscles of adult mice. Neuroscience 11, 241-253. [22]

Westall, F. C., Robinson, A. B., Caccam, J., Jackson, J., & Eylar, E. H. (1971). Essential chemical requirements for induction of allergic encephalomyelitis. Nature 229, 22-24. [16]

Westerblad, H., Bruton, J. D., Allen, D. G., & Lännergren, J. (2000). Functional significance of Ca2+ in long-lasting fatigue of skeletal muscle. European Journal of Applied Physiology 83, 166-174. [15]

Westerfield, M. & Eisen, J. S. (1988). Neuromuscular specificity: pathfinding by identified motor growth cones in a vertebrate embryo. Trends in Neurosciences 11, 18-22. [6]

Westgaard, R. H. & Lømo, T. (1988). Control of contractile properties within adaptive ranges by patterns of impulse activity in the rat. Journal of Neuroscience 8, 4415-4426. [18, 20]

Westling, G., Johansson, R. S., Thomas, C. K., & Bigland-Ritchie, B. (1990). Measurement of contractile and electrical properties of single human thenar motor units in response to intraneural motor-axon stimulation. Journal of Neurophysiology 64, 1331-1337. [12]

White, M. J. & Davies, C. T. (1984). The effects of immobilization, after lower leg fracture, on the contractile properties of human triceps surae. Clinical Science (London, England: 1979) 66, 277-282. [19]

Whittaker, V. P., Michaelson, I. A., & Kirkland, R. J. (1964). The separation of synaptic vesicles from nerve-ending particles ('synaptosomes'). The Biochemical Journal 90, 293-303. [3]

Wickiewicz, T. L., Roy, R. R., Powell, P. L., & Edgerton, V. R. (1983). Muscle architecture of the human lower limb. Clinical Orthopaedics and Related Research 179, 275-283. [1]

Widrick, J. J., Knuth, S. T., Norenberg, K. M., Romatowski, J. G., Bain, J. L., Riley, D. A., Karhanek, M., Trappe, S. W., Trappe, T. A., Costill, D. L., & Fitts, R. H. (1999). Effect of a 17 day spaceflight on contractile properties of human soleus muscle fibres. Journal of Physiology 516 (Pt 3), 915-930. [19]

Williams, G. N., Higgins, M. J., & Lewek, M. D. (2002). Aging skeletal muscle: physiologic changes and the effects of training. Physical Therapy 82, 62-68. [22]

Williams, P. E. & Goldspink, G. (1971). Longitudinal growth of striated muscle fibres. Journal of Cell Science 9, 751-767. [5]

Williams, P. E. & Goldspink, G. (1978). Changes in sarcomere length and physiological properties in immobilized muscle. Journal of Anatomy 127, 459-468. [19]

Williams, P. L. & Hall, S. M. (1971a). Chronic Wallerian degeneration—an in vivo and ultrastructural study. Journal of Anatomy 109, 487-503. [16]

Williams, P. L. & Hall, S. M. (1971b). Prolonged in vivo observations of normal peripheral nerve fibres and their acute reactions to crush and deliberate trauma. Journal of Anatomy 108, 397-408. [16]

Williams, P. L. & Langdon, D. N. (1967). Gray's Anatomy, 34th ed., pp. 62. Longmans Green, London. [2]

Willis, T. (1672). De anima brutorium. Blaeus, Amsterdam. [10]

Willison, H. J. & Kennedy, P. G. (1993). Gangliosides and bacterial toxins in Guillain-Barré syndrome. Journal of Neuroimmunology 46, 105-112. [16]

Willison, R. G. (1978). Preservation of bulk and strength in muscles affected by neurogenic lesions. Muscle & Nerve 1, 404-406. [6, 12]

Wines, M. M. & Hall-Craggs, E. C. (1986). Neuromuscular relationships in a muscle having segregated motor endplate zones. I. Anatomical and physiological considerations. Journal of Comparative Neurology 249, 147-151. [12]

Wise, A. K., Morgan, D. L., Gregory, J. E., & Proske, U. (2001). Fatigue in mammalian skeletal muscle stimulated under computer control. Journal of Applied Physiology 90, 189-197. [13]

Wísniewskí, H. M., Brostoff, S. W., Carter, H., & Eylar, E. H. (1974). Recurrent experimental allergic polyganglioradiculoneuritis. Multiple demyelinating episodes in rhesus monkey sensitized with rabbit sciatic nerve myelin. Archives of Neurology 30, 347-358. [16]

Wittwer, M., Flück, M., Hoppeler, H., Müller, S., Desplanches, D., & Billeter, R. (2002). Prolonged unloading of rat soleus muscle causes distinct adaptations of the gene profile. FASEB Journal 16, 884-886. [19]

Witzmann, F. A., Kim, D. H., & Fitts, R. H. (1983). Effect of hindlimb immobilization on the fatigability of skeletal muscle. Journal of Applied Physiology 54, 1242-1248. [19]

Wolpert, L. (1969). Positional information and the spatial pattern of cellular differentiation. Journal of Theoretical Biology 25, 1-47. [5]

Wright, E. A. & Spink, J. M. (1959). A loss of nerve cells in the central nervous system in relation to age. Gerontologica 3, 277-287. [22]

Wrogemann, K. & Pena, S. D. (1976). Mitochondrial calcium overload. A general mechanism for cell-necrosis in muscle diseases. Lancet 1, 672-674. [15]

Wuerker, R. B., McPhedran, A. M., & Henneman, E. (1965). Properties of motor units in a heterogeneous pale muscle (m. gastrocnemius) of the cat. Journal of Neurophysiology 28, 85-99. [12]

Xie, Z. P. & Poo, M.-M. (1986). Initial events in the formation of neuromuscular synapse: rapid induction of acetylcholine release from embryonic neuron. Proceedings of the National Academy of Sciences of the United States of America 83, 7069-7073. [6]

Yamaguchi, H., Takaki, M., Matsubara, H., Yasuhara, S., & Suga, H. (1996). Constancy and variability of contractile efficiency as a function of calcium and cross-bridge kinetics: simulation. American Journal of Physiology 270, H1501-H1508. [11]

Yan, Q., Elliott, J., & Snider, W. D. (1992). Brain-derived neurotrophic factor rescues spinal motor neurons from axotomy-induced cell death. Nature 360, 753-755. [18]

Yan, Q., Matheson, C., & Lopez, O. T. (1995). In vivo neurotrophic effects of GDNF on neonatal and adult facial motor neurons. Nature 373, 341-344. [18]

Yanagida, T., Esaki, S., Iwane, A. H., Inoue, Y., Ishimima, A., Kitamura, K., Tanaka, H., & Tokunaga, M. (2000). Single-motor mechanics and models of the myosin motor. Philosophical Transactions of the Royal Society of London [Biol.] 355, 441-447. [11]

Yemm, R. (1977). The orderly recruitment of motor units of the masseter and temporal muscles during voluntary isometric contraction in man. Journal of Physiology 265, 163-174. [12, 13]

Yin, Q., Kemp, G. J., Yu, L. G., Wagstaff, S. C., & Frostick, S. P. (2001). Expression of Schwann cell-specific proteins and low-molecular-weight neurofilament protein during regeneration of sciatic nerve treated with neurotrophin-4. Neuroscience 105, 779-783. [17]

Young, A., Hughes, I., Round, J. M., & Edwards, R. H. (1982). The effect of knee injury on the number of muscle fibres in the human quadriceps femoris. Clinical Science (London, England: 1979) 62, 227-234. [19]

Young, A., Stokes, M., & Crowe, M. (1984). Size and strength of the quadriceps muscles of old and young women. European Journal of Clinical Investigation 14, 282-287. [22]

Young, A., Stokes, M., & Crowe, M. (1985). The size and strength of the quadriceps muscles of old and young men. Clinical Physiology 5, 145-154. [22]

Young, J. L. & Mayer, R. F. (1981). Physiological properties and classification of single motor units activated by intramuscular microstimulation in the first dorsal interosseous muscle in man. In Progress in Clinical Neurophysiology: Vol. 9. Motor Unit Types, Recruitment and Plasticity in Health and Disease, ed. Desmedt, J. E., pp. 17-25. Karger, Basel. [12]

Young, J. Z. (1936). The giant nerve fibres and epistellar body of cephalopods. Quarterly Journal of Microscopic Science 78, 367-386. [9]

Young, M., Paul, A., Rodda, J., Duxson, M., & Sheard, P. (2000). Examination of intrafascicular muscle fiber terminations: implications for tension delivery in series-fibered muscles. Journal of Morphology 245, 130-145. [1]

Zacks, S. I. & Saito, A. (1969). Uptake of exogenous horseradish peroxidase by coated vesicles in mouse neuromuscular junctions. Journal of Histochemistry and Cytochemistry: Official Journal of the Histochemistry Society 17, 161-170. [18]

Zajac, F. E. & Faden, J. S. (1985). Relationship among recruitment order, axonal conduction velocity, and muscle-unit properties of type-identified motor units in cat plantaris muscle. Journal of Neurophysiology 53, 1303-1322. [13]

Zajac, F. E. & Young, J. L. (1980a). Discharge properties of hindlimb motoneurons in decerebrate cats during locomotion induced by mesencephalic stimulation. Journal of Neurophysiology 43, 1221-1235. [13]

Zajac, F. E. & Young, J. L. (1980b). Properties of stimulus trains producing maximum tension-time area per pulse from single motor units in medial gastrocnemius muscle of the cat. Journal of Neurophysiology 43, 1206-1220. [13]

Zerba, E., Komorowski, T. E., & Faulkner, J. A. (1990). Free radical injury to skeletal muscles of young, adult, and old mice. American Journal of Physiology 258, C429-C435. [21]

Zhong, H., Roy, R. R., Hodgson, J. A., Talmadge, R. J., Grossman, E. J., & Edgerton, V. R. (2002). Activity-independent neural influences on cat soleus motor unit phenotypes. Muscle & Nerve 26, 252-264. [19]

Index

Note: The italicized *f* and *t* following page numbers refer to figures and tables, respectively.

About the Authors

Brian R. MacIntosh, PhD, is associate dean of the graduate program and professor for the faculty of kinesiology at the University of Calgary in Alberta, Canada. MacIntosh is on the cutting edge of research in skeletal muscle and has published more than 50 papers and numerous book chapters in muscle and exercise physiology. He has been teaching undergraduate and graduate courses in these areas for 25 years and is a member of the Canadian Society for Exercise Physiology, the Canadian Physiological Society, the American Physiological Society, the American College of Sports Medicine, the Biophysical Society, and the Human Powered Vehicles Association. He is also an associate editor for the *Canadian Journal of Applied Physiology* and a former board member for the Canadian Society for Exercise Physiology.

Phillip Gardiner, PhD, is director of the Health, Leisure & Human Performance Research Institute at the University of Manitoba in Winnipeg, Manitoba, Canada. He is also an adjunct professor of physiology, a member of the Spinal Cord Research Center in the faculty of medicine at the University of Manitoba, and a Canada Research Chair, a position given to internationally renowned researchers. Gardiner is past president of the Canadian Society for Exercise Physiology and previous coeditor in chief of the *Canadian Journal of Applied Physiology.* He has published extensively in the area of neuromuscular adaptations and authored the book *Neuromuscular Aspects of Physical Activity.*

Alan J. McComas, MB, is emeritus professor of medicine (in neurology) at McMaster University in Hamilton, Ontario, Canada. McComas has more than 40 years of research experience in nerve and muscle. Among his accomplishments in research are devising a method for estimating the number of human motor units in human muscle, showing the importance of the electrogenic sodium pump in delaying fatigue, and carrying out early microelectrode studies of human muscle fibers. He has held named lectureships and is a member of the Society for Neuroscience.